Help for the Technology Shy

A COMMONSENSE APPROACH TO SAFE, PRODUCTIVE, STRESS-FREE TECHNOLOGY USE

— **Part II** —

MICHAEL GORZKA

Welcome!

Help for the Technology Shy - Part II

MICHAEL GORZKA

"Success is a journey not a destination..."

Published by Squirrels Drinking Coffee

Help for the Technology Shy - Part II

International Standard Book Number: 978-1-7349815-6-8

Table of Contents

"energy flows where attention goes"

Preface: What You May Have Missed

If you have not read **Help for the Technology Shy - Part I**, please put this book back on your nightstand until you do so.

Available in both print & digital versions, the first book in the Help for the Technology Shy series can be ordered from Amazon or directly from www.technologyshy.com.

In Help for the Technology Shy - Part I, we pleasantly discuss how to build a sound technology foundation which is essential for productive, stress-free contemporary technology use.

Other topics covered in Part I include…

- How to reduce technology anxiety
- Why you are not a dinosaur 🦖
- Rocket science 🚀 (not really 🙂)
- The most important thing to know about computers, tablets, smartphones, et al.
- And expecting the unexpected

Chapter 4: Your Getting Stuff Done with Technology Toolbox

This box contains everything you need to get stuff done with technology.

These are the things that you will need in the midst of technology use – **after** you have a *clear idea in your head* (or better yet on paper) what you want to do.

We have covered **mindful** use of *mouse clicks*, *screen taps*, and *keyboard presses* 🍿

The remaining tools which are *forms*, *menus*, *searching* and *Internet connection* are comprised of those first three:

"Productivity is never an accident. It is always the result of a commitment to excellence, intelligent planning, and focused effort." —Paul J. Meyer

The brain tool

The human brain is the captain of the *Getting Stuff Done with Technology toolbox.*

If the brain is a "bad steward" then consider your technology boat capsized!

Haphazard, willy-nilly use of any of the technology "tools" (e.g. mouse clicks, screen taps & keyboard presses) will run your Getting Stuff Done with Technology ship ashore faster than you can say Jack Robinson!

Unmindful use of the technology tools is the **#1 cause** of lost work, wasted time and technology-related angst 😬

As previously mentioned, our shiny devices are all as *dumb as a bag of hammers*.

Your technology use success depends on how much thought and planning you give to using them.

The internet connection tool

Connecting your device to the Internet will enable you to access a nearly infinite number of services and resources.

- some of which will be positive & beneficial
- some of which will be negative & counterproductive
- and many of which will be completely unnecessary

Practice *technology minimalism* (use only what is necessary) to save, time, energy and money – and to help you stay focused on your *objectives*.

Exactly **how** you can connect a device to the Internet will depend on your device, your physical location – because *context* will always be a factor during technology use – and how much you are willing to pay (if anything).

There are (at the time of this writing) three ways of connecting to the Internet:

1. through a data plan with a mobile carrier (i.e. a cellular connection)
2. through a wi-fi connection
3. through a wired connection (aka "ethernet")

NOTE: We cannot get into the nuts & bolts of data plans here (as it would be the most boring and out of date thing imaginable zzZ) but you can visit your Technology Tour Guide's *website* for his **current** cost-effective recommendations for data plans and other really good stuff, besides.

If you have a device that supports **both** of those connections, you can connect to free Wi-Fi in order to save money as it is easy to incur additional charges with cellular (i.e. paid data plan) connections.

How to connect a device to a Wi-Fi network (and through that network to the Internet)

The short answer (and is a Help for the Technology Shy mantra) is to *think about what you want to do and then look around for a way to do it.*

It might take some noodling around and following *visual cues* before you come across a Wi-Fi *icon* or *menu selection.*

There might be a Wi-Fi connection icon **somewhere** on your desktop or home screen and/or it might be within your device's *Settings*.

If you cannot locate your device's Wi-Fi connection utility, armed with the knowledge of the device you are using, you can borrow a device that is **already connected** to the Internet and do a *web search*:

Jumping through hoops (?)

Yes, depending on the organization offering the Wi-Fi, you may need to jump through hoops.

For example, your author's local library changed their public Wi-Fi login procedure to involve a web browser tab or window that automatically opened – **and needed to be kept open** – after the user agreed to the terms and conditions.

Many people would unmindfully **close** that web browser tab or window – an action which resulted in their Wi-Fi connection being lost.

The Wi-Fi password to join any public Wi-Fi network can change **periodically** (which would necessitate you having to reenter that password on your device).

There are people who get a bit discombobulated 😕 (if not outright angry 😤) when they are required to jump through hoops for Wi-Fi connection (**and many other things related to technology**).

Please keep in mind that well-meaning (and not so well-meaning) jackasses are often the culprits behind flawed and arcane procedures.

An organization's technology manager (who frankly may not be qualified for that position) can make some **poor choices** when it comes to "Wi-Fi gateway" products (and other things as well).

And the end users have to either make do with those choices – or find another Wi-Fi hotspot.

If you are uncertain as to how to connect your device to a Wi-Fi network, you can try asking a friendly staff member...

But please do not assume they are network experts or have any direct control or influence over their organization's Wi-Fi network.

Wi-Fi connection "gotchas" include:

- selecting the wrong Wi-Fi network
- paying for Wi-Fi access when free options are available
- entering the Wi-Fi password incorrectly
- assuming that the saved password on your device for a particular Wi-Fi network is still current

If your device will not connect to a Wi-Fi network, try force quitting your device's *Settings* app or restart your device 🍿

Armed with the knowledge of *which device* you are using (and the *operating system* that is running on that device) search the web for how to "force quit" an app on your device.

Unsecured Wi-Fi?

An unsecured network most often refers to a free Wi-Fi (wireless) network, like at a coffeehouse or retail store. It means there's no special login or screening process to get on the network, which means you and anyone else can use it.

"Screamin' beans" is an unlocked Wi-Fi network while the other three are **locked** 🔒 and require a password in order to join and subsequently access the Internet (i.e. "get online").

Is this biggie fries? 🍟 Are you at risk?

There are risks to taking a shower 🚿, crossing an intersection 🚦, walking out your front door 🚪 and many of the other things we do in real life.

Question: Does your hanging tough but pragmatic Help for the Technology Shy author worry about unsecured Wi-Fi networks?

The short answer is: "No."

Although no security expert, your Tech Wizard author **can say** he has been availing himself to unsecured Wi-Fi networks since Wi-Fi emerged in the late 1990s and has never experienced any adverse consequences as a result.

Your author also has never met anybody whose device has been compromised or "hacked into" through unsecured Wi-Fi access. So (at the time of this writing), he does not stress about them 🖊

Wired connections

Some devices can use a wired (faster) connection often called "ethernet" to access the Internet:

Even if your device does not have an ethernet *port*, it might still support a wired connection via an adapter – which is yet another good reason for knowing which *device* you are using.

Check your connection(s)

If *context matters* (and it does) then so do connections.

If you are using a wired device that is unable to connect to the Internet, check its physical connection.

The above photo shows an ethernet cable plugged into an ethernet *port*. Ports (or lack thereof) will come into play in the *Ed Loves Trains* chapter 🚆

Home Internet

Home internet usually involves an Internet Service Provider (or "ISP" for short).

The home internet service options you have available will depend on your physical location.

Home internet gotchas include lost Wi-Fi passwords, faulty *hardware*, unsecured networks, poor Internet service, and exorbitant costs.

Ask your friendly neighbors who their home internet service providers are, how much they pay and how satisfied they are with their ISPs.

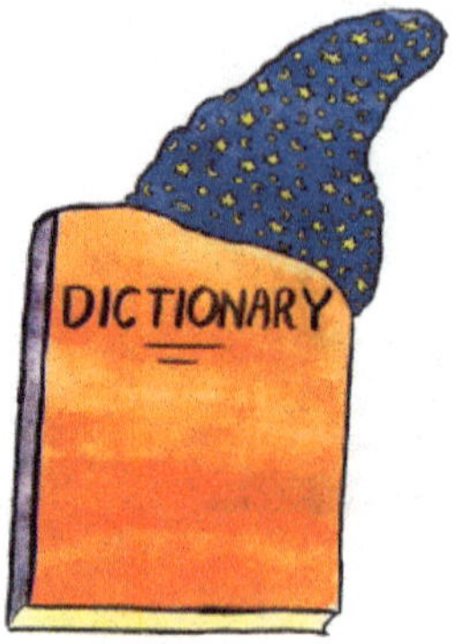

microcosm noun
mi·cro·cosm | \ ˈmī-krə-ˌkä-zəm \

in microcosm
: in a greatly diminished size, form, or scale

What is all this stuff? What does it cost?

Even if you do not have Internet at home and are not even considering being able to access the Internet from within the comfy cozy confines of your cloistered domicile, this section is still worth a read because researching, understanding and documenting the "nuts and bolts" of home Internet service is a microcosm of twenty-first technology use as a whole.

The first step is to gather some intel.

Make a cup of tea and casually embark on a no pressure, fact-finding mission ☕

Take notes (in your *technology notebook*). Write as neatly as possible. Take your time. Proceed methodically.

Low-key but productive procedures like this can provide a self-satisfying and tension-easing sense of **control** over your technology milieu.

- Check your bank and credit card statements 🏦 💳

- How much are you paying for anything related to technology and who are you paying it to?

- Are there cheaper options in your area? Are you paying for something that you do not need or use?

- Do you have any internet service hardware in your home such as a modem and/or a router? Or hey, maybe even some kind of a gizmo that acts as both?

Know what things are and what they do

It is important to know the correct terminology of things – for the purposes of documentation and getting good help if you need it.

For example, a router creates a network between the devices in your home, while your modem connects that network – and thus the devices on that network – to the internet.

If you do have (or think you may have) any such *hardware*, seek it out and examine it carefully.

Carefully write down in your *technology notebook* any information you see on it such as model name, model number, serial number, et al.

"The oldest and strongest emotion of mankind is fear, and the oldest and strongest kind of fear is fear of the unknown"
— H.P. Lovecraft, Supernatural Horror in Literature

Vagueness and abstraction can generate anxiety (and even shame) within adult technology users.

The very process of making unknown things **known** – that is getting a clear picture of what you are using and identifying the multimedia appliances you are living with – can help lower your technology anxiety level and empower you.

It may also make it easier for you to get *quality help* and possibly find better (and cheaper) options for what you want to do.

It comes down to taking control of your technology so it will not take control of you.

"It's great to learn
'cause knowledge is power"
— Schoolhouse Rocky

If your home internet has been set up by other people, talk to them 💬

You need to understand exactly what you have and what you are paying for because it may not fit your needs.

And even if these "other people" who set up your home internet service (or configured your devices) are 💯% trustworthy, they may not always be around to help you when something goes kaput or something unexpectedly changes.

If you have questions about your home internet service, get your ISP's phone number by hook or by crook or by phone book or by librarian.

Get the complete skinny on your internet service and write it down in your *technology notebook* and study it.

Call back and ask for clarification on anything you are not sure about.

(NOTE: Your Tech Wizard Tour Guide, "Tabby"🐱 & "Tough Cookie"🐭 are not really support specialists.)

The search tool

sleuth verb
sleuthed; sleuthing; sleuths
intransitive verb
: to act as a detective : search for information
transitive verb
: to search for and discover

You can search the entire World Wide Web…

…or just your device:

Searching the web 🔍

Searching the web can be your first step to complete any technology-related task or project.

You could for example, be curious about what one of the keys on your keyboard is for 🔍

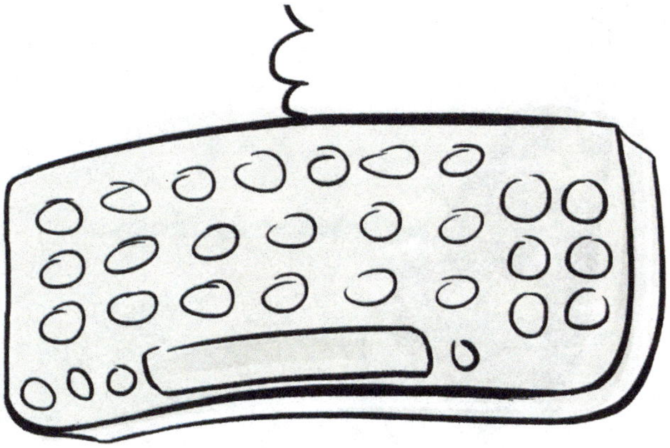

You might want to know how you can create emojis with your keyboard 🔍

You might be interested in setting up email on a tablet that you were just gifted 🔎

Can you video chat on your device ❓ 🔎

Example: How do you start an online business?

Example: How do you use physical materials in a digital presentation?

Web browsers

A web browser is an *application* that allows a user to display web pages on a device 🐿

And on web pages, you will find content including text, images and audiovisual materials.

By the way, a *website* is a collection of web pages, just as a book is a collection of paper pages.

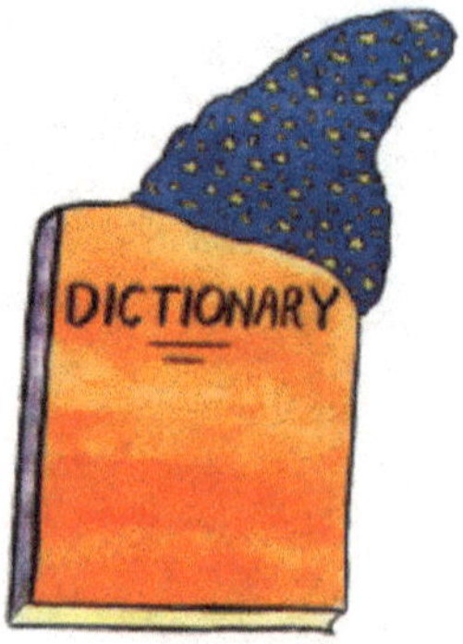

Web searching procedure

1. Connect your device to the Internet (if it is not connected already).

2. Start your device's web browser 🍜

3. Type what you want to find out into the web browser search box:

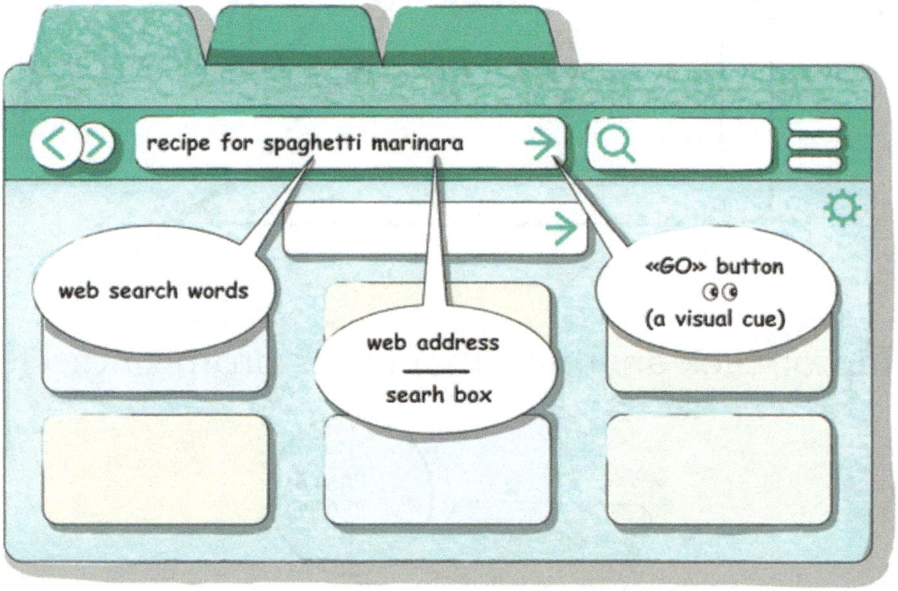

Remember to look for *visual cues* and **functionality**. For example, the above example browser window actually has two search boxes.

4. Peruse your *web search results page...*

5. Tap or click anything that looks promising:

You can return to your *web search results page* by clicking or tapping the web browser's "back button."

If you would like to "bookmark" a web page, look for a *menu selection* or a button that would enable you to do so 🤔 👀

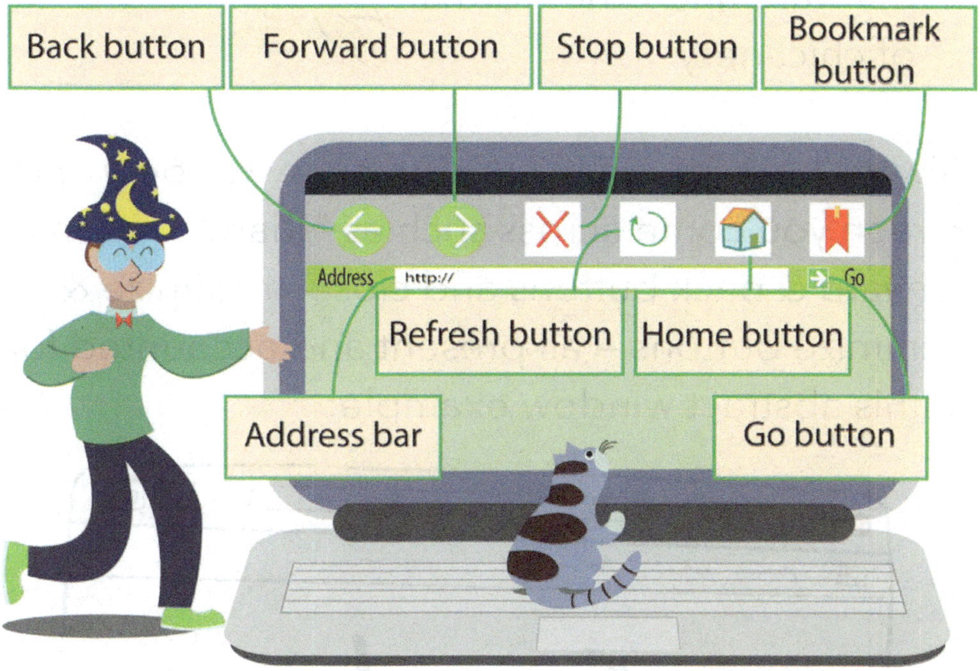

If your search results are not applicable to what you want to know or what you would like to do, try refining your search words.

Web searching "gotchas" include accessing:

- outdated information
- untrue information for your particular device and/or app
- sales pitches
- malware and other forms of chicanery

Always look for *visual cues* on every window and screen you come across such as "share" buttons, forward & back buttons and close, maximize & minimize buttons – all present and accounted for in this abstract window example:

Search Engines

Your web browser's search box is powered by a search engine.

Exactly **which** search engine will depend on your device and its web browser app(s).

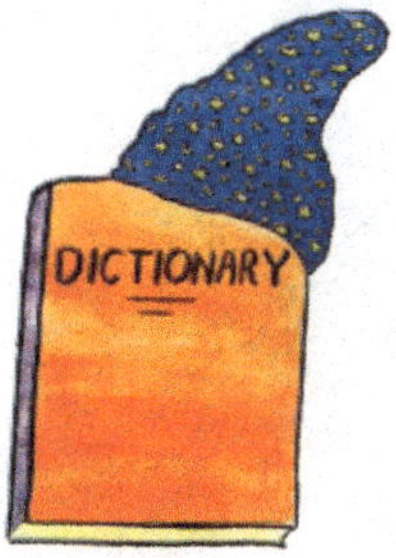

A search engine is a **service** that allows Internet users to search for content via the World Wide Web.

(fictional) search engine "homepage"

"Where's my stuff?"

This is truly one of your Tech Wizard author's "greatest hits."

He has been asked this question approximately 100 gazillion times, at least!

It starts **here** (i.e. from the moment you save a file and name it) and you need to know where "here" is...

Be aware, be apprised, be alert, whenever you save a file.

Did you...

- Save your file onto your device?

- Save it to an external storage device?

- Save it to a *cloud service*?

- Save the file on the device and **then** upload it to a *cloud service*?

You can **browse** for a particular *file* simply by clicking or tapping around your device's "filing cabinet" – which you got a general sense of as you were *noodling around your device*.

Or you can use the "search tool" to search for a particular file using words that you know are **in** a file or in the **name** of a file.

You could even search your device for **all** files that have a particular *file name extension* (such as **.doc**).

You may have to use *gestures* to reveal your device's search box.

Exactly **how** you can search your device will vary between devices but will likely involve a search box.

The important thing to know is that such functionality exists – you just need to make a nice cup of tea and locate it with *gestures* or by *noodling around*...

You can search for anything that has been stored "locally" on your device – such as *apps*, *files*, contacts, calendar events, et al.

You can search for anything that is stored on an external storage device (i.e. a *peripheral*) that is physically **connected** to your device.

You can search for anything in your *cloud* – **if** your cloud is connected to your device through an *app* (more about this in chapter 6 ☁).

"Those who dare seek, shall find what they are seeking for."
— Lailah Gifty Akita, Think Great: Be Great!

The FORMal tool!

It seems a bit odd that something so low-tech and well, mundane as filling out a form would still be so important in today's digital age – but it is!

Forms are very **powerful**!

There is frankly not much we can do with technology that does not involve a form at some stage of the proceedings.

An email message box is a form:

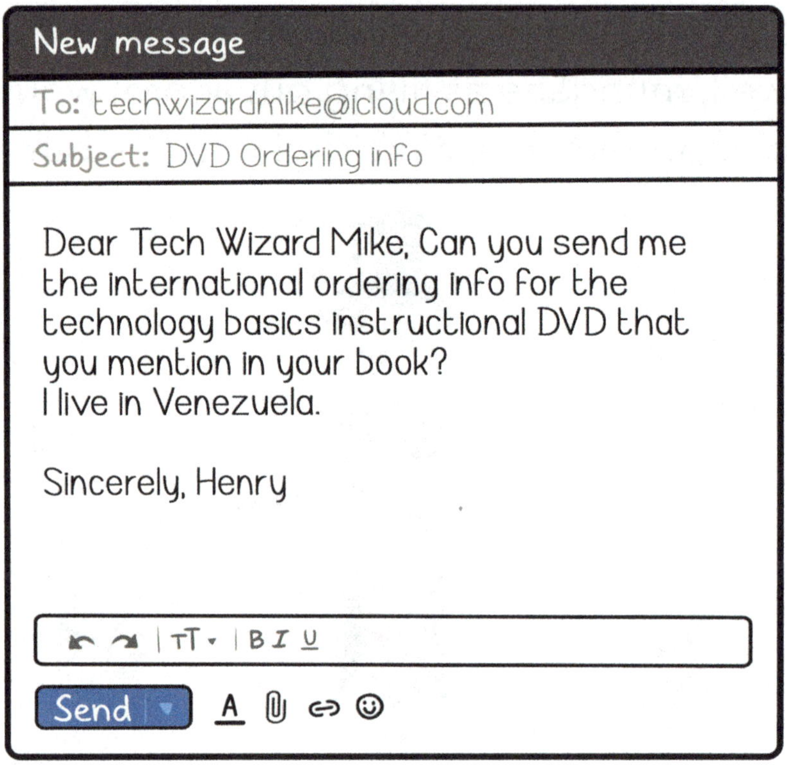

Please note that the above form also has **functionality**.

For example, the "curved arrows" are redo and undo buttons.

There is a formatting button (A), an attachment button ( ), a link button (to insert links to web pages within an email message) and an emoji button ( ).

When you have a chance, make a nice cup of tea and use the technology basics (🐭 ☝️) to *noodle around* with things like these and see what they do.

Filling out a form on a device is **conceptually** the same as filling out a paper form (and really about as complicated).

But even so, people often come to your affable Tech Wizard because they are unable to do certain things with their devices such as update their system software or download new apps 🍿

They are unable to do these things (and other important things, besides) because they did not **take note** of the usernames, passwords and security questions that they created for themselves via the forms they completed when they first turned on their new device.

We need to fill out forms on our devices the same way we fill out important paper forms – with thought, neatness and care.

As with many things related to technology, the skills needed to complete a form are few in number and rudimentary (i.e. click, tap, type and scroll).

It is how we apply them to the task at hand that matters.

Common Form slip-ups

(all of which have been committed by your Tech Wizard Tour Guide.)

1. Trying to rush through the process

The expression "haste makes waste" is **quintuply true** when it comes to completing forms.

Read each part of the form carefully before you click, tap or type **anything**.

2. Not noting the passwords and usernames they created and/or the answers to the security questions that they selected

Eeeek! This is where forms can truly come back to bite you! Carefully and *methodically* write these things down and be sure to understand what they are for 🤔

3. Trying to put a square peg into a round hole

Once again *context matters*! You (probably) wouldn't want to sing that unforgettable aria from Puccini's La Bohème while standing in line at the supermarket...

Nor would you want to type one of your passwords into a first name box:

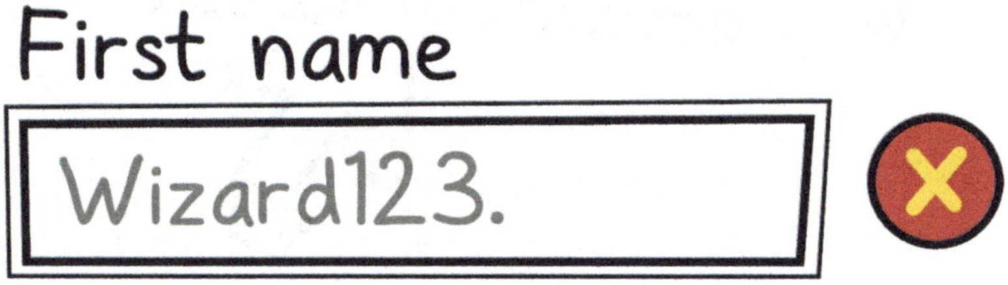

4. Pressing the Enter / Return key on the keyboard **before** the form is filled out:

Pressing the Enter key will attempt to **submit** the form.

It is the same as clicking / tapping a "Submit" button.

And attempting to submit an **incomplete** form will generate some kind of error message and you may have to start all over again.

If the form does not submit (or had an unintended result), look for *visual cues* as to why…

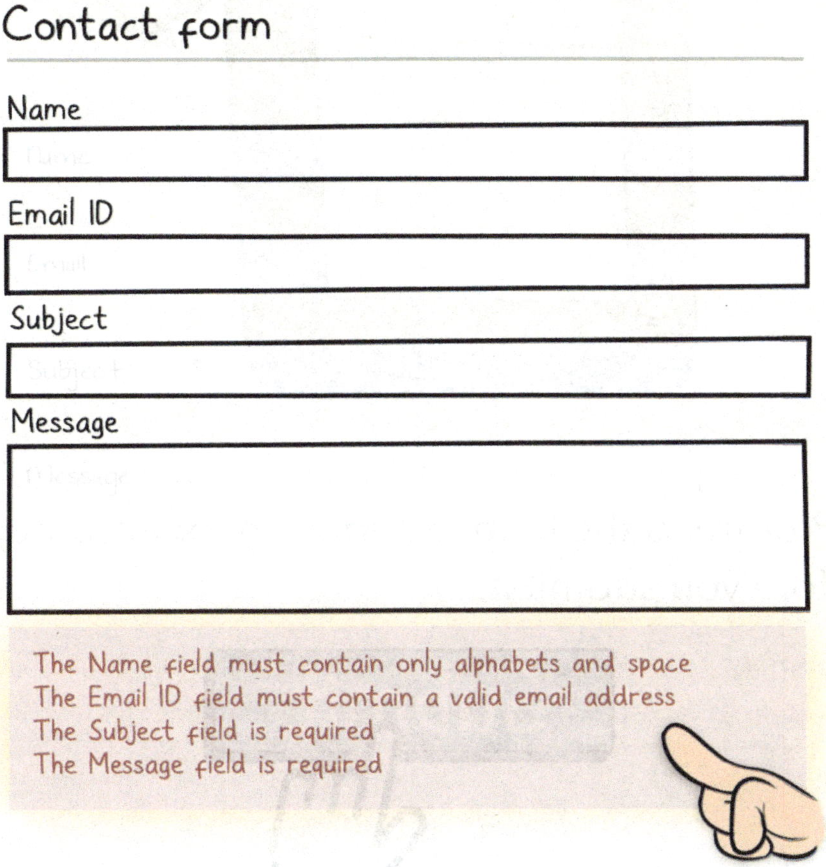

Things to remember about filling out forms

1. Proofread the completed form for accuracy before you submit it.

2. The blinking line is a text insertion point. It needs to be in any text box (including *web address / web search* boxes) that you wish to type into.

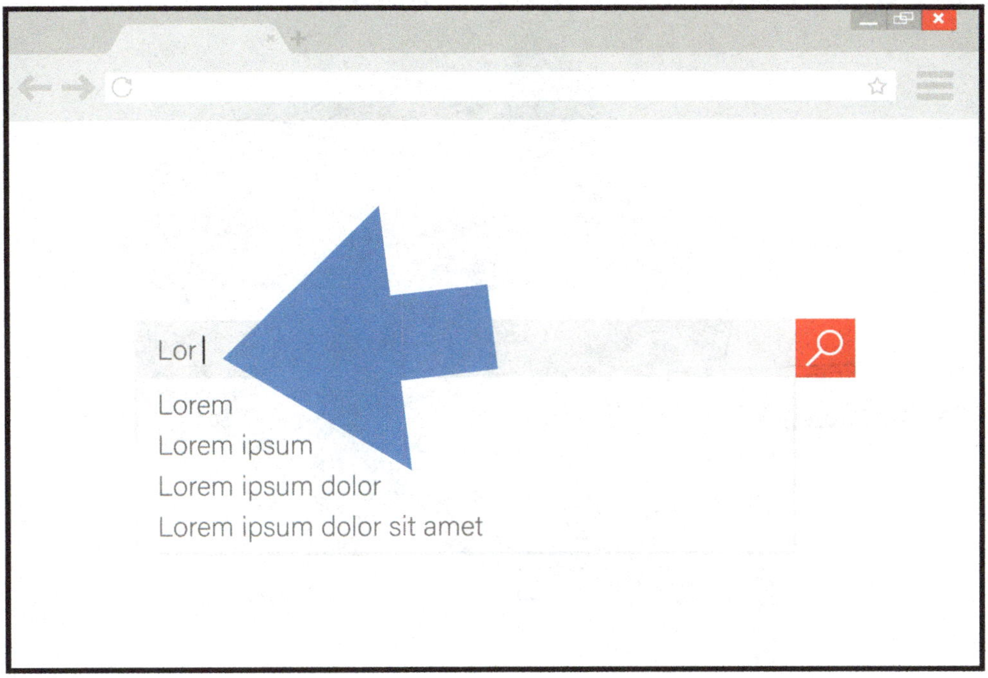

If you are using a physical keyboard with your device, you can press the "tab" key on your keyboard to move the blinking line (i.e. the point of focus) from one field on the form to the next.

3. Ask yourself "What is this form for?" Is it for a cloud account? Is it for device login? Is it to administer the device? It is so **important** to understand exactly what a form is for. If you are not sure, then get some *good help*.

We've all been here!

Staring at a login screen without a clue.

Being mindful (while taking notes) is the key, the key, the key.

4. Carefully write any usernames and passwords you create for yourself (+ any additional info) into your designated *technology notebook*.

5. If you can't see the entire form (or to make sure you can see the entire form), **scroll up and down** using your mouse, keyboard arrow keys, or fingers (depending on which device you are using).

The menu tool

Just like in real life, think about what you would like to have 🤔 Regardless of the device you are using, menus are often key to getting stuff done with technology – and they are usually pretty intuitive to use as they correspond (at least conceptually) to restaurant menus.

Menus provide "secret" functionality 🤫

Menus can be hidden away (until you summon them) or can be in plain sight on a screen – but in disguise 😯

In a nutshell, there are menus within applications....

There are "Start" menus...

There are "Settings" menus...

Query: How do you **find** these app menus, Start menus, and Settings menus of which you speak?

(hint: p. 137 of Help for the Technology Shy Part 1)

This mouse has two hands!

(for a very good reason)

Look for visual cues in *hardware* as well as *interfaces*. You can see that this mouse has two mouse buttons (and a "scroll wheel"):

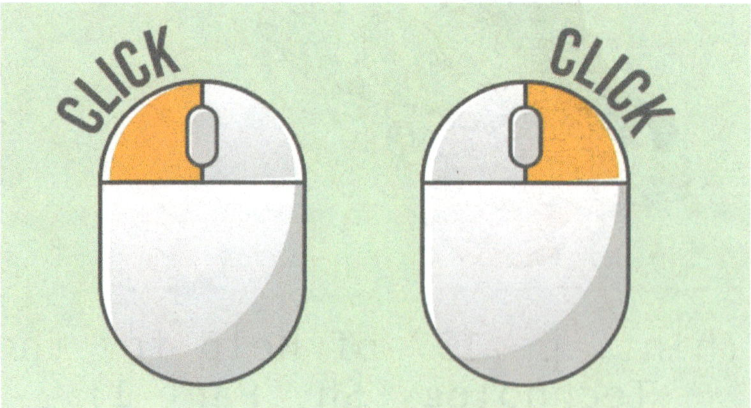

Clicking the right mouse button will cause a **context menu** (or "pop-up" menu) to appear:

The context menu is the secret agent technology tool because he is **hidden** until called.

Context menus can also be summoned with "mouses" that do not have a right mouse button.

Context menus can also be summoned on devices that only have a trackpad.

Context menus can also be summoned on touch screen devices.

The **selections** on a context menu will vary depending on which app is currently being used and what the tip of the screen arrow (or your finger or stylus) is currently over.

It is very important to remember that context menus **exist** and can be instrumental in getting stuff done with many devices.

Once you know *which device* you are using, you can easily find out how to summon a context menu on it by *searching the web*:

NOTE: You may have to add more details from your device to your web search…

Chapter 5: Think About What You Want to Do and Then Look Around for a Way to Do It

This is a simple but elegant and very **effective** formula for getting stuff done with technology.

For example, your author was hosting a library "Tech Meetup" on video editing...

"Step-by-step instructions for doing anything will often be right in front of you..."

"Sue" was very **intensely** taking notes before finally exclaiming "I can't write fast enough!"

In doing so, Sue exhibited a lack of calm and a total reliance on step-by-step instructions.

Neither of which are conducive to getting stuff done with technology.

Sue's anxiety stemmed from the fact that she thought she needed to have **everything** she needed to do in order to import, edit and share a video in writing as a series of *step-by-step instructions*.

The fact is, all ten pages of her hastily scribbled instructions could have been condensed into a single sentence:

Think about what you want to do and then look around for a way to do it.

Apply real life methods to your technology use...

For example, crossing a busy intersection is actually a pretty complex operation that we routinely accomplish in real life...

In order to achieve your current *objective* of crossing the street safely, you would look for *visual cues* (e.g. painted lines on the roadway, traffic lights and completely stopped cars).

You would proceed **mindfully** (e.g. look left, right and left again) before doing anything such as stepping off the sidewalk.

Approach desktops, home screens, and app screens the same way.

Quite often as you look at a screen, the "step-by-step instructions" will be right in front of you (so to speak).

One thing that was vexing Sue in particular was how to get the raw (i.e. unedited) video from a folder on her computer **into** the video editing app so she could work on it.

No written instructions were necessary for that particular task because there was an "Import Media" button in plain sight...

And there was an "Import Media..." *menu selection...*

Question: How did Sue get the raw video from her smartphone into a folder on her computer?

Answer: Once again that comes down to knowing *which device*(s) you are using.

Sue could have *searched the web* for how to get video from her smartphone to her computer.

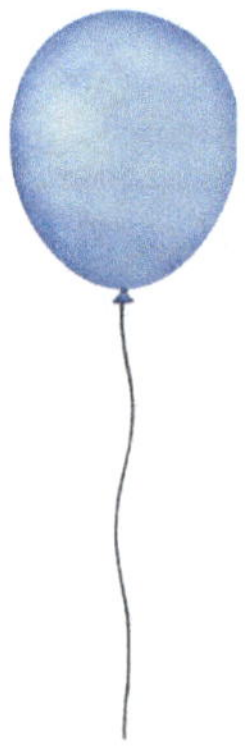

But Sue did not have to do that because the video she recorded on her smartphone automatically (albeit **eventually**) appeared on her notebook computer because the two devices were linked together via a *cloud* service 🍿

Sue broke down her project into the following action steps...

1. Record the video.

2. Get the video into her computer.

3. Edit the video.

4. Add title and end credits.

5. Save it.

6. Send the video to her friends.

(In addition to giving Sue a "map to follow", this set of written action steps made her feel more relaxed and in control of her technology use.)

Each of those action steps began with... You guessed it!

Think about what want to do and then look around for a way to do it.

Want to do something?

Reach into your *Getting Stuff Done With Technology toolbox* and try out your tools on the *app interface*...

- click or tap the buttons
- peruse the menus
- see what the *context menus* are

...in order to achieve your current action step (such as splitting a video clip):

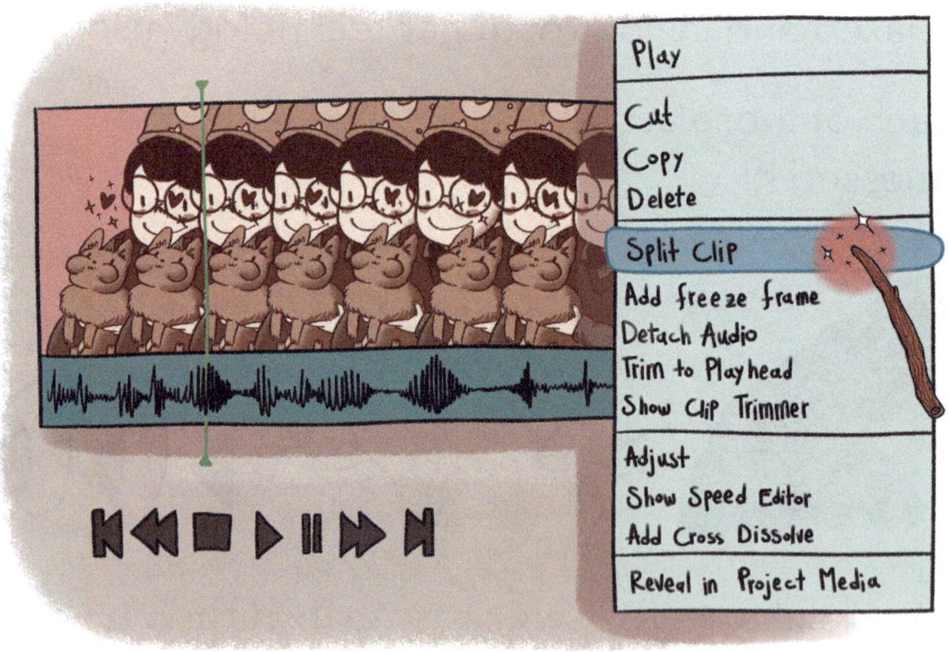

Chapter 6: The Cloud Chapter

If you choose to keep your *files* (e.g. photos and documents) and data (e.g. contacts and calendars) in a cloud, you are not really saving them in an actual, factual cloud ☁

Rather you are saving your files and data on a company's server:

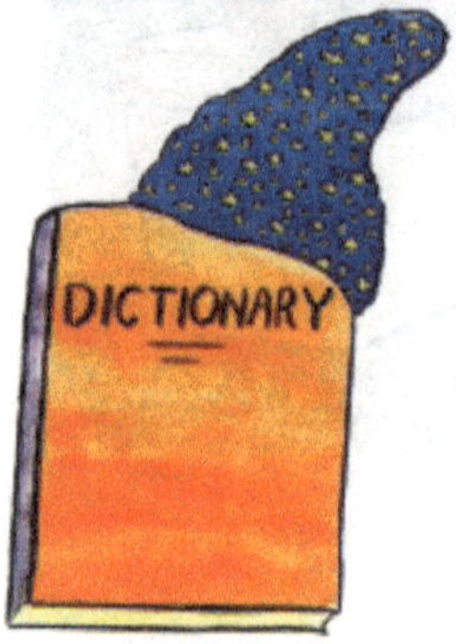

A server is a special computer that serves files and data upon request.

You can then access your files and data from many devices that can connect to the Internet:

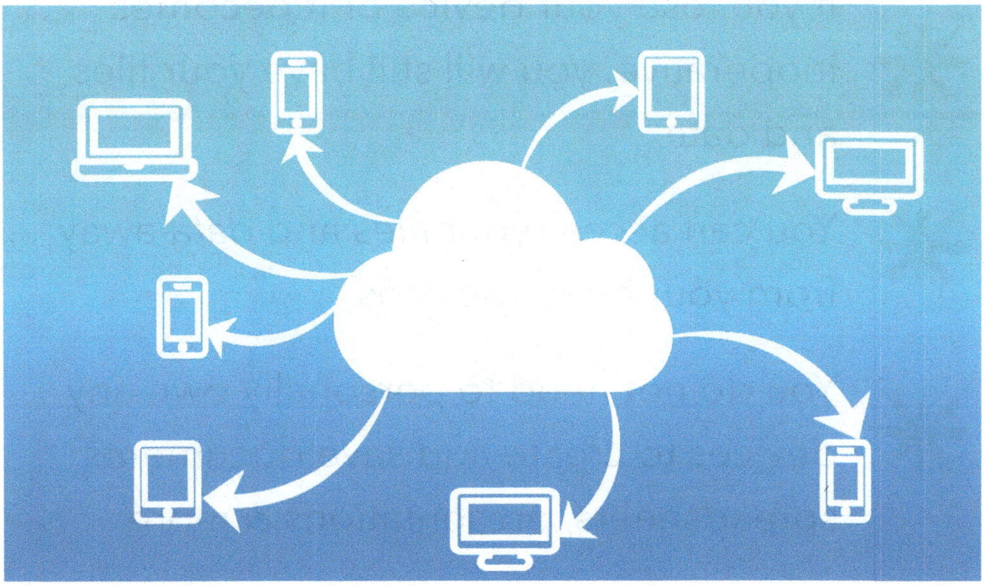

You will of course need your cloud account web address, *username* & *password* – all of which you diligently recorded in your *technology notebook*.

Email, by the way, is a cloud-based service that you can access through a *web browser* **and/or** through an *app* on your device.

Advantages of using a cloud service

- If you lose your device or it becomes inoperative, you will still have your files and data.
- You can access your *files* and *data* away from your home devices.
- You do not need to personally own any devices to create and save documents, spreadsheets, presentations & more.

Disadvantages of using a cloud service

- It may not be as cost effective if you need to store a large number of files or if you need to store files that have large file sizes.
- It can be slow as files need to be uploaded and downloaded through the Internet.
- The cloud service could go out of business (taking your data with it).

Your author recommends backing up your important files and data in **both** a cloud account **and** on an external storage device (which is a *peripheral*).

NOTE: This very chapter went *poof* because your author took his eyes off the ball 🙀

It disappeared from his *cloud* account without a trace and had to be redone from (almost) scratch – because it was **not** backed up on any external storage device.

Thankfully, the text and images were still in his Notes app and Photos app, respectively 😥

Tips for selecting a cloud service

If your device does not come with free cloud storage (or even if it does), you can *search the web* for "cloud storage" and/or "cloud services" and see what they have to offer in terms of free storage.

You can upload and download files and data between your cloud account and various devices (either yours or somebody else's) that can connect to the Internet.

For simplicity and ease of use, you can check your device's *App Store* for an *app* that you can use to connect your cloud storage account directly to your device...

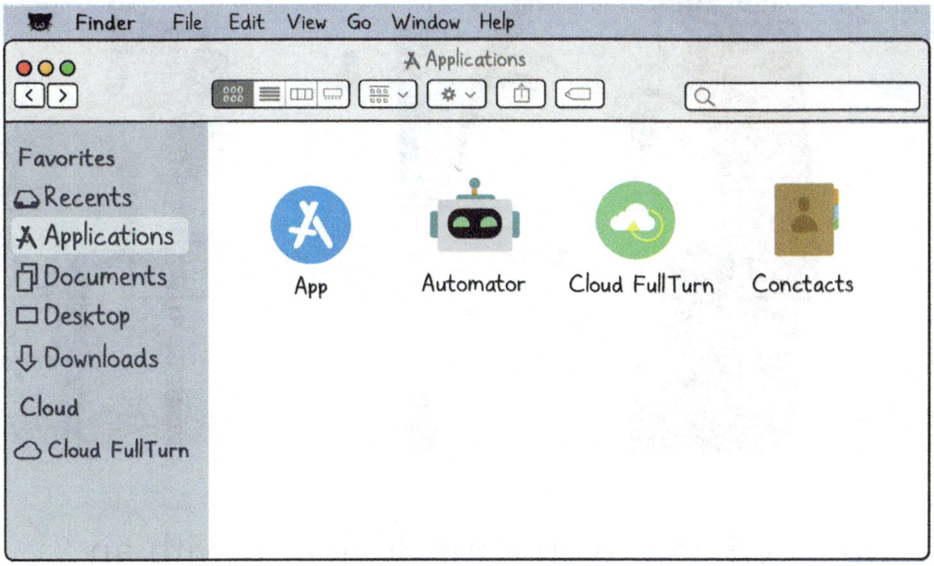

TIP: Make sure the app is from the same company that is offering the cloud storage 🍿

Chapter 7: Good Help Is Hard to Find...

If you need help with your device or with an *application* on your device or with a *cloud* service you can access through your device, then quality help is available from...

1. *searching the web*
2. online support communities
3. official customer support ☎️
4. real life human help

(You do need be able to separate the wheat from the chaff, though.)

Web searching

A good way to find out most anything is simply to type (or dictate) "How do I…" or "What is…?" into a *web search box*.

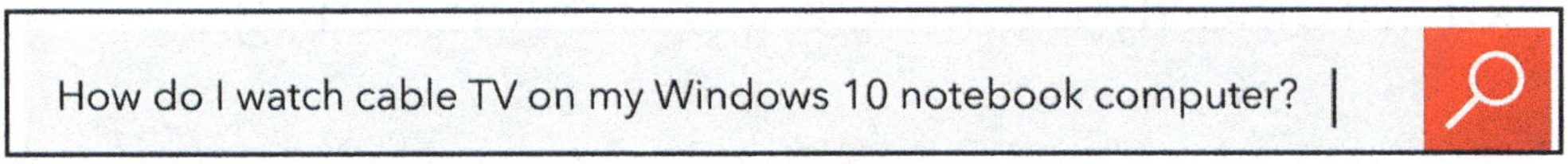

(This may be a solution or simply the first step to completing a project.)

If you have a technology-related issue or question, you should have it clear in your head (write it down, if that will help you stay on track) **before** you reach for your computer, tablet, smartphone, et al.

You should also have your *technology notebook* handy because you may also need to know…

- *which device* you are using
- the *operating system* that is running on that device
- the *app* that you are using on that device (if any)
- and the **version** of that app

Official support communities

You may very well find solutions and tutorials for your particular task / project by searching the web but online support communities can be extraordinarily beneficial (and even caring).

These official communities are overseen and administered by the manufacturer of a particular device or the creator of the device's *operating system*.

In such a forum, any assistance you receive will (likely) be from a real person who is knowledgeable and simply wants to help.

But once again, please be careful.

This is a lot of hoodwinking on the web so do not provide any financial or personal information.

Someone from an official online support community should not ask you to pay.

Official customer support

If your device or your device's operating system offers free customer support then by all means, call them:

You can *search the web* for your device's customer support phone number (chat and email may also be options) or visit your local library.

Once again, please do not provide your credit card number or any other personal information (including your address 🏡).

Real life human help

Real life human help in turn can be broken down into the following five categories...

A. someone who really knows what they are talking about
B. someone who can point you in the right direction
C. know-it-alls
D. well-meaning jackasses
E. nogoodniks

A. Someone who really knows what they are talking about

There are experienced, knowledgeable people who simply want to help (without any ulterior motives).

Your local library or adult community center may have such a person in residence. If they do, then by all means pay them a visit.

But again, make sure you can **clearly relate** what your technology-related challenge is and/or what your ultimate *objective* is.

Also keep in mind that people are often hired for positions (volunteer or paid) that they are not qualified for.

Look for signs of a "Know-it-all" and a "Well-meaning Jackass" – and read the "This Guy" case study at the end of this chapter.

B. Point you in the right direction

Celia said a wired connection was not possible because her smartphone's cable would not physically fit into the speaker audio jack.

(This was a shining example of incompatibility between "older" and "newer" technologies.)

Your author suggested that Celia could...

- Examine the speakers for a brand name, model number and *ports*.
- *Search the web* for devices that are compatible with those speakers (as Celia mentioned she had a number of "old" devices at home).
- *Search the web* for an **adapter** to make a wired connection possible between her smartphone and the yoga room speakers (which she was able to find).
- *Search the web* for speakers that are **compatible** with her smartphone (which became her preferred solution 👍)

This enabled Celia to take control of her technology use and make her own informed decisions.

C. Know-it-alls 🧐

Know-it-alls use technology (amongst other things) as a means to project their self-importance.

You cannot trust such people. Know-it-alls are generally not interested in affecting a positive outcome.

Their primary goal is often to promote the superiority of their chosen devices, *apps* and *operating systems*.

Look for someone who is thoughtful, methodical, open-minded and goal-oriented (as opposed to having character traits to the contrary).

Look for someone who takes the time to understand your issue and/or objective.

Try to verify or at least cross-reference what you are told by anybody.

D. Well-meaning jackasses

Such folks tend to know just enough to be dangerous (literally) and they have been the bane of your Tech Wizard author's existence.

They merrily pass along a lot of misinformation.

Well-meaning jackasses usually mean well (hence their name) but their often unsolicited advice can be extremely counterproductive to your *objectives*.

Well-meaning jackasses tend to be giddy and verbose (to use a kind word) instead of calm, thoughtful and methodical.

Such folks will reflexively offer quick and easy solutions (e.g. "just download an app!") when there are other considerations such as additional *hardware* and *cloud services* in order to develop a **workflow** to complete a particular task or project 🍿

A well-meaning jackass interrupted one of my tech help sessions and advised the person I was working with to clean her smartphone's delicate, electronics-filled charging port with a metal paperclip 📎 – an action which could have damaged the smartphone and generated an expensive repair bill.

Verify (preferably with an official support community or with *customer support*) when some random person enthusiastically tells you to do something.

E. Nogoodniks

Surrendering control of your device(s) to anybody – even a reputable person – is never the right course of action.

For example, my friend Helen wanted to use her notebook computer but was unable to do so because "Ted the Computer Guy" had not completed the *system updates* yet – **which could have easily happened all on their own**.

Do **not** try to take the easy way out.

Do **not** get into a "don't make me think" mode as it could lead to a very slippery slope and subsequently a very rocky road.

A very agreeable woman whom we will call "Marty" brought her notebook computer to your favorite Tech Wizard because it was doing some "very strange things" and would no longer function properly.

As it turns out, Marty had given her notebook computer to "This Guy" (she never knew his actual name) who had promised to make her computer safer, run more efficiently, and handle her computer's administrative tasks – for a fee 💸

This dirty, lowdown, lowlife, nogoodnik installed some nefarious *applications* on Marty's notebook computer which your author could not identify 🔎

Your Tech Wizard chronicler **could determine** that those *rogue apps* were basically running amok on Marty's computer – severely limiting its functionality.

"This guy" also changed the administrator *password* on Marty's computer…

What to do...?

1. The first order of business was for Marty to use another device (or a telephone) to contact her bank and credit card companies to let them know that her computer had been **compromised** (as Marty had been doing online banking and shopping from her notebook computer).

2. Next, Marty needed to regain **control** of her computer so we called the customer tech support number for Marty's device who were able to make that happen after a long arduous procedure.

Thankfully, Marty's device had good customer support so we were very lucky in this regard.

3. Now that Marty was once again the administrator of her computer, we were free to **remove** the nasty apps "This Guy" had installed.

4. Just to be on the safe side (and with the approval of customer support), we *factory reset* Marty's computer as there could have been remnants of the rogue apps on the computer.

5. All of Marty's files and data were in a *cloud service* so she did not lose any important *files* or *data* (e.g contacts, reminders, notes & calendar events) – but she needed to change her cloud account password.

Advertisements galore!

Ruth, a very calm, pleasant and **observant** person, had a feeling something was not quite right with her web searches.

For starters, she was suddenly seeing a far greater number of advertisements.

After viewing Ruth's *web search results pages*, it was apparent that her web searches were being hijacked 🙀

Ruth's computer had been infected with a "browser redirect virus."

Question: How did your attentive author know this at a glance?

Answer: Well frankly, a redirected *web search* is like pornography in that it is hard to describe but you know it when you see it.

To back this up, your Tech Wizard scribe showed Ruth what *web search results pages* were supposed to look like.

Carefully, your Tech Wizard Tour Guide walked Ruth through...

1. identifying the *adware* by comparing web searches on her notebook computer to web searches on another device
2. removing the adware (which turned out to be a quick fix)
3. researching exactly **why** Ruth's computer was so susceptible (actually wide open) to *adware* and other forms of *malware* (because Ruth's computer was *obsolete* and was no longer receiving *security updates* and **also** the *web browser apps* on Ruth's computer were not receiving security updates)
4. discussing her options (which turned out to be buying a new device and repurposing her *obsolete* notebook computer)

At the risk of sounding immodest, this was good help, folks!

Define "obsolete"

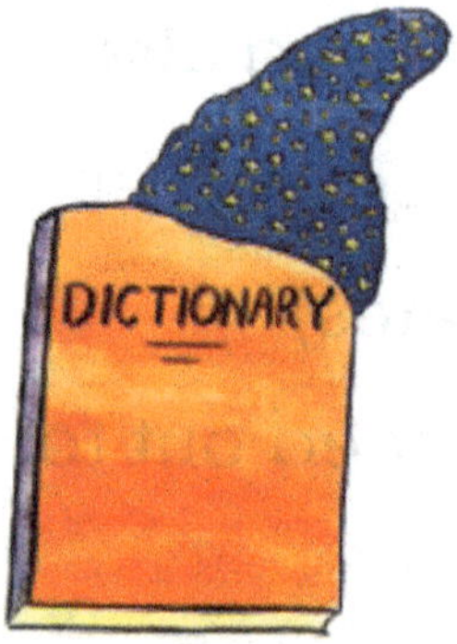


```
In technology the term
obsolete means the
hardware is no longer in
use, even though it may
still be in working
order.
```

Obsolete devices can still be put to productive use but you would not want to connect them to the Internet.

If something does not work properly on a device (e.g. your author's "Reminders" currently not showing up on his 7-year-old smartphone), it may be due to the age of the device.

Your author is not big on third party virus protection products but OS updates are essential for security.

Chapter 8: The Seven Deadly (Or at Least Counterproductive) Sins of Technology Use

The following "sins" (for lack of a better word) will derail your technology train just as sure as God made little green apples – Hence this book and *supplemental materials* 🖋

1. Lack of focus

2. Obstinance

Taking the "my way or the highway" approach to technology use (instead of getting *good help*, going with the flow and doing what works).

3. Clinging to *obsolete* devices

4. Negativity

The mindsets on this page are not technology challenges but actually technology non-starters.

5. Impatience

This is chomping the digital bit; engaging in a technology mad dash.

This is proceeding full-tilt boogie with zero regard to the very likely adverse consequences.

Be *methodical*. "Haste makes waste" is especially true when it comes to technology use.

6. Endless Looping

This is taking your eye off the ball. This is not allowing yourself to be **done** with a task or project by simply getting lost / bogged down in the technology-related nuts and bolts (i.e. losing sight of the *forest through the trees*).

7. Having unrealistic expectations

There is nothing remotely magical about contemporary technologies🔮

(Please keep in mind that our devices are all as *dumb as a bag of hammers*.)

When expectations are realistic, it is easy to experience success and feel personally valuable.

Realistic expectations have the power to make you happier.

Just for fun...

This chapter was originally intended to be "The Seven Deadly Sins of Technology Use" (but your Tech Wizard wordsmith did not want to get all biblical on you...)

Chapter 9: Case Studies

The Chilled College Girl

An agreeable and relaxed young lady who we will call "Lea" said she could not connect her notebook computer to the library's public Wi-Fi.

Sesame Street's "One of these things is not like the other" song often comes in handy when you are trying to figure out why something does not work on a particular device.

Other people were successfully connecting their devices to the library's Wi-Fi network, so why couldn't Lea?

In fact, your author had the **very same** notebook computer that this young lady was using and could connect to the library's Wi-Fi network without issue...

Your observant Technology Tour Guide then spotted something on Lea's computer screen which his notebook computer did **not** have...

When trying to figure something out, look for *visual cues* 👀

When asked about this "mystery app," Lea explained that a friend of hers had installed it.

Some friend 😒 (no doubt a *well-meaning jackass* or a *know-it-all*).

We *searched the web* for info on this mystery app and sure enough (according to her device's *official support community*), it was a form of *malware*. Boo!

We simply deleted this app from her computer and Lea was able to connect to the library's Wi-Fi network just fine.

Recap and Lea's keys to success

Lea was a girl on a mission! She simply wanted to get her work done and never took her eye of the ball.

Lea accepted things as they were 🕊️

After the Sesame Street "one of these things is not like the other" comparison and the research that we did on that *app* (which revealed it to be *malware*), Lea was all for **deleting it** – simply removing that obstacle from her path.

Such a positive mindset enables all of us to move **forward** toward our goals…

Rory the Writer

Obsolete devices and no longer supported apps can be problematic.

An aspiring novelist's manuscript has disappeared!

A cryptic message was received!

It looks grim but we are on the case!

(Sometimes technology use calls for some detective work 🕵️‍♀️)

A young fiction writer we will call "Rory" was housesitting when inspiration struck 💡

There was a bright **orange** computer in the kitchen...

Rory was unfamiliar with that particular computer but she inspected its outer case until she found a power button...

After the computer powered on, Rory noodled around its unfamiliar *interface*...

Armed with her sense of purpose, Rory used the mouse to open and close various *apps* on the computer **until** she found an *app* that she could use to create and format her manuscript...

Rory (not wanting to lose all of her hard work) diligently *backed up* her manuscript in progress onto **several** external storage devices.

Resuming work on a different device...

After leaving her housesitting post, Rory wanted to resume work on her manuscript, so she came into the library, sat down at a public computer and...

- followed the public computer login procedure
- inserted her mobile storage device into one of the computer's *ports*
- opened that storage device on the computer by following *visual cues* 👀

And then...

Rory reached into her *technology toolbox* and used the mouse tool 🐭 to **open** the document file on her external device by double-clicking it ("knock knock!").

The library's computer did not know what to do with the *file* (i.e. her manuscript) that Rory created on the **orange** computer she had access to during her housesitting gig.

So the library computer – and Rory herself – basically just threw up their hands.

And this is when your favorite Tech Wizard chronicler entered the picture...

Since Rory's file was an "unknown file type", we needed to find out exactly **which** *file type* it was so we could (hopefully) find an *app* to open it.

Rory opened her *Getting Stuff Done with Technology toolbox* and used the *mouse* and *context menu* tools to "get info" on her file.

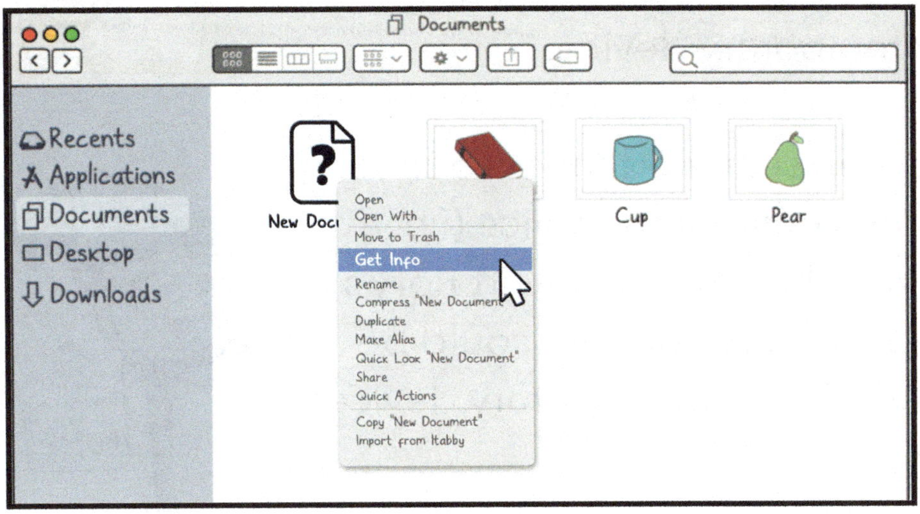

It turned out to be a "CWK file" and your author had no idea off the top of his head what that was.

This being the case, we *searched the web* for information on that *file type* and eventually learned that it could not be opened on **any** contemporary device.

We then searched the web for an online **conversion** service (i.e. a service that could convert Rory's file into a *file type* that could be opened on contemporary devices) to no avail.

As luck would have it, a library member had recently donated a *vintage* computer that had the same *operating system* as the **orange** computer and could run the app that Rory had used to create her manuscript.

The app itself had to be purchased from an online retailer (on CD 💿).

So Rory was **at last** able to open her manuscript on that donated vintage computer 🎉🎈🎊

Now that Rory could open her file, she needed to save it (or maybe "export" it) in a different *file type* that could be opened on contemporary devices.

Your author happened to know that the RTF file type could be opened on any contemporary device (including the library's public computers) but Rory certainly could have found this out on her own through trial and error.

Rory then backed up her manuscript file (now in a new *file type*) on several storage devices and in a cloud service.

Recap

Rory commendably accomplished her *objective* using the *hardware* (i.e. the **orange** computer) and *software* (a word processing *app*) that was available to her at that particular moment in time.

* But Rory had unknowingly used an *obsolete* computer.

* The *app* that Rory used to create and format her novel was no longer available on any contemporary device.

* There was at the time no conversion service available (online or brick & mortar) for the resulting *file type* of Rory's manuscript file.

But Rory ultimately succeeded because...

- She got *good help* (one of Rory's friends had recommended me to her).
- She was a very good listener and had no problem accepting the reality of her situation.
- She remained calm throughout and took the time to understand what she would need to do in order to reach her technology-related objective.
- She proceeded methodically toward her desired *visualized outcome*:

Shop 'Till You Drop, Arthur 🛒

"Arthur" wanted to be able to shop online from a popular Internet-only retailer that we will call "Shoppshack."

Arthur did not personally own a computer, tablet, smartphone or anything that could connect to the Internet.

But Arthur had heard that his local public library had computers that were *connected to the Internet* and he also heard that they were people there who could help him (at no cost 💰).

So in addition to his clear objective, Arthur had access to *hardware*, *Internet connection* and *good help* 🔑

Arthur also had **two** of the *technology basics* down pat which were clicking the mouse 🐭 and pressing keys on the keyboard ⌨️

He just needed some quick coaching on filling out web *forms* and *web searching*.

Arthur also needed a few prompts to follow *visual cues*.

When Arthur heard his new Tech Wizard friend say: *Think about what you want to do* 🤔 *and look around for a way to do it* 👀, he nodded his approval.

Reminder: The technology basics are very easy – but should be used in a calm and relaxed state of mind...

Arthur had a very clear *goal* and was not yet concerned with the technology nuts & bolts of online shopping – which were huge pluses for him.

Know and gather the necessary components to complete a project or reach a goal or realize a visualized outcome 🙌

Upon entering the Adult Services section of the library, Arthur immediately asked what he would need to be able to shop from this popular online retailer.

After some thought, your Public Librarian / Tech Wizard author replied…

🧩 a device (or access to a device) that is *connected to the Internet*

🧩 a *technology notebook* 📒

🧩 an email address 📧

🧩 an account with the online store that he wanted to shop at

Understanding what is needed...

As with the other examples in this book, we are going to paint Arthur's project in broad strokes; a more detailed and up-to-date video tutorial of Arthur's online shopping setup is available at www.technologshy.com/arthur.

Arthur's first puzzle piece: An internet connected device

Arthur's local library had Internet-connected computers that were professionally maintained and free from *malware*.

Arthur was very fortunate here as this is not the case in all libraries 🖊

Arthur's second puzzle piece: a designated technology notebook 📒

Your author recommended that Arthur acquire a *technology notebook* before doing **anything** on the computer.

The technology notebook is an integral part of the *technology foundation.*

As an added bonus, this notebook gave Arthur (a very technology shy individual) something physical to have close by. The notebook made him feel more grounded and in control as he worked toward his technology-related objective.

Arthur's third puzzle piece: Easy Email

Now that he had access to an Internet-connected device (i.e. a public computer) and a designated *technology notebook*, Arthur needed an email address.

Previously, Arthur had no idea how to sign up for an email address but it came down to…

- starting a web browser (which he found on the computer by noodling around 🍜)
- navigating to the email provider website that we will call "moogowoo mail" (by *searching the web* 🔍)
- methodically filling out a *form* to create his new email account 🤵
- writing down his email address, password and email provider web address in his technology notebook 📒

Working together, we broke down Arthur's primary *objective* (being able to shop online) into pieces and then broke down each piece (e.g. getting an email account) into **action steps**...

Getting good help!

Noodling around and getting acquainted

After establishing his email address, Arthur (in no particular hurry) noodled around the email web page (which is an *interface*) to familiarize himself with its functionality:

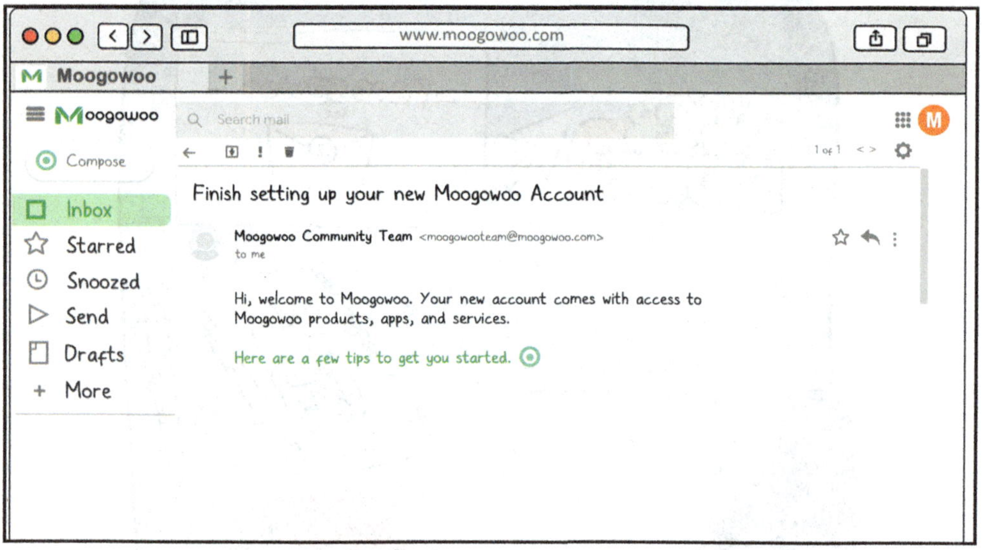

Review...

After establishing his email account and perusing its web page *interface*, Arthur **reviewed** the information about his email account he had carefully recorded in his *technology notebook*.

Stop, rest, digest, let it sink in... ☕

After a long enough pause to properly "digest" (so to speak) and congratulate himself on a job well done for completing this challenging component of his project, Arthur then began contemplating the next piece...

Arthur's fourth puzzle piece: online store account

- Arthur *searched the web* for his online merchant of choice...
- He looked around for a "create account" (or words to that effect) *button* or menu selection 🤔 👀
- Arthur filled out a *form* to create his account with that online storefront
- He carefully recorded his online store account information into his *technology notebook* 📒

Email verification

This particular online merchant required email verification:

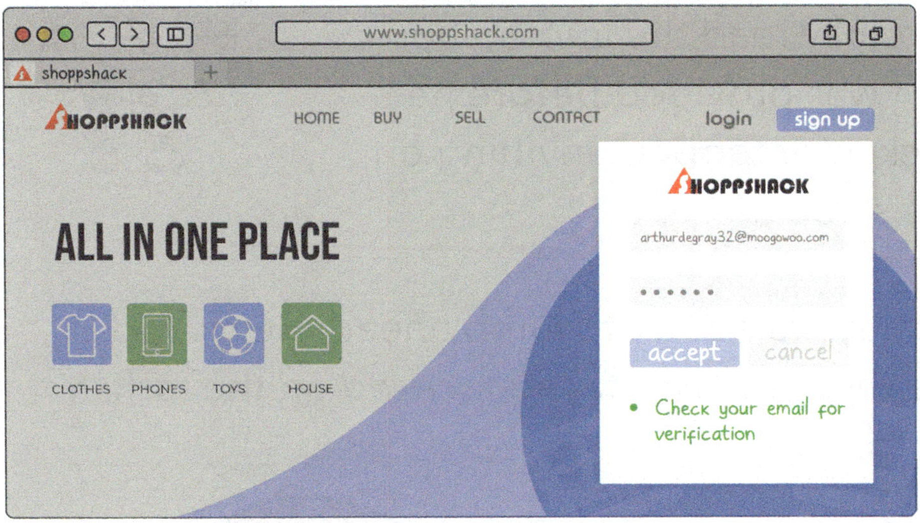

The actual wording and formatting of these messages is not important.

What **is** important is the fact that Arthur proceeded mindfully and *methodically*.

He **paused** on every new web page that presented itself and carefully read the information that was provided **before** he clicked or tapped anything on the screen.

Arthur did not try to jump ahead. He stayed in the moment. He stayed *focused* on the task at hand.

Arthur's "slow & steady wins the race" mindset served him especially well here 🐢 🏁

Arthur had his email web page open in one web browser "tab" and his online merchant account page open in another web browser "tab"...

Arthur **purposefully** opened a new tab so he could switch back and forth between the two but sometimes a new tab would **open on its own** after a web page link or button has been clicked (and we need to watch for things like this).

Arthur used visual cues to uncover functionalities in the apps and online services he was using. He then used those functionalities to create a productive workflow.

Ed Loves Trains 🚊

Ed is a trains enthusiast and has many interesting stories to tell about them.

This scholarly and low-key gentleman also has a considerable amount of information and insight regarding the social and economic impact of the railroads on the New England town that he has resided in all his life.

Ed's ultimate goal, his objective, his *visualized outcome* was to give a presentation on the history of trains in Maine.

Ed had a large collection of archival **physical materials** on his topic of interest including documents, newspaper articles, maps, photos and miscellaneous items such as train tickets and luggage tags.

He wanted to use selections from these materials in his *digital presentation*.

So what was Ed's first step?

With any technology-related objective or project, it is very important to understand what you want to do **before** you reach for any of your devices.

Projects are defined as any desired result that requires more than one action step[1]

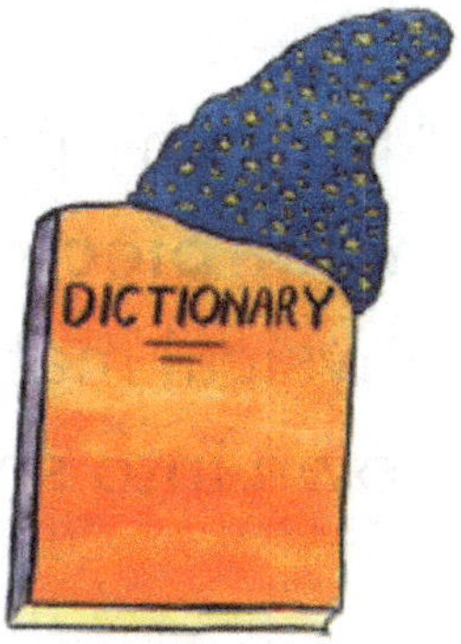

This could be a rather daunting project for anybody 🙀

But Ed wisely broke his *project* (i.e. his *visualized outcome* 🙌) into manageable pieces and proceeded to *focus* on each piece until that piece was in place.

[1] Allen David, Getting Things Done: The Art of Stress-Free Productivity (Penguin Books; Reprint edition, 2002), 37.

And as he proceeded, Ed realized that each **major piece** 🧩 (e.g. the digitalization of his physical materials) would have to be broken down into smaller pieces.

And he was okay with that 😊 🕊️

The key to Ed's success with this endeavor was his calm and methodical manner and not so much his technical expertise.

Because in **that** area (i.e. the technology nuts & bolts so to speak) *good help is not hard to find.*

Ed's first "action step" was to get his **physical** materials into **digital** format.

This first action step could be further broken down into…

1. Research digitalization options 🔎
2. If necessary, consult with people who know what they are talking about.
3. Make a cup of tea and consider those options (and possibly experiment with them).
4. Make a selection (keeping in mind that nothing will be engraved in stone).

digitalization
noun:
the conversion of text, pictures, or sound into a digital form that can be processed by a device.

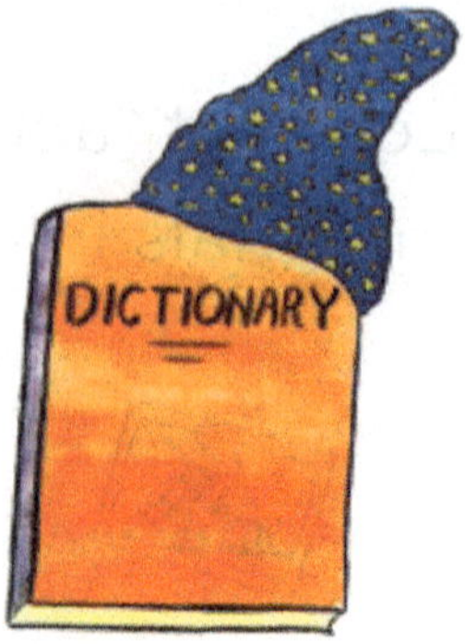

Ed was not yet familiar with the terms "scanning" or "digitalization" so he asked a few knowledgeable (but unpretentious) people who gave him the **terminology** necessary for a web search…

(**Note**: You may have to vary your search words to get applicable information for your particular task or project.)

Ed perused the *web search results pages* and quickly learned that he had at least several **options** for digitizing his physical materials which included scanning *apps*....

... and physical scanners:

After evaluating his **options** (and some experimentation with some scanning *apps*), Ed decided on a local reputable brick & mortar printing company which offered professional scanning services at very reasonable prices.

Ed's **informed decision** was based on the physical size of some of his materials (some of the maps were quite large and would not fit on a consumer scanner) and simply the laborious and time-consuming nature of scanning the large number of physical materials himself.

Subsequently, the printing company **scanned** Ed's physical materials and put them onto DVDs.

Action step 1 is complete ✅ On to step 2!

With the **digitalization** complete, Ed stored the physical materials in a safe place.

Next, Ed needed to **access** the scanned materials that were on the DVDs.

And this is where he hit a snag...

Ed's notebook computer did not have a DVD drive! He could not insert a DVD into it 🤨

Ed knew what he was using and knew what he wanted to do.

So he had *searched the Web* on how to "import" *files* from a DVD into his notebook computer.

This was a very sound strategy but it doesn't always work. And in this particular instance, Ed could not find an applicable solution for this particular challenge.

And here is where your steadfast Help for the Technology Shy chronicler entered the picture.

Before our meeting, Ed had thoughtfully formulated a very specific question:

"How do I get my files from these DVDs into my notebook computer?"

There will often be many options to complete any task or project including:

- an *app* that your device currently does not have
- other devices (yours or somebody else's)
- *human assistance*
- *cloud services* ☁
- *peripherals*

(This is why clarity of purpose is indispensable.)

For example, Ed could have borrowed a computer that had a DVD drive....

...and then "uploaded" his scanned files from his DVDs to a *cloud* account...

And then **downloaded** those files from his cloud account to **his** notebook computer:

Conceptually speaking, this is simply copying "stuff" from point A to point B to point C.

However, the large number and huge *file size* of the scanned images did not make this a practical option for Ed.

It would have been a slow, cumbersome process and *cloud storage* is (at the time of this writing) less cost effective than physical storage for large files.

And it is much faster to copy large *files* to and fro external storage media (such as a flash drive) than to and fro *cloud services*.

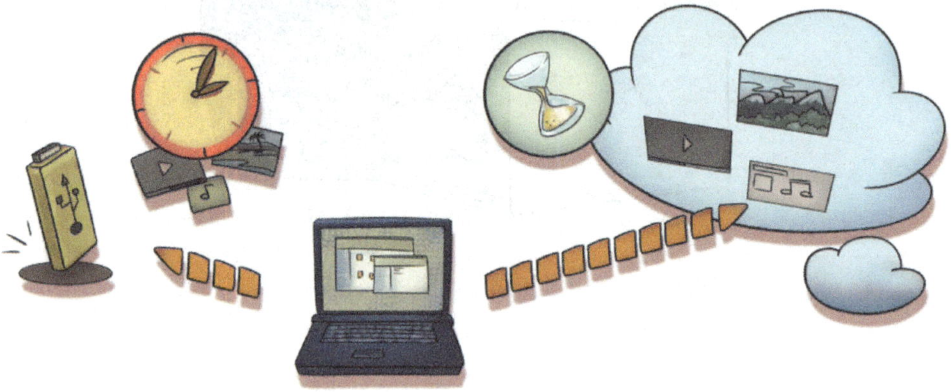

And as luck would have it, your favorite Help for the Technology Shy Scribe happened to have an external DVD Drive in his possession 😯

Ed's notebook computer and your author's "SuperDrive™" were **compatible** with each other (i.e. they could be used with one another).

But your author's SuperDrive™ could not connect **directly** to Ed's 2019 MacBook Pro – because that particular notebook computer did not have the right *ports*.

So Ed needed to purchase an **adapter** 🛒

The two devices could now talk to each other enabling Ed to access his scanned files on the DVDs 📀 through his notebook computer 💻

The scanned files are represented here...

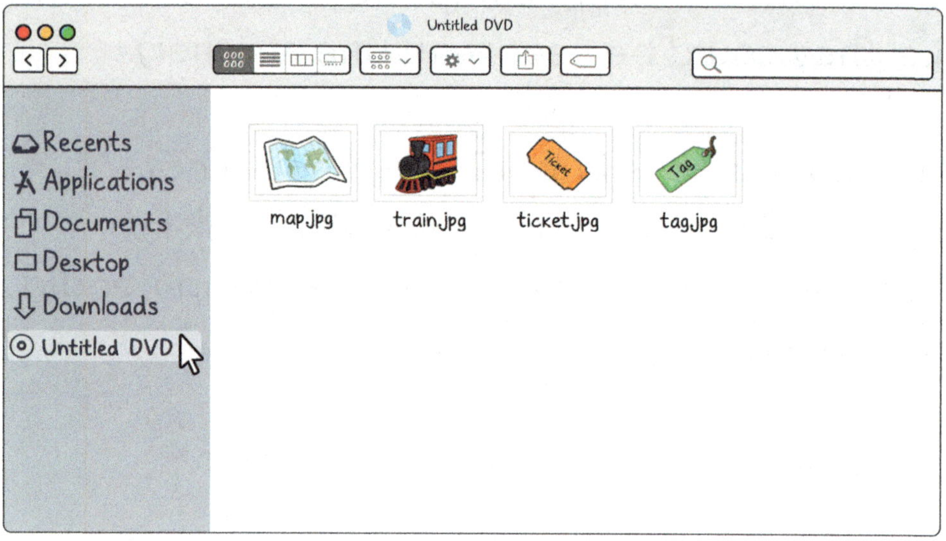

Whenever accessing your files (or simply when a new screen or window presents itself) look for *visual cues* such as...

- menus ("view" and "select all" might be most useful in this example)
- *breadcrumbs* (e.g. "back" & "forward" buttons)
- *buttons*
- names of *folders* and locations (e.g. "Untitled DVD")

Ed then thought **conceptually** about what he wanted to do (i.e. copy *files* from point A 📀 to point B 💻) and **then** reached into his *Getting Stuff Done With Technology toolbox* for the proper tools to make it happen.

In this example, Ed methodically...

1. **navigated** to the DVD location on his notebook computer (using the mouse tool 🐭 in combination with *visual cues* 👀 and *breadcrumbs* 🍞)
2. **selected all** of the images on the DVD (using the menu tool 👩‍🍳)
3. **copied** all of those selected images 👩‍🍳
4. **navigated** to the location on his computer where he wanted those images to be (e.g. his "Pictures" *folder*) 🐭 👀 🍞
5. **pasted** the copied items into this location 👩‍🍳
6. **made sure** the copied images were where he wanted them to be 👀

Using the technology basics + external storage media to create a workflow to accomplish a particular objective

And here is an abstract representation of Ed's workflow for selecting, copying and pasting his digitized materials from his DVDs into the "Pictures" folder on his notebook computer:

(Ed's conceptual understanding was instrumental to his success.)

Ed selects an app

The good news is there are **apps galore** that you can use to get stuff done on your device ✅

The not-so-good news is there are **apps galore** 🤨

Some apps will be shady (it is best to stay in your device's *App Store*).

Some apps may be **out of date** (and may not function properly on your device).

Some apps may be **poorly designed** and hard to use.

Some apps may be **defective** (your author once installed a simple "notes" app that proceeded to gobble up 90% of his device's resources 🐷).

Some apps will not be **suitable** for your task or objective (such as a cleric who had written his last will and testament with a spreadsheet app).

Ed had found a presentation creator *app* in his device's *App Store* that was well-reviewed and current…

After installing and starting that *app* on his notebook computer…

It came down to…

1. making a nice *cup of tea*…
2. taking in the app's *interface*…
3. following *visual cues*…
4. *thinking about what he wanted to do*…
5. and reaching into his *Getting Stuff Done With Technology toolbox*

And then (tools in hand) thoughtfully proceeding step-by-step to…

6. create his presentation file
7. add slides to his presentation
8. insert his digitized images onto his slides
9. save the presentation 🗄️
10. *back up* the presentation file ☁️

A condensed, abstract view of Ed's presentation creation workflow

Ed's Workflow 🚂

In this section, we are going to **summarize** (i.e. paint in *broad strokes*) how Ed made thoughtful and productive use of his files, apps, devices and cloud services.

After reviewing his **objectives** (which included being able to work on his presentation away from home)...

Ed used the *think about what you want to do and then look around for a way to do it* mantra...

and his *Getting Stuff Done With Technology toolbox*...

To

methodically...

Import **all** of the images from the DVDs into the Photos app on his home desktop computer...

Review and organize his images in a way that made sense to him and that would help him quickly find the images that he wanted to use for his current *project*...

Create his presentation on his desktop computer – and save that *file* to a *cloud* service...

Ed was then able to access that presentation *file* on his notebook computer:

Considerations... 🤔

It is always important to have the *files* that you need – when you need them.

If, for example, there were **additional** images Ed might want to add to his presentation when he is away from his home computer, he could upload them to his *cloud account* of choice before he leaves the house.

Or...

(And this is especially important if Ed is going somewhere without internet access.)

Ed could copy the images he might need from his desktop computer to a mobile storage device:

Once again mindfulness and conceptual understanding play key roles here. Ed used *visual cues* (including *breadcrumbs*) and the *mouse* and *menu* tools to **copy** the images from point A to point B.

File back ups (Better safe than sorry 😐)

At the end of each work session, Ed *backed up* his work in progress – and eventually his completed presentation – to an external storage device **and** to another *cloud service*.

It is well worth noting here that external storage devices are physical devices which can and do simply cease to function (turn into pumpkins so to speak 🎃).

Therefore, when it comes to important files, it is very important to **not** to have all of your eggs in one basket – because you could **lose** that basket.

For example, during the Autumn of 1997, a PhD student came into the library looking very distraught.

"Jenn" had put her entire thesis (+ supporting documentation) onto a floppy disk – and then lost it 🙀

Unfortunately, Jenn had not backed up her files onto any other computers or storage media and so all of her research and writing were lost 😿

This was a heart breaker but the real technology wizardry is learning from your mistakes...

Afterward: Where to Go From Here

Visit the Help for the Technology Shy website at www.technologyshy.com for supplemental materials including…

- step-by-step tutorials
- instructional videos 🍿
- technology tips & tricks
- recommended reading 📚
- ordering info for "Tabby's Guide to Thoughtful Tea Drinking" (i.e. the official help for the technology shy technology notebook)
- ordering info for the technology basics instructional DVD 📀

Ask your local librarian and/or bookseller to order Help for the Technology Shy books from Ingram.

Index

(NOTE: This index includes Help for the Technology Shy Part I)

A

B

F

G

H

I

J

K

L

M

N

O

P

R

S

T

This book (and Help for the Technology Shy Part I) were produced digitally on a *plethora* of devices and on multiple *operating systems*.

The fabulous illustrators / collaborators were engaged through upwork.com.

The case studies and other examples in this book were based on actual tech help sessions but the names and appearances of the people were changed.

Made in the USA
Monee, IL
02 November 2026

67381833R00116

THE VEDIC BIG BANG ©

A New Cosmology Scenario

Revealed in

the Ancient Veda Mantras of India

SHIV R. PAL PhD (Physics)

VOLUME 1

(Chapters 1 to 5)

THE VEDIC BIG BANG - A New Cosmology Scenario Revealed in the Ancient Veda Mantras of India

PRINT VOLUME 1: ISBN 978-1-9990979-2-9

Published by: Shiv R. Pal

For more information:
www.VedicBigBang.com.

E-mail:
shiv@vedicbigbang.com

Table of Contents
VOLUME 1

About VOLUME 2

The book is in two parts, VOLUME 1 (430 pages) and VOLUME 2 (340 pages). The Chapters 1 to 5 comprise the VOLUME 1 and the Chapters 6 to 11 are in VOLUME 2.

The Chapter 5 of VOLUME 1 presents the unprecedented definitions of entities and processes that serve as the cause and effect for the advent and formation of the Vedic universe and prepares the stage for this author to formulate the "Vedic Cosmology Principle" in Chapter 6 in VOLUME 2.

The "Vedic Cosmology Principle" is based on 12 RigVeda mantras from its "*Bhavavrittam" suktas* (hymns) and 5 *shloka-couplets* from the ancient book of Manusmriti's first chapter, entitled the "*Srishti Utpatti*". These mantra references lead to 28 physics deductions. The RigVedic term "*Bhavavrittam*" (discussed in Chapter 6) means "*The History of Creation of the Universe*" or "*A description of the Birth of the Universe*" and the term "*Srishti Utpatti*" means "*The Birth of the Creation* (universe)". Supplementing the 28 deductions in physics with 40 additional Veda mantra references, the author has developed the "Vedic Cosmology Scenario" in Chapters 7 to 10 in VOLUME 2.

The Chapter 11 presents numerous intriguing facts and features pertaining to the Vedic formation of the solar system including the birth of the Earth and the other planets. The VOLUME 2 is embellished with colour graphics delineating the stages of the primordial matter-ovate, first in the incubation period of 4.32 million years, developing "heat" and "rotation" and then exploding into uncountable matter-ovates forming the universe and the solar system. The titles of the six Chapters in VOLUME 2 are as following.

Chapter Headings of VOLUME 2

To
my wife and our son - my best friends.

Preface to the Vedic Big Bang

Professor Carl Edward Sagan in his writings and in the famous TV series, 'The Cosmos', praised ancient Vedic thought on cosmology. He recited the English translation of a Veda mantra and remarked, '*most sophisticated ancient cosmological ideas came from India*'. I have attempted to bring together the advanced, intriguing cosmological ideas from a large number of the ancient Veda mantras that lay the foundation for a Vedic Cosmology Scenario for the formation of the universe. My humble effort would further strengthen Carl Sagan's enunciation. Of the wide range of subject areas covered in the Vedas, the emphasis in my book has been restrained to invoking the mantras relevant to physics and the Vedic Cosmology Principle that reveal the process of the formation of the universe. All this hidden knowledge permeates the ancient Veda mantras.

When I began writing this book in 2008 there were no modern books addressing Vedic Cosmology from a physics perspective. It took me many years of intense Vedic research and cosmological analysis to complete this book entitled "The Vedic Big Bang". I therefore chose thorough, comprehensiveness rather than speed of publishing.

This author (Shiv R. Pal) named the Vedic creation process "The Vedic Big Bang"© because it sets forth the birth of the universe out of an impetuous explosion in the enormous primordial ovate of brilliantly hot matter named हिरण्मयमाण्डं (*hiranmayam-aandam*) the literal meaning of which is – "the golden-embryo" or "the ovate/spheroid with hot golden brilliance". There are several similarities and numerous differences between the ancient Vedic Cosmology Principle developed in Chapter 6 and the Big Bang Hypothesis in modern cosmology; both separated by a time-gap ranging in millennia.

An overview of the status of contemporary cosmology with an emphasis on the Big Bang Hypothesis is presented in Chapter 2 in simple terminology at the level of a general reader. This author raises numerous questions on the enormity of the explosion in a dimensionless point called the 'space-time singularity' at infinitely high temperature. The 'universe out of nothing' from the dimensionless space-time singularity is counter intuitive and in violation of the simple logic affirmed by the ancient sutra (1/43) - नावस्तुनो वस्तुसिध्दिः (*Naavastuno vastusiddhi*) of Rishi Kapil's Saankhya Darshan (Philosophy) which means - "*Nothingness cannot be the cause of something*". The contrasting Vedic Big Bang mechanism uses the pre-existing indefinable *Asata*-state of matter (Nature) as the "material-cause" of the real definable *Sata*-state of matter that forms the universe. The universe is the ensuing 'effect' of the pre-existent "material-cause". The Vedic Big Bang is 'something out of something'. This book presents a new perspective on the deficiencies of the modern Big Bang Hypothesis including the mathematical exercise in the hypothetical Cosmic Inflation to accelerate the universe and the un-harnessable Dark Energy to expand it. One of the Big Bang theory's creators, Paul J. Steinhardt of Princeton University (USA) stated (Scientific American, April, 2011) – "*....But some of the theory's creators ....are having second thoughts. As the original theory has developed, cracks have appeared in its logical foundations.....*"

The Vedic Big Bang Scenario in Chapters 7 to 9 is not translated into a mathematical framework, however, this Veda-based qualitative scenario works with real physical entities, real physics processes and the real by-products of matter and energy, including the force responsible for the expansion of the universe (discussed in Chapters 8 and 9) unlike the hypothetical Dark Energy in modern cosmology. Developing a mathematical model of the Vedic Big Bang would be straightforward when taken up.

This author, a physicist by training, found in his Vedic studies a treasure-trove of mantras laden with the principles and mechanisms pertaining to the advent of the universe and also found thoughts rich in the cognate subjects including physics, cosmology and astronomy. The four Vedas: RigVeda, YajurVeda, SaamVeda, AtharvaVeda, known as "*the books of knowledge of the Aryans***", are the fountain-head of the mantras. Sanskrit, a language with perfect grammar, is the language of the Vedas and the subsequent Vedic Literature. Each term (word) in Sanskrit is a derivative of a root or roots, the etymological analysis of which yields the meaning of the term and reveals the properties and nature of the entity the term stands for. The term "*arya*" originates from the Veda. In spiritual context it means the *eeshwar-putra* (progeny of God) and in the social context it means "*a person with noble qualities*". From ancient times the inhabitants of the Indian sub-continent have been known as 'Arya' and the sub-continent has been known as 'Aryavart' (the place of the Aryans**). The term Hindus for the Aryans of India and the term Hindustan for Aryavart is a later identification owing to invaders from outside India.

Utilizing the vast Sanskrit vocabulary the Vedas are composed in mantra formulations in unique and mathematically perfect metered poetry. That itself is mind-boggling. The RigVeda alone has 10,521 mantras in it, grouped into 1,028 *suktas* (hymns or sections) with 209 *devatas* (subjects) credited to 354 *rishis* (sages), among whom are 21 *rishikas*, the female *rishis* (Ved Shastra Talika, by Pundit Jiyalal Varma, 1958).

Based on 12 RigVeda mantras from the "*Bhavavrittam*" *sukta* and 5 *shlokas* of Manusmriti from its first chapter on "Birth of the Universe", this author (Shiv R. Pal) developed the "Vedic Cosmology Principle" in Chapter 6. From these mantra references, the author draws 28 physics deductions. These deductions with the additional 40 mantra references enabled developing the Vedic Cosmology Scenario in Chapters 7 through 10. The rishi (sage) appropriately entitled the RigVeda *sukta* (hymn) as the '*Bhavavrittam*' because the term (discussed in Chapter 6) means "*The History of Creation of the Universe*" or "*A description of the Birth of the Universe*".

This author has further strengthened the foundation of this unusual book on ancient cosmology by complementing the mantra references with the sutras and statements from the precious ancient Vedic Literature (Darshans, Smritis, Upanishads, Brahmana books and so on); all anterior to a time-milestone on February 18, 3102 BCE. The book presents the etymology of the important terms in Vedic cosmology, leading to the fundamental definitions of entities serving as a 'cause' in the formation of the universe.

The author takes several steps for translating mantras from Sanskrit to English. A mantra is first presented in Devnagari script (the name of the Sanskrit script) followed by its transliteration. After separating conjugated words, all the words of the mantra are listed with etymological root-based *yaugik* meanings, keeping in mind the context of the *sukta* (hymn) and the mantra. Then, following the instructions emphasized by Swami Dayanand Sarasvati (1825-1883 CE) in his interpretive translation of RigVeda and YajurVeda in the Hindi language and the etymology of terms in the book of Nirukt by Rishi Yask, this author presents the composite mantra translation in English. All Sanskrit terms in the book are italicized. The Vedas being theistic, at times the mantra gives credit to the "efficient-cause", the Creator, for a physical phenomenon in the "material-cause", the Nature. This is reflected in the mantra translation. However, even if the Creator is separated from the subject matter of the mantra, the valuable physics lesson of the mantra still persists.

The major participating entities in the Vedic creation process of the universe such as Time (*Kaalah*), Space (*Antariksh*), Nature (*Prakriti*) and so on, including the definition of the fundamental building-block of *Prakriti* (Nature), the smallest unit of matter termed as *parmaanu*, are all defined with clarity. The Nature in the Vedas is an inert entity without the virtue of inspiration to create. It is the entire matter/energy of the universe; the universe is the Nature. The Nature in Vedic physics, at its most fundamental level is divided into a triplet of fundamental particles named *sata, raja* and *tama* with diverse properties. All definitions are in Chapter 5.

The abstract entities, Time (*Kaal*) and Space (*Aakaash*), are thought to be real entities in the Vedic Cosmology, same as in the modern Big Bang cosmology. The difference is that these entities in Vedic Cosmology are eternal in nature through the perennial cycle of creation-dissolution of the universe whereas in the modern Big Bang Hypothesis they are said to be first-time created at the time-zero of the Big Bang. They are used in the Big Bang algorithm but are least defined or not defined in terms of what they actually are, what mechanism created them with what ingredients and what is at their fundamental structure.

A general principle in Vedic Philosophy stipulates that all created entities are quantized at their fundamental level. Therefore, the onus is on the modern Big Bang Hypothesis to define the entity being used. The mantra formulations in Vedic Cosmology render considerable commentary on the nature of these entities (discussed in Chapters 1 and 5, Volume 1). The entities such as the Time (*Kaal*) and the Space (*Aakaash*) are common to the two cosmologies; however, one finds that their Vedic definitions are superior in uniqueness and depth to the definitions available from the modern cosmology. The large number of Vedic synonyms for these entities represent their specific properties and add to our understanding of the uniqueness of these entities. The specificity in a Vedic thought at times is mind-boggling. It is mind-boggling to read in the RigVeda mantra (10/56/5) that the universe is expanding. The full mantra is in Chapter 9 in Volume 2, however when one reads that the universe with "*...its self-propelling force expands greatly the celestial objects from their minimum* *positions ....in togetherness ....*" one has no choice but to put such mantras into the physics frame of reference. In the Vedic mechanism the force of expansion is real and not hypothetical as the un-harnessable Dark Energy of modern cosmology.

The Sanskrit term for the advent of the universe is *'srishti'*. The closest term for 'srishti' is 'creation'; the etymological analysis signifies that the term 'srishti' has no religious connotation and stands for the advent of the universe with all the substances created in it, animate and inanimate. It has no relation to the creationist-argument. Implicit in the term '*srishti*' however is that "*the universe is not an ever-existing phenomenon, it has come to be at a given point in time*" and its active state of matter during the creation phase will dissolve into an inactive state leading to its non-creation phase termed as '*pralaya*'. In the Vedic thought the *'srishti'* and '*pralaya*' are part of the eternal cycle with long-term frequency; they are the day and night of the creation-dissolution process.

To prove that the Veda mantras and the sutras of Darshans are extremely ancient and were not influenced by the modern thought and by the thoughts of other known ancient civilizations the author postulated a 'Reference Hypothesis' in Chapter 3 stating, "***The Vedic references on cosmology are ancient and precede the 5,000-year period***". The hypothesis is proven by establishing a realistic timeline deduced from the historicity of Vedic Literature and validated by the ancient records available even today in three different calendars that have been in active use since India's distant past.

The antiquity of the Vedic references used in this book takes one anterior to the time-milestone at February 18, 3102 BCE, which was not established by a common calendar initiated at the coronation of a king or introduced after a famous person in history but it marked the onset of a calendar that is concomitant with astronomy. This date in the Common Era Calendar is the onset of the Vedic sub-Calendar called the Kaliyug-Calendar within the Vedic Srishti-Calendar (long-term Natural Vedic Calendar).

This book provides a glimpse into the ancient Vedic civilization in India going past 5,000 years. The overview of the Vedic Literature in Chapter 4 exhibits its richness. The religion, science and philosophy go hand-in-hand with the concept of the form-less unitary all-pervading consciousness *Brahmn* as the Creator, with the inert *Prakriti* (Nature) serving as the material-cause of the universe. In addition to the conterminous Creator and the quantized Nature, the universe is also populated with the quantized units of consciousness called the *aatmas* (souls) whose interaction with matter is observable in the living-beings bound by the principle of free-will, karma and the system of re-birth.

The brief historical overview of the Vedic Civilization presented in Chapter 3 begins with the kings of the Sun Dynasty in the distant past of the time-milestone in 3102 BCE who ruled from Ayodhya including King Ram. The kings from the Moon Dynasty ruled from Hastinapur (Delhi). This book gives information about the years, months and days of rule of each king in the Lunar Dynasty, starting from the coronation of the King Yudhishthir in 3140 BCE after the great war of Mahabharata. The Aryan dynasty at Hastinapur continued till 1191 CE when the last Aryan King Yash Pal was forcibly replaced by a Moslem invader from outside India. Parallel to Yudhishthir dynasty was the Nand dynasty of Kashmir kings. The historical account is compiled in the book entitled "Raj

Tarangini" written in 1148 CE by the court historian Kalhan. The remarkable continuity in the succession order of the head-priests at Shankaracharya monasteries, beginning in October 492 BCE, is available in three different ancient Indian calendars up to date. These data provide the proof for the Reference Hypothesis stated above.

Max Muller and the other Western Indologists of the nineteenth century, who translated the Veda from Sanskrit to English, arbitrarily shrank and distorted the ancient history of India without providing a solid reference in support. This assault did not spare even the Vedas and the Vedic Literature. That erroneous history is still being taught by the contemporary scholars. Therefore, the Chapters 3 and 4 in Volume 1 of this book are critical of Max Muller. His inadequate preparation in the Sanskrit language and grammar and his bizarre mistranslations of the RigVeda mantras are pointed out. The writings of Max Muller's contemporary Swami Dayanand Sarasvati and Pundit Gurudatta Vidyarthi, the great scholars of Sanskrit, are invoked to point out erroneous mantra translations by Max Muller and others. To disprove Max Muller's Aryan Hypothesis and the Aryan Invasion Theory, the author has tied some historical loose ends invoking the state-of-the-art data from satellite and ground-truthing instrumentation under the "Extinct Sarasvati River Project" and the national genome projects of the Government of India.

Sooner or later this unique book was to happen because when a contemporary physicist (like this author) reads in Rishi Kanaad's ancient Vaisheshik Darshan *sutra* (5/2/3) - अपां संयोगाभावे गुरुत्वात् पतनम् (*Apaam sanyog abhaave gurutvaat patnam.*), which means - "*The rain drops in the absence of the union (with air) fall down due to* ***gurutva*** (gravitation)", this author has to let the underlying fact be known to all, that the knowledge of gravitation (***gurutva***) is far more ancient than 1666 AD when Newton may have seen the apple fall from a tree. The inescapable fact is that Rishi Kanaad's statement on gravity is neither a fortuitous one nor a fleeting one because, to explain the intended phenomenon of gravitation at a deeper level of inquiry, he took up two other examples from the daily life, exhibiting the effect of gravity. The first example was that of a *moosal* (heavy pestle) falling in the *okhal* (mortar) under gravity (5/1/7) when released by hand. The second example was that of an arrow in-flight with the additional force imparted to it by the string of the bow, falling gradually under gravity (5/1/16, 17, 18). He also emphasized gravity as the अदृष्टकारितम् (*adrishtkaaritam*, 5/2/1, 5/2/2) of the Earth, the invisible action (karma) of the Earth, an erudite deduction indeed!

The Vedas are not just focused on cosmology. They deal with considerable astronomy beginning with the mantras on the formation of the solar system (Chapter 11) and the meaningful naming of the 28 constellations and 9 planets; the Earth and the Moon included. The Vedic prescribed solar system formation is quite similar to one of the modern theories of the formation of the solar system. The *rishis* had developed tremendous insight into the characteristics and orbital motions of planets. They named Jupiter as *Guru*, meaning *'the largest'* and Saturn as *Shani* – meaning '*the slow-moving*'. The seven rays of the Sun are given seven distinct names and the Sun is designated as the controller of the solar system. Considerable discussion is presented on the Sun, the Earth and the Moon. The mantras state that the Earth and the planets orbit around the Sun and the Earth is round and not the centre of the universe. A mysterious statement in the Mantra Brahmana (1/5/13) states - '*That black substance on the surface of the Moon is there from the interior (heart) of the Earth'*. It is astonishing that the modern astronomy has not yet suggested the Earth's interior as the source of the black-maria regions of the Moon, predominantly on its front surface facing the Earth. Based on the age estimates of the lunar samples, this author has explained the possibility of the veracity of the Mantra Brahmana statement.

When a Veda mantra assigns a definite period of 4.32 million years for the primordial matter-conglomerate of the universe in the ovate shape (*Hiranmaya-aandam*) to have developed tremendous heat energy and rotational momentum before it exploded (Big Bang) creating *koti-koti* (hyperbole for uncountable) smaller *aandams* (ovates), which are said to be the cause for the *mahaan* (great) celestial bodies, a physicist cannot label such a mantra as the byproduct of faith. The number 4.32 MY being in cosmological scale and associated with the raw material of the universe in the shape of a brilliantly hot spheroid (*Hiranya-aand*), he has no choice but to put the mantra into the physics realm. The mantra thus is a part of the Vedic Cosmology Scenario.

Mind-boggling Cases in Vedic Physics

In the development of the Vedic Big Bang Scenario several deductions are drawn. It makes one wonder how the *rishis* (sages) of the Veda mantras reached such conclusions in physics, so long ago. Of the numerous mind-boggling cases found in the Vedas and discussed in this book, two are cited here that have strong relationship to physics.

Rishi King Manu in his book entitled Manusmriti has divided the time-scale from the largest time-unit of 4.32 billion years to the smallest time-unit of *nimesh* (178 milliseconds). A *nimesh* being the blink-time of a human eye, Manu could not have derived this unit without a rigorous experiment in human physiology and an accurate time measuring methodology. The state-of-the-art blink-time measuring experiments carried out by other researchers (discussion in main text) with computerized timing with the optical imaging system, provided the average blink-time of a normal human eye to be 175 milliseconds. This is very close to Manu's definition of *nimesh* measuring 178 milliseconds. This is not an ordinary result.

The RigVeda mantra (1/123/8) states that the distance of morning twilight illumination due west of the sunrise point is 30 *yojan* (390 km). This is mind-boggling because using the ray-tracing geometry and knowing that the troposphere contributes to twilight, one finds that 390 km is highly accurate (details in Chapter 5). It is mind-boggling because it could not have been determined without the *rishi* having the knowledge of the daily spin rate of the Earth and its circumference and radius. These cases are amazing because all this took place anterior to the time-milestone at 3140 BCE!

Shiv Raj Pal

Acknowledgements

Acknowledgement by an author of a book is generally about expressing indebtedness to those who have played a role, directly or remotely, in helping to enhance the substance of the book. About indebtedness, the ancient commentary book on YajurVeda, the Shatpath Brahmana states (1/7/2/1), - "Soon *as a human-being is born he/she is indebted to the devas, rishis (sages), pitars and the other humans of the society*". The *devas* in this context are the inert physical systems of Nature, that are essential for sustaining the universe and 'selflessly' enabling the living creation in it. Just by the virtue of taking birth in this creation a human-being becomes indebted to Nature (*devas*). The Veda advises humans to become conscious of the environment and refrain from misusing and harming the Nature's systems. Conserving Nature's resources and minimizing wastage are the payments towards the debt the humans owe the *devas*.

Humans are indebted to the *rishis* (sages) who selflessly impart Vedic knowledge through the Veda mantra formulations. This author is indebted to the Vedic *rishis* for the mantra formulations in the four Vedas and the *sutras* in the Darshans (Philosophies), Upanishads and Smritis that form the basis for the physics and cosmology in this book. The author is especially grateful to Rishi Swami Dayanand Sarasvati (1825-1883 CE), the great scholar of Vedas and the Vedic Literature who revived the interest in the Vedas and gave an impetus to the society to be proud of the Vedas and the Vedic knowledge. He emphasized to use the right tools of grammar and methodology for understanding, translating and interpreting the Veda mantras. The author has benefitted immensely from his Veda translations in the Hindi language and his guidance for unravelling the meanings of the mantras using the '*science of word formation*' in Rishi Panini's *Ashtaddhyaayi* grammar and the ancient etymology of Vedic terms in Rishi Yask's Nirukt and Nighantu. Utilizing the Veda mantra references this author has become increasingly aware of the irredeemable debt that he owes to the Vedic *rishis*.

The author wishes to express his gratitude to the late Pundit Jiyalal Varma of Kota, Rajasthan, the author of a Vedic resource book entitled "Ved Shastra Talika" ("A Key to Vedas and Philosophy") who was a mentor to this author for the Vedic knowledge. Pundit Jiyalal ji discussed with the author the intricate concepts of Vedic Philosophy when the author was of an impressionable age during his school, college and university years while pursuing education in physics. In 1961 Pundit Jiyalal ji from his large book collection gifted to this author a book entitled "Ved Vidya Nidarshan" written in 1959 by Pundit Bhagvaddatta and asked this author to read the book and explore the science thought in the Vedas. The author cherishes this inspiring book and has referred to it in several places in this book.

The author is grateful to Acharya Udayaveer Shastri (1894-1991 CE), a great Vedic philosopher of our times, a great Vedic historian and a scholar of Sanskrit language, with a large number of books to his credit, for his interpretations of the mantras and the *sutras*. Referring to the historical data arduously researched and presented by Acharya Udayaveer Shastri in his remarkable book entitled "Vedant Darshan Ka Itihaas" (History of the Vedanta Darshan, 1991) has tied some historical loose-ends for this author's "Reference Hypothesis" (in Chapter 3). Deep gratitude is expressed to the late Acharya Gyaneshvar ji of the Vanprasth Ashram of Rojad, Gujarat, who generously gave his time for discussions on Vedic topics.

Humans are indebted to the *pitars* (parents and forefathers) because they contributed to establishing a civilized society. Under the Vedic karma principle and the rebirth system one is considered to be blessed if one takes birth from 'good parents and in highly developed society'. My parents are the *pitars* who followed their *dharma* towards the society they were born into and brought up the next generation with their best effort, responsibly. I am indebted to my late parents.

Even though my parents were unsophisticated village folks with education at a basic reading-writing level, they had high reverence for education. Not having a school in my small farming village and discovering that I had the potential for acquiring higher education, my parents left no stone unturned to walk me to a school wherever possible in the countryside even during monsoon rains and hot summers. Poverty did not stop my determined parents to send me to the city for education at the primary and middle-school levels. The well-deserved educational scholarships and the small income from the private tuitions that I offered in the community, made my continued education viable, unburdening my parents from the financial strain.

I am deeply grateful to my parents for inculcating the values and the discipline that took me forward, step by step, through successes and setbacks. Their words of wisdom disguised in the form of a simple 'village-saying' or an ancient proverb have left an imprint on my inner-self and the same words echoed different meanings under differing circumstances of life. When I was small and walked by my father's side he told me: "*Shiv, there are two possible benefits of looking at the ground ahead of you while walking rather than looking at the sky; you may find a rupee* (money) *and you will never step on a snake*". Their simple words of wisdom always had a much deeper meaning. I will remain indebted to my parents.

My wife, a Ph.D. from the University of Toronto, a Special Educationist, kindly accepted my request to edit the manuscript. A talented singer with the music degree of Sangeet Visharad in North Indian Classical Music, with three albums to her credit, she is generous with her time and teaches meditational music on a voluntary basis at a charitable cultural institution. With her busy schedule she devoted valuable time to help refine the manuscript. I am grateful to her for spending uncountable hours on editing and for continuing to be a tower of strength for the family.

After retiring from research in physics, I spent considerable time on the studies of the Vedas and Vedic Literature. Based on my notes I prepared several public lectures and discourses. Our son helped me with the logistics of the presentations. He encouraged me to compile the Vedic ideas and the copious notes into this book. I am grateful to him for his persistent encouragement which led me to write this book. He has closely followed the progress of the book and was always helpful in resolving any technical difficulties. He has been a steadfast friend through all the challenges that I faced.

Shiv R. Pal

कालो ह भूतं भव्यं च पुत्रो अजनयत् पुरा ।...

(*"Kaalo ha bhootam bhavyam cha putro ajanyat pura"....*)

" *Time (Kaal) the son is its own father* (the past) *and will be born as its own son* (the future)"….

(AtharvaVeda, 19/54/3).

CHAPTER 1: Introduction to the Vedic Big Bang

"...few of us spend much time wondering why nature is the way it is; where the cosmos came from, or whether it was always here; if time will one day flow backward and effect precede causes; or whether there are ultimate limits to what humans can know...and why there is a universe." (From Carl Edward Sagan's introduction to "A Brief History of Time" by Stephen Hawking).

The queries, such as in Sagan's quotation, -...*why nature is the way it is; where the cosmos came from.... if time will one day flow backward..."* are certainly not recent products of the human mind. Owing to the overwhelming developments in science and technology of the modern era, the tendency is to think that such queries, if raised in previous civilizations, must have remained at a primitive level. However, the documented facts preserved in the ancient Vedic Literature suggest that such queries are millennia upon millennia old and they were dealt with the thought system that was highly advanced.

If one raises the blind on the window of the present civilization and makes an attempt to take a peek upstream at the other shore of the unidirectional 'river of time' one will witness much more going on there than just the indulgence in the primitive queries about the universe. If one were to make a sincere effort in focusing on the blur of the distant past, one would find that there is an early civilization there which uniquely owns an abundance of well-focused thoughts in science and philosophy. The added excellence associated with this ownership is that these ancient thoughts are well-documented, well-preserved in their originality and available to the present civilization. These thoughts are not expressed in a primitive language either; instead they are well-articulated in amazing and unique poetry formulations with mathematically precise metered scales, in a most sophisticated language that is rich in vocabulary and grammar. This language is Sanskrit. It is the language of the four Vedas, '*the sacred books of the Aryans (Hindus)*. These extraordinary Vedic thoughts in mantra formulations are preserved unadulterated, without even a slight change of a syllable since they were first generated in the minds of the *rishis* (sages) in the distant past of the Vedic Aryan Civilization of India. The remarkable feature of this civilization is that it is not just a few thousand years old but it embodies unknown multitudes of eons from the far antiquity. Sanskrit continued to be the language of the vast post-Vedic literature that expounded the thoughts contained in the Veda mantras, which laid the foundation for a high level of spirituality, philosophy and the knowledge-base covering different realms of human development, including the science of the cosmos.

Contemporary linguists and literary scholars have distinguished Sanskrit to be the most ancient and the 'mother of all modern languages'. It is said to be the language of literature that is the oldest in the human library. Sanskrit for the Hindus (Aryans of India) is an endowment from the ancient *rishis* and is revered as the perfect *vaigjyanic bhasha* (scientific language) to express *vigjyan* (science). A NASA scientist, Rick Briggs (at Roacs, NASA Ames Research Center, Moffet Field, California) has expressed the same opinion differently in an article published in the Artificial Intelligence Magazine (Vol. 6, No.1, Spring 1985). He stated:

"... *Understandably, there is a widespread belief that natural languages are unsuitable for the transmission of many ideas that artificial languages can render with great precision and mathematical rigor.*

But this dichotomy, which has served as a premise underlying much work in the areas of linguistics and artificial intelligence, is a false one. There is at least one language, Sanskrit, which for the duration of almost 1,000 years was a living spoken language with a considerable literature of its own. Besides works of literary value, there was a long philosophical and grammatical tradition that has continued to exist with undiminished vigor until the present century. Among the accomplishments of the grammarians can be reckoned a method for paraphrasing Sanskrit in a manner that is identical not only in essence but in form with current work in Artificial Intelligence. This article demonstrates that a natural language can serve as an artificial language also, and that much work in AI (Artificial Intelligence) *has been reinventing a wheel millennia old...."*

A natural deduction from what has been stated above is that if the Vedic people developed a perfect language with superiority over other languages, they must have expressed thoughts in that language on different subjects that should have some value even today. The science thoughts on nature and cosmology are of direct relevance to this book. The queries and questions on the universe, similar to those in Sagan's statement, have been raised in the Veda mantras and the answers are also furnished in the form of the mantra formulations. These queries are not just fleetingly mentioned in the Vedas; they are deliberately grouped under chapters and sections with well stated subject headings that cover various aspects of the universe from its inception to its establishment and functioning. There is positive affirmation in the mantras that the queries are not left at the level of mere verbalization but the answers are sought through serious thought process. The most fascinating aspect of the queries and answers in the Vedas is that they being the product of the far antiquity of human civilization were not influenced by the present age of science and philosophy which commenced only a few eons ago. Carl Sagan has elegantly put forth the un-aging queries, touching the areas of physics, astronomy, cosmology, philosophy and even spirituality, all in one sentence quoted above. Regarding Vedic thought, Sagan in his writings had expressed that "*most sophisticated ancient cosmological ideas came from India".*

The other name for the matter and energy of the universe spread out in all directions in the form of effulgent cosmos is Nature. Cosmology is the science of cosmos. The matter of the universe is seen and detected in its subsistent and non-subsistent, observable and non-observable matter and energy forms. On a grand scale of things the Nature is not alone by itself; it is embedded in a real entity called Time, the entity that Sagan in the above quote is also talking about. It is an invisible non-subsistent entity that is not a form of energy, not a part of the matter of the universe and is not created by the matter of the universe. Among his queries Sagan has not included yet another real entity called Space, which according to its Vedic definition also pervades through the universe encompassing even the biggest celestial objects and penetrates through each minutest of minute matter particle of Nature. The Space by the Vedic definition spreads out up to the edge of the universe, wherever that edge might be. As expressed in a Vedic expression - *the Space serves as an overall container of the entire universe*. A common container is outside the matter it contains such as a glass around a volume of water. But, the Space entity of the Vedas is a unique container - it is outside the matter it contains as well as it permeates through the matter that it contains. That the Space is an entity, that it is curved and that all objects are embedded in it is also a concept in modern Big Bang cosmology, however, it is not thought to be permeating inside matter. The Space in Vedic cosmology, like Time, is also an invisible non-subsistent entity which is not created out of matter or energy and it is not created out of Time either.

To answer questions such as – when, how and *where the cosmos came from*, the cosmologists resort to explaining matter (Nature) and its behaviour. They soon find out that the universe cannot be created by matter (Nature) alone; the Time and Space entities are also needed. The Time-entity is needed to create the universe in and the Space-entity is needed to put the created universe in. The irony is that the modern Big Bang cosmology is not able to explain how these two entities were created in the first place. Modern cosmologists are certain on the advent of this beautiful universe at a given point in its distant past but are not certain on whether the Nature itself pre-existed before that distant past. In this regard, this author could add a few more queries to Sagan's list: did the Nature pre-exist before the beginning of the universe or is the present universe its first birth, did the Time and Space pre-exist or were they born for the first time along with Nature, if they were born for the first time what necessitated their birth, what are they made of, what mechanism caused them and why are they different from each other and from Nature itself? The reader may find it interesting to note that in spite of overwhelming progress in the science of cosmos, the jury is still out on modern cosmology to furnish satisfactory answers to these queries.

Hidden in the effulgence of the vast Nature are certain all-pervading non-transient rules that Nature has imposed on itself that govern the behaviour of its matter/energy and the functioning of the universe. The physicists through observations, laboratory measurements, mathematical deductions and logic have uncovered these rules and called them the 'laws of Nature'. Other terms used for Nature's laws are 'physics laws' or 'physical laws'. Associated with these non-transient laws of Nature are invariant physical constants that together define quantitatively the behaviour of matter. They are also termed as Nature's constants or physics constants. Because the Nature's laws and mathematical values of physics constants do not change with time and locale the adjective of 'invariant' is associated with them. For this reason they are also known as the 'universal constants'. If the laws of Nature were transient and the physics constants were variable, the smooth functioning of the universe will be in doubt because the universe and its matter would behave differently at different times and in different parts of the universe. Such a scenario is the recipe for chaos. Observation and logic led the ancient Vedic *rishis* to formulate in the AtharvaVeda mantra (10/8/7) that the entire universe functions under एकनेमि (*ek-nemi*), the same law (one principle). Nature chose order over disorder. The presumption that the universe has followed the same laws of Nature since its birth and that it will continue to do so in future, allows a physicist to extrapolate in both directions on the arrow of Time to recreate the universe's past and predict its future as well. This extrapolation in both directions is the science of cosmos. To perform this extrapolation a cosmologist invokes physics laws in a complex mathematical framework and attempts to explain the observed phenomena of Nature. It is the reverse interpolation right to the first moments of the creation that explains how it all began.

The term, 'creation' throughout the book has no religious connotation. It has no relation to the contentious issue between contemporary cosmologists and creationists or to the ongoing debate on creationism, grand design or Darwinism. It is as neutral as in – the creation of water requires one Oxygen atom and two Hydrogen atoms.

The term 'creation' in the Vedas includes the advent of the universe and all substances in it, animate and inanimate. Pertaining to the physics laws and constants, one must grasp the fact that there is extreme criticality in their existence and the values. They are critical in the sense that they are just right and even the minute alteration in them would send the present universe on the course to chaos or create altogether a different universe. Nature, within this criticality in its laws, has allowed us conscious-beings to be born at a given point in its history and among them allowed the intelligent humans to be born to comprehend the knowledge of its existence and its functioning. Nature allowed us to observe its diversity in the form of various phenomena embedded in it and allowed us to become cognizant of its all-pervasive rule of "cause-and-effect" that leads us to discover the laws that govern it. If a phenomenon like the universe is time-dependent and if Time is seen as a non-reversing entity, one could conclude that what existed before may have a connection to what exists later. This connection is obvious in physical phenomena where pre-existence of something enables the existence of something else. The pre-existent substance therefore is the cause for the subsequently enabled substance. In the specific example of a mother and child, the mother is the cause for the child who is born later in time. When one looks deeper into this law of Time one finds that after all 'what came before what' and 'what caused what' has led to the discovery of all of Nature's laws that we know of and will lead to the discovery of all those laws yet to be discovered. Knowing Nature's laws has in turn enabled us to predict Nature's behaviour.

To be governed by laws is inherent in all aspects of Nature. Like Nature, the living-beings also set some rules and follow them for smooth functioning and predictability in their lives. While no one can tell how much deliberation went on before a rule of Nature matured into its final critical form, the rules set by living-beings have certainly evolved over time and out of intellectual analysis and logic; an example being – 'telling the truth is good for all'. We do not see any sign of evolution in Nature's laws. The laws and rules of humans generally are not perfect and need tweaking every now and then whereas the laws of Nature are perfect and are non-transient and invariant from the beginning. Whereas the thoughts generated in the human mind are responsible for inventing a rule, the thoughts are also responsible for the needed tweaking of the rules. It is the thought generating machine or a force with its exclusive existence in living-beings, decidedly sitting somewhere in the head, that generates all thoughts and is not found in the inert confines of the inert Nature. If our laws need to evolve to reach a level of 'telling the truth is good for all', why is it that Nature's laws are perfect from the very beginning? Did Nature have thoughts to tweak its laws? Answers to such queries have not come forward and are not easy.

How are thoughts created by the thought-producing machine in a living-being's head, has also been an age old query. In modern times differing possibilities are proposed and explanations are furnished for what generates thoughts. It is said that the brain creates thoughts. It is also said that the brain is only a tool but it is the mind that initiates all thoughts. A lot of research has gone on and still continues to determine what the mind is and how a thought is initiated in it. A lot has been learnt about the subject, however, the present status of this subject is generally summarized in a sentence - it is not yet understood how the mind works. The ancient Vedic thought on this subject takes us a step beyond the mind in that the thought creation is credited to an entity, whose presence enables consciousness in a living-being. All thoughts and actions that follow in a living body are thought to be the effect of this cause, the quantum of consciousness. This quantum of consciousness is not a physical machine made up of atoms and molecules; it is the operator beyond a physical plane that generates thoughts in the mind which delivers them to the brain for further processing and shaping. Whatever the mechanism of thought creation, the ability to initiate action in the physical body resides in the brain. The brain also has the ability and built-in functions to analyze the thoughts that are generated, sets rules based on the thoughts it had generated earlier and also adjusts, changes or drops the pre-established rules later.

We create rules and change them for greater perfection. So, the question remains - when and how did Nature perfect its laws? When we observe Nature and discover its unchanging laws, it makes us wonder if it also had its own mind that was used to evolve its laws over time and came up with a perfect set of laws to create the universe. Or could it be that a seemingly strict follower of all physical laws, the Nature itself was captivated by rules that were extrinsically imposed on it because it had no mind of its own or it had no choice to break them or adjust them? A modern cosmologist's response to these questions is likely based on un-provable hypotheses that he introduces and renders the expected common answers: no one is doing the thinking for inert Nature; this is just the way Nature is; it is by chance that our universe has this perfect set of laws; the other universes have their own sets of laws. One thing is certain. When a modern cosmologist uses a hypothesis for the creation of the universe, his model has no provision to create Nature's laws first. He follows the laws that already existed in Nature.

Just as the observation of Nature by modern scientists led to the discovery and formulation of the laws of Nature, it must have been the observation of Nature and its behaviour that led the ancient sages of the Vedas to express their thoughts in mantra formulations. As is evident from the formulations found in the Veda mantras, the deductions in Vedic cosmology must have been based on the profound study of Nature.

To express a thought with clarity one needs not just the vocabulary but the right vocabulary. The right vocabulary and its richness reflect the degree of thought investment that was put in to delineate and explain a subject or a phenomenon. A phenomenon in Nature can have more than one property. To express differing properties one needs to generate differing vocabulary. This is what is found in the Vedas. This is what tells us with certainty that a lot of thought investment had taken place to understand and to explain Nature and the process of the creation of the universe itself. The *rishis* of the Vedas gave not just the needed terminology for the subject of cosmology; they created a system to create generously the synonyms to represent differing properties of important entities in the creation.

In the Western philosophy and languages one word, 'nature', derived from *natura* in Latin, which means 'essential qualities or innate disposition', serves to represent all aspects of the Nature-entity that Carl Sagan was referring to. The Vedas use distinctive terms for Nature to represent its specific stages, properties and functions. Whereas the Sanskrit term *prakriti* is the most common synonym for Nature, one finds in the Vedas a large number of synonyms for *Prakriti*. This author has listed in Table 5.1 of Chapter 5 in this book some 39 synonyms of Nature out of many more found in the Veda mantras. Each name is based on the 'root' of the word-formation which reflects the property of the entity. Out of the large number of synonyms found in the Vedas for the important objects of Nature this author has listed 20 for the Sun (Table 11.4 in Chapter 11, Volume 2), 33 for the Earth (in Tables 11.6a, b in Chapter 11), 22 for the Space (Table 5.6 in Chapter 5). This alone suggests that the Vedic Civilization in its distant past had invested quite heavily the mental power to understand Nature and its actions. They also uniquely defined the abstract entities like the Time and the Space. The solid evidence of this is well preserved in mantra formulations in Sanskrit, the language of the Vedas.

The names of the four Vedas are: RigVeda, YajurVeda, SaamVeda and. AtharvaVeda

The science of word formation in Sanskrit is based on *dhatu* (root) system. Therefore the term used for an entity itself delineates its property and ascertains what the term stands for. Several of the Vedic terms for the Nature-entity in Table 5.1 of Chapter 5 reveal that it is an inert entity, same as in modern cosmology. The examples of the synonyms reflecting this specific property are: *Achetan* (an entity without consciousness), *Agjyaan* (a non-knowledge-entity), and *Andham* (without the ability to see itself and around it to give itself a direction or one that is blind to knowledge). There are several more synonyms indicating the same. The uniqueness of these Vedic synonyms becomes obvious from the meaning of the terms. A detailed discussion on Vedic-defined Nature is in Chapter 5; here it would suffice to say that the Nature for the Vedas is a conglomerate of inert matter causing the physical universe without having the ability and initiative to create its own rules. It is *Andham* and therefore a slave to the physical laws imposed on it extrinsically.

Even though Nature is an inert entity in Vedic as well as in modern cosmology there are differences in how it is perceived in the two cosmologies. The example of an earthen pot-maker is usually cited in the Vedic Philosophy. The inert lump of clay on the fly-wheel does not transform into a pot without the extrinsic intervention of the pot-maker's thought, energy and hands. For modern cosmology, the clay, the pot-maker, the pot-maker's thoughts, energy and hands are all together part of the same thing – the inert Nature. In modern cosmology the laws of Nature and its behaviour are self-generated and self-implemented, without having regard to how, by whom and when all the laws were created in such perfection. In this scenario the advent of the conscious-beings in the universe is just another act of Nature and the appearance of the mind in a living body is just another accident in Nature. A living-being is found to be enslaved by Nature, not having the freedom to increase its mind's capacity to any level it wanted. On the one hand the living-beings, humans in particular, derive from Nature the ability to understand it and even to make educated guesses regarding when and how 'it all came to be' and on the other hand the Nature disallows us from fully understanding it by putting hidden bounds on *what humans can know*, as Sagan put it. The fact remains that while a better understanding of the universe has been developed in the present times, the ancient queries, some of which being the same as in Sagan's quotation, have not been answered fully yet.

Modern cosmologists generally believe that Time and Space are real entities and without them one cannot explain the universe's creation and behaviour. This author has differentiated the 'entity' from its 'measured value' by using respectively the upper and the lower cases of the first letter of the term. Thus the terms Time and Space represent the Time-entity and the Space-entity and the terms 'time' and 'space' represent their measurements. The same style is followed for the other entities as well.

The fact about Nature is that a physicist could capture a piece of it in some form or the other (matter or energy) and save it in his laboratory for one to see, feel, smell or taste and the fact about Time and Space is that he cannot cut or capture a piece of Time or Space to save it. This raises a natural question – if one is not able to capture the Time and Space entities, do they exist in reality or are they mere illusions of mind? A modern physicist uses them for interpreting Nature's phenomena. It would be of interest to know at what point in the history of civilization a physicist learned that Time and Space are real. One has to admire those who thought about the real existence of the abstract entities, such as Time and Space and established their importance in the formation of the universe. When one explores into the Vedic mantras one finds that the concepts of Time and Space have been as old as the Vedas. If Vedic *rishis* learned about the reality of Time and Space they must have used these entities to enhance their knowledge, science and philosophy.

Concept of Nature, Time, Space and Big Bang are Ancient

Modern cosmology's trinity (Nature/matter, Time, Space) is needed to explain how the universe came to be. In his theory of relativity Einstein said that space and time were in a continuum. The mathematical meaning of his 'space-time continuum' is quite deep; however, its simple meaning is that one needs both entities to understand the universe. Although a modern cosmologist has no idea what these abstract entities are made of, it is universally accepted that both, Time and Space, are two separate entities completely different from each other in nature, function and usage. If both are different entities then it is counter intuitive to think that one would affect the other and the two dissimilar entities of abstract nature would form a single 'space-time continuum'. Considering that the modern Big Bang Hypothesis emphasizes that the Time and Space took birth first time at the time zero of the Big Bang the Einstein's 'space-time continuum' in this author's perspective is more like conjoined-twins with different DNA, born of different mothers, but got attached at birth only by the skin! Since no one has witnessed *where the cosmos came from, or whether it* (universe) *was always here* and what Time and Space actually are, any proposition on this by the most brilliant of brilliant cosmologists at best remains a hypothesis.

There have been some hypotheses suggesting that the universe was always here and will continue till eternity. Presently, the most accepted hypothesis of the universe is the Big Bang Hypothesis. In this hypothesis the trinity of modern cosmology is said to be born for the first time out of dimensionless Nature's womb, the 'space-time singularity' that gives birth to the universe in an explosion called the Big Bang.

The universe being created out of the dimensionless singularity means that the entire matter of the universe was concentrated at one dimensionless point before the Big Bang took place or was concentrated at one point at least during the first moments before it spreads. The Big Bang singularity at time zero is said to be at an extremely high temperature, of the order of 10^{32} °K. The number with 10 raised to 32 is equal to the number with 32 zeros after 1, which in common units is - one hundred million-trillion-trillion degrees Kelvin. The temperature of Sun's centre in comparison is only at $1.6x10^{7}$ °K (16 million °K).Such a high temperature is neither observed in Nature nor can it be created in a most advanced modern high temperature physics laboratory.

We observe in Nature that the temperature is a property of matter. Therefore the temperature of space-time singularity is the temperature of the matter in the singularity. That the matter existed in a dimensionless singularity is by itself counter intuitive thought and that the matter of the entire universe, estimated to be 10^{60} kg, came out of a volume-less point, is a strange thought by all standards.

One can think of the temperature as equivalent of the energy in a system, however, the nature of temperature is that it does not exist without matter and cannot be sensed in the absence of matter. The Matter is primary and the temperature is secondary. Therefore, the dimensionless space-time singularity before matter is created should logically be cold, leave alone being at trillions of degrees of temperature. The matter in the first moments of the Big Bang appears in the so-called Planck Era when it is said to be at the highest temperature; the Planck Era duration being only $0.539x10^{-43}$ seconds. The minus sign before the exponent 43 means that it is less than 1 second. The Planck Era therefore lasts only for a period of 0.539 seconds divided by a number with 43 zeros after one. The universe during this period grew from zero to one Planck-length of $1.6x10^{-35}$ meters across. More details on the Big Bang Hypothesis are in Chapter 2 under the title – "The Present Understanding of the Universe: A Critical Scrutiny of Big Bang Hypothesis".

One could ask a physics riddle in three different ways: "in what zero-dimensional entity is hidden away the entire universe?", "what entity was made up of two ingredients which are not in it?", "what entity exists in mathematics but not in physics?" The answer to all three riddles according to the Big Bang Hypothesis is – the 'space-time singularity' which by definition was dimensionless. This in common language – was not there! It is not a real physical condition that can be created in a physics laboratory or observed anywhere in the present universe; however, it can certainly exist in a mathematical construct of a cosmologist. As mentioned before, three entities are simultaneously born in the Big Bang Hypothesis; the Nature-entity (matter/energy), the Time-entity and the Space-entity. Not considered in the Big Bang Hypothesis or not clear from it is the possibility of matter of the universe existing before the Big Bang and whether Time and Space entities also existed before the matter appeared? The other limitations of the Big Bang Hypothesis, discussed in Chapter 2, forbid this hypothesis from acquiring the status of a principle in physics.

To the surprise of most, the Vedas have a definite scenario of their own for the creation of the universe. The Vedic scenario is formally encased in the mantras in specific sections of the chapters, embellished with appropriate titles. Vedas have dealt with the subject of cosmology in considerable detail. The term "Vedic Cosmology" will be used in this book to differentiate the ancient cosmology in the Vedas from the modern Big Bang cosmology. With regard to the creation mechanism and the sequence of the created byproducts the Vedic Cosmology is remarkably different from modern cosmology. On the other hand the modern Big Bang Hypothesis has considerable semblance with Vedic Cosmology as well. Based on a large set of Veda mantras and other Vedic references (RigVeda, 10/129/1-7; 10/72/8-9; Manusmriti, 1/5, 6, 14-16; Shatpath Brahmana, 10/1/3/1; 11/1/6/2; 11/1/1, 2) this author has drawn 28 physics deductions in Chapter 6 that set the stage for developing **the Vedic Big Bang** scenario in the Chapters 7 to 10. The author named the Vedic creation process as "Vedic Big Bang" because it sets forth the birth of the universe out of an impetuous explosion in the enormous primordial ovate of brilliantly hot matter named हिरण्मयमाण्डं (*hiranmayam-aandam*), the literal meaning of which is – "the golden-embryo" or "the ovate/spheroid with hot golden brilliance". The similarities between the Vedic Big Bang developed in Chapter 6 and the modern Big Bang Hypothesis presented in Chapter 2, both separated by a time-gap ranging in millennia, are mind-boggling. Based on different sets of Veda mantras the Vedic Big Bang Scenario and the Vedic Big Crunch Scenario are developed in Chapters 7 to 10.

The definitions of the major entities taking part in the Vedic creation process (Nature, Time, Space, …) and the analysis of the Sanskrit terms pertaining to the creation process of the universe in Vedic Cosmology and Vedic mechanics, form the body of the Chapter 5.

The universe in Vedic cosmology begins from the indefinable *Asata*-state of Nature (matter) held in shrunken Space; albeit not with zero dimensions as in the Big Bang singularity. The *Asata*-state of matter by definition being completely devoid of energy or dynamic activity, there is no temperature in it. After all, the temperature in matter is due to dynamics in its components. The Matter in such an indefinable state, without energy (dynamics), can therefore come together to confine in a relatively small space. This space occupied by Nature in its dissolved state is termed as '*tuchchhaya*' (RigVeda, 10/129/3). The term '*tuchchhaya*' stems out of the 3rd mantra in the Vedic Cosmology Principle stated in the 129th *sukta* of RigVeda's 10th chapter. It represents the adjective for Nature in the *Asata*-state and RigVeda has exclusively used it for *Prakriti* (Nature). The meaning of the term '*tuchchhaya*' is '*one that is confined in small space*' or '*one that is in diminutive state*'. The ordinary meaning of the term is - 'relatively very small'. It is amazing that the Vedic universe begins with matter in diminutive volume (*tuchchhaya*) and the modern Big Bang begins with a dimensionless 'space-time singularity'. The physics reality necessitates that the indefinable *Asata*- state of matter without energy will confine in diminutive volume (*tuchchhaya*) and not in zero space of 'space-time singularity'. The process of creation of the universe in Vedic Cosmology is the transition of matter from the *Asata*-state to the *Sata*-state the definable matter state suitable for forming substances and objects of Nature.

The term '*tuchchhaya*' being the geometric adjective for the Nature in its inactive state one can by reduction apply it to the Space-entity that holds the inactive Nature. The Space-entity in the Vedic creation scenario therefore does not ever disappear in a space-time singularity. The function of the Vedic Space-entity is to provide home for the matter and energy that forms the universe. Unlike the modern Big Bang cosmology where the existence of the Space-entity begins in the singularity, the state of Space in the Vedic creation changes but its existence continues eternally through the cycle of creation-dissolution. The important outcome of this comparison is the relationship of Nature to its container - the Space. On this, Swami Dayanand Sarasvati (1825-1883 CE) a monk and a great nineteenth century Vedic philosopher and reviver of the Vedas, expressed with clarity of thought in his book "Satyarth Prakash" (publ.1881 CE) that the matter of the universe (Nature) and the Space are two distinct entities and that the Space was the place for the matter to exist. And all this he expressed on the basis of the Vedas as back as the late 19th century, well before the hypothetical concept of the all-pervading 'ether' of the Western science and philosophy. Swami Dayanand Sarasvati's statement differentiating between Nature and Space came precisely 35 years before the birth of the 'space-time continuum' in Einstein's Theory of Relativity. The basis for Swami Dayanand's statement is not his own or from a near past, he only stated the Vedic Philosophy established in the far antiquity beyond Mahabharata about 5,000 years ago.

One difference between the two cosmologies is in the initiation process of the creation itself. The initial Nature in the modern Big Bang Hypothesis is dimensionless yet hot and is its own energy to trigger the explosion at the time $t = 0$ second of the Big Bang. The Vedic Big Bang on the other hand is initiated at the Time $t = t_n$ seconds where time t_n refers to the nth creation riding on the eternally flowing Time-entity. This means that the present creation according to the Vedic Creation Principle is one of the infinite creations in the past and is one of the infinite creations to come in future. The Vedic creation-dissolution of the universe is a perennial cycle of the state of matter transiting from *Sata* to *Asata* and back to *Sata*, with the life-span of each phase according to the Veda is not thousands or millions of years but in billions of years.

The other main difference in the two cosmologies is that the Vedic Big Bang takes a period of one *deva-yuga* of 4.32 million years (MY) to develop the required heat energy in its primordial हिरण्मयमाण्डं (*hiranmayam-aandam*), the golden-embryo, before it explodes into *koti-koti* smaller *aandams* (ovates of matter); the *koti-koti* being the superlative for 'uncountable'. The details on this scenario are deferred to Chapters 7 to 10.

This author feels that it is the Nature (matter/energy) that is understood better by modern cosmology than Time and Space; both eternal entities in Vedic physics (philosophy). The reason is that the matter of the universe interacts with the knowledge senses directed by the entity called Mind. The Mind in a living-being according to the Vedic creation is a byproduct of matter at extremely minute level. On one hand the entire universe finds itself immersed in the Time and Space entities and on the other hand both entities offer themselves to matter at extremely minute levels. To answer a query like, *if time will one day flow backward*, one has to first know what Time is. Since the modern cosmology received the gift of 'space-time continuum' in 1916 CE from Einstein's Theory of General Relativity, small sectors of the Time-entity with minute pieces of the Space-entity are utilized by modern cosmologists in the field equations to hypothesize the advent of the universe, yet they do not seem to understand what is the nature of these two entities.

Modern physics has discovered that the fundamental nature of Nature is quantized. By quantized is meant that matter at its most fundamental level is an individual quantum, the ultimate unit of matter. These quanta or corpuscles of matter are known as quarks which serve as the smallest building blocks of matter. In addition to quarks there are other stable and unstable matter particles that physicists have detected in their laboratory measurements. The quarks are not found free in Nature because they are all consumed in forming the neutron and proton that form nuclei of atoms. A neutron, the electrically neutral nuclear particle, is formed when one 'up-type' and two 'down-type' quarks come together and a proton with a positive electrical charge is formed when two 'up-type' and one 'down-type' quarks come together. The Hydrogen atom, the smallest atom, is formed by a proton (two 'up-type' and one 'down-type' quarks) and a negatively charged electron orbiting around it. More triplets of quarks in a nucleus form heavier atoms. A quark therefore could be considered as the smallest indivisible particle of matter. A proton having a diameter of 10^{-13} cm (one-tenth of one trillionth of a centimeter) occupies an unimaginably small space. Even though the fundamental particles are extremely small they are separate entities. Thus the quantized nature of matter makes the entire Nature an entity that is discontinuous at its fundamental level. The energy form of matter, as in a ray of light, is also quantized into individual packets of energy called photons.

It is of profound importance to state here that the Vedas eons ago stated that the Nature (matter) at its fundamental state is quantized with its smallest quantum called *aapah*. It further stipulated that all *aapah* are not one kind but they are three distinct types owing to three different properties: *satah, rajah* and *tamah.*

When the three properties are emphasized they are addressed as *sato-guna*, *rajo-guna* and *tamo-guna*, where the suffix term *guna* means 'property' (details in Chapter 5). Whereas the general term for the fundamental components of matter is *aapah*, the smallest particle of matter is termed as *paramanu*, the last stage of matter. Therefore a Vedic *paramanu* is either a *sata* particle or a *raja* particle or a *tama* particle. A *paramanu* by definition in Vedic concept is an entity not divisible further. More than one *paramanu* form an *anu* just as more atoms form a molecule. The Sanskrit term *anu* stands for the antonym of 'fat'. This means that an *anu* is not fat yet with extra matter in it. It is implicit in an *anu* that it is a three dimensional substance and a unit by itself. The euphonic combination of *param* + *anu* forming *paramanu* therefore refers to the last stage of *anu*. Because it cannot be divided further a *paramanu* is a *nitya* entity - one that does not change with time. These unchanging smallest units of matter make up the entire *Prakriti* (Nature) and therefore the Nature is also a *nitya* entity.

With regard to the physical shape of a *paramanu*, Rishi Kanaad in Vaisheshik Darshan (7/1/20) has stated the nature and geometry of a *paramanu* stating - नित्यं परिमण्डलम् (*nityam parimandalam*). This translates into - "*eternal and spherical*". It is *nityam* because it will remain non-divisible forever is Nature and the sphericity will remain its geometry. The fact is that if a spherical particle is attached to another spherical particle the resulting particle cannot be spherical. Therefore, a non-spherical shape cannot be a particle's last state. It is interesting to note that the Greek word 'atom' also means – 'one that is indivisible'.

All these Vedic definitions are not the invention of this author. They are preserved in the Vedas and the Vedic Darshans (Philosophies) since the antiquity of human civilization; the Vaisheshik Darshan of Rishi Kanaad in particular. Remarkable in the ancient Vedic science is that a trio of matter particles forms a fundamental unit of functioning matter, something similar to what is in modern physics.

One thing glaringly obvious here is that when the matter's fundamental state is defined in a Vedic scripture, it has little to do with religion, spirituality or worship of a Creator. It is pure science. It may not be commensurate with the contemporary physics, but the Vedic definition of matter is not ordinary. It makes one wonder as to what led the *rishis* of the Vedas to come up with the fundamental division of matter into a triplet that was so close in approach to one in modern physics eons later with three quarks forming a nuclear particle. Details on the Vedic definition of matter and its states are in Chapter 5.

The modern physics/cosmology is not fully clear about the nature of Time and Space. There are numerous questions on these entities that remain unanswered. To come up with a solution to the Big Bang field equations a cosmologist begins his calculations by inputting small slices of time and space equivalent of Planck-time and Planck-length respectively. The cosmologist calls this extrinsic insertion of time and space in the equation as the birth of Time and Space entities. The fact is that in this birthing of the universe out of the dimensionless space-time singularity the new matter of the universe is created according to the laws of Quantum Physics but under what laws are created the time and space?

A cosmologist understands from previous deductions in physics that the Planck-time is the smallest division of time where the meaning of time begins; however, the cosmologist does not really know whether Time is a continuum or a quantized entity and whether this first moment of Time is a single quantum of Time or just a slice of time with many quanta hidden in it. After the Time is presumably born in Big Bang a cosmologist is not sure how it grows; quantum by quantum accumulating to 13.8 BY till now or in a continuum with its beginning in Big Bang.

Physicists are in the same boat with the Space-entity because the Big Bang cosmologists surmise that the Space-entity like the Time-entity is also born at the onset of Big Bang. According to the Big Bang Hypothesis and on the basis of the astronomical observations of the size of the observable universe, it is now estimated that, corresponding to the age of the universe of about 13.8 BY, the Space has stretched from zero to well beyond the observable limit of the universe about 8.8×10^{23} km across (0.88 trillion-trillion km) and it still continues to expand. Physicists do not know what the Space-entity is made of, the entity that suddenly appeared out of nothing at the birth of the universe and thereon continued to stretch in all directions to great distances without snapping. It is not certain yet whether it is a continuum or is composed of small quanta.

The Time and Space entities in the Vedic thought are eternal because an eternal entity according to the Vedic philosophy has no 'cause' for its birth. No 'cause' for the creation of Time or Space is known. The logic behind this Vedic principle is that if an entity is created as an 'effect' of a 'cause' the entity cannot be a continuum and anything that is created has the transience in its nature. Therefore the Time and Space in the Vedic Cosmology are continua. This also means that the transient universe comes into existence and goes into dissolution phases, again and again.

The Time and Space as the primary participants in the creation of the universe are common factors in the Vedas and in modern cosmology; however, there are distinct differences between the two perspectives as well; eternal versus born in Big Bang. The birth of the universe in Vedic Cosmology is only an event in the perennial cycle of creation and dissolution with Time continuously flowing forward and the Space expanding initially during the creation-phase of the universe and shrinking back to smaller confines at the universe's dissolution. At dissolution, the entire matter of Nature returns to its indefinable concentrated state where Space is still needed to hold it and Time is still needed to flow in one direction to bring the next event of creation.

Thus, the Time, Space and Nature in the Vedic Cosmology never return to a condition of 'nothingness' of the space-time singularity of the modern Big Bang Hypothesis. Based on the Veda mantra references, a detailed discussion on the concepts of Time, Space and Nature is presented in Chapter 5 of this book. After defining the fundamental nature of matter (Nature) a brief discussion on the definitions of the Time-entity and the Space-entity is preempted from Chapter 5.

Time-entity in Vedic Literature

The universe is a mind-boggling phenomenon, even for a cosmologist or an astronomer. It occupies vast space with '*bbbillions-and-bbbillions*' of stars in it, as Sagan used to put it with tremendous emphasis on the letter 'b' in the word billions! Numerous questions on the phenomenon of Time have been raised but all answers may never come owing to human limitations as Sagan had also wondered – '*... whether there are ultimate limits to what humans can know...*' Implied in Sagan's earlier quotation - '*...if time will one day flow backward and effect precede causes...*' is that the 'cause-and-effect' is a time-dependent phenomenon with the 'cause' preceding the 'effect'. Therefore turning an 'effect' back to its 'cause' could happen only when time is an entity and could be reversed too. Even in rewinding a movie and seeing the characters going backwards the time pertaining to all other events and substances has moved forward. The Big Bang Hypothesis is a forward model with time starting at t = 0 sec and moving forward.

As in the Vedas and the Vedic Philosophy this author sees the 'cause-and-effect' as an all-pervading principle in Nature. It is an all-pervading phenomenon because every object of the universe, every event in the universe and the entire universe are immersed in unidirectional Time-entity and bound by it. Simply put - all events in the universe ride on Time while the Time keeps moving forward.

Is 'cause-effect' a recent tenet recognized in human civilization? Certainly not! In the Vedic Cosmology, the cause-effect principle is a pre-ordained character of Nature, the latter being an inert conglomerate of fundamental particles serving as the 'material-cause' of the universe. The cause-effect principle is taught by the Vedas through mantras, directly as well as indirectly. The joint term 'cause-effect' corresponds to the joint term कारण-कार्य (*kaaran-kaarya*) in Sanskrit. The cause-effect (कारण-कार्य) principle is briefly discussed later under the sub-heading - "The Vedic *Kaaran-Kaarya*: Cause-and Effect-Principle".

One develops the specific feeling of first and next at different levels of awareness. When one ponders over a specific event of the past with eyes closed, one gets the cognizance of 'first and next' in that event. When one uses the eyes and observes a sapling grow one tends to store in memory the feeling of 'first and next' and 'earlier and later'. This feeling is created because of the existence of Time. The development of the basic feeling of 'first and next' does not require high intellectual potential in a living-being, however, to develop the feeling of 'first and next' while looking for the relationship between the first and next is an intellectual pursuit. When this intellectual pursuit establishes a definite relationship between the first and the next, such that the existence of the next is out of the first, the mind recognizes the preexistence of the principle of cause-effect. The cause-effect is inherent to Nature owing to Time and its recognition and application to Nature's phenomena is the source of knowledge. It strengthens the logic system of the individual through the *differentiation between the knowledge and fallacy*'. Its application has no bounds. It is the application of cause-effect to the creation of the universe that leads to the development of the science of cosmology and when applied to the Nature and our place in it, it leads to the development of philosophy. The degree to which it is applied in a society indicates the degree of intellectual pursuit in that society.

The ancient *rishis* (sages) of the Vedas and the Sutras (Darshan-philosophies) have since time immemorial comprehensively emphasized the logic of cause-effect to understand the concepts of Time, Space and the creation of the universe itself. A society that explored the application of cause-effect in both realms of the universe, animate and inanimate, is no ordinary society. Admittedly the hearsay of the excellence of a society does not make it an extraordinary society either. The evidence of a society's high intellectual level is reflected in the wisdom contained in the literature of the society. Mantras after mantras in the Vedas, *sutras* after *sutras* in the Darshans (Vedic Philosophies) and the *shlokas* after *shlokas* in Smritis, Upanishads and Brahmana books contain rich discourse on science and philosophy that includes the phenomena of Time, Space, Nature, Consciousness and other concepts difficult to grasp. The Vedic sages explored in depth these phenomena primarily because these phenomena have been the primary participants in the process of creation.

The Vedas emphasized that the Time, Space, Nature (matter) and Consciousness are different from each other, are not consumed by matter, are not consumed by each other and they do not consume matter. The matter is utilized by consciousness that emerges as the living bodies of living-beings, but it is not consumed; the matter of the universe remains conserved. The Time-entity for the Vedas and the Vedic *rishis* remains unaffected by the presence of matter in it, the Time-entity flows forward in one direction and the Space-entity expands from its *tuchchhaya* (diminutive, RigVeda, 10/129/3) volume before creation to its capacious volume like a natural un-packing of a flower from a microscopic bud to a full-bloom.

As mentioned earlier the most common Vedic synonym for Nature is *Prakriti* and the common synonyms for Space-entity are *Antariksh* and *Aakaash.* The Vedic term for the Time-entity is *Kaalah* or simply the *Kaal.*

AtharvaVeda, the fourth Veda, invokes the virtues of the *Kaal*-entity presenting 15 mantras in two *suktas* (hymns). Several of these are discussed in Chapter 5. A glimpse of how the *Kaal* was viewed by the ancient Vedic *rishis* is exemplified here by referring to just three part mantras from the AtharvaVeda. The literary decoration with parable and simile is obvious from these mantras:

> *"Kaal the son is its own father* (the past) *and will be born as its own son* (the future)...". (AtharvaVeda, 19/54/3).

> *"Kaal is like the potent un-ageing horse in perpetual motion (flow) on which ride the* ***sapt-rashmi*** (seven-colour-rays of the Sun) *and the* ***sahashra-aaksha*** (*thousand-eyed*) *Sun....*" (AtharvaVeda, 19/53/1).

> *"In this creation none is older than Kaal... "* (AtharvaVeda, 19/53/4).

These part mantras alone convey several physics aspects of the Time-entity. The first mantra, a marvel in metaphor, states the three sectors of the Time-entity. The first and the last mantra reveal the eternal nature of the Time-entity.

Everything in this creation is transient and aging. It is therefore a highlight of poetic expression when the Veda in the second mantra compares Time with an un-aging, unstoppable (powerful) horse that moves-on perpetually. The riders on this horse are the observable time-dependent natural phenomena such as the "thousand-eyed Sun" (***sahashra-aaksha*** a poetic hyperbole) that sees all around itself and the sunlight with its beauty in seven colours (***sapt-rashmi***). The seven colours in the solar spectrum and the Sun being all-illume all around is today's scientific reality stated in the Veda thousands of years ago. We see the brilliant Sun and the beauty created by its light in a rain-bow but the Veda recognizing the solar rays and the Sun as time-dependent phenomena is remarkable in itself. Anything with a dynamic nature, which includes everything in the universe, is time-bound. As we know now that the solar rays take about 8 minutes to reach the Earth, indicating that they are not a phenomenon of action-at-a-distance.

For Carl Sagan's query, "*If time will one day flow backward...*", to come true the solar rays must become static first and all cyclic orders in Nature should cease. We do not observe any sign of time-reversal in our vicinity or in the astronomers' observation of the far reaches of the universe. We do not even know what it would look like or feel like if Time started to reverse. We may as well reconcile with the fact that *Kaal* (Time) the "*potent un-ageing horse in perpetual motion* (flow)" will keep moving forward.

Some time ago when asked by a host on Canadian Broadcasting Corporation, '*what is Time*', the visiting guest, Professor Roger Penrose, the eminent British mathematician/cosmologist answered – "*...to understand Time we have to understand what Mind is and what consciousness is...*". The fact is that the Big Bang field equations have no parameters or variables to deal with the phenomenon of Mind or Consciousness. Therefore, the modern cosmology with all its complexity in modeling the universe would not be able to shed light on what consciousness is! This is why this author remarked earlier that the modern cosmologists use Time and Space entities in their model but they do not understand what they actually are!

The Vedic philosophical perspective on consciousness is a discourse to be rendered in another book; however, a brief commentary on the phenomenon is contextually included at places in this book. It may be mentioned here that this author has questioned the basic logic of the evolutionary biologist, Dawkins, who defines consciousness in terms of a "*useful model*" of a conscious-being.

When modern cosmologists refer to 'time reversal' and the 'effect preceding the cause', the context is the Time undoing sequentially what it has done this far in the creation of the universe. Logic requires that for it to reverse the Time must have a beginning. In the Vedic thought the Time is *anaadi* (eternal), as implicit also in the AtharvaVeda mantra (19/53/4) cited above. The Nature (matter) in Vedic thought is also eternal but its matter transits from *Asata*-state of dissolution to the *Sata*-state of creation. The reversal from creation to dissolution and vice-versa takes place without reversing Time. Because all matter of the Nature in the Vedic creation-dissolution cycle is conserved and only its state changes, like in a reversible chemical reaction, the universe reverting from creation to dissolution is not time reversal. The discourse on cause-effect is enriched with simple deductions in Darshan (philosophy) *sutras* based on the Veda mantras, such as '*the effect is always different from its cause*', '*the effect is coarser than the cause*' and several such.

A detailed discussion on the tremendous progress in time measurement by modern physics is beyond the scope of this book. However, on how deep the Vedic scholars have gone in measuring and resolving the Time entity is presented in detail in Chapter 5 and briefly in what follows. The definition of Time from the perspective of practical application of Time is furnished in the ancient book on practical astronomy entitled 'Surya Siddhant'. The book of Surya Siddhant, better known in Western astronomy as 'The Sun Principle', is a treatise on the indigenous practical astronomy developed by the ancient Aryans of India. It is in the form of *shloka* composition; a *shloka* in Sanskrit is a metered poetic couplet expressing a specific thought. There are numerous *shlokas* in Surya Siddhant on Time and its division as well. The first *shloka* on the basic nature of the Time-entity is the 10th *shloka* in the first chapter of Surya Siddhant. Its English translation is as follows:

"*Time is in two classes. First, the cause for creation and dissolution* (being eternal) *is unknowable and the Second is knowable. The Second-class is in two categories – one is moort* (comprehensible and measurable) *and the other is sookshma* (immeasurably minute)." (Surya Siddhant, 1/10).

Implicit in the first part of the *shloka* is that the Time-entity itself being eternal remains an unimaginable entity, because only the created objects and phenomena in the universe have the property of transience, form, function and expanse. It is the form, function and expanse of objects that enable their measurement and at least enable the imagination of their existence and enormity. In the Vedic thought, both kinds of physical phenomena, *moort* and *sookshma*, are deemed as real. The term *moort* means - one having the form and size, enabling its detection, observation or measurement and therefore it is the subject of the knowledge-senses (unaided or aided). The term *sookshma* on the other hand means - one that is subtle, beyond the senses and only the subject of cognitive faculties. An example of *sookshma* time in modern physics is a Planck-time, 10^{-43} seconds (10th of a million-trillion-trillion-trillionth of one second) which is not a subject of the senses and cannot even be measured directly in a laboratory. Thus the Time-entity has been recognized in the Vedas as a real entity accessible in the realm of measurability and inaccessible in its realm of immeasurability wherein it is eternal and extremely minute.

Modern physics deals with the, *moort* and *sookshma* classes of time, the former in practical applications and the latter in theoretical applications. It is shown in Chapter 5 that the Aryan time measurement units are derivatives of science application and quite elaborate as mentioned above, covering a range from 4.32 BY down to the practical unit of *nimesh* (1 *nimesh* = 0.177778 sec = 8/45 sec) in 18 divisions of the time scale. In some ancient astronomy texts, for example in the 16th and 17th *shlokas* of 'Siddhant Shiromani', written by Bhaskaracharya in the 12th century CE, there are two more finer divisions of the time unit of *nimesh*. The 30th part of a *nimesh* is 1 *tatpar* which is 6 milliseconds and 30,000th of *nimesh* is defined as a *truti*, where the term *truti* stands for the minute division of time representing the "uncertainty" or the "accuracy" in time measurement (see also Surya Siddhant commentary by Mahaveer Prasad Shreevastava, 1987CE). A *truti* in Vedic astronomy therefore is about 6 micro-seconds (six one-millionth of a second). The physics of the derivation of the practical time unit named *nimesh* is also discussed in Chapter 5. Several paragraphs devoted to this practical time unit show that it could not have been arrived at without accurate experimentation in physics and human physiology and all that was accomplished by a Vedic *rishi* in the far antiquity of the Vedic civilization.

The Space - entity in Vedic Literature

The Vedic concept of the Space-entity based on Vedic mantra references is also discussed in detail in Chapter 5 under the sub-title of "*Aakaash-tatva*: The Space-entity". The Space-entity of modern science is the *Aakaash*-element of the Vedas. This invisible entity is a real entity in the Vedas as also in the modern cosmology. An example of how the Space is viewed in the Vedic discipline is exemplified from Swami Dayanand Sarasvati's statement in his book "Satyarth Prakash" (1881 CE). What he stated in the Hindi language is translated into English by this author, as:

"...Reality is that the Aakaash (Space-entity) *is not a created entity* (it is eternal)*, because otherwise without its pre-existence where could the Prakriti* (Nature) *with its paramanus* (atoms) *stay....?"*

Swami Dayanand Sarasvati's above statement is based on the Veda mantras. It is obvious from this statement that the *Aakaash* (Space-entity) in Vedic Philosophy is eternal and must pre-exist the formation of the atoms of Nature and its function is to provide a place for their existence. It is also obvious that the preexistent *Aakaash*, responsible for holding the matter of the universe, which is *Prakriti*, is a different entity altogether from *Prakriti* and that *Prakriti* at its fundamental level is made up of corpuscular *paramanus* (fundamental particles).

The practical Vedic units for space measurement are tabulated in Chapter 5 with emphasis on the unit of *yojan* (1 *yojan* = 13 km). The distance unit of *yojan* in differing applications appears in several Veda mantras, either independently or in association with the time unit. The RigVeda mantra (1/123/8) relates the unit of *yojan* to the motion of the light swath of the morning-twilight sweeping the Earth as the Earth turns; the morning-twilight being a dynamic physics phenomenon in atmospheric optics. The mantra states:

The morning Twilight (Usha) with its immutable occurrence is born daily in the vast home of Varun (atmospheric water-air-ocean system) and moves swiftly demonstrating its qualities in एकैका (*akaika*)*, in concatenation, up to 30* ***yojan***". RigVeda (1/123/8).

Among several physics observations in this mantra it specifically states that the morning twilight, the diffused sunlight (indirect sunlight) successively keeps illuminating the dark-side of the west of the sunrise point with a light swath extending up to a distance of 390 km (30 *yojan*). Because the Earth rotates around its axis the dynamic spread of twilight moves westward of the sunrise point, nonstop. A *rishi* must have carried out optics measurements to determine this number. Additionally, to estimate this distance the *rishi* had to know the circumference of the Earth and the speed of spin of the Earth. Knowing these parameters in the far antiquity of the Vedic civilization is remarkable in itself.

Now we know that a point on Earth's circumference moves with the speed of 464 meters/second, which is faster than the speed of sound. Therefore, the twilight swath successively illuminating and moving forward at this rate not only implies that the Earth is round but the *rishi* also knew what its circumference is. A complete discussion on why 390 km twilight swath in the Veda corresponds to the calculated tropospheric height of 11.95 km is in Chapter 5 under the sub-heading – "The Case of Vedic *Usha*: the Morning Twilight", delineated with a ray-tracing diagram. The twilight is the diffused sunlight scattered from the tropospheric layer. From the 26 RigVeda mantras on *Usha-devata* (Twilight) several virtues of the morning twilight are listed that glaringly exhibit physics thought. To conclude, the rate of spread of Twilight west ward on the surface of the Earth could not have been arrived at without a rigorous experiment in time, space and atmospheric optics. And all that happened in the far antiquity of the Vedic Civilization!

The other example that boggles one's mind is the ancient Vedic calculation by some *rishi* for the speed of light represented in terms of *yojan* and the time unit of *nimesh*. It is also amazing that, the speed of light determined by this *rishi*, discussed in Chapter 5, is extremely accurate from present standards of unit measurements.

As mentioned before that this author has collected 22 of many synonyms for the Space-entity. This means 22 differentiating properties of the Space-entity. This author could safely surmise that if asked a contemporary cosmologist to list all properties of the Space-entity he would have difficulty exceeding 22.

The Vedic cosmology presents with dexterity the characteristics of the Space-entity that have not been imagined as yet in modern times. *Pushkaram* is a Vedic synonym for the Space (*Aakaash*). Rishi Yask's ancient book Nirukt (5/3/2/2) provides the etymology for the term *pushkaram*. It relates to Nighantu 48th (special collection of Vedic terms). *Pushkaram* is derived from a single *dhatu* (root) but is a multi-meaning term. Three out of five meanings of *pushkaram* are furnished here:

- *Space is* ***pushkaram*** *because it does poshan (provides care) of all substances of the universe, animate and inanimate, by benignly providing space for their existence.*
- *Space is* ***pushkaram*** *because it rises above matter, like a lotus flower rising out of mud, yet remaining beautiful and unblemished.*
- *Space is* ***pushkaram*** *because it has the property of pushpate, one that swells like a flower.*

These statements have physics implications: that the Space is essential for the existence of matter and therefore for the existence of the universe itself; the Space is an entity separate from the matter of the universe (lotus and mud); the matter does not react with Space or change Space's constitution (does not *blemish* it) and the Space expands (*pushpate*) as the universe expands (expanding universe an eons old Vedic thought, mantra references in Chapter 9). The remarkable physics properties of Space in these statements and a dozen additional different properties of the Space-entity in differing contexts are discussed under the sub-heading "The Space-entity (*Aakaash-tatva*)" in Chapter 5 entitled "The Primary Vedic Entities in the Creation of the Universe". The discussion is supported by about 15 references from Nirukt (2/3/1), Bhagvadgita (13/32), Shatpath Brahmana (11/1/6/22), Vaisheshik Sutra (2/1/20), and the Vedas. Several of these properties of Space have not been explicitly identified or explored by modern cosmology.

The Veda has addressed the Space-entity with the adjectives such as the "Flower of the Creation". The Space after creation does what a flower does when it is created by a plant. This simile could be understood by most. A bud before efflorescence begins at a single point, folded with efficacious mastery to occupy very compact space with all the petals in an irreproducible and irreversible packaging. This compact space of the bud contains the ingredients of its own little universe, waiting to burst open with a bang. When the bud blooms into a flower, it is like an event of a little big bang, displaying its ingredients while drawing unseen energy from its plant and rendering a variety of unique shapes of its components. The bloom spreads from a *tuchchhaya* (diminutive) space into a relatively larger space revealing a full-fledged creation of its own.

This phenomenon starts with a microscopic three-dimensional bud in the shape of an ellipsoid or an ovate which announces the advent of a flower in the diversified form and splendor of nature with colours and fragrance.

***Pushkaram* property of Space**. The two blue dots on the opposite sides of the tip of the compact bud (light green) correspond to the two blue dots far apart on the opposite petals of the opened flower. The expanded flower (pink) and the bud are in proportionate size.

When we observe its advent and observe its previously hidden components and ingredients, we take it upon ourselves to define it by creating appropriate nomenclature such as sepals, petals, filaments, pollen and so on in the likeness of stars, super-novae, quasars, solar systems, galaxies, black-holes and a host of differing formations in the vast universe. Thus, it becomes obvious as to why the Space acquired the adjective of a flower in the Vedas. In terms of the cause-and-effect, each Vedic word for each object, an action or a phenomenon in Nature, is the effect, the cause of which is the root of the word with specific properties.

The *rishis* in the Vedas and in the post-Vedic literature explained complex subjects with simple examples, parables and similes. Invoking the simile of a flower for the Space-entity is most appropriate and clever. This author draws parallels between the blooming of a flower and the spreading of the cosmos in Space. The author followed a bud till it bloomed fully into a flower. A tiny green bud and the pink flower are shown in the picture. Two ink dots on the opposite side of the bud correspond to the two ink dots on the opposite petals of the flower. The bud and the flower in the picture are in actual relative size and on blooming, the flower occupied 40-50 times the space of the bud, in just a few days. The separating of the dots relative to each other as the flower bloomed reveals the way the galaxies separate from each other increasing the space between them. A Vedic *rishi would* not draw a simile between the *Aakaash* (Space) and a *pushp* (flower) and call Space a *pushkaram* without knowing that

Aakaash expands like a flower. The question remains - how did the *rishis* know that *Aakaash* expands? Amazing!

Another Vedic term *vapushkar* for *Aakaash* (Space) means '*Aakaash keeps its form even when in use by a substance of any form*'. The implication of *vapushkar* property of Space, pertaining to space-time continuum and warping of Space near massive celestial bodies and bending of light rays is discussed with new explanation in Chapter 9.

The swelling of the Space-entity after the advent of the universe, carrying the matter of the universe with it, relates to the "expanding universe". The author presents numerous mantras stating that the universe is expanding. As an example the RigVeda mantra (10/56/5) states with clarity that the *Antariksh* (Space-entity) was spreading and along with it spread out the cosmos; which in modern cosmology is the same as the expansion of the universe.

Considering that Rishi Brihaduktho Vaamdevya formulated the mantra stating that the universe is expanding, one could surmise that the *rishi* must have been an assiduous astronomer observing the constellations night after night and observed the cosmos separating from each other. The fact is that he must have found, like a contemporary astronomer with most powerful telescope, that the galaxies and stars in our observation range do not show apparent daily or yearly separation as a part of the expansion of the universe. This is because the expansion in the locally observed constellations is not discernible. Yet the *rishi* said that the Space was expanding. So what was the meaning of all this? Let us discuss this a bit further.

We are not talking here about the planets of the solar system changing their position night after night – we are talking about the stars and the galaxies outside our solar system. Observing the stars of the night-sky at their respective positions, night after night, would not lead anyone to detect the property of expansion in the universe. If it weren't for the Doppler red-shift effect in the spectrum of the faraway galaxies, modern astronomy could not conclude that the universe was expanding. It was the red-shift in the spectrum of faraway galaxies reaching the telescope and not the actual motion of the galaxy that led Edward Hubble in 1929 to conclude that the galaxies were flying away at high speeds. But the fact also is that Rishi Brihaduktho Vaamdevya concluded in this RigVeda mantra that the invisible entity *Antariksh* (Space) was spreading and that the cosmos were spreading with it. Therefore, the mantra and the Vedic thought in it are remarkably intriguing. Rishi Vaamdevya must have a basis for labeling the universe as "expanding". The *rishi* must have had a sound theory or carried out experiments in astronomy, not necessarily measuring the red-shift. No matter how one interprets this mantra (and several more); the Vedic principle contained in it is no different from one of the most important findings of modern astronomy that the universe is expanding. It is intriguing!

After detecting the red-shift caused by the Doppler Effect between moving objects (observer and the galaxy), the Big Bang equations were corrected, making the expanding universe as the basic feature of the Big Bang Hypothesis. A word-for-word translation of this mantra is presented under the sub-heading "Expanding Universe a Vedic Discovery Eons Ago", in Chapter 9 on Vedic Big Bang scenario. A brief discussion on the red-shift is included in Chapter 2 and its cause in the faraway galaxies, from a Vedic perspective, is presented in Chapter 9.

The concept of Space in the Vedic civilization is as old as the RigVeda itself. The Veda mantras have been the basis for the interpretations and elaborations in the post-Veda literature on Vedic philosophy. A section of the post-Veda literature is called the Brahmanas or the Brahmana Books. Yet another section of the literature is in *sutra* formulations called the Vedic Darshans (philosophies). Rishi Kanaad, a post-Veda philosopher in the far antiquity of the Vedic civilization expounded the subject of Space in great depth in his Vaisheshik Sutra. Again, the details on this are in Chapter 5, however, a brief philosophical perspective on Rishi Kanaad is presented here.

Rishi Kanaad states the *lingam* (defining-sign) of the Space-entity (*Aakaash*) in the Vaisheshik Sutra (2/1/20):

"The property permitting the 'exit from and entry into' is the defining sign of Aakaash" (Vaisheshik, 2/1/20).

Implicit in this simple formulation, *exit from and entry into*, is that the Space is unaffected by objects moving in and out of it. It is an astute deduction by Rishi Kanaad for an abstract invisible entity. This defining-sign (*lingam*) of the Space is delineated in the example of a ball (in Chapter 5) thrown by a pitcher where the ball on its trajectory successively enters the *Aakaash* (space) at one point and exits to enter at the next point without distorting the space left behind. Vaisheshik by Rishi Kanaad is one of the oldest Darshans that precedes the Mahabharata War which according to the Indian calendars, when converted to the Western calendar, ended in 3140 BCE (proven by this author in Chapter 3).

Krishna, well after Rishi Kanaad and just before the onset of the Mahabharata War preached the philosophy of life to Prince Arjuna of Hastinapur (near Delhi). Krishna's sermon to Arjuna at the battle-field of Kurukshetra (near Delhi) is well-known as the book of Bhagvadgita with 18 chapters in it, which in fact is a part of the book of Mahabharata. Bhagvadgita has fascinated some of the recent scientists, including Albert Einstein, Albert Schweizer and J. Robert Oppenheimer, just to name a few.

A *shloka* couplet from Bhagvadgita's 13th chapter also relates to Rishi Kanaad's basic definition of the Space-entity and goes beyond it, drawing a parallel between the Space (*Aakaash*) and another entity that is also not bound by matter. The translation of the Bhagvadgita's *shloka* reads thus:

> "*Just as the all-pervading Aakaash (Space), owing to its minute nature, does not attach to any object in it, the soul-entity in the living being also does not attach to the matter of the physical body.*" Bhagvadgita (13/32).

The main clause in this Bhagvadgita *shloka* pertains to the soul-entity, wherein Krishna emphasizes the soul's most important property of non-attachment to matter. Implicit in Krishna's expression is that the soul-entity and the physical body (matter) are two separate entities and the soul and the Space are two different entities. The most important observation in the *shloka* is the parallel that he draws between a specific property of the soul-entity and that of the *Aakaash* (Space-entity), the latter being a subject in cosmology. The Space owing to its minute nature does not attach to any object embedded in it. The primary difference between the two according to the Vedic philosophy is that the soul is a single quantum of consciousness and therefore localized at a given point in time and space, whereas the inert Space-entity is conterminous through the universe. The common property between the soul-entity and the Space-entity is the property of non-attachment. In other words, the soul-entity is neither bound by the matter nor is it consumed by matter. This raises a natural question - if the soul does not attach to the living body how can it function in it?

The answer to this question lies in the example of a driver in a car. The driver is a conscious-being and the car is an inert entity. The driver desires to hold the steering-wheel, to start the ignition, to put the car in gear and to press the gas-pedal with his foot. This transforms the inert machine into an apparent animate object. The driver's hands and foot remain detached respectively from the steering-wheel and the gas-pedal. The driver after driving stops the car, stops the engine and steps out of the car. The car turns into an inanimate object. Krishna's explanation of non-attachment of the soul to the body is in the likeness of the driver's non-attachment to the car. It is this property of non-attachment, common to the soul and the Space that Krishna explained with clarity to Prince Arjuna. Modern cosmology gives no consideration to the imperceptible phenomenon of consciousness and therefore to the concept of the soul, however, it cannot do without the imperceptible Space-entity. One could still pose a non-philosophical question. Since no one has seen a soul and no one has seen the Space then how could Krishna know what a soul is and what the Space-entity is? Obviously there are no physics answers to these questions. Even a physicist who believes in the existence of Space-entity will not be able to give exact answer to - what is the Space-entity?

Rishi Kanaad's 'defining-sign' of the Space-entity and Krishna's complementary elaboration of its property of minuteness, its non-attachment with matter and non-consumption by matter, are remarkably brilliant deductions. The phrase in modern cosmology - 'the space is curved', did not escape the careful consideration of the *rishis*, although it may have slightly a different meaning in the ancient Vedic cosmology. The Space-entity for this property is adorned by the adjective, *teeryak*. The literal meaning of the term *teeryak* is – '*one that has directionality like the path of an arrow*', which is always curved. Simply put, the Vedic 'Space is curved' but it is not curved by the gravity of a massive celestial body. The discussion on this aspect of the Space being quite involved, it is taken up in Chapter 5 under the sub-heading of '*Antariksh*: the Vedic Space is Curved', where several mantra references are invoked (AtharvaVeda, 10/2/24; 10/2/25; ….).

Not just the Space-entity (*Aakaash*) as a "general-cause" or the Nature-entity (*Prakriti*) as the "material-cause" of the universe but other entities and processes partaking in the formation of the universe are also uniquely defined and described in the Vedas.

Rotating and Expanding Universe – an Ancient Vedic Thought

The force of gravity due to mass in matter pulls the universe inward. To have the universe expanding, a force opposite to the force of gravity was to be conceived. Not having such a force in existence the cosmologists had to imagine that a background energy field, ubiquitous through the universe, must exist to explain the continued outward expansion of the universe. Not knowing what this energy could be, it was named as Dark Energy. From there on the Big Bang Hypothesis shifted from the trinity formed by Nature, Time and Space to a quadrate with the fourth corner formed by the Dark Energy. Because the universe is said to be presently expanding at an accelerated rate, the effect of the Dark Energy exceeds the effect of gravity. This necessitated Einstein's original field equations for a static universe to change to include the expansion of the universe.

Not having a confirmed mechanism for the production of the Dark Energy and not having a convincing common physics logic to support its existence a rhetoric question was put forward - if not the Dark Energy what else is expanding the universe? Hypothesizing the Dark Energy aside, it is not clear whether the Dark Energy pre-existed the Big Bang or it is a byproduct of the Big Bang. If it appeared with the Big Bang what mechanism created it? If it is a quantized entity like other known radiational energy forms, what kind of quanta would constitute it? If it is a continuum, is it discontinuous beyond the last galaxy at the edge of the universe? The answers to these questions have to come from the Big Bang cosmology without invoking new hypotheses. Are all other potential possibilities exhausted to explain the observed expansion of the universe? The Big Bang Hypothesis has answered many questions related to the observations of the phenomena in the observable universe and as discussed in Chapter 2, it has also raised many questions.

The question - if not the Dark Energy what else is expanding the universe – is answered uniquely in the Vedic Big Bang Scenario developed in Chapters 8, 9. Startling to the contemporary cosmologist it may be, but the eons old fact in the Vedic Cosmology, preserved in the form of mantra formulations, is that the universe is expanding. (RigVeda, 10/56/5; RigVeda, 10/72/8; RigVeda, 10/129/5; ....). The explanation for the expansion of the universe, as identified by this author, comes from other mantras (RigVeda, 7/63/2; RigVeda, 10/89/2; ...) where, in addition to the translational (linear) expansion of the universe, the major expansion is due to the rotation of the universe as a whole. That the entire universe is rotating around an axis termed as *Skambha* is a unique physics thought in the Vedic Cosmology. It is the centrifugal force of rotation that creates the tendency of throwing the matter of the universe outward causing the expansion of the universe. The ancient Vedic thought that the universe is expanding because it is rotating has led to several new deductions in new cosmology, discussed in Chapter 9.

The moving objects generally are in two types of motion: linear and angular. In a linear motion the object moves from one point to the next along a straight line whereas in angular motion the object moves on a curved path. All celestial bodies are on angular motion around their own foci - the Moon around the Earth, the Earth (and other planets) around the Sun and the Sun (and other stars) around the centre of our galaxy, the Milky Way Galaxy. Our galaxy may be rotating around its own focal point. According to the Vedic thought the Milky Way Galaxy and all other galaxies of the universe are rotating around a common axis named the *Skambha*. The Veda states with clarity that the rotational motion is not only a feature of the celestial objects that rotate around their own respective local centers of rotation, but they all together are also in rotation around the centre of the universe. This feature of the Vedic Cosmology is quite different from anything in the modern Big Bang cosmology.

The relevant Vedic mantra references, implicitly and explicitly state that the rotational momentum (mass x rotational velocity) in the objects of the universe is an endowment of the inherent rotation in the primordial Nature (RigVeda, 7/63/2; ...). In the Vedic scenario of the creation of the universe, both, the translational (linear) and rotational (angular) momenta within the primordial Nature are imparted to the matter of the universe at the time of the Vedic Big Bang. It is the rotational energy, not thought of in the modern Big Bang Hypothesis, that could be adding its own effect to the observed red-shift in the faraway galaxies.

In the Vedic scenario, with gravitation to contract the universe and the centrifugal force of rotation to expand it, one does not need a hypothetical Dark Energy. The observed red-shift in the spectrum of the fast receding galaxies, in the presence of this universal rotation, could partly be an artifact created by the translational and rotational components of the rotating universe. This effect in red-shift is further discussed in Chapter 9.

According to such a Vedic scenario, the source of rotational momentum in all galaxies, solar systems and in all celestial bodies is primarily due to the original rotational momentum in the primordial blob of Nature, whose original shape is defined by its Vedic names, *Maart-aand* (Mortal-ovate) being one. The Sanskrit meanings of the terms used in Vedic Cosmology, depicting the primordial universe's shape, nature and action are discussed in detail in Chapters 7 to 9. It is also discussed there that the Vedic scenario for the formation of galaxies in an ever expanding universe makes a better sense as well. The galaxy formation is the result of matter accumulation under gravitation in a given region of space. In a linear expansion of the universe, as in the Big Bang Hypothesis, one should really be surprised that even a single galaxy was formed when matter was only separating away vigorously.

The galaxy formation mechanism in the modern Big Bang Hypothesis is based on the hypothesis that the galaxies are the result of the original irregularities in the matter distribution of the primordial universe, that were formed right in its first moments during Planck-time of 0.539×10^{-43} second. The mathematical manipulation aside, it is not intuitive and not very convincing that in the modern Big Bang scenario with tremendous explosion at one point with super high radial speeds in matter, the matter density nodes that existed in the unimaginably short time period and in the negligibly small space would not fall apart with time as they flew away with tremendous force.

The Vedic Principle formulated in Chapter 6 and a large number of additional mantra references, lead to the development of the Vedic Cosmology Scenario developed in Chapter 7 to 10. Discussions on rotation and expansion of the universe, logic for galaxy formation and the age of the universe are a part of this scenario. The Vedic age of the universe is not as high as the age estimate in modern astronomy; however, a remarkable feature of the Vedic age estimate of the universe also runs in billion years.

The Theistic Argument

In modern Big Bang cosmology the two opposite forces, gravitation and Dark Energy, pervading through each other affect matter of the universe, one contracting it and the other expanding it, but without one affecting the other. In addition to the force of Dark Energy, there are four fundamental forces in Nature discussed under the sub-heading, "Fundamental Forces of Nature" in Chapter 2. All forces have the inherent nature to pervade through each other without destroying each other. What mechanism created these four forces and what mechanism created the non-destructive property in each force is another mystery of Nature. Then, who is to vouch that in addition to the force of Dark Energy and all the other known physical forces, yet another all-pervading force-field could not or did not pervade through the universe with consciousness its additional virtue? A failure to detect a force using instruments made of matter does not prove its absence.

The Vedas being theistic, the mantras at times give credit to another kind of all-pervading force with consciousness its additional virtue. Present cosmology is carried out with half a dozen physical forces including the force of Dark Energy, each playing their own specific roles. Then a consciousness force could also coexist to play its own role. In the Vedic concept of creation of the universe, the entire universe with all its matter, energy and different physical forces is embedded in an all-pervading Consciousness Field. It is this force that is considered to be governing the behaviour of the universe in both its realms - matter and consciousness. It is the Consciousness Force that is the source of all the laws in animate and inanimate realms. Most mantras with science thought in them are without a reference to the Consciousness Force, however, when a mantra with science thought in it gives credit to the Consciousness Force; one could set aside the Consciousness Force without losing the benefit of the science thought in the mantra. Therefore, the theism in the Vedas does not negate knowledge extraction from the mantras and does not affect developing the scenario for the creation of the universe.

The Consciousness Field is Veda's 'efficient cause' of the universe with the matter (Nature) being its 'material cause' and the Time and Space being its 'general causes'. Based on the discussion thus far, especially on the Vedic definitions of Time and Space, it is evident that there was considerable thought investment by the Vedic *rishis* in the realm of cosmology and the entities that form the set of the 'material-cause', the 'general-cause' and the 'efficient-cause'. Veda mantras delineate with clarity that the thought investment has led to knowledge development in the areas of physics, astronomy, cosmology, philosophy and even spirituality; the last two still remain outside the framework of pure sciences. The *rishis* also developed enriched knowledge base in the area of the relationship between consciousness and matter. They gave consideration to the metaphysical energy fields created in the first moments of the universe that lead to the development of functions in conscious-beings, the functions such as the Intellect, Ego and Mind. These entities, the product of Nature, are the subject of another book. A brief comment on Mind is as following.

In trying to understand the Mind-entity the contemporary scientists find that it is a very complex phenomenon in living beings. Mind according to Vedic Philosophy is a quantum of energy that operates inside and outside of the living being. Owing to its minuteness it cannot be detected or captured. Anything not measurable by physical means is generally said to be metaphysical or super-natural and is generally labeled as 'may exist in reality or may not'. In the Vedic concept, the Mind-entity arises in the living unit of matter owing to the presence of the consciousness corpuscle present in that living unit. The Mind is an entity in the realm of matter and is a companion of the soul; the latter is an eternal non-matter quantum of consciousness. Both coexist in the preferred locale of the physical body that has a limited transitory life-span: metaphorically prescribed to be 100 years for humans. The soul-entity is said to be a perpetual traveler of the universe, stopping at rest-areas (living bodies), gathering knowledge about the world with the help of companion instruments, the Mind being its most active instrument.

Recognition and Dissemination of New Thoughts

Complexity exists all around us. The complexity is in the nature of Nature. Living-beings, a byproduct of Nature, are riddled with complexity in their nature and in their physical construct. It is in their nature to try and unravel the complexity that surrounds them. When a human-being uncovers a secret of Nature it is not kept a secret. It is disseminated for the benefit of others. Others examine it, evaluate its truthfulness, reject it or accept it. Recognition of the process of thought generation and dissemination has been ongoing since the existence of living-beings and more so in the humans in developed societies and civilizations. When the truth is reached and recognized as such by a society, it becomes a precursor for discovering more hidden truth.

What is 'truth' and how do we know if it has been reached? The Vedic definition of truth is contained in the Sanskrit alphabets that form the Sanskrit term for 'truth'. This term is सत्य (*satya*). The word '*satya*' (सत्य) is derived from the *uss dhatu* (root) and *krit pratyaya,* leading to its meaning – '*that which is'*. The word analysis of the term *satya* (सत्य) itself points out that it is a conjugation of two words *sa* and *iti* (स + इति) which together mean "*that's it, period*" or '*that only*' or "*nothing else in it*". Truth has no structure or fine-structure and is without alternatives. It does not associate with probabilities and exclusion principles. Often times it looks as though the truth was reached until a new thought comes along, necessitating reconsideration and reiteration. The सत्य (*satya*) or the truth by Vedic definition is an invariant in time and space.

At times when the complexity of the creation of the universe in both realms, the animate and the inanimate and the complexity of many phenomena embedded in it are well understood by some brilliant and well-known scientists, it may seem that the final truth was reached. At times the enthusiastic scientists present their discovery as the 'final truth' and declare that they have reached सत्य (*satya*). The human limitation at times causes confusion between *satya* and what seems like *satya*. An example of this is the 'final' declaration by one of the most brilliant cosmologists of our time, Stephen Hawking. Two parts of this 'final' declaration are in the following that he stated while discussing his new book entitled "The Grand Design" co-authored with Leonard Mlodinow.

(A) *"The scientific account is complete. Theology is unnecessary…"*

(B) *"God is not necessary and that the universe is deterministic with free-will in living-beings only an illusion".*

Let us examine these statements and determine how close both statements are to the 'truth', the सत्य (*satya*). (The author's discussion pertaining to Stephen Hawking was written before his passing). A discovery by a scientist and its experimental confirmation by him/her as well as by other scientists certainly deserves to be elevated to the category of 'a final declaration'. However, the fact remains that the *scientific account* of the universe, as claimed in (A), is not complete yet, because if it were complete, all the Big Bang cosmologists would be out of work! The fact that they are gainfully employed strengthens the logic that the thought process on the creation of the universe has not come to finality yet and the *scientific account* has not been completely put through a final scientific test yet, where a 'hypothesis' should morph into a 'principle'. The Big Bang Hypothesis, on which Stephen Hawking based his declarations, is still thronged with a large number of hypotheses, some of which do not even make commonsense in physics.

Let us for a moment put aside *theology*, *God*, *free-will in living-beings* and put them all in the category of *illusion* as Hawking suggested in (B) and look at the remainder of Hawking's above statement (A) where he declares – "*The scientific account is complete*". To check the validity of this statement one need not resort to a theological argument because that may not yield the desired results; instead let us invoke the assertions from some high caliber cosmologists and their latest research.

Julian Barbour, a British physicist, has authored and coauthored several books and carried out theoretical research in quantum gravity. Numerous other physicists have published papers based on his theories. Barbour and the physicist Bruno Bertotti developed a technique called "best matching" for deriving gravitational equations directly from astronomical measurements, which without the use of the 'space-time continuum' described gravitational effects as accurately as Einstein's general relativity. The "best matching" also explains that the not well-understood Dark Energy in the Big Bang Hypothesis, responsible for the accelerated expansion of the universe, was unnecessary. Based on Veda mantra references, this author will discuss in Chapter 9 that the universe is expanding and that the mechanism for this expansion is not based on the hypothetical Dark Energy but it is due to the centrifugal energy naturally stored in the rotating universe itself.

Based on his book entitled "Does Time Exist?" a preamble to Professor Barbour's public lecture on June 6, 2012 at the Perimeter Institute for Theoretical Physics in Waterloo, Ontario, Canada, stated:

"Many attempts to create a unified theory of the universe using relativity and quantum mechanics suggest that time as we seem to experience it does not exist- it may be only a well-founded illusion. The idea of timeless universe can be traced to Plato and his insistence that only being is real, while becoming is an illusion. In this talk Prof. Barbour will explore how the Wheeler-DeWitt equation of quantum gravity suggests the fundamentally timeless nature of the quantum universe. He will also raise unresolved mysteries of our conscious experiences, and why these might provide insight into how a fundamentally timeless universe may be perceived as intensely temporal".

Hawking's conclusion that "*The scientific account is complete*" is based on the Big Bang cosmology, which is based on Quantum Mechanics to create matter, the role of which lasts only for one Planck time of 0.539×10^{-43} second, a period of 0.539 seconds divided by a number with 43 zeros after one. After that the newly born universe is taken care of by Einstein's relativity. The Big Bang cosmology in turn cannot exist without the space-time continuum and without the mysterious expanding force, the Dark Energy, but the Dark Energy is not needed in Barbour's approach. A slice of time as small as 0.539×10^{-43} sec creating all matter of the universe 10^{60} kg at one dimensionless point in space means that the matter is produced at the rate of 10^{60} kg/10^{-43} sec = 10^{103} kg/sec. This could happen in a mathematical equation but is beyond the reality of what one observes in Nature. This is exceedingly beyond commonsense and one would think that the Big Bang Hypothesis is pseudo-science. Hawking has to have Time and Barbour says that it is an illusion. Therefore without suggesting who is right, a comparison between the two approaches leads one to think that Hawking's declaration is premature. Let us analyze further the logic in Hawking's two statements (A) and (B).

Hawking's conclusion, '*theology is unnecessary'* has not been scientifically proven yet. Any conclusion without a scientific proof is in the category of a 'belief'. Therefore the conclusion by a modern cosmologist claiming to have reached the सत्य (*satya* = truth) that *'theology is unnecessary'* is a belief similar to the belief of a theologian who concludes without proof that 'God is necessary' for the creation of the universe. The Vedic term सत्य (*satya*), which stands for 'that only', applies as much to a theologian as to a cosmologist.

Stephen Hawking's statement (B) *"God is not necessary..."* also falls in the same category as his first statement. As argued before, a unique all-pervading force with consciousness as its additional virtue could exist in the universe and go undetected by physical means. Because the non-detectability of a phenomenon is not the proof of non-existence of the phenomenon, Hawking's statement (B) is not on a sound logic. Considering the discussion in the previous pages one cannot claim that the modern cosmology (Big Bang Hypothesis) has reached the *satya* (truth). The fact is that truth in Big Bang Hypothesis is sought based on several un-provable hypotheses and the fact is that the same basis is being used for disproving the least understood phenomena such as the possible fifth force that theists address as God. The reality is that many consider the phenomena such as consciousness, Mind and spirituality as least understood but real. Thus, the सत्य (*satya*) has not been reached yet. However, whatever the level of proximity of the emphatic statements to the truth, one useful outcome of this is that it moves the discussion on cosmology forward and contributes to the story of cosmology. One could say that all ideas are good if they are the stepping stones towards the discovery of *satya.*

While presenting the historical review on cosmology in the first chapter of his earlier book entitled "A Brief History of Time" Stephen Hawking followed the tradition for scholarly publications; acknowledging and referring to the work of other scholars. He started with the Greek philosopher Aristotle (384 - 322 BCE) who suggested that the Earth was a stationary round sphere rather than a flat plate and all other celestial bodies including the stars moved around it in circular orbits. Aristotle also theorized that everything was made out of four elements - earth, air, fire and water. Hawking then referred to the famous Greek astronomer Claudius Ptolemy (87 - 150 CE) who elaborated Aristotle's idea by putting forward a cosmological model of the universe wherein the Earth was still the centre of the universe around which the universe rotated. His ideas dominated till about 1500 CE.

Hawking takes the reader forward with brief stops at well-known milestones on the onetime thorny path of astronomy and cosmology. Big names like Nicolas Copernicus (1473 - 1534 CE), Johannes Kepler (1571 - 1630 CE) and Galileo Galilei (1564 - 1642 CE) are referred to who observed and asserted that the Earth and the other planets revolved around the Sun. Then other milestones came a bit faster which revolutionized the subjects of physical sciences, cosmology in particular. Sir Isaac Newton's (1643 - 1727 CE) remarkable discoveries of gravitation and laws of motion changed the scene of science. Even these were found to be only approximate when Albert Einstein (1879 - 1955 CE) put forward his theory of relativity.

In addition to the references of the individual personalities of yester years in science and philosophy, Hawking in his story of cosmology also invoked the Jewish/Christian/Muslim religions for their theories of the creation of the universe. However, the much older Vedic religion having had coherent philosophical principles and the *"most sophisticated ancient cosmological ideas"* as Carl Sagan had put it did not get any mention in Hawking's book. The reasons for the absence of Vedic references in Hawking's book could have been that he was not aware of the Vedic religious tradition; that the language barrier did not permit him direct scrutiny of the Vedas; that he went along with the impression created by the available English translations of the Vedas by Western authors and Sanskritists that the so called 'books of knowledge' had no knowledge in them. Mediocrity in understanding the Sanskrit language, particularly the Vedic Sanskrit and grammar, by Indologists and Sanskritists, has certainly been responsible for creating confusion about the Vedas. Whatever may be the cause for the absence of recognition or even a mention of the ancient Vedic effort in developing the knowledge on cosmology, the effect certainly is that the Vedic astronomy and cosmology do not get the rightful place in the modern story of the universe.

One gathers an impression from Hawking's story of cosmology that the sophistication in astronomy and cosmology began from Aristotle in the fourth century BCE leading to the present realistic understanding of astronomy and cosmology. Is it really the case that a better and more coherent understanding of astronomy and cosmology did not exist before Aristotle? To prove that astronomy at a level superior to the earth-centric astronomy of Aristotle did exist before Aristotle, one has to show some authentic references dating prior to the fourth century BCE, indicating that the universe was not earth-centric. It would exhibit an even higher degree of sophistication in astronomy if such references indicated that the Earth was a dynamic celestial body in orbit around the Sun. The fact is that this is exactly what is preserved in the mantra formulations of the Vedas written eons before Aristotle. Therefore there is a need in the modern society for revisiting the story of cosmology and readjusting the historical milestones.

This author's effort in this book is to present the Vedic version of the story of cosmology from the onset of the creation of the universe leading to the formation of the cosmos, with specific attention to the solar system – the planets, the Moon and the *mother* Earth. All parts of the Vedic story of cosmology are based on Veda mantra references. Chronologically these Vedic references are from eons and eons before Ptolemy in the second century CE and before Aristotle in the fourth century BCE. The astronomical deductions from the mantras, that the Earth is a sphere, that it is a dynamic system and is in an orbit around the Sun, are presented in detail under the sub-heading of "*Prithvi:* the planet Earth" in Chapter 11 in Volume 2 entitled "Functioning Vedic Universe". A short discussion on it is advanced here.

The Vedic Earth - a Sphere, in Motion, Not the Center of the Universe

It would be acceptable to Western Indologists, Sanskritists and historians that the Vedas preceded the Greek thinkers cited above. Even Max Muller (1823 - 1900), who is responsible for publishing the erroneous chronology of the Vedic history, said that the Vedas were assembled in their final form by at least 1,500 years BCE. Hindus have always believed that their Vedas are extremely ancient and no date can be assigned to them and now it is accepted by most Vedic scholars that the Vedas were written in the far antiquity of the Vedic Civilization before any book of Sanskrit Literature got written. Whereas an exact period of the advent of the Vedas is not available, most scholars of Sanskrit accept that they were written before the Mahabharata War. Based on astronomy and literature there is now ample proof, discussed in detail in Chapter 3 entitled "Historicity of Vedic References : The References on Cosmology Precede 5,000 Years" that the Mahabharata War ended around 3140 BCE with the coronation of the Emperor Yudhishthir of Hastinapur (near Delhi). With his coronation began a working calendar named Yudhishthir-samvat (calendar) within the long-running Aryan calendar called Srishti-samvat. The Yudhishthir-calendar remained in use well after the Vikram-calendar introduced in 57 BCE by the Emperor Vikramaditya of Avantika (now Ujjain).

Written records commencing from the 5th century BCE have been independently kept and are still preserved in the four Shankaracharya monasteries, geographically separated by over a thousand kilometers from each other. They were first established by Shankaracharya himself between 499 BCE and 485 BCE. These records are reproduced and discussed by Acharya Udayaveer Shastri in his book "Vedanta Darshan Ka Itihas" (History of Vedanta Darshan) published in 1991 ACE. The beauty of these records is that the events at the monasteries were recorded independently following different ancient calendars - Yudhishthir-calendar, Srishti–calendar (Natural-calendar of the Aryans), Vikram-calendar and Shalivahan-calendar. At some monasteries more than one of these calendars were simultaneously used which established a relationship between the onsets of these calendars. The last two calendars commencing in a well-known and well-established period in India's history are presently in use in the Indian society. The written records at the monasteries validate the onset of these calendars. Based on the hard data available from the old hand-written manuscripts, discussed in Chapter 3, it is concluded that the Mahabharata War, its time period, the onset of Yudhishthir-calendar in 3140 BCE, the account of the war (barring recent insertions) in the book of Mahabharata are a historical reality.

Hastinapur (now Delhi) continues to be India's main capital since before the coronation of Emperor Yudhishthir in 3140 BCE and the onset of the Ydhishthir-calendar. From Yudhishthir onwards, the rule of indigenous Aryan kings continued at Hastinapur, without a break, for 4,332 years till the Muslim invasion of India in 1192 CE. In addition to the social and historical significance of the new-year celebration in two of the most recent calendars, they are in close synch with the Natural-calendar of the Aryans called the Srishti-calendar, all beginning with the winter-solstice. The onset of the Srishti-calendar's *Kali-Yuga*-epoch was on February 18, 3102 BCE which is based on the astronomy of the solar system. Details on the Srishti-calendar and its epochs are provided in Chapter 5. The date February 18, 3102 BCE serves as a time 'milestone' in the Vedic history. Except for the books on Hindu mythology named Puranas, eighteen in number, the Vedas and post Veda books in Sanskrit Literature were written in an unknown period anterior to this 'milestone'. Thus, all Vedic references in this book including those pertaining to the planet Earth, are very ancient and anterior by eons to Aristotle's time.

The Vedic references being the marvels of thought-sophistication, rich in science and philosophy, they ought to find a place in the modern science arena, regardless of how old they are. They cannot be viewed merely as a part of some ancient religion. The sophistication in thought content and its expression in sophisticated Sanskrit vocabulary should caution one not to lump Vedic Cosmology with the Earth-centric astronomy of Aristotle and Ptolemy. The reader will note from the cited Veda mantras that in spite of a big time gap between the Vedas and the developments in modern science, the content in these Vedic references could contribute toward building a suitable partnership with the modern astronomy and cosmology. A full section in Chapter 11 is devoted to the planet Earth where the Vedic terminology emanating out of the mantra references leaves no doubt in the mind of a reader that the Vedas have as scientific a view of the Earth as modern science. Based on the Vedic references a detailed discussion on the phenomena of the created and functioning universe is presented in Chapter 11 which includes the process of the formation of the solar system, the meaningful nomenclature of planets and some details of their characteristics with emphasis on the Sun, the Earth, the Moon, the Venus and the Jupiter. A brief account on the planet Earth is included here.

There is considerable discussion on the planet Earth in Vedic mantra formulations and discussions in the vast Vedic Literature. These references cover the origin of the Earth, its early state, its geology and its shape. It is remarkable to note that the Veda mantras refer to several of its properties: *gurutva* (gravitation), sphericity, the state of its interior, three types of its motions and so on. According to its specific properties the Vedas have addressed the Earth with a large number of meaningful Sanskrit terms. Compilation of Earth's synonyms is found in millennia old Nighantu, the ancient topical dictionary of specific Vedic terms included in the book of Nirukt authored by Rishi Yask. The importance of the book of Nirukt was emphasized earlier in the present text for etymological interpretation of Vedic terms and mantra translation. Cross-referencing by the scholarly in Sanskrit Literature places the book of Nirukt in a historical period anterior to 1900 BCE. Rishi Yask's interpretation of the Vedic terms and mantras related to the planet Earth is quite ancient and unaffected by later Hindu punditry resorting to giving *rudhi* (colloquial/non-*yaugik*) meanings of the Vedic terms. The book of Nirukt certainly belongs to a historical period at least 1500 years prior to Aristotle and Ptolemy from whom the sophistication in Western modern astronomy began. Nirukt's interpretive commentary on the Vedic terminology and interpretive translation of Veda mantras and Nighantu's word assemblage of the Vedic terms takes us beyond 1900 BCE and the inundation of the book of Mahabharata with Veda mantra references takes us towards the Vedas, well before the milestone at February 18, 3102 BCE.

In the context of his interpretation of specific Veda mantras on the topic of the Earth and the Vedic terms used for the Earth, Rishi Yask in the first section of the second chapter of Nirukt (2/2/1) has listed 21 property-based synonyms for the Earth. In addition to Rishi Yask's list of 21 synonyms, this author has collected about a dozen more from the Vedas and the Vedic Literature. The richness of the Vedic terminology, with a total of 32 synonyms for the Earth, is exhibited in Tables 11.6a and 11.6b in Chapter 11 in Volume 2. Nowhere else, not even in the current state of science development in the society, does one find such rich, defining terminology for one object, the Earth.

The Earth, throughout the Vedic literature, starting from the oldest book of RigVeda down to the five thousand year old references in Sanskrit, is undoubtedly a round spherical body. A RigVeda mantra (10/58/3) states that the Earth slopes down in all four directions. The mantra's relevant first line and translation are as follows:

यत्ते भूमिम्[1] चतुर्भृष्टिं[2] मनो जगाम् दूरकम्। (RigVeda, 10/58/3)

Transliteration: "*Yatte bhoomim*[1] *chaturbhrishtim*[2] *mano jagaam doorkam*".
Mantra Translation:

"O man! Your *munn (mind) travels far on the Earth*[1] *that slopes down in all four directions*[2]" (RigVeda, 10/58/3).

The first superscripted term is भूमिम्[1] (*bhoomim*) which means 'the Earth'. The second superscripted term चतुर्भृष्टिं[2] (*chaturbhrishtim*) is the conjugation of two terms, *chatur* and *bhrishtim*. The term *chatur* means 'four' and *bhrishtim* means 'falling' or 'dropping down'. The composite meaning of the three terms भूमिम्[1] चतुर्भृष्टिं[2] (*bhoomim chaturbhrishtim*) is – '*the Earth that drops down in four directions*' or simply put '*the Earth that slopes in all directions*'.

The mantra stipulates with remarkable clarity that the Earth is not a flat plate; instead it slopes down in all directions. With this deduction in a Veda mantra, it should not take much for a scholarly *rishi* to conclude that the Earth is round. If any object slopes down in all directions, the object has to be a sphere. Need one emphasize more?

As in the above RigVeda mantra the terms *Bhoo* or *Bhoomi* for the planet Earth are the most common in the Vedic Literature. The name rendering its geometry is formed with the noun and an adjective as "*Bhoo-gol"*. The noun prefix *bhoo* means 'the Earth' and the adjective *gol* means 'spherical'. If translated in the syntax of Sanskrit, the term *"Bhoo-gol"* means "Earth-spherical". In English expression with the adjective preceding the noun, it is "Spherical-Earth".

The various adjectives assigned to describe the Earth in the ancient Vedas, followed by the Vedic Literature including the 18 mythological Puranas written two to three millennia ago, all state that the Earth is a spherical entity. In addition to mythological interpretation of the Vedic Literature, the Puranas also include Aryan history and philosophy. The Earth throughout the Sanskrit Literature has been addressed as *andaa-kaara* (spheroid or ovate) and *chhatraa-kaara* (umbrella-shaped or curved like an umbrella). The Earth therefore, in the Vedas and the subsequent Vedic Literature, was never a "flat plate".

In addition to the above mentioned adjectives associated with the shape of the Earth, the Earth is also called as *koorm-prastha-nibhaa* which in the order of words translates into, *turtle-back-like*. Similarly, the Earth is also called a *koorm-prashtha.* Once again the meaning is - 'shaped like the back of a turtle' or 'like the turtle-shell'. These adjectives convey two most important properties of the Earth: that its shell was hard like a tortoise-shell and that it sloped down in all directions like the back of a tortoise. These similes and adjectives are a unique product of the Vedic science thought. It provides a first glimpse into the physical observation of the shape of the Earth.

There are a large number of mantras in the four Vedas that relate to the Earth, however, the AtharvaVeda, the fourth Veda, has mantras and *sukta* (section or hymn) devoted specifically to describing the properties of the planet Earth. These mantras will be invoked in Chapter 11 with complete analysis of all terms. However, a mantra from RigVeda (the first Veda) is presented in the following.

"This Earth in space repeatedly goes around (आ अक्रमीत्[1]) *in a stable orbit, around its progenitor* (पितरम्[2]) *the Sun, holding its water forward".* (RigVeda, 10/189/1).

Of the two important terms, आ अक्रमीत्[1] (*aa akrameet*) means '*repeated motion*' and पितरम्[2] (*pitram*) means '*the father or the progenitor* or *the protector*'- all meanings apply.

The compact character of the Veda mantra reveals in just a few words several scientific observations about the Sun-Earth system, including that the Earth is born out of the Sun, that it is not stationary, that it repeats its orbit around the Sun, that it does not leave its water behind, that it is in space and so on. In this author's perspective, these are un-ignorable Vedic scientific facts about the Earth. It is obvious that the age-old Vedic science is in league with the relatively new-born (five century old) modern science. Needless to say that the superiority and sophistication one finds in Vedic science is remarkable and exceeds the earth-centric astronomy of Aristotle and Ptolemy.

From the Vedic mantra references on Earth and the brief discussion presented above, it becomes clear that the Vedic view of the Earth is straightforward, that the Earth is spherical and secondary to the Sun in the way a progeny is secondary to a progenitor. Implicit in the Vedic statement is that the Sun, apparently a small disc of fire traversing the sky, is primary and the Earth, with an apparent vast horizon around the observer, with large trees and mountains included, is secondary. This observation is also remarkable and mind-boggling. The angle subtended at an observer's eye by the apparent diameter of the Sun's disc is only about half a degree (32.7' minutes of an arc) which at times is seen even smaller than the apparent diameter of the Moon (34.1'), yet the Veda did not say that the Earth was born out of the Moon or that the Sun was born out of the Earth. The Earth in the Vedas is addressed as a feminine entity. The Vedic view of a father-daughter relationship between the Sun and the Earth is obvious. A large section in Chapter 11, entitled, "*Prithvi*: the planet Earth", is devoted to the Vedic perspective of the Earth.

Earth's Gravitation - a Vedic Concept

Mantras about the Earth's gravitation are discussed in Chapter 11 under the sub-heading "Earth's Gravitation: Veda and Vaisheshik Darshan". As an example a *sutra* (aphorism) from Rishi Kanaad's Vaisheshik Darshan, states –

अपां संयोगाभावे गुरुत्वात् पतनम् (*Apaam sanyog abhaave gurutvaat patnam.*), which means –

"*The rain drops in the absence of the union* (*with air*) *fall down due to* ***gurutva*** (gravitation)".

The fact is that the term *gurutva* in the *sutra* is a Sanskrit term for gravitation and nothing else. The underlying fact of the *sutra*-aphorism is that the knowledge of gravitation (***gurutva***) is far more ancient than 1666 AD when Sir Isaac Newton may have seen the apple fall from a tree. Rishi Kanaad's period being well beyond the book of Mahabharata at 3140 BCE the knowledge of gravity existed millennia before even Aristotle. Newton of course gave gravitation a mathematical expression known as Newton's law of gravitation discussed briefly in Chapter 2.

It may be mentioned here that Rishi Kanaad's statement on gravity is neither fortuitous nor a fleeting one, because to explain the intended phenomenon of gravitation at a deeper level of inquiry, he devoted several *sutras* to the subject. To exhibit the effect of gravity under differing conditions he took up two other examples from daily life. (A) A *moosal* (pestle) free-falling in the *okhal* (mortar) under gravity (Vaisheshik, 5/1/7), when released by hand. He also explains that the upward motion of the pestle is owing to a different force imparted by the hand. (B) An arrow in-flight, with a different force imparted to it by the string of the bow, falling gradually under gravity (Vaisheshik 5/1/16, 17, 18). He explains the forward motion of the arrow invoking incremental initial steps as in calculus.

He also associated gravity to Earth's अदृष्टकारितम् (*adrishtkaaritam,* Vaisheshik 5/2/1, 5/2/2) the invisible action (karma) of the Earth; an erudite deduction indeed! In the same context he formulated the *sutras* for several other functions in mechanics that relate to gravity and to the potential energy converting into kinetic energy (motion) – *nodan-vishesha* and *udsan vishesha* of Vaisheshik Sutra (5/1/9, 10; 5/2/1).

In the RigVeda mantra (10/189/1) cited above, the Sun was addressed as the पितरम् (*pitaram* = progenitor) of the Earth around which the Earth orbits in आ अक्रमीत्[1] (*aa akrameet*), the '*repeated motion*'. Let us now include here a fascinating example with science thought showing how the Vedic *rishis* have viewed the पितरम् of the Earth. This example is from one of the earliest interpretive commentary books on RigVeda, entitled "Etareya Brahmana" authored by the ancient *rishi* Etareya.

The Sun Never Sets

The oldest books after the Vedas are several specific commentary books on the Vedas, known as the Brahmana books written eons before Rishi Yask and the great Grammarian Paanini. These books include interpretation, elaboration and instructions for the applications of principles in Veda mantras for improving the life of the individual and the society. There is one primary Brahmana book on each Veda. Additionally there are a few more under the Brahmana title. The common English term *brahmin* for an individual with scholarly pursuit or for one born into Brahmin-caste has no relation to a Brahmana book. Since Veda mantras abound in science thought the Brahmana commentaries are replete with science thoughts as well. All Brahmana books belong to a period that is in the distant past of the Mahabharata period (3140 BCE). Brahmana books have valuable material on cosmology. The oldest among the Brahmanas is said to be the Etareya Brahmana of the RigVeda, authored by Rishi Etareya. The word *etareya* conveys that he was born to the mother named Itara. An extraordinary reference to a scientific deduction about the Sun appears in Etareya Brahmana. This reference states:

> "*When the Sun sets in one region of the Earth it creates day in another region of the Earth and that the Sun actually never sets*".

In the enlightened society of today, one could say that this statement about the Sun by Rishi Etareya is no big deal because with TV and internet information everyone knows that the sunset in one region of the Earth is sunrise in another region. But one has to appreciate the fact that the deduction by Rishi Etareya was drawn in the far antiquity of a society commonly labeled as 'primitive'. What Rishi Etareya stated is the observation in astronomy and physics and therefore this author labeled it as 'a scientific deduction'.

The context of this reference in the Etareya Brahmana is a specific worship ceremony that a worshiper performs with mantra recitation. This specific ceremony should begin and end between the sunrise and sunset. Rishi Etareya advises in Etareya Brahmana (iii, 4, 6, 44) on how to perform this worship ceremony. The relevant part of it is summarized thus:

"The worshiper should begin worship with the day and end with the day just like the Sun begins the day at sunrise and ends its day at sunset. The voice being the instrument (for mantra recitation) *the worshiper should do recitation in strong sound and maintain it through the completion of worship or even enhance the sound in the voice towards its third* (last) *phase, if possible."*

"In the third phase, owing to the doubt that the Sun might set before the completion of the ceremony, the worshiper need not rush through the mantra recitation and service because –

स वा एषनकदाचनास्तनेति नोदेति॥ इति ।

(*sa vaa esh na kadaachanaastneti nodeti. iti.*)*"*.

The last part of Rishi Etareya's instruction to the worshiper presented in Sanskrit, as it is in the book of Etareya Brahmana translates thus:

"*It* (the Sun) *neither ever sets nor it ever rises*"

This alone is a profound observation in science. Rishi Etareya did not stop there. He elaborated upon the meaning of the above statement using four more separated statements. What he said after this is a lesson in ancient physics (only translation provided here):

"*When humans consider it set, the Sun ends the day in their region and it is said to have* विपर्यस्त *(viparyast =inverted). Thus, in the past region* (of Earth) *it creates night and in the future region of Earth it creates day*."

"*And when humans say – 'it rises in the morning', in reality it ends the night in their region, is* विपर्यस्त *(viparyast = inverted) and begins the night in the region which is to come in future."*

"*This Sun never ever sets*".

"*In conclusion, it never sets* (नहवै कदाचन = *na ha vai kadaachan*). *One who understands this knowledge, acquires coexistence with the Sun* (achieves brightness, enlightenment), *acquires its brilliance* (in intellect) *and is born* (in next birth) on planets with brilliance".

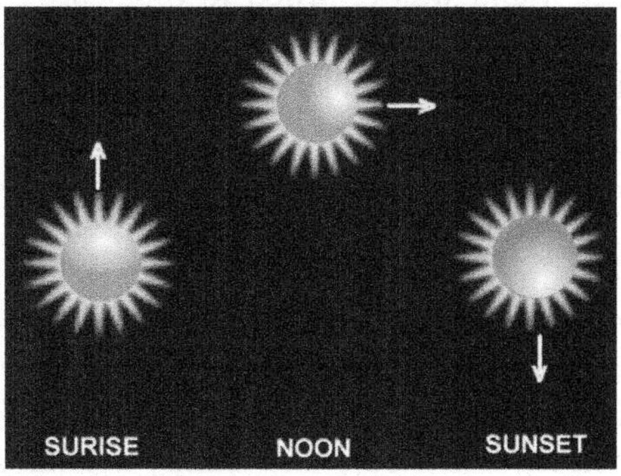

The term विपर्यस्त (*viparyast*) in the above two statements stands for 'it sets inverted', thus representing Sun's appearance to an observer at the time of sunset with respect to its appearance at sunrise. In the context of the Sun being inverted,

Rishi Etareya is talking about the Sun's disc, the upper edge of which at sunrise becomes the lower edge at sunset. He further emphasizes that the upper edge at sunset inverts at sunrise. Rishi Etareya's statement is depicted in the diagram by this author with three positions of the Sun; at sunrise, noon and sunset.

Aside from his concluding remarks, Rishi Etareya's emphasis that the Sun never sets is as good today as it was when he wrote the Etareya Brahmana such a long time ago. He clearly means that the Sun is always there in space illuminating one region of the Earth or the other and that the sunrise and sunset are the human definitions of the beginning of the day and the beginning of the night respectively. This also implies that it is the same Sun that is seen inverting its disc when it is in the west at sunset to when it is in the east next day at sunrise.

Rishi Etareya through this simple lesson in physics wanted to tell the worshiper not to get bogged down in blind belief that the prescribed ceremony should end abruptly at sunset, because, as he said with emphasis that '*the Sun never ever sets*'. At the end of his discourse when Rishi Etareya states that the person achieving brilliance in this life is "*born on brilliant planets* (in next birth)", he made two points: that the rebirth system is part of the Vedic Philosophy and thatthere are planets in the universe that are brilliantly lit for life to exist on – what are today's exo-planets.

This was from a voluminous Brahmana book written by a *rishi* in the far antiquity of Indian civilization. It has 285 sections with a large number of statements in each section distributed in 40 chapters. It is now available in modern fine print. This author's copy of the Etareya Brahmana has more than 1400 pages. Based on mantra references from the Vedas and Brahmana books, this author has presented many more scientific details on the Sun, under the sub-heading of "*Soorya*: the Sun" in Chapter 11 entitled "Functioning Vedic Universe".

In contrast with this sample of ancient Vedic science, anterior to Mahabharata, one has the foundation of Western science and philosophy established in the 5th century BCE Greece when a pre-Socratic Greek philosopher, Heraclitus of Ephesus (535 - 475 BCE) believed that the Sun 'took birth every day' and was a giant bowl orbiting the Earth wherein the sea water turned into fire at sunrise. At the Sun set, the fire in the bowl turned into water again and was returned to the sea. At night the empty bowl travelled under the Earth and was ready for another orbit around the Earth. Therefore according to Heraclitus, the Sun was renewed every day. Not in the Vedic thought!

Aristotle's Philosophy versus Vedic Philosophy

In Aristotle's philosophy everything in the universe was made up of four elements: Earth, Air, Fire and Water. The Vedic elemental philosophy has five elements; the additional element being the Space (*Aakaash*). In spite of the difference in the two elemental philosophies the meaning and function of elements in Vedic philosophy are altogether different from Aristotle's elemental philosophy. Vedic Philosophy emphasizes that the *panch-maha-bhootas* (five-great-elements) are essential for the sustenance of life on Earth and they relate to the five knowledge-senses in conscious-beings. This is dealt with in another book. It suffices here to say that the Indian Philosophy was well ingrained in the

Indian society well before Mahabharata (3140 BCE) and there was a fair bit of exchange of philosophy and mathematics between India and Greece around the time when Alexander the Great invaded India in 326 BCE, a period contemporaneous to Aristotle (384 - 322 BCE). Aristotle probably did not know that the fifth element *Aakaash* (Space) was also essential for the existence of substances.

Hawking in his earlier book has narrated an interesting story of an elderly lady with her Earth-Tortoise theory. This and many such stories have their origin in ancient cultures and civilizations. Such a story may have had full meaning when told first and may even have its origin in a science principle. Given enough time within a given civilization, a story with meaning may turn into a legend or myth. The story also travels from one civilization to another and from one continent to another. The travel changes the story's composition. Dilution with time is an inherent nature of information. Before we get to Hawking's story of the elderly lady in England and establish a Vedic connection to her Earth-Tortoise theory, let us first talk a bit about the nature of information itself.

Entropy of Information

In the 'knowledge' exchange process between A and B the 'knowledge' is first converted by A into corresponding 'information' which is transferred to B who converts it back into the corresponding 'knowledge'. Because A and B are dissimilar human-beings the knowledge at A is different from the knowledge at B.

Most effective medium of information transfer has been the verbal communication. The information transfer in this methodology is local and at the speed of sound. The modern information technology has changed the speed of information transfer from the speed of sound to speed of light. Therefore the information transfer now is quite instantaneous, global and more accurate. Even then the conversion of information to knowledge renders a differential between the knowledge at A and B. This could be viewed as the knowledge at A being the 'cause' and the knowledge at B its 'effect'.

Under the Vedic cause-effect principle, *'the effect is the vikaar of the cause'*, where the term *vikaar* represents the order of state lower than the one existed in the original cause. It could be expressed in physics terminology as - '*the effect is a degenerate state of its cause*'. Simply put, the knowledge at B is at a lower order than the knowledge at A. This is like the principle of entropy in thermodynamics according to which the 'disorder' in a system always increases.

The factors responsible for introducing 'disorder' in the information-system (knowledge) are time, locale, language etc. Before the modern technological development took place the information took hundreds or thousands of years to cross the borders of the countries. Although slower, the information did travel in those days too. Just as the cause-and-effect is an underlying principle in the functioning universe where cause precedes the effect and just as the physics principle of entropy is inherent to physical systems in nature where disorder increases with time, a simple information-entropy theorem could be formulated as:-

"The entropy of information increases with time"

The increase in information-entropy is also subject-specific. For example the information-entropy increases faster in the realm of history of a society than in science. This difference is owing to the fast changing rules governing a society compared to the non-variance in the rules of science. The older the history of a nation the higher is the degree of uncertainty for ascertaining the 'true' history and historical events of that nation as well as the 'truth' about its civilization. Just as in studying the physical system that we call 'universe' a Big Bang cosmologist begins with some seemingly logical assumptions on its present state (present entropy) and goes back in time to predict its beginning in a space-time singularity, a historian goes back in time to put together a society's history, based on the information of the present and in the near past. The common process in the two situations is the extrapolation back in time.

For any extrapolation, one needs reliable data points on the timeline of the history itself. Such data points in the case of the history of a society, a culture or a civilization generally are in the form of hints, verbal accounts, stories, references in the books as well as the facts preserved in the written records; the latter furnishing high reliability. There is always a probability for misinterpretation. This is equivalent to the increase in information-entropy. A higher degree of entropy is introduced in the information when a historian attempts to write the history of a civilization with his mediocre command of that civilization's language, literature and culture. This in fact has been the situation when the Vedas, Vedic Literature and the Vedic Civilization's history were interpreted by some Western Indologists and Sanskritists of the past couple of centuries. Reliability of these historical deductions increases when the historical records prepared by the historians of the past belonging to the civilization are interpreted by competent present historians from the same civilization. History will be more reliable when it is based on the writings of the historians of that culture who lived closer to the historical events in that culture, establishing the validity of the history and events.

The Vedas, Vedic knowledge, Vedic philosophy and Vedic history, having their roots in the far antiquity of the Vedic Civilization of the Aryans in India, have suffered from increased information-entropy owing to the fact that a long time period has elapsed since their advent. In this process the information about the value of Vedic thought has gone through an unnaturally high entropic change. The entropic change seems to have shaped the narration of Hawking's story of the elderly lady and her Earth-Tortoise Theory.

Hawking's Story of the Elderly Lady and her Earth - Tortoise Theory – How myths get Created!

It is in human nature to want to know more about the universe and then to tell others about it. How best to tell for most to understand has always been a challenge for all civilizations. So that others can understand, the persons with relevant knowledge have resorted to examples, similes and metaphors. No matter how complex the knowledge is, it comes down to the person of knowledge turning into a story teller. Hawking, Sagan and some other popular science writers are good story tellers and have helped common people to understand a lot about the universe. However, the answers to basic questions about the universe continue to lie in the realms of complex mathematical formulations and concepts in quantum mechanics. This too must be brought to the level of a narration for the benefit of the others.

In telling the story of cosmology, Stephen Hawking in his earlier book entitled 'A Brief History of Time' narrated an interesting story as:

> *"A well-known scientist (some say it was Bertrand Russell) once gave a public lecture on astronomy. He described how the earth orbits around the sun and how the sun, in turn, orbits around the center of a vast collection of stars called our galaxy. At the end of the lecture a little old lady at the back of the room got up and said: "What you have just told us is rubbish. The world is really a flat plate supported on the back of a giant tortoise." The scientist gave a superior smile before replying, "What is the tortoise standing on"? "You are very clever, young man, very clever" said the old lady, but it is turtles all the way down!"*

This author feels that under the humor of the earth-tortoise story is hidden some science with its origin in ancient India because similar stories have been prevalent in the Indian countryside. It can also be said that the stories related to the Earth and the creation of the universe that are narrated by the farmers sitting around a fire began as a lesson in science contained in the Veda mantras. Slowly, as the millennia passed, the information-entropy did its job and the science lesson changed into the earth-tortoise myth narrated by the old lady. Loss of the source of information (story) with time is natural, as in Hawking's story; "*...some say it was Bertrand Russell*" indicates a hint of uncertainty. By the time Hawking told the story in his book there are three stories in one - first is the old lady's earth-tortoise story, second is the story of the encounter between the elderly lady and the well-known scientist and the third is about the first two stories that Hawking is telling us in his book. Obviously the 'information-entropy' has been in play beyond just the A and the B mentioned above.

As a child this author had heard the stories related to the Earth, the Sun, the constellations and the cosmos from elderly farmers sitting around a fire, continents away from the stage of Hawking's story in England. Some said that the Earth was supported by some extraneous entity, some said that it was supported on the back of a turtle exactly as in Hawking's story of the elderly lady and some said that it was supported on the hood of a cobra. Others said that it was supported on the horn of a bull. The farmers even had their own cosmological theories for the creation of the universe. When this author acquired higher education, including the study of the Vedic and Puranic literature, the source of the village myths became obvious. The 18 books of Puranas are the source of Hindu mythology.

The ancient Vedic Literature is vast and the references to the word 'turtle/tortoise' would be many. The author would address just a few pertinent to the subject at hand: the 'earth-tortoise story'.

The Sanskrit word for tortoise is *koorm*. The earliest appearance of the word *koorm* is in RigVeda. It is in the name of the *rishi*, Koormo Gaartsmado Gritsmado Va (son of Gritsmada) who is credited for the revelation of 35 mantras in three *suktas* (hymns or sections) with 31 other full *suktas* credited to his father Gritsmada. These mantras, revealed by both of these ancient *rishis*, are on the subjects (*devatas*) that relate to the Sun, the air system (atmosphere), the Earth-Space system, and the energy and so on. Therefore, when a reference of these subjects is made in Sanskrit Literature it is natural to state that the particular knowledge regarding the Sun and the Earth is based on the revelations of Rishi Koorm, who in fact is a scholarly man and not a turtle.

This is something similar to an expression in modern cosmology for the phenomenon of gravitation. Often we read - '*Gravitation is better explained by Einstein's theory*'. We know what it means and it is unaffected by the information-entropy to interpret it differently. But this expression would have a different meaning in an imaginary scenario thousands of years from now if a cataclysmic event wiped-out the present civilizations on Earth, gone with it the entire science development. If the German language was still alive after the cataclysm and someone read the above mentioned sentence, he could translate it as - "*gravitation is better explained by one-stone theory*", because the compound word '*einstein'* could be interpreted as *ein + stein* which means *one-stone*. Or it could also turn into "*gravitation is surrounded by stone"*, because *einsteinen* means '*to surround with stone*'. With time and even without a cataclysmic event a Vedic statement - '*knowledge of Earth is based on the revelation of Koorm* (the *rishi*)' could turn into a myth that the Earth was on a *koorm* the turtle. Therefore this could be a source of the earth-*koorm* or the earth-tortoise myth that this author also heard from the farmers.

The Vedic Literature that followed the Vedas has used the word *koorm* for a tortoise and as an adjective for describing the shape and properties of objects. In the book of Mahabharata, the compound word *koorm-prashthaka* is used for the lid of a dish, for it is in the shape of a turtle's back. The other words with *koorm* as prefix are: *koorm-kalp* (an ancient period of time), *koorm-chakra* (a geometrical circular drawing), Koorm-Purana the 15th of the 18 Puranas that are laden with mythological Hinduism.

Pertaining to the subject at hand, the 15th Purana with the title - 'Koorm-Purana' is on the reincarnation of Vishnu as a tortoise. The term *vishnu* in the Vedas stands for the '*creator and the upholder of the universe*'. The Puranic interpretation said that the Earth is upheld on Vishnu the Koorm (tortoise). This is one other source of the myth that the Earth was resting on a tortoise.

Two adjectives for the Earth, *koorm-prastha-nibhaa* meaning "*like-turtle-back*" and *koorm-prashtha* meaning "*hard and sloping in all directions like the shell of a turtle*" were presented earlier. These adjectives for the Earth, delineating the scientific observation of the Earth by Rishi Gritsmada, could also be the source of the earth-tortoise myth. Instead of the intended meaning of the earth-tortoise simile for the spherical Earth, the possibility of the simile taking the form of a myth was high and the possibility of it turning into folklore was there. There certainly was a possibility for such folklore to reach the lecture hall of Bertrand Russell in England. In my village, since the time I heard the earth-tortoise theory first, independent India progressed and the education also progressed. The younger generations have brought new knowledge to the villages. The earth-tortoise theory has faded away along with the supporting tower of tortoises.

Another example of a Puranic story built on the distortion of a Veda mantra is that of the Earth sitting on the head of a cobra. A Vedic mantra states that the Earth is *'sarpvati'*; the actual meaning is that the *'Earth moves like a serpent'* or that the '*Earth is in serpentine motion*'. While the sliding of the Earth in space is a scientific fact emanating from the Veda, the misinterpretation of the mantra could create a myth.

Sanskrit generates synonyms with each name representing a property and Sanskrit also has multi-meaning words such as '*indra'*. The word *indra* stands for God in the spiritual and creation realm, it stands for the Sun in the physical and cosmological realm and it stands for a king in the social realm. Not following this important nuance could lead to creating a myth!

The Earth in the Veda is described as *agni-garbha* (pregnant with heat/fire); a simple observation of volcanic eruption could lead to such an adjective - a fact. The Earth in a myth is supported on the horn of a bull - not a fact. This myth could have come from mistranslation of the two RigVeda mantras (1/164/47 and 1/146/2) wherein the Sun is addressed as *ooksh* and credited to be the cause for the rain on the Earth (a scientific fact in the Vedas). The two RigVeda mantras (1/164/47, 1/142/2) state-

> "*O man, know the sequence of cause-effect for the Earth getting wet by rain. The rays of the brilliant Sun cover all waters of the Earth, attract (raise) water in space and send it in flight to wet the Earth"*. (RigVeda, (1/164/47).

> "*The Ooksh* (Sun), *bigger than the Earth, is responsible for creating clouds and sprinkling water drops* (in rain) *and thus the Earth surface gets impregnated for the birth of vegetation*". (RigVeda, 1/142/2).

The word used for the Sun in the last mantra is *ooksh*. There are several other mantras, RigVeda (5/47/3, 10/31/8), that have used the *ooksh* synonym for the Sun. The mantras (RigVeda, 1/142/2) states clearly that – "... *the Sun is bigger than the Earth...* the *Earth is in space and is supported by the attraction of the Sun...*".

Since *ooksh* the Sun is said to be supporting the Earth, there is a high likelihood of a myth originating out of these mantras - that the Earth was supported by an ox. An interesting point to note here is the similarity in the sound of the Vedic term *ooksh* impregnating the Earth's surface and the word ox impregnating a cow. One wonders, how the word oo*ksh* travelled to other languages still sounding more or less the same. The Earth-bull relationship is a scenario in some other religions also.

Hawking and the Universe Out of Nothing - Contradicts the Saankhya Philosophy

Prof. Hawking in his recent book, 'The Grand Design', coauthored with Prof. Leonard Mlodinow, a physicist at Cal Tech in USA, touched the subject of theism/philosophy. Let us look into the logic of the statement well-publicized in the media:

"*...Big Bang was merely the consequence of the law of gravity and God no longer has any place in theories on the creation of the universe due to a series of new developments in physics...*".

To say that *'God no longer'* has any place in theories on the creation of the universe may imply that at some point in time the God was part of some cosmology theories and now God is taken out of the theories or that the theories with God are now proven wrong. It is common knowledge that neither the religions have ever attempted to prove God through mathematical equations nor physicists have attempted in their equations and experiments to prove or disprove God. The reality is that the physicists cannot provide a direct proof of some of the entities that they believe exist, the Time and Space are such entities. If God never was a part of any Big Bang field equation then to say that "*God no longer has any place in theories*" has no strong logic.

Even the part statement – "*Big Bang was merely the consequence of the law of gravity...*" has no solid logical basis when the fundamental questions related to the Big Bang Hypothesis raised throughout the earlier text in this chapter and more raised in Chapter 2 remain unanswered: why is there mass in matter and the gravitational property in mass, when did Nature perfect its laws, what are the Time and Space made of and so on. The Big Bang Hypothesis itself is not based on pure mathematical equations either. They are a parameterization in a mathematical construct which once setup could churn out a desired universe. Or churn out as in Multiple Universe Theory as many as 10^{500} (500 zeros after 1) different universes. With a large number of questions outstanding can one conclude that the gravity alone created the universe?

An example of an inconceivable situation in Big Bang Hypothesis, as pointed out earlier, is the Cosmic Inflation Era when the matter of the universe is made to rush out at un-natural and un-imaginably high speeds of 3.33×10^{23} times the speed of light (details in Chapter 2), the speed of light being 3×10^{8} m/sec. Nothing in the universe is supposed to travel faster than light. A weak explanation for this strange behaviour of matter came from Hawking's co-author Leonard Mlodinow. Mlodinow at a CNN panel discussion said that the '*time does strange things*'in those early moments of the universe.

On September 10, 2010, the host on the CNN television, Larry King, held a panel discussion on Hawking and Mlodinow's new book entitled "The Grand Design". The panel included the well-known authors: Leonard Mlodinow, Deepak Chopra and Father Robert J. Spitzer. Prof. Hawking did not appear on the television but his answers to Larry King's questions appeared in the videos forwarded by Prof. Hawking. It was one of the most captivating panel discussions on CNN. Regarding the Time-entity, Mlodinow stated:-

"Well, the question of time is also one that you have to be careful about; ...time can become part of space. Time and space affect each other. As you go back in time, when the universe is very, very small, time ceases to have the meaning we think of it as having today. You can't really go back -- you can't follow time backwards like it's a railroad track and get to the station and that's the beginning. So that question of what's time zero is a difficult question to phrase." **(script from CNN)**

This statement pertains to the Time-entity, yet it does not deal with the nature of Time. The first part of the statement suggests that as one goes closer to the beginning of Time (in the Big Bang Hypothesis) it becomes a part of Space and both start to affect each other. Mlodinow's statement raises more questions than provides answers. When cosmologists do not know yet what Time and Space are made of at their fundamental level, can one claim that the Time and Space are affected by each other when Time approaches zero? Do tiny parts or corpuscles of Time or Space, if they are made of corpuscles, combine to form something else or do they consume each other to affect each other? What exactly is this effect that Mlodinow is referring to? Why is Time so small when the universe was small and what physical, chemical or biological mechanisms make Time to grow? The matter is created under the laws of Quantum Physics but under what law or laws the Time or Space are created?

If Time and Space affect each other when Time was close to zero, should they not affect each other in the same way now when they are still growing and embedded into each other at each and every minutest possible point in the universe? These are simple questions and their answers ought to be simpler than – '*So that question of what's time-zero is a difficult question to phrase*' as in Mlodinow's above statement. Even now after 13.8 BY, the age of the universe, one is not able to cut a piece of Time to measure it separately and save it, away from the main Time. The same goes for Space. If two tiny pieces of Time and Space cannot be cut and tied together in a physics laboratory of today, to form a tiny 'space-time continuum', it seems futile to think that Time and Space in the physics laboratory of Nature formed a tiny 'space-time continuum' at time-zero of Big Bang when "*...time can become part of space. Time and space affect each other*", as stated by Mlodinow.

In Mlodinow's railroad track metaphor that – "*You can't......follow time backwards like it's a railroad track*", a station is implicitly conceived at the "space-time singularity" from where the train started but cannot go back to. What effect makes it an irreversible process? If the track in the metaphor were a Canadian Pacific Railway stretching over 22,500 km through about 1,300 stations, all stations are reachable! What if there were many stations on the other side of the so-called last station named "space-time singularity"? Unless proven otherwise the Time-entity could even be an eternal entity flowing from well beyond the station called the "space-time singularity". The same goes for the Space-entity. Therefore, a logical statement would have been that one is not able to follow time right close to t = 0 second owing to the limitation in the Big Bang field equations wherein the mathematical algorithm becomes unstable. The fact is that nothing physical should happen to the Time-entity itself at the beginning of the assumed Big Bang. Therefore, to say that the '*time does strange things*' near t

= 0 second is not based on sound logic. Also to say – "*time ceases to have the meaning we think of it as having today*" is also not based on sound logic because there is no proof to such statements. There is some degree of circularity in Mlodinow's argument because the Time is dealt with in specific equations with a pre-thought of specifically ending it into the space-time singularity at t = 0 sec; but that is happening only in the equation of a cosmologist and not in reality.

The Big Bang Hypothesis first works with Quantum Physics to create matter for a negligible part of the universe's age and then works with Einstein's Theory of General Relativity for the rest of the universe's long life of 13.8 BY and continues beyond that. In this process there is no overlap (no unification) between the Quantum Physics and the Theory of General Relativity. Just for fun, if one were to take a reverse journey right up to the 'space-time singularity' at time zero that would be like tracing back the history of the universe. In order to take up this reverse journey, the caravan of time, space, matter, energy gravitation and Dark Energy, the 'camel-tied to the tail of the camel', follows the rules of General Relativity for most of its journey. This caravan carries the cargo of General Relativity back in time towards the first moments of the Big Bang, however, before reaching the space-time singularity the cargo will be forced to pass through the torturous 'time-warp' in the desert of Cosmic Inflation! Then the cargo of General Relativity must be left at the doorstep of the Planck Epoch just 10^{-43} seconds away from the birth of the universe!

A camel of General Relativity is too big to enter the minute nursery of Nature named the Planck-time regime where the nurse named Quantum Mechanics works and would not permit even the tail of the camel to pass! Physics has yet to develop a conciliatory relationship between Quantum Theory and Gravitational Theory (General Relativity). Effort is being made in Particle Physics (Quantum Mechanics) through developing a Unified Field Theory to reconcile the two. There has not been measurable success in this area yet. Contrary to Hawking's statement, the fact remains that the General Relativity is not even effective in the Quantum Mechanics regime near the space-time singularity and the law of gravity not being operational in Quantum Mechanics is not even the initiator of the early universe in Planck-time epoch. Therefore, to state emphatically that "*Big Bang was merely the consequence of the law of gravity...*" is not strictly logical.

Hawking's much quoted other statement is:

"*Because there is a law such as gravity, the Universe can and will create itself from nothing. Spontaneous creation is the reason there is something rather than nothing, why the Universe exists, why we exist. It is not necessary to invoke God*"

A similar emphasis at the panel discussion on CNN came from Mlodinow:

"*...One of them is this concept of nothingness. According to quantum theory, there is no such thing as nothingness. You can have nothingness in quantum theory. But from that, things will arise. You'll pop into existence and out of existence again. So I know that's hard to understand and hard to accept....*".

When Larry King asked - So there can be a nothing? Mlodinow replied-

"*So there's nothing, but nothing is not stable. Things come and go from nothingness.*"

Hawking's above premise hinges on the existence of the 'law of gravity' and the 'spontaneity of matter creation out of nothing'. According to Quantum Physics the Higgs particle imparts the mass to other matter particles just created and the mass in turn creates the property of gravity in matter. Therefore, in the sequence of creation the spontaneous creation of matter, including creation of Higgs particles that impart mass to other particles just created, is primary and the creation of mass and law of gravity is secondary. An implicit assumption in this logic seems to be that God is incapable of spontaneity. The theists have always believed the *spontaneity* as an attribute of God. On the other hand the law of gravity is not known to act with *spontaneity*. Regardless of whether God exists or not, Hawking's premise at face value is not logical.

According to the Vedic Big Bang (discussed in Chapters 6 through 9 in Volume 2), the properties of matter, including mass, are invoked when the *Asata*-state of undefined matter transits into the *Sata*-state of defined primordial matter. The *Sata*-state of matter forms a primordial matter conglomerate in ovate shape named *Hiranya-garbh* (Golden-embryo) which over a Veda prescribed period of 4.32 million years (one *deva-yuga*) develops enormous heat and rotational energy. Then the 'spontaneous' Big Bang took place with a tremendous explosion in the Golden-embryo. That it was a spontaneous explosion is stated in the RigVeda mantra (10/129/5). The *sukta* (hymn) 129 is entitled "*Bhavavrittam*" meaning the "*History of Creation of the Universe*" –

"*At the onset of creation, there is* वितत: *(vitatah) the instantaneous-expansion of the rays of* रेतोधा: *(retodha) the fundamental seeds of matter omnidirectional, interlaced transversely all-around, existing above as well as below…*".

The Vedas being theistic, वितत: *(vitatah)* the *instantaneous-expansion* is credited to the Creation Energy Field addressed as God. The spontaneity therefore is not a necessary condition for eliminating God from the creation process of the Big Bang Hypothesis.

A reference to God in these pages may give an impression that this author is making an effort to defend the concept of God. This is not the case. The effort here is to show that a sensible serious consideration to the subject of cosmology exists in ancient Veda mantras as far back as one can go in the ancient civilizations on Earth where God happens to be a part of that theistic frame of reference. The author's emphasis in this book is on "material-cause" the Nature (matter/energy) along with the "efficient-cause" (God) and the "general causes" such as the Time and the Space. The aim is to bring in mantra references on cosmology, just as a cosmologist would bring in references from the work of other cosmologist, an atheist or a theist. The emphasis is on the thought contained in the mantra rather than on pedant. If one so wishes, God could be dropped from the Vedic scenario because what will still be left are the matter (Nature-entity) and the physical mechanisms that create and shape the universe. The Vedic scenario is not like – let there be universe and the universe was there. For example, another reference (Shatpath Brahmana, 11/1/6/1, 2) states that the brilliantly-hot golden embryo made of primordial matter in ovate shape was in *paryaplav* (rotation around an axis while in floatation) for a period of one *deva-yuga* measuring 4.32 MY before it exploded spontaneously. This Vedic scenario (in Chapters 8, 9) is as neutral without invoking God as physics is.

Many Veda mantras pertaining to Vedic Big Bang deal with physics without invoking God. The 'material-cause' (Nature) of the universe still remains the same whether it is in a Veda mantra or in modern cosmology. The bulk of the above mantra delineates the instantaneous expansion of the rays of fundamental matter particles going in all directions. The brilliant use of the Sanskrit term representing *interlacement* of all matter (rays of matter) further emphasizes that in spite of the sudden explosion in Nature, all matter remains connected. This is a fact of the universe, emphasized in modern cosmology as well. The all-pervading attractive force of gravity owing to the mass in matter of the universe ties together all the mass of the universe.

In the second part of the mantra cited above, the reference is made to both, the conscious-being Creator and the inert Nature with its synonym *Svadha*, which means – a '*self-existent entity*', an eternal entity. The Vedic terms for Nature in its primordial matter and energy phase are adorned with specific adjectives: *Hiranyam-aand* (Golden-ovate); *Hiranya-garbh* (Golden-embryo); *Maart-aand* (the Mortal-ovate); *Braham-aand* (Brahmn's ovate) and *Prajapati* (Progenitor and governor of the population, the cosmos) (RigVeda, 10/127/9; 10/168/3; Shatpath Brahmana, 11/1/1, 2). This rich terminology itself begins to delineate the creation scenario of the early universe, which has several similarities with the modern Big Bang Hypothesis.

The process of spontaneous explosion in the primordial matter leading to the advent of the universe therefore is not a necessary condition limited to an atheistic reference frame only. The विततः *(vitatah)*, *instantaneous-expansion* in the RigVeda mantra (10/129/5) and several other mantras presented in Chapters 7 to 9 represent only a physics process without religious connotations. The important point here is that the Veda put forward the process eons before the modern Big Bang. When one is dealing with a physical process in Nature, *it is not necessary to invoke God* as Hawking stated, but when one wishes to look into 'cause-effect' a theist traces back to the 'efficient-cause' and 'material–cause' whereas an atheist traces back only to the 'material-cause' and the God is replaced by "spontaneity".

The statements of Hawking and Mlodinow, "*Universe can and will create itself from nothing*" and "*... You can have nothingness in quantum theory. But from that, things will arise*" imply that an 'effect' need not have a 'cause'. Therefore effectively these statements are in contradiction with the age-old deductions in philosophy, Eastern as well as Western, wherein the principle of 'cause-effect' is seen as a natural principle.

A dictum under the "cause-effect" principle in Vedic Philosophy stipulates that –

"What is being created cannot be a cause of itself".

Therefore if Time, Space and Nature (matter), all three entities with different characteristics and properties, are created for the first time in the Nature's experiment called Big Bang, then creating all three without a 'cause' or 'causes' means an indulgence in pseudo-science!

There is some discussion pertaining to this subject in contemporary published literature. One example is an article in the British Journal for the Philosophy of Science (pages 649-668, 45, 1994) by Quentin Smith. He has addressed a question- "Did the Big Bang Have a Cause". He refers to William Lane Craig's contention from *The Kalam Cosmological Argument* (Harper and Row, pub. 1979, p. 145) that- '*everything that begins to exist has a cause of its existence*' and finally Smith concludes thus:

> *"...there is no sound Kantian-style argument that the big bang singularity has a cause. I am not aware of any other even remotely plausible a priori argument that the big bang has a cause, and so I think it is reasonable to conclude that the prospects for such arguments are poor. ...Thus we reach a general conclusion: there is no philosophy of big bang cosmology: that the universe began to exist without a cause."*

Quentin Smith refers to Immanuel Kant (1712-1804 CE) the well-known Kaliningrad (Russia) philosopher with German background. Kant is considered to be a German philosopher of high scholarship. It is also well-known that Kant had deep interest in Indian philosophy as well as in Sanskrit language. According to the Soviet scholar Theodore Stcherbatsky (1866-1942 CE), a Russian Indologist, Kant derived his doctrine of the '*categorical imperative*' from none other than the Hindu philosophy, which is the same as the Vedic philosophy. The Nobel laureate Ravindranath Tagore and Pundit Jawahar Lal Nehru praised Stcherbatsky for his scholastic achievements in Hindu and Budhist philosophies and his fluency in the Sanskrit language. There were well-known writers and philosophers who followed Kant, such as August Wilhelm von Schlegel (1767-1845 CE) and Friedrich von Schlegel (1772-1829 CE). Both brothers pioneered the Indo-European linguistics and comparative philology in Europe.

Kant's doctrine, the '*categorical imperative*', states:

> *"It is impossible for anything to begin to exist and that every change is an alteration in the state of some permanent substance".*

Kant formulated this principle about two hundred years ago. Contrasting with this is Quentin Smith and most modern cosmologists who argue that the Big Bang happened out of nothing, without a 'cause'. If for a moment one puts aside Smith's argument related to Big Bang and looks at Kant's doctrine from the Vedic frame of reference, one is certainly up for an unanticipated surprise.

If this author were to take the reader into the distant past of the Vedic Civilization well beyond the historical milestone of 5,200 years ago and open the first chapter of Rishi Kapil's Saankhya Darshan, the 43rd *sutra* therein states,

नावस्तुनो वस्तुसिद्धिः |(*Naavastuno vastusiddhi*) (Saankhya Darshan, 1/43)

Sutra translation:
"Nothingness cannot be the cause of something". (Saankhya Darshan, 1/43)

Rishi Kapil's above statement is in the context of the creation of the universe. The Vedic Creation Principle is based on RigVeda's 129th *sukta* of its 10th chapter and it is formulated in Chapter 6 of this book. As expressed earlier the universe comes into existence owing to the transition of the state of matter (Nature) from *Asata*-state (indefinable state) into the *Sata*-state (definable state). A direct implication of this is that something as real as the universe is the 'effect' with its pre-existing 'material-cause' being the *Asata*-state of matter. Thus in the Vedic creation "nothingness" is not the cause of the universe. This is also Kant's emphasis in contradiction with those contemporary cosmologists who emphasize that the universe came out of nothing.

The most common term for Nature in Vedic Philosophy, including in Rishi Kapil's Saankhya Philosophy, is *Prakriti*, however contextually he has also used the term *Pradhaan* which means '*the primary-material-cause*'. Another expression of the *Asata*-state of Nature is the 'causal state of matter' which has a real existence.

Those arguing that the entire matter in Big Bang or any other piece of matter was created out of nothing, have to first disprove Rishi Kapil's principle that *"Nothingness cannot be the cause of something"*. Particle physicists argue that the particles and anti-particles appear all the time in their experimental chambers out of nothing and disappear into nothing. The observation of new particles in experimental chambers may be valid but their argument of 'out of nothing' is not on a sound basis because their definition of 'out of nothing' is quite arbitrary. Conjectures such as - 'the Big Bang had no cause', are in contradiction with the Saankhya Darshan's *sutra* नावस्तुनो वस्तुसिद्धिः (1/43) and therefore with Kant doctrine as well.

According to the Veda there is no place in the universe with a perfect vacuum. A so-called vacuum chamber of a physicist may be completely devoid of atoms of gases and other particles but it is still populated by the ubiquitous *aapah* matter, the most fundamental particles. In the creation process of the Vedic universe the entire universe is first filled with 7 *deva*-systems (discussed in Chapter 8) before the Vedic Big Bang takes place in the compact so-called 8th progeny of Nature named *Hiranyamaandam* (Golden-embryo). One of the 7 *devas* is the ubiquitous *Aapah-devata* forming the background field of most fundamental particles in triplet *sata, raja* and *tama* with differing properties (details in Chapter 5). They are so minute that they cannot be evacuated completely from a vacuum chamber. The term '*aapah*' means '*present everywhere*'. Then there is a part of the Space-entity, gravitational field and the Dark Energy (if it is real) in the same vacuum chamber; none of which could be evacuated out. Thus there is not a thing like '*nothing*' anywhere. Before claiming that the phantom particles appear out of *nothing* and disappear into *nothing*, a physicist has to think of all-pervading '*aapah*'. A statement such as the '*matter created out of nothing*' is not logical because the matter is the 'effect' of a 'cause' that could be obvious or hidden.

A *sutra* (aphorism) in philosophy does not come out of a hat. It first springs up instantaneously, revealed as a thought in the mind of a *rishi* who ponders over it before formulating it. A quantum of thought may appear spontaneously in the mind but it does not vanish like a matter-antimatter particle pair of Quantum Physics annihilating in the next moment. The quantum of thought is expressed, correlated with observations, tested by logic and argued extensively, sometimes over centuries by other brilliant thinkers and ultimately the formless thought incarnates into an observable and readable aphorism. *Sutra*-aphorisms of Vedic Darshans are the ancient pearls of wisdom intercalated on the thread of brilliance and they never age. For the brevity of the text a lot more relevant material from the Darshans cannot be invoked here. In the context of cause-and-effect, only a limited number of *sutras* from Saankhya philosophy should suffice as following.

In relation to the creation of the universe, Rishi Kapil in his Saankhya philosophy, taking advantage of the power of the Sanskrit language, brilliantly formulates a concise *sutra* on the 'material-cause' and the 'cause-of-a-cause',

मूले मूलाभावादमूलं मूलम् | (Saankhya, 1/32)

Transliteration: "*moole moolaabhaavaad-moolam moolam*".

Word Translation: मूले (*moole*) = in the mool, मूलाभावात् (*moolaabhaavaat*) = from nonexistence of *mool*, अमूलं (*amoolam*) = is without *mool*, मूलम् (*moolam*) = *the mool.*

Sutra Translation:

"The mool is without mool due to nonexistence of the mool of the mool".

(Saankhya, 1/32)

In this outstanding intellectual display of the power of Sanskrit grammar, through the forms of the single term *mool*, Rishi Kapil has formulated a universal principle. A first read of the English translation of the *sutra* may not make sense because the same word *mool* is being used as the subjective-noun and the objective-noun. It is difficult to extract the actual meaning being conveyed by a sentence if its grammatical logic is unclear. The fact is that the *sutra* of Rishi Kapil with four parts (see transliteration) is a complete sentence and the *sutra* statement makes full sense right away for someone with Sanskrit background even if the actual meaning hidden in the *sutra* may require further explanation. It is the power of the Sanskrit language that enables the single term *mool* in four variations of grammatical conjugations to convey the logical meaning of the *sutra* statement. To extract the meaning hidden behind the words one requires some background-reading of the interpretive commentary and the knowledge of word formation. Therefore this author feels that a brief discussion on Rishi Kapil's *sutra* is in order here.

What is this *mool* that Rishi Kapil has so beautifully used in this *sutra*? The *rudhi* (ordinary/colloquial) meaning of the Sanskrit word *mool* is the root of a plant. The sprouting of a seed begins with a root which supplies nourishment to the plant. Thus the process of the root being the feeder to a plant, ranging from a micro-plant to a gigantic tree, is universal. Therefore implicit even in the ordinary meaning of *mool* is that a substantive root of a tree is the 'cause' and the substantive tree is the 'effect'.

The English usage of the word 'root' in a two-worded common expression - the 'root-cause', also conveys a similar meaning wherein 'root' is the adjective and 'cause' is the abstract noun. Its application for a tree may sound awkward if expressed as, the 'root is the root-cause for tree's existence', albeit that is a fact. A tree's existence begins from the first molecules of its root and its root continues to be the cause for its existence. Cut the visible part of a fully-grown tree and the probability is that it would sprout again, but if its root is cut off, its death is inevitable. This is why the root of a tree is the cause for its 'existence'. Obviously there are two kinds of effects caused by a 'root'– subsistent and abstract. The subsistent effect is the three dimensional physical tree and the abstract effect is the very 'existence' of the tree.

As mentioned before, all words in the Vedas and *sutras* have *yaugik* meaning (root-based, abstract), not the *roodhi* meaning that represents concrete objects. The 'existence' of a tree is the effect of a cause which is the root of the tree (*mool*). The 'existence' is the *yaugik* meaning of the effect. On the other hand the three dimensional substantive tree is the effect of a cause which is the root of the tree (*mool*). Rishi Kapil's *sutra* is to be treated in the *yaugik* realm where no substantive causes or effects are involved. Both meanings of *mool* are applicable and necessary to understand the soul of the *sutra*. The usage of *mool* in this *sutra* is for the material-cause of creation which is the Nature itself, which Rishi Kapil has addressed as *Prakriti* or *Pradhaan*. The rephrased English translation of the above *sutra* in familiar words is:

"The Nature is without a cause due to nonexistence of its cause" (Saankhya, 1/32)

The straightforward meaning of this statement is that the Nature is not caused by something else; it is without a cause. This by definition makes it an eternal entity. Because the Nature is the entire matter forming the universe, the Nature is universe's 'material-cause' (the *mool*). Matter being an inert substance the Nature in the Vedas is devoid of consciousness. The consciousness in the Vedic perspective is altogether a separate, non-matter, eternal quantum of consciousness that creates the observed consciousness in the otherwise inert conglomerate forming a living body.

The *mool* in Rishi Kapil's above *sutra* is not the quantum of consciousness (individual soul) because the *mool* of the Saankhya *sutra* (1/32) is the 'material-cause' of the universe, the Nature. Rishi Kapil in other *sutras* of his Saankhya Darshan has also dealt with the quanta of consciousness that serve as the cause for consciousness in living-beings. The *sutras* mentioned above, Saankhya (1/32, 1/43) and several more were established by Rishi Kapil to lead to the principle of eternity applied to certain entities in the creation process of the universe. The other two inert eternal entities serving as the 'general-cause' in the process of creation are Time and Space entities. The eternal entities that have no cause for their existence are addressed as *anaadi*. A detailed discussion on eternal entities is in Chapter 5.

The author of yet another Darshan (Philosophy) Rishi Kanaad, who gave the definition of *paramanu* the last stage of matter and the revealer of the property of gravity in Earth, deals with the subject of *kaaran* and *kaarya* (cause-and-effect) using his own unique perspective for terminology. Two relevant *sutras* are discussed below.

कारणाभावात्कार्याभाव: | (Vaisheshik, 1/2/1)

Transliteration: *"kaarn-abhaavaat-kaarya abhaavah"*.

Sutra Translation:

"In the absence of a cause the effect is non-existent".

(Vaisheshik, 1/2/1)

Then in the subsequent *sutra* he stated the same principle completing the logical loop:

न तु कार्याभावात्कारणाभाव: | (Vaisheshik,1/2/2)

Transliteration: *"Na tu kaarya-abhaavaat-kaarana-abhaava"*.

SutraTranslation:

"The absence of effect does not prove the absence of cause".

(Vaisheshik, 1/2/2)

These principles related to the cause-and-effect are remarkably concise, yet they convey a complete meaning. In the context of the universe's creation out of Nature these two *sutras* together mean that if there was no Nature (cause) there would not be the universe (effect) and conversely the absence of the universe does not mean that the Nature (cause) was non-existent. For example in the dissolution state of the Nature when the matter is in the indefinable *Asata*-state confined in *tuchchhaya* (diminutive) space (RigVeda, 10/129/3), it is there but non-observable. This means that if the universe (effect) is absent, its cause is not necessarily absent. Rishi Kanaad's second statement (1/2/2) has a direct implication for the modern Big Bang cosmology wherein the birth of the entire universe is hypothesized from a dimensionless space-time singularity. It is counter-intuitive to think that a dimensionless space-time singularity served as the womb for 10^{60} kg of matter. If Rishi Kanaad's perspective prevails the Big Bang logic is in serious trouble. If Big Bang assumes that the universe was absent in the singularity at time zero it fails to prove that its cause (matter = Nature) was also absent in it. If the Big Bang Hypothesis emphasized that the universe first appeared as matter-antimatter out of nothing, then that 'nothing' is the cause of the universe. This argument is also in trouble, this time owing to Rishi Kapil's Saankhya *sutra* (1/43) that "*Nothingness cannot be the cause of something*".

It is interesting to note from the second part of Kant's statement cited earlier in this section, "*…that every change is an alteration in the state of some permanent substance*". In the present context of the universe his '*permanent substance'* is the eternal Nature-entity, the *Prakriti* of the Vedas and Darshans. It may be noted that the Kantian maxim that relates to the preexistence of *permanent substance*, provides considerable value to the philosophical argument on the creation of the universe. It is concurrent with the eons old Vedic view expressed in the Upanishads and by the great *rishis*, Kapil and Kanaad.

It was mentioned earlier that the truth (*satya*) is invariant and non-transient. A conscious-being seeks truth because the truth leads to increase in knowledge. If a perceived truth brings with it some 'doubt', think that the truth is not reached yet. Doubt is the trait in conscious-beings that plays a vital role in uncovering the truth.

Doubt (संशय = *sanshaya*)

The term for doubt in Vedic Philosophy is the संशय (*sanshaya*), defined by Rishi Kanaad in his Vaisheshik Darshan (2/2/17). Till the *sanshaya* (doubt) about a substance or a phenomenon exists in the mind of a person, the person cannot claim to have true knowledge of that substance or that phenomenon. Doubt blurs the mind.

To remove doubt one must understand what doubt is and how doubt arises. The great Vedic philosopher Rishi Kanaad brilliantly defined the *sanshaya* (doubt) as:

सामान्यप्रत्यक्षाद् विशेषाप्रत्यक्षाद् विशेषस्मृतेश्च संशयः ।

(Vaisheshik, (2/2/17)

Transliteration: "*Saamaanya-pratyakshaad vishesha-apratyakshaad vishesha- smriteshcha sanshayah.*"

Sutra Translation:

"*Cognizance of ordinary nature of a substance (dharma, properties), non-revelation of its extraordinary nature (properties) and retrospection of its specific nature (properties) is 'sanshaya' (doubt) of the substance*". (Vaisheshik, 2/2/17).

Notice that in Rishi Kanaad's definition of *sanshaya* (doubt), the *sanshaya* is part of the substance. When the substance goes out of the mind of the person the *sanshaya* (doubt) of that substance also goes with it. The doubt arises in the mind due to the incomplete knowledge of the substance. Doubt will not arise if there is complete absence of the knowledge of the substance (or entity). This means that doubt about a substance arises when the knowledge of its ordinary properties is already there but there is ignorance about its extraordinary properties that are not obvious and the mind recalls the possibility of some unusual properties that the substance may possess.

Based on Rishi Kanaad's *sutra* (2/2/17), this author draws two straightforward deductions:

(1) "*Doubt about a substance is generated by incomplete knowledge of that substance*", and

(2) "*Complete knowledge of a substance neutralizes all doubts about that substance*".

This chapter began with Professor Carl Edward Sagan's quotation with half a dozen questions about the universe including the question - *why there is a universe*? This author has also raised a large number of other questions in this chapter on the universe and on the modern Big Bang Hypothesis. When there are unanswered questions about an entity, the entity is not absolved of the संशय (*sanshaya*) the doubt that it evokes. So is the situation with the universe. It would be difficult to remove all doubts about universe because it is complex, vast and unfathomable.

As we continue to learn about the ordinary properties (Rishi Kanaad's *saamaanya*) of the universe and learn about its extra-ordinary properties (Rishi Kanaad's *vishesha*) the doubt about the universe will diminish. But this cannot happen without the *retrospection of its specific nature (properties).*

Other Considerations in the Contemporary Big Bang Hypothesis

Some contemporary Big Bang cosmologists have added to the *sanshaya* (doubt) about Big Bang Hypothesis. They believe that Time must have existed before Big Bang. Gabriele Veneziano published an article entitled "The Myth of the Beginning of Time" in Scientific American, May 2004. In the article he stated- " *Strings Theory suggests that the big bang was not the origin of the universe but simply the outcome of a preexisting state*". In this scenario the Big Bang is not the beginning of Space and Time. He further stated that "*The new willingness to consider what might have happened before the bang is the latest swing of an intellectual pendulum that has rocked back and forth for millennia"*. Through the rigors of mathematics, some cosmologists as well as Paul J. Steinhardt at the Princeton Center for Theoretical Science at Princeton University, USA, are now working to prove the validity of the cyclic universe where a perennial cycle of expansion and contraction gives birth and death to the universe. The basic idea in this new development in cosmology is the same as in the ancient Vedic principle with the universe in the cyclic state, clearly laid down in the Vedic mantras. The contemporary cosmologists had to accept that if Time existed before the Big Bang of the present universe the Space also must have existed and therefore the other sequential Big Bangs must have occurred before the present one. Then who can arguably deny that the Nature, Time and Space are eternal entities? This already is the Vedic Big Bang Scenario. The extension of logic would render that the Big Bang was a recurring phenomenon. This leads to the matter being the material-cause and the Time and Space being the associative causes. In such a theory by contemporary cosmologists, the efficient-cause (the Creator) is out and the mathematics is in and the rest is the same as the Vedic Cosmology Principle. The argument in this situation is that if the creation is cyclic and if it is going to happen one after the other, where is the need for a Creator?

A long answer to this question needs a separate discussion; however a short point is made here. The Nature follows the law of gravity owing to the mass in matter. The mass in matter according to the Big Bang quantum mechanics came from Higgs particle in a later step of creation. So the mass in a matter particle, say an electron, is extrinsic to electron. Therefore for a moment not worrying about the other properties of the electron, an electron has two things in it the negative charge and the mass imparted by a Higgs particle. Because the mass is extrinsic to what the electron was, it should be separable from the matter of the electron. But a physicist is not able to separate mass from an electron or any other piece of matter. Then the question is – what is mass and when did it evolve to have attraction to other mass to create the law of gravity. In a Big Bang all things happen instantaneously and there is time for sequential events but there is no time for evolution of any of the four forces in Nature and there is no time for evolution of all created particles. So, first time within the first moment (Planck-time) all matter diversity and all laws of Nature turned out to be just right with the criticality in their values. Can a physicist create only the property of mass in the laboratory or separate the mass from a matter particle? Can a physicist observe the Nature resorting to these actions now and indulge in evolving the process for further perfection? Because there is no sign of Nature evolving and arriving at the critically balanced Nature`s laws and the laws do not seem to be the outcome of a random draw, the only answer seems to be that the inert Nature is not alone and its different properties are extrinsically put into it.

From a theistic Vedic perspective, because the properties in matter are not seen developing in Nature (not evident from the behaviour of matter), they are extrinsic for someone to own them. The Veda credits the Creator for two sets of laws, *satyam* and *ritam*, which govern two realms of the creation - physical and spiritual respectively (discussed in detail in Chapter 5). A Creator is needed to infuse all matter properties to the inactive indefinable *Asata*-state of matter at the point of the commencement of the creation. Because He is all-pervasive each fundamental particle gets triggered with *satyam* and *ritam* laws.

Hawking, Grand Design and the Veda Mantra

The Grand Design proponents have argued that the Nature's laws together with the values of solar mass, Earth-Sun distance and other factors are so critically balanced that life exists on Earth and small variation in them would not allow life to exist on Earth. They credit this to divine design, popularly known as the Grand Design. Because life thrives on Earth and presumably not on other planets of the solar system the proponents of Grand Design said that the Earth is uniquely created by the Grand Designer. Hawking wrote his latest popular book entitled - "The Grand Design", wherein he stated:

> *"The 1992 discovery of a planet orbiting a star outside our own Solar System is a turning point against Isaac Newton's belief that the universe could not have arisen out of chaos... That makes the coincidences of our planetary conditions -- the single Sun, the lucky combination of Earth-Sun distance and solar mass -- far less remarkable and far less compelling as evidence that the Earth was carefully designed just to please us human beings".*

If the Grand Design implied that the Earth is special and the only planet in the universe to have life on it, the Grand Design with exo-planets discovered now in thousands has shot itself in the foot. Hawking's basic argument seems to be that if other solar systems may have Earth-like planets with life on them, then where is the need for God to specially design the Earth?

Hawking's argument is meaningful only if life existed on some exo-planets. The Grand Design continues to persist because life on exo-planets has not been proven yet. From the Vedic perspective on life on Earth and on the other planets both sides are only half right!

If the Earth-like exo-planets are already discovered in large numbers in other solar systems, within our reach with a telescope, it could be generalized that life exists throughout the vast universe. This would strengthen Hawking's argument that the Earth is not unique. But at the same time it puts the argument in jeopardy because if life exists on exo-planets, the possibility is that by now the intelligent-beings on them have proposed their own Grand Design. This makes the Grand Design argument to continue, now universally. The proponents of Grand Design do not come out a winner either because the exo-planets with life on them make the Earth a just another planet with life on it. So both are half right!

There is a positive side to this for the followers of the earth-centric Grand Design. If the astronomy had not discovered exo-planets with the possibility of life on them the believers of earth-centric Grand Design would not have had the benefit of learning that the Master Designer is capable of creating planets throughout the universe that give opportunity for living-beings to take birth on them and do their karma.

This argument does not make a difference to the followers of Vedic Philosophy because the Vedas have already enlightened them through mantras stating that the exo-planets exist throughout the universe that are inhabited by living-beings. There are numerous mantra and *sutra* references in the Vedic Literature that explicitly state that there are planets, seen and unseen that are inhabited by living-beings. Let us begin with a YajurVeda mantra (17/2) stating that living-beings are also born on the planets in the universe other than the Earth. The YajurVeda did not come up with such a pronouncement to counter Hawking or the proponents of Grand Design, it has always been there since the ages. Therefore in Vedic science the 1992 discovery of the exo-planet, as stressed by Hawking, is hardly a turning point.

The first part of the long YajurVeda mantra (17/2) teaches to count in a decimal system from 1 to 10^{18} (billion-billion) giving nomenclature for all divisions: 1, 10, 100,....till 18 zeros after 1. It is of interest to note from this *chhanda* (long mantra) that a decimal system for counting was established in the far antiquity of the Vedas. The second part of the mantra relates to the astronomical bodies categorized as 'seen' and 'un-seen':

Mantra Translation:

"*...When you think of counting bricks, think of the vast Space and those worlds out there that are observable* (धेनवः लोके = *dhenavah loke*)[1] *and non-observable* (धेनवः अमुष्मिन् = *dhenavah amushmin*)[2] *to which also these numbers apply, where you may be in rebirth* (अमुत्र = *amutra*)[3]*"* (YajurVeda, 17/2, last part of mantra).

This mantra could not have made it clearer that there are exo-planets out there in the universe where humans (life-systems) are born. For further elaboration on the translation of this part mantra this author has emphasized three crucial Vedic terms. The first superscripted term धेनवः लोके (*dhenavah loke*)[1] has two words in it wherein the first word *dhenvah* serves as a pluralized adjective to the second word *loke*. The singular *dhenavah* is *dhenu*. The root meaning of the term *dhenu* is the same as that of its other synonym, गो (= *go*, *gau* or *gavah*) which in the ancient word assemblage, Nighantu and Yask Rishi's Nirukt (2/2/1, 12/1/7) stands for an astronomical body, the Earth included. The properties of the object named *dhenu* or "*go*" represents the property of motion, 'one that has gone far' and the one that is in orbit around its progenitor (further discussion is in Chapter 11). The term *dhenavah* stands for planets. The specific property represented by the term *dhenu* for a planet is one that is 'livable' and which 'provides nourishment' to one who lives on it. In other words it has the property of 'one that supports and is in motion or is in orbit'. The term *dhenavah* in the context of the mantra represents those planets that provide nourishment to substances and living-beings on them. Implicit in the term 'nourishment' is that the planet provides all inert substances such as water, air and matter to sustain life.

The term *lok* means 'terrestrial world where *lokas* live'. The pluralized term *lokas* mean all living-beings. Thus the term *lok* stands for '*the abode of living beings*'. In addition to the word *lok* frequently appearing in the Sanskrit Literature it is also in the active vocabulary of the present Indian society. Educated and uneducated people alike, use the compound word *prithvi-lok* where *prithvi* means the Earth. Therefore *dhenvah* the pluralizing adjective to *loke* means '*the orbiting worlds that are observable and are the support for life*'. The expression *dhenavah loke* in the terminology of modern astronomy would mean - 'on the observed exo-planets'.

The mantra not only takes us outside of our solar system but it takes us to the region of space beyond our observation. The second pair of superscripted terms धेनव: अमुष्मिन् (*dhenavah amushmin*)[2] means - '*on the orbiting worlds that exist but not seen*'. It is the term *amushmin* that puts the terrestrial bodies in the category of 'non-observable'. Even when more of the presently non-observable exo-planets become observable in time with the help of powerful telescopes, there will still be uncountable exo-planets that would remain un-observable (*amushmin*). The two Vedic categories of exo-planets, the seen (*dhenavah-lok*) and the unseen (*dhenavah-amushmin*) will continue to be the feature of the vast universe. The Vedic *rishi* did not miss out the fact that the universe is not only what one observes but it is unfathomably large with unobservable celestial objects beyond the human ability to observe.

The fact that the *rishi* of the mantra puts planets in two categories, the seen and the unseen, indicates that it is not a fleeting reference or a chance observation. To address an illuminated dot in the night-sky with the term *lok,* which by definition is like the Earth, is no ordinary observation. Most planets of our solar system, discussed in detail in Chapter 11, are counted and named in the Vedas and the ancient Vedic Literature. So it is important to note that the *rishi* of the YajurVeda mantra (17/2) cited above did not say that "... *you may be in rebirth* (अमुत्र = *amutra*)[3]" on the planet *Mangal* (Mars) or on *Brihaspati* (Jupiter) and so on which are *dhenavah-loka*s seen in the night-sky. Obviously the *rishi* had made a reference to the seen and the unseen exo-planets capable of sustaining life.

Implicit in the YajurVeda mantra (17/2) cited above is the understanding that the emergence of life and consciousness on planets with terrestrial makeup that are capable of nourishing substances and living-beings, is a rule throughout the universe than an exception. The *rishi* certainly is not limiting the advent of life and consciousness to our Earth alone. A synonym for the Earth is *Vasudha* (*support for vasus*). The term *vasu* means "*living-being inhabitants*". Swami Dayanand Sarasvati referring to the ancient commentary book on YajurVeda (Shatpath Brahmana, 14/6/7/4) answered a question in the chapter 8 of his book Satyarth Prakash (publ. 1881) stating – "*Yes, living-beings inhabit celestial bodies throughout the universe and that is why celestial bodies acquired the name vasu (including the Earth)*". He said this over a century ago before modern astronomy discovered exo-planets.

Those emphasizing earth-centric Grand Design are to understand that the Earth according to the Vedas is not alone in the universe to sustain life. The Earth no doubt has been sustaining an unimaginably high diversity of life-forms for billions of years; however, it is not sufficient to accommodate all quanta of consciousness of the universe taking birth in all possible forms of their bodies. Those emphasizing the discovery of life on distant exo-planets light-years away from the Earth and pitching it against Grand Design hypothesis are to understand that the discovery of exo-planets and even a certainty of life on them do not prove that the advent of the universe was an accident in the gravitational field. If anything, the *discovery of a planet orbiting a star outside our own Solar System is a turning point* for all to recognize that life and therefore consciousness is also a ubiquitous property of the universe. With exo-planets supporting life, the earth-centric Grand Design believers will have to extend their Grand Design to the entire universe as it is in the Vedic perspective.

One unique feature obvious from the above Vedic reference is the 'rebirth system' followed by the quantum of consciousness, the soul. Implicit in the rebirth system is that the quantum of consciousness called the soul is a non-perishable entity that causes life and consciousness in matter, the matter otherwise being inert. The Vedas and Vedic philosophy have expounded on the subject and have asserted that the soul is a non-perishable conscious entity. Of numerous references at hand a reference on the nature of the soul (*Atma*) from Bhagvadgita (2/20) is presented below:

> *"Neither it* (*Atma* =soul) *is born nor it ever dies. It is not that if it came once* (to take birth in a body) *it will never come back again. It is eternal, unchanging perpetual and ancient. It is not killed by killing the body, it lives on"* (Bhagvadgita, 2/20).

The emphasis in this and the *shloka* that followed (Bhagvadgita, 2/23) implies that the tools made of matter neigther can capture or destroy a soul nor can they subject it for testing and measurement. The best proof of soul's existence is that its presence in the physical body gives rise to consciousness in otherwise an inert conglomerate of limbs and organs. Under the rebirth principle, taking birth is possible on the Earth or on another planet if the planet is capable of sustaining life (YajurVeda, 17/2). Therefore, a sage may have experienced being born on another planet before formulating the above mantra. In the long history of the universe we all may have been born on other planets but cannot recall owing to our limited spiritual development in personal consciousness unlike the self-realization that a sage might have had. This is another explanation from the Vedic perspective for a belief in the existence of धेनवः लोके (*dhenavah loke*)[1] and धेनवः अमुष्मिन् (*dhenavah amushmin*)[2].

The Vedic definition of the quantized consciousness of the second-kind called *Atma-tatva* (individual souls) and the conterminous consciousness of the first-kind the supreme, omnipresent, unitary spirit, named Brahmn in the Vedas, God in English, are discussed in Chapter 5. Under the Vedic principle, all souls of the universe are ubiquitously present (Saankhya, 1/114; Vaisheshik, 3/2/21) through the universe and not limited to the Earth only.

There is hardly any difference in the Vedic principle of ubiquity of life in the universe and the thought contained in the statement of one of the discoverers of the new exo-planet Gliese 581g orbiting around the star Gliese 581, two hundred trillion kilometers away in the constellation of Libra (Vishakha in Vedic astronomy). Announcing its discovery on September 29, 2010, Steven Vogt, a professor of Astronomy and Astrophysics at the University of California at Santa Cruz stated:

> "*Personally, given ubiquity and propensity of life to flourish wherever it can, I would say, my own personal feeling is that the chances of life on this planet* (Gliese 581g) *are 100 percent. I have almost no doubt about it.*"

Whether a conscious-being soul creates life as stated in Vedic Philosophy or life is a onetime accident in the chemistry of the planet before the hypothetical evolution takes over, as assumed in modern biology, the *ubiquity and propensity of life* is turning out to be an increasingly accepted fact regarding the universe. The possibility of life on exo-planets and life's ubiquity through the universe encourages one to think again about 'life on Earth began as an accident'. If life were an accident on Earth, it has to be an accident on all those exo-planets that support life. Then, it is hardly an accident and rather it is a phenomenon with a root-cause. The theory suggesting the accidental start of life on Earth would therefore remain to be an un-provable hypothesis. This is the real turning point. As this author generalized above using a mantra reference and as stated by Professor Steven Vogt and subsequently by other astronomers, life seems to be sustained throughout the universe. Then the *discovery of a planet orbiting a star outside our own Solar System* should have been *a turning point.* Life being ubiquitous in the universe suggests that the advent of life on Earth is not an accident. This generates the possibility of the forces and fields of differing nature yet to be discovered that could have been the participants in the process of the creation of the universe.

For a truth-seeker physicist, searching for answers to questions regarding our universe, the concepts of the conterminous consciousness (the first-kind, Supreme Consciousness), the quantized consciousness (the second-kind, souls), the karma principle leading to the birth and rebirth of a soul on this planet and its instantaneous transmigration to other planets, ought to all remain on the table for discussion and study.

The Vedic explanation for how the *rishis* arrived at such mind-boggling deductions relates to both, physics and philosophy, to be discussed in another book. Limiting this author's suggestions to the present reference frame of physics, the deductions of the ancient *rishis* could have had a theoretical basis if not the astronomical. But why would a *rishi* teach about large numbers first up to 10^{18} then talk about the universe that is observable to have धेनवः लोके (*dhenavah loke*)[1] and the universe that he was not able to observe but knew it was there to have धेनवः अमुष्मिन् (*dhenavah amushmin*)[2] the planets with life on them that were non-observable? This mystery remains!

The author concludes this section with one more reference on this topic. Rishi Etareya stated in the book of Etareya Brahamana (iii, 4, 6, 44) that the person (person's soul) who understands the physics knowledge about the Sun acquires coexistence with the Sun and "*...is born* (in next birth) *on planets with brilliance*".

The Vedic "*Kaaran-Kaarya*": "Cause-Effect" Principle

The cause-effect principle and its application has been discussed a number of times in the earlier pages. Under this section the Vedic perspective of 'cause-effect' is presented with formal definitions of its components and its applicability in both realms – material (physical) and spiritual (metaphysical). The consciousness in Vedic philosophy is seen in a realm separate from the material realm. The two realms make the Vedic cause-effect (*kaaran-kaarya*) principle a universal principle. As mentioned before, all physics laws governing the matter of the universe come under Creator's laws addressed as *satyam* and all laws governing the spiritual realm come under Creator's *ritam* laws (RigVeda, 10/190/1). As an example, the law of gravity is under *satyam* and the law of rebirth, that determines spiritual evolution of a quantum of consciousness (soul) leading to the consequential rebirth comes under Creator's *ritam* law. Because they are Creator's laws to govern the universe in both realms they are invariant. The gravity cannot be switched off when one is falling under gravity and bad karma cannot go unpunished.

Vedas are inundated with mantras that explicitly point to the *kaaran-kaarya* (cause-effect) in differing situations in physical and spiritual realms. Relevant parts of some mantras are listed below:

> "*... those scholarly who understand with clarity the causal-Nature (matter) coming into the creation-phase giving birth to all substances and reversing into the causal-state dissolving all substances of the universe, cognize both* (सह वेद उभयम् = *saha-ved-ubhayam*) *simultaneously...*" YajurVeda (40/11).

The term उभयम् (*ubhayam*) means 'both' and stands for the *kaaran* and the *kaarya* - the 'cause and effect'. The RigVeda further makes it explicit in the following mantras:

> "*...the Nature (universe) in its state of existence was hidden in its diminutive causal state of matter...*" RigVeda (10/129/3).
>
> "*... Rishis in their contemplation comprehend cognitively the binding principle of causal relationship that the cause for Sata the definable (state of matter) lies in the Asata the indefinable (state of matter).*" RigVeda (10/129/4).
>
> "*...fundamental seeds of matter are the cause for the great celestial objects. In this process, the Svadha* (Nature) *is subordinate* (material-cause) *and the Creator's inspiration is paramount* (efficient-cause)*".* RigVeda (10/129/5).

The three main 'causes' in the Vedic philosophy are formally defined in the following with their equivalent English names. They are similar to the components of the 'causes' in the Western philosophy.

Nimitta-kaaran : Efficient-Cause

The *Nimitta-kaaran* in Western terminology is equivalent to the efficient-cause or the first-cause or the prime-cause. The implied meaning of the word *nimitta* includes '*a cause with motive to create a kaarya*' where *kaarya* means the created '*action*' which is the 'effect'. Since the *Nimitta-kaaran* does not consume itself in creating the resulting effect, this cause is also addressed as the associative-cause.

Upaadaan-kaaran : Material-Cause

The *upaadaan-kaaran* is the same as the material-cause in Western philosophy. The other Vedic term for this cause is *samvayi-kaaran* (intimate-cause). The implication here is that whether the resulting 'effect' is a substance, an action or a property, the *upaadaan-kaaran* (or *samvayi-kaaran*) is matter and its forms, whatever the resulting form of the 'effect' may be.

Saadhaaran-kaaran : Ordinary-Cause

In some contexts this cause is also addressed as the associative-cause because this cause is not consumed in creating the effect. It does not initiate the action (cause) but is needed to enable the action which causes the effect. The *Saadhaaran-kaaran* could be a substance or an entity such as the Time-entity or the Space-entity that is neither the material-cause nor the efficient-cause. Between the efficient-cause and the ordinary-cause the efficient-cause is the initiator of the effect which may necessitate association of the ordinary-cause. The *saadhaaran-kaaran* (ordinary-cause) is different from the material-cause but owing to its associative nature it is sometimes viewed as the efficient-cause.

The cause-and-effect is present in all processes in all directions of the universe. From the onset of the creation of the universe, a cause leads to an effect and what is produced in the effect serves as the cause for the next state of matter, thus setting a chain of cause and effect. Cause-effect is a pre-existing rule in Nature and we see it all around us. Contrasting this, there is no specific logic that would suggest that the universe, which is the action (effect) of Nature, came into being without its cause. Also the observed fact on the 'arrow of time' is that the cause precedes the effect and observing the chain of cause-effect leads to increased knowledge about Nature.

If $2x10^{30}$ kg of matter was not available about 4.6 BY ago at a given time and locale of the universe our solar system would not have come into existence. Therefore, this matter was the *upaadaan-kaaran* (material-cause) for the solar system. Of course, the preexistence of this mass alone is not sufficient to create the solar system and the orderly shape that it is in, with one of its planets called Earth in such a remarkable balance that intellectual beings and the rest of the complex living systems could exist on it. Simultaneously with the material-cause existed other associative-causes like – gravitation as the inherent property of matter, the solar system's position in our galaxy with its initial linear and angular motion and so on and so forth. Therefore, in Nature we always see something preexisted to serve as its material-cause. When this observation is extended to the actions of the living-beings one finds that the living-being serves as the *nimitta-kaaran* (efficient-cause), at a smaller scale though, to create new things (the effect). The initiative in the case of a living-being is his own and he draws the matter for the material cause from Nature. Now, to gain a bit more insight into how the Vedic *rishis* used the subtle principle of cause-and-effect to gain knowledge about the creation, let us invoke an example from the distant past of the Vedic Civilzation.

The *kaaran-kaarya* (cause-effect) discussion is by no means a recent revival in Hindu Philosophy (the ancient name being Vedic Philosophy). The process of the creation of the universe in the Vedic Philosophy has always been thought in terms of cause-and-effect. The Veda has even assigned a Sanskrit term *maatali* for the *rishi* with specific knowledge of creation, space, time and the cosmos. Simply put, a *maatali* is a cosmologist. In the distant past of India was a Vedic *maatali,* Rishi Shvetashvatar, the author of the Shvetashvar Upanishad. His disciples while putting together a list of entities enabling the creation of the universe invoked the subject of cosmology and asked their guru Rishi Shvetashvatar - which single entity is the most important in the process of the creation of this universe?

काल: स्वभावो नियतिर्यदृच्छा भूतानि योनि: पुरुष इति चिन्त्यम् ।

(Shvetashvar Upanishad, 1/2)

Transliteration: "*Kaalah swabhaavo niyatiryadrichchhaa bhutaani yonih purush iti chintyam*"

Sutra Translation

> *"It is our thought provoking curiosity as to which among: kaalaha* (Time), *svabhaava* (self-creation), *niyati* (preset rules of Nature or pre-existing order in Nature), *yadrichhaa* (to manifest by accident or by chance), *bhootaani* (classes of matter), *yoni* (Nature) *and Purush* (the Creator), *is uniquely important for the creation of the universe?"* (Shvetashvar Upanishad, 1/2)

Even though this question, posed by the students of Rishi Shvetashvatar, belongs to the far antiquity of the Vedic civilization, its essence is as fresh and as scientific as the questions posed in modern science and philosophy. The list in this question is inclusive of subjects and entities that modern science deals with and those discussed in modern philosophy. Whether a follower of modern science or philosophy agrees or not with Shvetashvatar's answer is altogether a different matter, the comprehensive character of this question is certainly a rarity in modern times.

Even though both, science and philosophy, relate to all that is in the universe, the apparent disconnect between modern science and modern philosophy keeps the two in different compartments. For example in this Upanishad statement, the term *niyati*, the preset rules of Nature or the pre-existing order in Nature, is well understood by the modern science and it has successfully discovered those rules through theory and experiment. The modern science has also learnt that these invariant rules of Nature govern the matter of the universe from the time it was created. However, the modern science and modern philosophy work independently; with the exception of the Natural Philosophy which would probably repeat what modern science has hypothesized and discovered. One connection between the science and philosophy is through the Anthropic Principle in cosmology where the critical and 'just right' values of Nature's laws and constants are thought to be a possible cause for the advent of living-beings on Earth. The cosmologists do not like to think of this as a pre-ordained rule of creation and therefore explain away the Anthropic Principle, with a mathematical proof without invoking a physics proof. This mathematical proof is the Multi-universe theory with the underlying argument that our universe is not special; it is just one of the 10^{500} possible universes out there.

Answering the question posed by his disciples, guru Shvetashvatar imparted in-depth knowledge to his students and concluded by answering that among all the possible causal entities for the creation of the universe, the uniquely important cause is the *nimitta-kaaran* (efficient-cause) and this *nimitta-kaaran* was the *Purush*, the unitary all-pervading conscious Spirit, popularly addressed in the Sanskrit Literature as the *Brahmn*. The word *purush* means '*the one pervading all*' and is used for the Creator of the universe. In the context of the creation of the universe the efficient and material causes (Creator, Nature) appear as a duo throughout the Vedic literature– *Purush-Prakriti*, *Brahmn-Svadha*, *Paramatma-Prakriti* and so on.

From the above statement of the Upanishad it would seem that there are 7 causes listed by the students of Shvetashvatar where *Kaalah* (Time-entity), an important participant in the creation of the universe serving as the *saadhaaran-kaaran* (ordinary-cause), is included but another item of great importance, the Space-entity, was excluded or missing. In fact the Space-entity is included in the fifth term *bhutaani* and therefore there are a total of 11 items in the set presented by the Upanishad statement. In the '*vibhakti*' rule of Sanskrit grammar the singular term *bhuta* changes to *bhutau* in double-plural and it changes to *bhutaani* in plural (more than two) as in the Upanishad statement. The term *bhutaani* has five different items in it. The general term *bhuta* stands for matter. The plural usage of it in the term *bhutani* represents the five classes of elements of Vedic Philosophy, the *panch-maha-bhootas* (five-major-elements), which were briefly discussed earlier in this chapter contrasting with Aristotle's philosophy having only four elements. The first of the five *bhutani* of the Vedic Philosophy is the *Aakaash* (Space); not included by Aristotle in his philosophy.

They are part of the students' list because the existence of life is not possible in the absence of even one of the five *maha-bhootas*.

It becomes obvious from the question of Shvetashvatar's students that the discussion on philosophy, cause-effect and recognition of the entities of Nature, which is the creation itself, had gone on before the period of the Shvetashvar Upanishad. This is why the students in the statement of Shvetashvar Upanishad came up with the list of 11 important entities that are observed to be the causes in the process of creation. The concept of Time (first item in students' list) and Space (part of the fifth item in their list) as entities playing a role in the creation and sustenance of the universe are well established in the ancient Vedic science/philosophy. Notice that *Purush* the Creator and *Yoni* the Nature appearing in the same list as two distinct entities is indicative of the fact that in the Vedic philosophy they certainly are two different entities. This is unlike in pantheism where everything is Nature (matter) or in Advaita Vedanta Philosophy of non-duality propounded by Aadi Shankaracharya where everything is *Purush* the Creator. The inert Nature for the Vedic Philosophy remains as the material-cause (*upaadaan-kaaran*) and the conscious-being Creator as the efficient-cause (*nimitta-kaaran*).

Similar philosophical thoughts are expressed throughout the continuity of the Vedic Philosophy. Someone posed a question to the great Vedic philosopher and author of the 19th century, Swami Dayanand Sarasvati, whom this author has cited a number of times - '*How many types of causes are there for the creation of the universe*?' Based on the Darshans, Upanishads and Smritis he elaborated the answer in his book Satyarth Prakash. A translation of his answer from the Hindi language to English is presented below. The bracketed items are this author's insertions:

> *"Three: one nimitta, second upaadaan and third saadhaaran* (efficient, material and ordinary). *Something will come into being due to the nimitta kaaran* (efficient-cause) *and nothing will come into being without it. It* (efficient-cause) *itself does not come into being but transforms other. The second upaadaan* (material-cause) *is the one without which nothing can come into being, a change in it is a change of its state to generate another - it degenerates into another. The third saadhaaran-kaaran* (general or ordinary cause) *is saadhaaran-nimitta-karan* ('ordinary-efficient-cause') *that coexists in association with the primary-efficient-cause. Of the two efficient-causes, one 'primary efficient' is the Paramatma* (the Creator of the universe) *who creates the universe out of universe's material-cause, sustains it, dissolves it and keeps it in order.*
>
> *The other cause* (ordinary-efficient-cause) *takes the created substances from Creator's universe and variously transforms them for its utilization, is the conscious-being souls. Prakriti* (Nature), *the material-cause is made up of atomic matter* (fundamental particles), *is known as the primordial building material for the construction of the universe, is inanimate and inert and is neither capable of transforming itself into something specific nor is it capable of demolishing itself* (from Satyarth Prakash, Ch. 8, 1881 CE)*."*

It is obvious from the above quotation that the cause-effect and differentiation between the specific causes and effects have been a central theme of the Vedic Philosophy. In another context when a scholarly person expressed doubt on the harmony among all six Vedic Darshans (Philosophies) on the cause of creation, this Vedic scholar par excellence, Swami Dayanand Sarasvati, summarized it in the book of Satyarth Prakash, that all six Darshans complemented each other on the question of the creation principle of the universe. He wrote:

> "*In Meemansa* (Jaimini Rishi's Meemansa Darshan) - *there is no 'effect' in the universe without desired action; in Vaisheshik* (Kanaad Rishi's Vaisheshik Darshan) - *nothing is created without time*; *in Nyaaya* (Gautam Rishi's Nyaya Darshan) - *nothing could come into being without its material cause*; *in Yog Darshan* (Patanjali Rishi's Yoga Darshan) - *nothing can happen without knowledge, wisdom and thought*; *in Saankhya* (Kapil Rishi's Saankhya Darshan) - *creation is not possible without togetherness of causal entities*; *and in Vedanta* (Vedvyas Rishi's Vedanta Darshan, same as the Brahm Sutra) - *nothing can be created if the Creator does not create it"*.

It becomes abundantly clear from Swami Dayanand Sarasvati's answer that the subject of the creation of the universe in Vedic thought is indigenous to India. Also the Veda mantras are the basis for the philosophical principles in all six Darshans, a fact made clear by the *sutras* or from the direct reference of Veda cited by the *rishi* in his Darshan. The six Darshan treatises written in different time epochs prior to Mahabharata War (ended in 3140 BCE), in six differing realms of philosophy, by six different *rishis* have also served as a methodology for the dissemination of Vedic knowledge. The Vedic knowledge in the form of mantras, the statements of Brahmanas, the *sutras* of Darshans and *shlokas* of Smritis and Upanishads was handed down directly to deserving minds and was presented to the common people at a simpler level with examples and similes. The example of the 'pot-maker and the earthen-pot' is often cited in the interpretive commentaries on the Vedas and Darshans. The example of the 'pot-maker' to differentiate between the causes perhaps is as old as the Darshans and Upanishads. Let us for a moment indulge in this ancient example.

When an earthen-pot is created from the clay (sand) drawn from the earth, a human being pot-maker is the *nimitta-kaaran* (efficient-cause or first-cause), the clay is the *upaadaan-kaaran* (material-cause) and the implements like *chaak, dand* (spinning-wheel, stick), water, place (space), time and several other factors that enable the creation of a pot are the *saadhaaran-kaaran* (ordinary-cause or associative-cause).

A pot-maker does not make pots without purpose. If pots had no utility the pot-maker would not have the desire to make pots. He makes pots to fulfill the need of the others. The Creator creates the universe for all souls to enjoy the creation and do karma. Although the villagers are not involved in pot-maker's preparation of the raw material and pot-creation, his association with the villagers is the main cause for him to make pots. Therefore while the pot-maker is the *nimitta-kaaran* (efficient-cause) in this scenario, the people of the village serve as the *saadhaaran-kaaran* (ordinary-cause). The pot-maker's example delineates a glimpse into the simplicity of explanation in the ancient Vedic Civilization.

A pot is created 'because' of the pot-maker's 'desire to create' and the villagers' 'desire' to be fulfilled. A physicist could setup a set of equations to model the pot-maker's process but the physicist would not know how to work-in the parameter named 'desire' in his complex set of equations. Desire is a metaphysical or a spiritual entity present in conscious-beings. It is absent in clay molecules, because, if desire existed in clay molecules, they could assemble themselves into a pot! With reference to the cause-and-effect Swami Dayanand Sarasvati also stated in Satyarth Prakash,

> "*...the cause is subtler than the effect and in the sequence of creation the preceding entity is the cause of the next*".

The Karma Principle

The Karma Principle is a vast subject to be taken up in a separate book. Contextually this author presents here a glimpse into the subject.

The cause-effect in the Vedic perspective pervades through both realms – the consciousness and the physical. The quantum of consciousness (soul) cannot carry out any action without a physical body. All 'actions' (karma) performed by a physical body are inspired by the soul in that body. Therefore, the soul is responsible for all karma, good or bad. Good karma raises the level of spiritual evolution of the soul and a bad karma brings it down. The soul has the freedom of action (karma) but has no freedom to reward itself for good karma or punish itself for bad karma. To have universality and uniformity in accounting for the consequence of a karma the Vedic Philosophy has put it under the *ritam* law of the Creator. This basically is the karma principle of the Vedas. Let us now consider some logistics of the Karma Principle.

The cause in a physical cause-effect chain is consumed in creating the effect. In a metaphysical situation (related to consciousness) the cause may continue to create effects without getting consumed. A conscious-being may generate a thought (cause) in his mind. A thought is a non-material and non-substantive action. However, the thought could be implemented using the physical body to create a tangible object (effect). The conscious-being has the 'free-will' to repeat the process without apparent consumption of the thought producing machine, the mind. Therefore the cause-effect has to be seen differently in the physical and the spiritual realms.

In the case of an aggressor the mind of the aggressor is the cause and the assault by his physical body is the effect. This kind of karma and its consequence are settled by the laws of the society by putting the aggressor behind bars. Thus the physical body of the aggressor is getting the punishment and not the thought producing system, the mind of the aggressor. The thought creation is at a metaphysical level and the implementation of thought is at a physical level. Therefore for complete justice, the punishment ought to be levied at both - the metaphysical and the physical levels. It is straightforward to determine and apply the punishment at a physical level but it is difficult to quantify and apply the punishment at the metaphysical level. So, does the aggressor get away with only the physical punishment and the victim is left to suffer an irreversible loss?

Also if the aggressor is not caught in the act of committing the crime, the societal system is not able to levy any punishments on him. If the aggressor gets away unpunished or bribes his way out of a punishment the people call it as society's unfair system and a theist may even call it the unfair system of God. For a truly just system the aggressor may escape society's law but he should not be able to escape the *ritam* law of the Creator under which the spiritual level (soul's evolution) of the aggressor is automatically brought down (punished). The aggressor may face the consequence of this change when his karma catches up with him in this life or when he takes a new birth.

All quanta of consciousness in the universe come under the law of karma and the balance between the good karma and bad karma determines the level of evolution of a person's consciousness. Depending on the level of bad karma the consequence could be the next birth as a low level-life form. Because the cause-effect in the Vedic Philosophy is seen encompassing the physical as well as the spiritual aspects of the universe, the law of karma is accepted as a just and universal principle. The cause-effect in the Vedic Philosophy is also expressed as the *karma-phal* (action-result) principle. This preset law of creation displays the differences between different living-beings. The level of consciousness in a living-being relates to his/her accumulated karma (actions). Because the law of karma applies uniformly to all and a conscious-being owing to free-will has control on his own karma but has no control on the *karma-phal* (result), the law of karma is considered to be just and fair.

The karma balance-sheet according to the Vedic Philosophy is part of the person's consciousness which the person cannot shake off. Also, the person has no freedom and control to delete unwanted items from the karma balance-sheet! The record of his good and bad karma remains with his consciousness and is a part of that living-being's nature. The cause-effect principle in the physical realm and the karma-*phal* (karma-consequence) principle in the spiritual realm are an integral part of the Vedic Philosophy because without it one would have difficulty explaining why conscious-beings have different starts in this life, right from birth, even before they have had a chance to do good or bad karma. A full exposé answering all questions pertaining to the karma principle and the rebirth system is beyond the scope of this book.

About the Vedas and the Mantra Translation

The Chapter 4 entitled "A Brief Review of Ancient Vedic Literature – the Source of References on Cosmology" presents a brief discussion on the vast Vedic Literature and its branches. The term '*veda*' means the '*revelation of knowledge*', When the term *veda* is used as a noun it means the '*book of knowledge*', which includes the four Vedas (Rik, Yaju, Saam, Atharva) without any label of a prophet attached to it. The term is not limited to representing knowledge in a specific area but implicit in it is knowledge pertaining to all realms of creation; animate and inanimate. The general Sanskrit term for 'knowledge' is *gjyaan* and the term for specific knowledge is *vigjyaan*. It is the prefix *vi* to *gjyaan* that changes the meaning from 'knowledge' to the 'specific knowledge' or to the 'special knowledge'. The term *vigjyaan* in common usage of the word means 'science' because it reveals specific knowledge of a given phenomenon in a given subject. One finds in the Vedas and the subsequent books on the Vedic philosophy many adjectives for knowledge in specific realms. A *sutra* from the ancient Vedic Philosophy expressed thus – इति ह विज्ञायते (*iti hh vigjayate*) meaning *'this is deduced from scientific knowledge'*, where the scientific knowledge in Vedic terminology is defined to be the knowledge acquired by practical means and confirmation. Therefore, examples of Veda mantra references, with added details deferred to forthcoming chapters, are worth examining from the science perspective.

The two basic states of matter (*Sata* and *Asata*), the Vedic concepts of Space and Time, the birth of the universe with a bang, the Earth being round, having the property of gravitation, being in space and revolving repeatedly around the Sun and so on are just a few examples presented in this introductory chapter that are topics of modern science as well. This author having a background in physics has found the Vedic knowledge fascinating from the scientific perspective and therefore the effort in writing this book.

The knowledge content in a Veda mantra is presented in Sanskrit terminology in poetic expression with specific meters. The Sanskrit script is also known as the *Devnagari*-script. In the earlier times the Vedas were learnt by heart and hand-written in indelible black ink, on *bhoj-patra*, the thin bark of the Himalayan birch tree. The universality of thought is the unique feature of the mantras and they cover a wide range of subjects. The four Vedas together contain a total of 20,416 mantras in the realms of the physical and the spiritual creation, including cosmology, the subject of this book. The term "Vedic Cosmology" frequently used in this book represents cosmological thoughts and cosmological principles contained in the mantras as well as the cosmology scenario developed in this book by this author based on the Veda mantra references.

This book also includes contextual general science thoughts and subjects related to the physical phenomenon in the universe. The development of the Vedic Cosmology Scenario by this author (Chapters 6 through 10 in Volume 2) is based on the relevant mantras from the Vedas and the ancient commentary books on the Vedas, supplemented by the references from the books on Vedic Philosophy. The post-Veda ancient books, together, are addressed as the Vedic Literature.

This author takes several steps for the composite mantra translation from the Vedic Sanskrit into English. First the mantra is presented in *Devnaagri* (Sanskrit script) followed by the transliteration of the mantra. Then all euphonic conjugations between the words are delineated and individual words are listed with their English meaning. The Sanskrit language being rich in science terminology, a Sanskrit term having multiple meanings is translated using English terms that convey the closest representative meaning in the context of the mantra. For the sake of brevity at times the direct English translation of a mantra without a detailed translation of the Sanskrit terms is presented. The translation of each term from Sanskrit to English is primarily based on the interpretive translation of the RigVeda and the YajurVeda by Swami Dayanand Sarasvati (1825-1883 CE), wherein he has provided term-by-term translations in the Hindi language. From his Hindi translations this author has synthesized the English translation. The English translation of mantras from the AtharvaVeda and the

SaamVeda are based on the Hindi translations of well-known Sanskrit grammarians and Vedic scholars commissioned by the Arya Pratinidhi Sabha, a world body of all Arya Samajs.

An Arya Samaj is a society with the aim of promoting the Vedas, the Vedic Literature and the Vedic Culture. The first Arya Samaj was initiated by Swami Dayanand Sarasvati in Bombay (Mumbai) in 1875. For the word analysis and the mantra translation this author has frequently referred to the Hindi interpretation of Nirukta by Pt. Sitaram Shastri (Parimal Publications, Delhi, 2005) and Nighantu the ancient word assemblage for specific Vedic terms. Whenever necessary the author has used two Sanskrit-English dictionaries: "The Student's Sanskrit-English Dictionary" by V. S. Apte (publ. New Bharatiya Book Corporation (India) and the "Sanskrit-English Dictionary" by Monier-Williams (publ. Clarendon Press, 1899).

The very first Veda, the book of RigVeda alone contains a total of 10,521 mantras in 1,028 *suktas* using about 153,792 words. A *sukta* represents a collection of mantras specific to a *devata* (subject). Therefore, a *sukta* is a section within a chapter which the Indologists termed as a 'hymn'. Sanskrit, the language of the Vedas, is rich in grammar and vocabulary and is universally accepted as a highly developed ancient language. The richness of a language can be judged by the dictionary of that language. The English language, enriched with words from Latin, Greek and other European languages has an impressive vocabulary of about 86,000 words. It is well-developed to express and explain complex thoughts and phenomena in most subjects, including physics and cosmology.

When M. Monier-Williams (1819-1899) published his Sanskrit-English dictionary in 1899 it had 180,000 Sanskrit words and by no means was it a complete dictionary.

The Sanskrit language, according to an authority on the Vedas, with a title of Vedalankaar, Professor Subhash Taneja, Head of the Sanskrit Department, Rajasthan University, India, has over 3,683,800,000 words in its dictionary; not a single word borrowed from any other language. His reference is the well-known ancient Sanskrit dictionary called "Shabd Kalpdrum" and the number of words mentioned above is based on the computer count of the words in that dictionary. This dictionary is available in fine print in five thick volumes in the page size of one foot by one and a half feet. Professor Taneja, however, insists that this computer count is an under-estimate. Such immense vocabulary containing nearly four billion words can only stream out of an extraordinary language that is predisposed with the scientific methodology for word-creation and word usage with a seamless grammar developed by extraordinary people of an extraordinary culture and civilization. The linguists know that Sanskrit is an extraordinary language.

Mistranslations of the mantras by ill-prepared Sanskritists and Indologists of the nineteenth century have caused extensive damage, misconceptions and misunderstandings about the Vedas. Some Western scholars of the past few centuries learned Sanskrit and enthusiastically translated Veda mantras into English and other languages without learning the proper rules of Sanskrit grammar and its science of etymology that is prescribed specifically to understand, interpret and translate Veda mantras. Their ignorance led to generating and disseminating prejudice and misconceptions regarding the Vedas.

Professor Max Muller (1823-1900 CE) of Oxford University, England, translated the RigVeda shabbily with weird meanings of Vedic terms.

Based on mantra references these misconceptions are dealt with in Chapters 3 and 4 emphasizing that the Vedas have no rituals, no multiple gods, no human history, no mythical characters and that the Vedic poetry is unique, complex and not easy to emulate. Vedas offer amazing science thought for understanding a complex subject like cosmology. The post-Veda literature (Sutras, Upanishads, Smritis, Brahmanas and so on) is the result of the in-depth study of the Vedas by the ancient *rishis*.

The mantra formulations in the Vedas are highly organized in chapters and sections with accreditation to the *rishi* responsible for the formulation. Large number of references from the Vedas and Vedic Literature are presented throughout this book. It will become obvious that the *rishis* have dealt with the subjects of cosmology, astronomy, general science and philosophy with brevity and methodically.

A natural question, the answer to which is still being sought by the modern scholars of Indology is - how old are the Vedas? Many among the modern scholars have guessed erroneously but no one has convincingly determined the period for the advent of the Vedas. A reasonable chronological hierarchy of literary works in Vedic Literature, beginning with the Vedas in the unknown ancient times, till the book of Mahabharata and Bhagvadgita (3140 BCE) is presented in Chapter 4. In Chapter 3 entitled- "Historicity of Vedic References: The References on Cosmology Precede 5000 years", the author initiates a hypothesis and proves that the Vedas and all the references from Vedic Philosophy and literature used in this book, for building the story of the Vedic Cosmology, are anterior to an important time-milestone in human history at February 18, 3102 BCE. As mentioned earlier, this date is the astronomical onset of the *Kali-yuga*, the present epoch of the *Srishti*-calendar of the Aryans, discussed further in Chapters 3 and 5. Therefore, the Vedic references being ancient in the real sense of ancientness, Vedic thought had not come under the influence of cultures and writings of civilizations exterior to India and has certainly not been affected by modern scientific development which is relatively very recent.

Based on new genetic research on Indian population, carried out under the national genome projects and archeological finds it is established in Chapter 3 that the Aryan Invasion Theory and Aryan Hypothesis concocted in the late 19th century by Max Muller et al. was a hoax. The detailed genetic research on the DNA of the Indian populace with state-of-the-art scientific tools under a multinational research collaboration yielded the fact that the Aryans of India are indigenous to India (see Sahoo et al. in Proceedings of the National Academy of Sciences of the Unites States of America, Vol. 103, no. 4, 843-848, January 24, 2006 and Sengupta et al. in The American Journal of Human Genetics, Vol. 78, 202-221, February, 2006).

To be specific, these papers concluded, along with many more that followed, that the age of the accumulated microsatellite variation (of DNA) in the majority of Indian haplogroups exceeds 10,000-15,000 years leading to the fact that the Indians have been on the Indian sub-continent indigenously for an extremely long time. In terms of geological epochs, the Indians in India have shaped the distinctive South Asian Y-chromosome (paternal) landscape at least since the pre-Holocene and Holocene-era exceeding 10,000 - 15,000 years going back into the Pleistocene epoch extending from 12,000 years to 2.5 million years. Therefore, the Indo-European expansion under the Aryan Invasion Theory, hypothesized earlier by Max Muller et al. is proven wrong. The genetic research has indisputably validated the antiquity of the indigenous people of India and therefore their corresponding ancient Aryan culture which includes the Sanskrit Literature and the four Vedas.

Age of the Universe

Hawking in his earlier book remarked that - "*According to a number of early cosmologies and the Jewish/Christian/Muslim tradition, the universe started at a finite, and not very distant, time in the past*". This finite time was about 5,000 to 6,000 BCE. This is negligible compared to 13.8 BY estimated in contemporary astronomy. To match this big number, the numbers provided in the scriptures were re-interpreted by saying that God's one day was much bigger than Earth's.

Hawking's statement does not include the Vedic Creation Principle, now discussed by this author in detail in Chapters 6 through 11, Volume 2). The universe according to the Vedic cosmology principle did start in a distant past, about 2 billion years ago. This number assigned to the age of the universe by the Veda is closest to the estimate by contemporary astronomy and is in the same order of magnitude (in billions). A remarkable fact about this number is that it was assigned to the universe by the Veda in the distant past of the historical milestone at 3140 BCE. It is an extraordinary ancient result.

Some of the first estimates of the age of the world in the modern era came from counting Biblical genealogies. John Lightfoot in 1642 CE estimated that the world was created 4,940 years ago and according to James Ussher's estimate in 1650 CE it was created 5,654 years ago. Then the scientific methodology for age estimate of the Earth began. Based on the cooling rate of the Earth, Buffon in 1760 CE estimated Earth's age as 75,000 years. From 1831 CE to 1890 the scientific methods yielded the age in million years. In 1907 to 1947 the age of the Earth was estimated to be about 2 billion years. As this author understands it the Vedic age of the universe and the Earth are treated near simultaneous. Thus the Vedic estimate of the creation about 2 BY was on par with the modern era scientific estimates. There are no conversion factors between the Creator's and Earth's year because the Vedic estimate is based on astronomy and the Aryans have been tracking it since an unknown period.

The definition of a day or a year is based on a given frame of reference. The above numbers on the age of the universe are in Earth's reference frame. The one year of Jupiter (Jupiter's one revolution around the Sun) for example is 12 years of the Earth. Likewise the 'cosmic year of the Sun' in which the Sun revolves once around the centre of the Milky Way Galaxy, carrying the entire solar system, is 225 million Earth-years. Therefore, God's one day could be defined in Earth's reference frame with a multiplier and a rationale attached to it.

For helping mathematical manipulation of cumbersome numbers, deriving a conversion factor is a brilliant invention of the human mind. The practice of using conversion factors is not a new phenomenon in human history. The science of determining a conversion factor in Vedic tradition is extremely old. The examples from Manusmriti not only exhibit the science of the conversion factor, but go beyond a simple mathematical multiplier, delineating clearly Manu's grasp of physics/astronomy as well as of global geography.

Manu realized that the definition of a day on the Earth itself is not absolute. A day (sunlight period) and a night (dark period) form Manu's *aho-ratra*, which is a 'day' in the present society even though it includes a night as well. A day at the Equator (day + night) is 24 hours. At the geographic pole there is a bright and a dark period, each six months long. Therefore, one day at the Pole is one year at the Equator. To differentiate a day at the pole and a day at the equator, Manu picked the terminology from the Veda and addressed the polar-day (one year of Earth) to be a *deva-day* or a *devic-day*. Several *shloka* couplets (Manusmriti, 1/65, 1/67) define the conversion factors, stating as "*One year of humans is one devic day-night…*" (at the pole). It points to the fact that Manu knew the difference in geography between the pole and the equator. Introducing about three dozen terms for the time units, Manu has provided the conversion factors for the time starting from 4.32 BY down to the time-unit of *nimesh*, where one *nimesh* is exactly 8/45 second (177.778 milliseconds). This is discussed in detail in Chapter 5 under the sub-heading of 'Measure of Time-entity'. The conversion factors are presented in Tables 5.2 through 5.5. The time measurement of *nimesh* is also discussed in Chapter 5 under the sub-title as "The Case of Nimesh: A Mind-Boggling Situation". The accuracy in extracting it from the human physiology, the blink-time of the normal human eye-lid, is unbelievably high. It is this small unit of time that was in practice in the Vedic society and continued to be a unit in practical Vedic astronomy for accurate ascertainment of the position of star in a constellation.

Even if there is a mismatch between the age assignment in modern astronomy and the Vedic creation age, the Vedic system has worked within a preset premise. This premise, to be discussed later, has the astronomical basis. According to the calculation method used for the Vedic age of the universe (Earth), which is also the method that enables the calculations of important dates in the history of the Vedic civilization, this author estimates the onset of the Vedic clock for the present creation to be 1,972,949,118 years ago (about 2 BY) in 2018 CE. In fact, by studying the well maintained practical luni-solar calendar of the Aryans, the *Srishti-samvat*, one can even work out the age of the creation, right to the day, hour and minutes. For some unknown astronomical reasons, the Aryans introduced a long-term sub-harmonic within the same Aryan calendar with the frequency of 306.72 MY that they called *manvantar* and gave 14 different names to them covering the entire creation period of 4.32 BY (see Table 5.4 in the Chapter 5). The *manvantar* relates to the periodic Big Flood which destroys most of the life and civilization on Earth, rejuvenating the creation again with a new beginning and new order. The last Big Flood was the 6th since the creation. More discussion on this is in Chapter 5.

The references pertaining to the numerous physical principles show that it is not just a chance factor that happens to assign 4.32 BY for the total age of the creation-epoch. The present section is concluded by presenting one of the several mantras that provide quantitative information on events and the early history of the universe. This RigVeda mantra provides an estimate of the period that separated the first appearance of vegetation and humans on the Earth. Rishi Bhishgaathrvvan is the revealer of this mantra and its translation is:

> *"I know with certainty the names and places of 107 medicinal vegetation of deep red color that were born first in three deva-yuga periods before the birth of humans".*(RigVeda, 10/97/1).

Based on the RigVeda mantra (10/72/3) it was pointed out that it took one *deva-yuga* period of 4,320,000 years from the onset of the creation of the universe for '*vegetation with root system*' to first emerge on the Earth. This mantra (RigVeda, 10/97/1) essentially states in quantitative terms that the vegetation on the Earth appeared first and a period of three *deva-yuga* (12.96 MY) elapsed since the beginning of the universe for the humans to appear on the Earth. The modern evolutionist may dispute the time period mentioned in these mantras, however, the basic science-thought of the mantra remains indisputable - that the vegetation appeared in this creation before the humans and that the time-lapse between the two events was millions of years. What is also obvious from the mantra is its compactness expressing a science-thought with total absence of any religious overtones. The thoughts in the two mantras, having the association with the specific time periods for the advents of '*vegetation with root system*' and humans, 8.64 MY apart, exhibit the finesse needed in a science-thought, particularly when the specific time period runs not in just the thousands but in millions of years. Such mantras put the Vedic thought closer to the ideas of modern science and remarkably distant from the creation time-lines in other non-science sources. This is substantiated further in the forthcoming chapters where many such mantras containing science-thoughts are presented and the Vedic cosmology scenario for the creation of the universe is delineated.

CHAPTER 2: Present Understanding of the Universe – A Critical Scrutiny of the Contemporary Big Bang Hypothesis

Developments in Contemporary Cosmology

Time-to-time brilliant ideas have jumped out of the human mind that has changed the course of societies. Why the ideas jump out of the mind is only minimally understood by the scholarly who study mind. Since human beings are spread all around the Earth the new ideas are jumping out of the human mind all the time all around the Earth. Some ideas are pursued further, developed and implemented by the person who gave birth to them or by the other members of the society. Depending on the nature of the idea and the nature of the society the implementation of ideas may take generations. The persistence in pursuing the ideas builds the knowledge base of the society and leads the society on the path of progress. Therefore, the new advance in human knowledge in a society generally is not in isolation and not completely independent.

All new ideas need further nurturing by others or they will not take root. The implementation of the accumulated knowledge-base by the previous generations of a society serves as the precursor for the advancement of knowledge for the generations to come. The humans having existed on the Earth for millions of years as suggested by modern science and the human erudition still being far from the level of omniscience, repetition of similar ideas in different locales thousands of years apart is highly likely.

In the case of the scientific development of the past several centuries when tremendous strides have been taken, it is several generations of scientists and engineers that have contributed to the present level of advancement around the globe. One scientist's theory, invention or methodology has served as the base for the next level of development. Newton's idea of gravitation and the universal gravitation-constant that he introduced, was there since 1687 CE before becoming a part of Einstein's Theory of General Relativity that Einstein put forward two centuries later in 1916 CE. Einstein's Relativity Theory was there for Georges-Henri Lemaitre to propose in 1927 CE his theory for the origin of the universe; his theory being the Big Bang Hypothesis. That first story of the birth of the universe was by no means completely convincing. The framework of the Big Bang Hypothesis needed to invoke Quantum Mechanics that was already put forward in 1900 CE by the German physicist Max Planck (1858-1947). A hypothesis being a new idea that jumped out of the human mind faces hurdles in its implementation. Overcoming these hurdles may require implementation of the old and the new ideas. Implementation of such ideas may lead to the inclusion of more hypotheses within the preexisting primary hypothesis. This actually has been the situation in developing the contemporary cosmology theories. However, a general consensus among cosmologists working within the framework of a hypothesis leads to a much needed consolation - that if the hypothesis explained 'most' phenomena of the universe it must be deemed as a 'good' hypothesis and therefore be elevated to the status of a 'physics principle'. In the field of cosmology it is not just the Big Bang Hypothesis that has aspired for the status of a 'physics principle', there are several other hypotheses and methodologies that also have while cosmologists continue to work in that direction. In such a situation the hypothesis that explains 'more' phenomena of the universe with least number of sub-hypotheses in it is said to be more accurate and therefore most acceptable.

Even though none of the present hypotheses in cosmology is proven to be a 'physics principle' yet, several have made great strides in explaining the physical phenomena and the overall behaviour of the universe. The fact remains that as yet not all questions are answered by the existing hypotheses and instead some have even led to new questions. As of now no cosmology theory is complete and the answers to numerous fundamental questions in cosmology remain outstanding.

To review all developments in cosmology one has to attempt to bring in important details from as far as possible in the modern history of cosmology. This has been done by prominent physicists, astronomers and cosmologists through innumerable articles, journal papers and the published books. Therefore, for the sake of continuity and completeness the bare essentials of modern cosmology, that have direct relevance to the age-old Vedic Cosmology Principle and the functioning of the universe as documented by the ancient thinkers in the Vedic Civilization, are presented here.

A limited sample of the science ideas in Veda mantras and the scientific thought in Vedic philosophy was presented in Chapter 1 and their antiquity is established in Chapter 3. There is a span of undetermined millennia between the two cosmologies and therefore the ancient Vedic Cosmology could not have borrowed ideas from the modern cosmology. Yet this author finds similarity between the basic ideas of both disciplines. Once we are through the Chapters 7 to 10 in Volume 2 of this book the reader would appreciate the numerous similarities between the ancient Vedic and the modern approach to physics; cosmology in particular. These similarities are not just in a few fleeting mantra references on cosmology, they are spread through the different subject areas covered by the Vedas.

The ancient Vedic concepts of Nature's fundamental nature and the Vedic definitions of the entities essential for the creation of the universe seem so fresh and young that reading them in the perfect metered unique poetry of the Veda mantras and pondering upon them makes one wonder when exactly in time the first foundation stone of cosmology was laid in the human civilization? There is no time, locale or DNA restriction on the human mind to pop up a brilliant idea. Therefore it is not necessary that brilliant cosmology ideas popped up only during the past several centuries in given locales on the Earth and that the ancient ideas had to be completely dissimilar from modern ideas. There are major differences between the two as well. Even these differences, pointed out in the forthcoming chapters, would show that they are enlightening enough to generate new thoughts for modern cosmology and may even give it a nudge out of its present groove to create new thinking in the subject.

When we follow the differing cosmology principles or hypotheses for the creation of the universe, the common factor is the matter of the universe, the Nature. Cosmology looks at the behaviour of this matter which appears in the form of the fundamental quanta, the energies and forces at their subtle levels and the colossal cosmos at their macro level. The attempt in cosmology theories is to explain and predict Nature's behaviour on both sides of the timeline. This includes announcing how it all started, how it was, how it will be and how it will end, if it ends. Cosmology has to tell us what the matter is, was it there before the trigger of the Big Bang or was it created out of nothing? Several such questions have been raised in Chapter 1 as well.

To describe the modern cosmology hypothesis it has specific terms and definitions for forces, energies and the state of matter such as matter-antimatter, virtual and real particles and so on. The Vedic creation of the universe also begins with a bang. It could only be considered adequate and worthwhile if it had at least the necessary and appropriate terminology to describe it. Several terms of the Vedic Cosmology were introduced in Chapter 1 and its richness will become obvious in the chapters to follow. Recall that the Vedic universe in the creation phase (*srishti*) is when the matter is in the definable state (*Sata*-state) and it is in dissolution phase (*pralaya*) when matter goes back to its indefinable state (*Asata*-state). To show the clarity in Vedic thought on the existence of the two states of matter in the two phases of Nature, a part-mantra from RigVeda's *Bhavavrittam sukta* is invoked here:

> "*Preceding the srishti (universe in creation phase), it is not that 'Asata', the indefinable matter was not in existence, it is the 'Sata', the real definable matter that did not come into existence. Then neither the 'rajah', the observable matter and energy existed nor existed 'Vyom', the Space spread out yonder in all directions, the place for cosmos....*" (RigVeda, 10/129/1).

The Nature in Vedic cosmology is exclusively either in dissolution phase or in creation phase. No inter-mingling of the two states of matter ever exists at the same time. The Vedic premise is that if the inert Nature in inactive *Asata*-state of matter was left alone, it will never begin the creation phase of the universe, because, after all what would trigger any change in the *Asata*-state of matter? Consequently the Vedic argument is that an extrinsic trigger of energy was needed to send the inactive *Asata*-state of matter to the active *Sata*-state of matter. This is why the Vedic creation is theistic wherein the switching of matter from one state to another is instantaneous, as fast as a thought. The needed extrinsic thought-energy (mind over matter) comes from the all-pervading Consciousness. A related question on modern Big Bang Hypothesis remains outstanding – what triggers the dimensionless 'space-time singularity' to instantaneously create all matter of the universe, 10^{60} kg of it, within 10^{-34} seconds?

The above mantra points at the emblems of the creation phase; the *rajah* representing the observable matter and energy in the form of the cosmos and the effulgence of the universe and the spread out *Vyom*, the Space-entity. The *Vyom* by Vedic definition being the Space-entity its function according to the Vedas is to hold the objects of creation. Therefore, the spread out *Vyom* with observable energy and cosmos embedded in it is yet another emblem of the creation. The Vedic thought in the mantra basically clarifies that all these emblems of creation phase did not exist in its dissolution phase. Philosophically expressed, the matter in *Asata*-state is not into 'being' and while the matter in the *Sata*-state of the universe has come into 'being', it becomes observable in the form of *rajah* the observable matter and energy in the cosmos. In Vedic cosmology the Nature never comes to an end into nothingness, only its state changes from colossal (*Vyom*) in creation to *tuchchhaya* (diminutive) in dissolution.

The mantra emphasizes that the *Vyom* (Space) did not exist then as *"spread out yonder in all directions"* but it did exist in a diminutive confinement. The conterminous Vedic Space-entity with the property of 'stretchability without snapping' being an 'eternal' entity, never disappears into nothingness as a 'space-time singularity' of the modern cosmology. More insight into these Vedic concepts is to follow in Chapters 5 and 6.

What is being said in these mantras about the state of matter (Nature) and the Space-entity is in many ways similar to what is being said in modern cosmology, the difference being in the terminology and definitions of the states of matter. With regard to the geometry of the universe in its *Asata*-state when the *Vyom* is not yet *spread out yonder in all directions* should also not sound strange to a contemporary cosmologist. Even though the Vedic principle with its own space-time confinement of Nature, before it transits into the creation phase, is different from the 'space-time singularity' of the modern Big Bang with no space or time in it, the Vedic space holding the Nature in *Asata*-state is addressed as *tuchchhaya* which means diminutive relative to what it is now as the vast *Vyom*. Thus, in both mechanisms, the modern Big Bang and the Vedic Big Bang, the universe begins from diminutive Space and extends into the vast spread-out *Vyom*. This happens because the expansion of the universe is a pre-requisite in both mechanisms.

More about the Vedic Creation Principle is revealed in Chapter 6 with full interpretation of all mantras of the *Bhavavrittam sukta* (129th) of RigVeda's chapter 10. There are specific statements and considerable terminology that exists in these mantras that help create a scenario for the creation process of the universe (in Chapters 7 to 9). Here it would suffice to say that both states of matter at their fundamental level, the *Asata* and *Sata,* are in discrete form made up of fundamental particles called *paramanu,* that cannot be divided further, that serve as the ultimate tiniest building blocks of matter/universe. The *paramanus* in the creation phase enter into their active phase.

CHAPTER 2: Present Understanding of the Universe – A Critical Scrutiny of the Contemporary Big Bang Hypothesis

There have been attempts to theorize the creation of the universe with and without a Creator. A big name in Christian Philosophy is St. Augustine (354 - 430 CE) who put forward the principle of '*ex nihilo*'. According to his *ex nihilo* principle, God created the universe out of His mere omniscience and omnipotence without any pre-existing matter or physical energy. St. Augustine believed that nothing except God existed before the creation of the universe. This means that no preexisting material-cause of the universe was needed to create the universe. St. Augustine's *creatio ex nihilo* (creation out of nothing) sounds very much like the universe out of the Big Bang's space-time singularity, which by definition is a 'nothing'. The difference however is that in the modern Big Bang Hypothesis St. Augustine's God is out and the Quantum Mechanics followed by Einstein's set of differential equations are in. Contrasting with these theories is the Vedic creation principle where the material-cause of the universe is the Nature-entity manifesting in its diverse forms as matter, energy and forces from its preexistence in another state. Obvious from the above part-mantra from RigVeda (10/129/1), to be elaborated in Chapter 6 later along with other mantras in Chapter 6, is that in the Vedic cosmology, though God is the efficient-cause of the universe the preexisting matter is its material-cause; there is no place for '*ex nihilo*' principle of St. Augustine where 'nothing' except God existed before the creation. In the context of the creation of the universe, the phrase of the RigVeda mantra (10/129/2), आनीदवातं स्वधया तदेकं (*aneedvaatam svadhaya tadekam*) states that "… *Present then* (before the creation) *with Svadha the eternal Nature is the One and none other like Him…*". This means that present before the creation was the Creator (*one of a kind*) with *Svadha*; where *Svadha* is the Vedic synonym for Nature; 31 in Table 5.1 in Chapter 5. In the Vedic *kaaran-kaarya* (cause-effect) the Creator is the 'efficient-cause' and *Svadha* (Nature) is the 'material-cause'.

Now let us jump from the '*ex nihilo*' principle of the 5th century church arena to the 17th century science arena and comment briefly on the Clockwork Universe proposed by Sir Isaac Newton.

Clockwork Universe Theory of Newton

Sir Isaac Newton (1642-1727 CE), is a big name in science. His interests and expertise covered many branches of science and philosophy - physics, mathematics, astronomy, alchemy, natural philosophy and theology. He published a remarkable book in 1687 CE, called 'Philosophiæ Naturalis Principia Mathematica' that changed people's thinking and brought transformation in science. It is this book wherein he described his theory of gravitation and presented his famous three laws of motion that established the foundation for Classical Mechanics. He conducted many experiments of fundamental nature and discovered several laws of Nature. He showed that there was universality in natural laws that governed the motion of heavenly bodies and he advanced people's knowledge by putting forward ideas such as the conservation of momentum and energy. Newton's three laws of motion are well-known. Newton is credited for discovering gravitation and for formulating the 'law of universal gravitation'. Recall that with reference to Rishi Kanaad and his ancient Vaisheshik Sutra eons prior to Newton a brief discussion on *gurutva* (gravitation) was included in Chapter 1 under the sub-heading of "Earth's Gravitation - a Vedic Concept". The Sanskrit term *gurutva*, as treated by Rishi Kanaad in his Vaisheshik Sutra, **stands for nothing else but 'gravitation'.**

Newton's Law of Universal Gravitation:

> *"Gravitational force between two masses is proportional to their product and inversely proportional to the square of the distance between the masses."*

The gravitational force between two masses is due to the natural property of matter-matter attraction. The law of gravitation can easily be understood from the following diagram (Figure 2.1) where two static masses M_1 and M_2 with uniform density are placed at a distance R from each other.

Newton's law states that the mathematical relationship between the force of attraction F between the two masses M_1, M_2 separated by the distance R is given by the equation,

$$F = G\frac{M_1 M_2}{R^2}$$

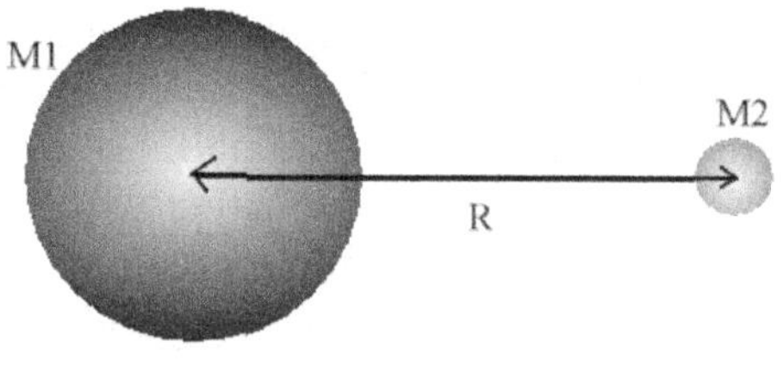

Figure 2.1: Two masses M1 and M2 separated by a distance R.

Here G is the constant of proportionality between the two sides of the equation. It is called Newton's Universal Gravitation Constant. The term universal is applied to G because its value is non-variant throughout the universe, be it the gravitational attraction between the Moon and the Earth or between two far apart massive galaxies. The distance R between the masses being in the denominator and squared, it is apparent from this equation that when the masses are separated further the force of attraction F decreases faster. If the masses weigh 1 kilogram each and are separated by a distance of 1 meter, the force F = G, the Gravitational Constant itself. The value of G is about 6.673×10^{-11} $m^3kg^{-1}s^{-2}$ (meter-kilogram-second unit). In practical terms this strength of the force is extremely small. One would not even detect any apparent pull between 1 kg masses. For this force to become discernible at least one mass has to be enormous, such as the Earth weighing 6.0×10^{24} kg. The gravitational force F on a 1 kg mass on the surface of the Earth with its radius being R= 6378.137 km is about 9.84 $m^3kg^{-1}s^{-2}$. This number looks small but in comparison with the force between the two 1 kg masses, it is about 150 billion times in comparison. Owing to the enormity of the force due to Earth's gravitation the pull on a small mass becomes discernible when the mass is let go from hand – it is attracted towards the surface of the Earth. If Earth's surface-hardness permitted the mass to keep moving, it would end up reaching the Earth's centre where the gravitational pull on the mass is the same from all directions.

When the distance R between the two masses M_1 and M_2 does not change with time the state of the masses is said to be static. When the state of an object changes from static to dynamic, it is said to be in motion. A moving mass possesses two distinct properties, the momentum and energy. Consider a mass M kg (kilograms) moving with a velocity v m/s (meters/second) in a linear direction, its linear momentum is the product of its mass and its velocity, Mv (kg.m/s). Therefore, a small object with high velocity can have as much momentum as a big object with low velocity.

In addition to the linear motion in an object it could have an angular motion on a curved path, the curvilinear motion; like Earth's orbital motion around the Sun. The motion on a curved path is said to be angular motion because the object in motion keeps changing its angle with respect to a fixed point; the Sun in the example of the Earth. If the Earth's angle with respect to the Sun is changing at a rate of ω radians/sec (Greek letter ω is Earth's angular velocity around the Sun. Therefore its angular momentum is Mω (kg.rad/s).

The other term in mechanics is the object's energy owing to its motion. It is called the object's kinetic energy. Without going into the details of mechanics the relationship between the kinetic energy E_{KE}, its mass M and its velocity v or ω is given by the equations,

$$\text{Linear } E_{KE} = \frac{1}{2}Mv^2, \qquad \text{Angular } E_{KE} = \frac{1}{2}M\omega^2$$

The terms such as linear momentum, angular momentum, kinetic energy, gravitational force and gravitational field will be mentioned frequently in the forthcoming chapters. The universe is held together due to the property of gravitation in matter pulling all matter inward.

Newton had deep faith in God and did not exclude the Creator from the creation (universe). His contribution to the philosophical interpretation of Biblical texts is considerable and he believed that the Universe and the physical laws that governed it were created by God. He put forward the "Clockwork Universe Theory". The implication of this theory was that the whole universe was like a clock wound up by God and it continues to run like a perfect machine with preset laws of physics enabling its perpetuity. He did not elaborate on how matter existed before this creation or what the underlying mechanism was for this clock (universe) to emerge. As it is observed by the scholarly that the "Clockwork Universe Theory" was merely a philosophical statement. Factually, the basic philosophical idea that God was the Creator and the Prime Mover of the universe and that He did this without any external interference in its operation, was around for centuries prior to Newton. Newton's era being the period of 'enlightenment' in the Western philosophy where the 'age of reason' dominated, the clockwork universe theory of Newton was well respected because his laws of gravitation and motion were able to predict the dynamic behaviour of the solar system and the Nature itself as understood during his time. Newton knew that if the universe and its course were already predetermined it could lead to atheism because God in a pre-wound clock-universe was redundant. Several factors seemed to work against Newton's principle for the creation of the universe, the free-will factor in living-beings generally violating physical laws, being the most talked about at the time.

When the two laws of thermodynamics were established, they did not help the "Clockwork Universe Theory". The two laws of thermodynamics are:

(1) *Energy is conserved*, and

(2) *The total entropy (disorder) of any isolated thermodynamic system tends to increase with time, approaching a maximum value.*

The second law of thermodynamics overruled the clockwork principle of the universe, because there was no room for disorder in a universe that was pre-wound by God. Also, along came the 'Uncertainty Principle' in Quantum Physics, which said that nothing in the universe can be predicted with complete certainty. Newton's "Clockwork Universe Theory" did not proceed further.

Accumulation of knowledge leads to inheritance of knowledge. The modern science is richer in knowledge owing to Newton's contribution to that pool of knowledge. Even though his "Clockwork Universe Theory" had no sound mathematical or scientific footing, the current science is indebted to Sir Isaac Newton for advancing the knowledge through his discoveries in different areas and laying a foundation stone for building modern science. In the post-Newton era, while considering new cosmology theories, Newton's idea of God as the Creator and the Mover of the universe was left behind and also left behind was his Clockwork Universe Theory. Yet his laws of motion and law of universal gravitation and much more that he contributed to physics continued to be extremely valuable. Due to the Clockwork Universe Theory *'the baby was not thrown with the bath water'*. Rather the baby was only the beginning of what was to come and what was yet to be known. Present theories of the universe, be it the Continuous Matter Generation, the Big Bang or the Strings, they all owe Newton for laying a foundation stone of cosmology and enriching the inheritance pool of knowledge in the Western civilization.

The question arises whether the foundation stone laid by Newton is the first for the building called cosmology? Certainly not! There have been foundation stones laid earlier but not necessarily part of the present building. They belonged to an entirely different building in a completely different locale. This locale was in the East. The foundation of that building is still preserved in the form of mantra formulations. Although some stones of the building have gone to other regions of the Earth, still traceable by linguists and historians, the foundation stones have remained behind. These foundation stones are enriched and preserved in the form of the Vedas, a quick commentary on them was provided in the previous chapter. The *rishis*, *munis* (sages, the thinkers) of the Vedic Civilization laid these foundation stones and built on them unique literary monuments in the far antiquity of the human civilization. They did so not to gain recognition, praise or fame but to leave behind this literature in Sanskrit as their gift to humanity for advancing knowledge and to encourage further development.

The *rishis* of the Vedas have stated the principles and the laws that may have implicit assumptions, which may not be suitable for mathematical manipulation or may be non-provable in a physics experiment, but they certainly are at a high intellectual level. They have given consideration to the ingredients and forces necessary for the creation of the universe and based their deductions on the universality of 'cause-and-effect'. A glimpse of this was presented in Chapter 1 under the sub-headings of "Some Cause-Effect Specifics in Vedic

Philosophy" and "The Vedic *Kaaran-Kaarya*: Cause-and-Effect Principle". Disagreements on the mechanism of creation are a certainty but modern cosmologists would not have disagreement with the Vedic principle that creation is not possible without the real existence of matter.

Immeasurability of an Entity

Sometimes we come across statements in the science arena that are true only to a point. At times a criterion applicable to a specific physical system in one realm is generalized for differing realms. For example a statement such as - *'Only the things that can be observed or measured have their existence in the universe'* is not strictly true. The entities like Time and Space are real but not observable and what we measure on Time and Space are their intervals and not their absolute quantities because their beginning is not known. The other limitation is that one is not able to cut a piece of these entities to put aside for later perusal and experimentation. When one moves from physical to metaphysical realm one comes across other forces and phenomena that are incomprehensible; consciousness being one that is immeasurable with tools made of matter. Therefore, owing to these limitations one tends to conclude that the consciousness phenomenon may not even exist.

Here is an interesting conundrum pertaining to immeasurability. Let us revisit Newton's formula for gravitation. Assume that all parameters on the right side of the equation are known. The gravitational constant G is about 9.84 $m^3kg^{-1}s^{-2}$, small mass M_2 is 1 kg, the Earth's mass M_1 is 6.0×10^{24} kg and Earth's radius R is 6378 km. When the small mass M_2 is on Earth's surface the distance R between the two masses is the same as Earth's radius and the calculated force on M_2 is approximately 10^{13}, directed towards the centre of the Earth.

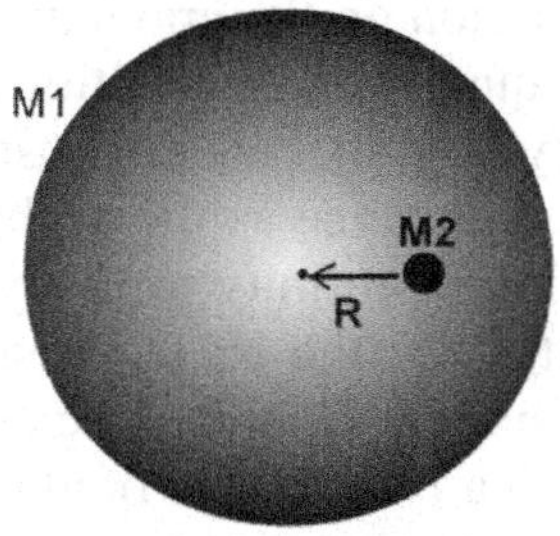

Figure 2.2 A

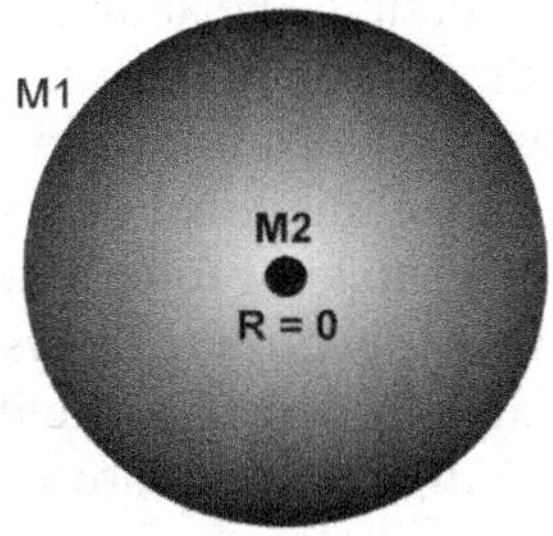

Figure 2.2 B

Now let us move the small mass M_2 inside towards the centre of the Earth (Figure 2.2A). Because R is in the denominator of the equation, as R decreases the force F on M_2 must increase. Finally if the M_2 is placed at the centre of M_1 the R becomes zero (Figure 2.2B). Dividing by zero means the force F = ∞ (infinity). Also the mass of the Earth is in all directions of M_2 the force must be pulling M_2 equally in all directions. This also means that there is no net directed force on M_2 pointing in a given direction. Since the two masses have not changed, therefore the force F cannot be infinity. This is a case of **immeasurability**.

We will now do one final hypothetical experiment with the masses M_1 and M_2. Let the mass M_1 be the mass of the universe, 10^{60} kg, as suggested by cosmologists. Let M_2 be any small mass. The mass of the observable universe is said to be spread out into a sphere of radius 4.4×10^{23} km. For all practical purposes the boundary of such a sphere is at infinity and the mass M_2 for all practical purposes is situated at the centre of the observable universe, as in Figure

2.2B. The contemporary astronomers also suggest that mass of the universe is uniformly distributed all-around. Therefore, a small mass M_2 embedded in such a universe will stay put wherever one puts it and will not experience any net gravitational pull to the left or right or up or down. Considering the large dimension of the universe, moving M_2 from one point to the other will not change the outcome either. An observer of the mass M_2 would quickly conclude that as far as the gravitational force on it is concerned, the mass M_2 was not gravitationally interactive with the rest of the universe even though the mass of the universe does have its own gravitational field. Or at least it is as good as not interacting with the rest of the mass of the universe. Since the observer is unable to measure a change in gravitational pull on M_2 the observer is unable to use M_2 as a probe to infer mass M_1 of the universe using Newton's equation even if other parameters in the equation were given. This leads the author to formulate a simple physics *sutra*!

"The true knowledge of an entity is possible if the knowledge gathering instrument is able to interact with the entity."

This requires one to make sure that one is using the right tool to detect the right entity. The above discussion on the gravitational field tells us that when one deals with infinitely large quantities and entities one runs into the limitation imposed by 'immeasurability' and the modern cosmology is not immune to it. To visualize this, treat the entire universe as the probing mass M_2, embedded in yet another field of force extending to infinity well beyond the universe. The universe with all its physical laws sitting at the center of such an unknown field may not be able to probe that field. The physical universe and its laws may not even be the right instrument to probe such a field. If this field is of metaphysical nature or it is a non-physical pure consciousness, probing it with physical means will be an exercise in futility. Therefore the author draws another physics *sutra*:

"*Futility of detection of an entity is not a proof of the absence of the entity*".

In addition to the *gurutva*-field (gravitation) owing to matter of the universe the theistic Vedic philosophy/physics gives consideration to yet another all-pervading conterminous energy field to which applies the 'immeasurability'. It is said to be a formless, Supreme Conscious Being (Creator), all-pervading energy field that binds the universe together. Just as a human-being with the inherent property of consciousness has the ability to create thought, alter the chemistry and physics of his own physical body, alter matter and physical events within his area of influence and create tangible things that did not exist before, a Supreme Conscious Being with His area of influence encompassing the entire matter of the universe could trigger an event such as the creation of the universe. This is the Vedic premise for the existence of the 'efficient-cause' of the universe which does not apply to an atheistic perspective of modern physics/cosmology. The cosmologists under the Multi-universe hypothesis have theorized the existence of 10^{500} universes outside ours, forming so-called a 'tree of bubbles' (universes). Because there is no positive proof of Multi-universe hypothesis, the immeasurability applies more to Multi-universe Hypothesis than to anything else.

The fact is that none of the infinite and semi-infinite entities, in the realms of physics or metaphysics, is measurable. Regarding the immeasurability of an entity let us consider the case of electromagnetic radiation (light).

Fast Particles – Photon and the *Munn* (Mind-entity)

When one sees a car at its maximum speed one thinks that the car must have been at rest first, then began moving and gradually accelerated to attain a maximum speed. On the other hand when a photon (quantum of light) is produced by an 'excited atom', the atom does not lay the photon next to it like a hen lays an egg! The newly born photon does not remain next to its mother-atom, instead it instantly moves away from the atom. One would think that initially the photon must have been at rest before picking up speed to attain the speed of light, a speed which, according to Einstein's General Relativity, is the maximum attainable by any object in the universe. The fact is that the maximum attained speed of the photon is under measurability but its acceleration, if it exists, is under immeasurability. A photon is thought to shoot out from the emitting atom at the speed of light. One cannot imagine that a photon, like all other objects, attains the maximum speed without some initial acceleration. The maximum speed of the photon in vacuum is 300,000 km per second and a bit less in the air and in the other media. To understand how fast a light photon travels one could compare it with something fast and familiar like a 747 passenger jet propelled with four jet engines at a speed of about 900 km/hour. While it may take about nonstop 45 hours for the jet to complete one round of the globe of the Earth, a photon of light in the same time would have gone around the Earth 21,630 times.

The question, as to why a photon unlike all other objects, has the property to travel from its birth to the death with one speed without initial acceleration, is not an easy one. A photon of light is considered to be a mass-less particle which when shoots out at this speed will maintain its speed, if not hindered by an obstacle or by the medium itself. Once on its way, this packet of energy never gets tired. A natural question therefore is - where does a photon continuously get the energy replenished for its travel to infinity? As it is travelling, nothing from outside is put into it and nothing goes out of it. The photon instantly attaining speed at its source without any sign of initial acceleration in it is also not because the parent atom or a force within the photon itself gives a tremendous initial push to the photon. This speed of light is the same in any direction.

It may be mentioned here that there is a hypothetical particle that is said to travel faster than a photon – it is called the "tachyon".

In 1962, Bilaniuk, Deshpande, and George Sudarshan first proposed the possibility of particles moving faster than light. For such particles they used the term "meta particle". In 1967, Gerald Feinberg gave such a hypothetical particle another name "tachyon", derived from the Greek word *tachys* which meant "swift". The name tachyon is now used for a particle moving faster than light.

The existence of tachyons will be in violation of the known laws of physics because nothing should travel faster than a photon. The research on tachyons however continues. Scientists at the European particle accelerator announced in 2011 that they had detected a tau-neutrino particle that had travelled faster than light; however, later they announced that it was not a tau-neutrino; rather it was a misdetection owing to *a faulty element of the experiment's fibre optic timing system*!

There is yet another particle in the metaphysical realm, considered in the Vedas and the Vedic philosophy, which not only is said to travel faster than light, it does not need acceleration and reaches its destination instantaneously. Its description in the Vedic philosophy fits into the modern physics expression – "action at a distance". It is called the "*munn*" (pronounced as 'fun'). It is associated with the conscious-being and resides in the *anteha-karan*, the interior-sense-function in the brain of the living-being. A *munn* is associated with consciousness and not with inert matter or energy. Numerous mantras and sutras refer to the speed of *munn*. The mantra references specifically draw its comparison with a photon of light (*mareechee*). The *munn* quantum has instantaneity to acquire infinite speed. It interacts with matter in limited regime without being affected by matter.

That a *munn* is minute, mobile and a unitary entity, one for each conscious-being, has been stated time and time again in the Vedas and Vedic philosophy. So that a Vedic reader understands with clarity that the *munn* is a real dynamic entity and that it has a wide outreach and deep entry, YajurVeda (34/1-6) has devoted six successive mantras on *munn*. The RigVeda in addition to several other mantras in other *suktas* has devoted (10/58/1-12) a full *sukta* with its *devata* being *Munn Aavartanam*. All 12 mantras in this *sukta* are in praise of the *munn* and together they create a picture telling us that all the subjects mentioned in these mantras are in the reach of the *munn* and that after exploration of the spiritual inner world of a living-being and exploration of the exterior physical world, it returns back into the *anteh-karan* (interior senses) of the body where it belongs.

This author's interest in including a metaphysical Vedic concept, the *munn*, in this section is its relationship to physics. First let us look at the sixth of the 12 mantra of RigVeda (10/58/6), which is related to the *munn*'s speed and then summarize the thoughts contained in the remaining eleven mantras,

यत्ते मरीची:[2] प्रवतो[3] मनो[1] जगाम दूरकम् ।
तत्त आ वर्तयामसीह क्षयाय जीवसे ॥(RigVeda, 10/58/6)

Transliteration: "*Yatte mareecheeh*[2] *pravato*[3] *mano*[1] *jagaam doorkam* |
Tatta aa vartyaamasiha kshayaaya jeevse ||"

Mantra Translation:

"*O man, your munn*[1] *travels afar conquering*[3] *(pravato*[3]*) over mareechee*[2] *the fast-moving-light-rays*[2]. *I reward you your munn*[1] *to dwell in your living body all through your cyclic birth process*". (RigVeda, 10/58/6).

The term मरीची:[2] (*mareechee*), in a Sanskrit dictionary and in mantra interpretations by ancient Sanskrit scholars, stands for '*particle of light*' which is no different from the physics term - 'photon'. The meaning of मरीची:[2] as a '*particle of light*' has not entered into the Sanskrit dictionary after the term "photon" was coined in physics in 1916 or after the corpuscular nature of light radiation was recognized in physics. The Sanskrit term *mareechee* is as old as the Vedas and its meaning '*particle of light'* is also as old as the Vedas. Therefore the Veda treating light as a discretized phenomenon, by itself is remarkable. In whichever way one looks at the term *mareechee,* it represents a physics definition of light photons. At this point, this author is tempted to say that the Vedic *rishis* through the mantras have presented deep insight of a phenomenon in physics - mind-boggling indeed. There is much more to this mantra than just the definition of a photon. Please read on.

In addition to the term *munn*[1], the other important term in the mantra is प्रवतो[3] (*pravato*). The elemental term प्रवन् with verb conjugation yields the term प्रवतो[3] (*pravato*); the *munn*[1] being the subject. Based on other mantra interpretations and on the ancient dictionary of Vedic terms, a translation of प्रवतो[3] is '*one that conquers*'. The other same root-based translations are - '*one that gains on*' as in a horse race, or '*one that aims at*' or '*one that targets something*'. That something in the mantra is *mareechee* the photons of light and the conqueror is the *munn*[1]. At this point the simple logic becomes obvious. When an archer aims his arrow at a moving target, for the arrow to 'gain on the target', the arrow has to travel faster than the target itself or the whole exercise of shooting an arrow is meaningless. In this mantra the chaser is the *munn*[1] and its target is *mareechee* – the light photons. One does not attain victory over an adversary without catching up to the adversary. Therefore, the logical conclusion of the RigVeda mantra is that the *munn*-entity of the Vedas, associated with the consciousbeing, travels faster than light.

The implication of the second line of the mantra is that the *munn* of a person repeatedly goes out and comes back to the same person without getting lost in the universe and that the Creator endows it cyclically to the conscious-being every time he/she/it is born through the cycle of birth-death-rebirth.

The most important find here is that the *rishi* chose light rays to compare the speed of the *munn* and not the wind or a fast running horse or a flying arrow. To choose light, the fastest known entity in physics, for comparing it with the speed of *munn,* is simply mind-boggling.

There is no equipment or experiment in the physical sciences that could measure the speed of this metaphysical real entity called *munn*. The question here is – how can something like the *munn* with the unusual properties defying physics laws exist, claiming to be residing in the head of the living-being and not be detected? This question is meaningful when opined in the frame of reference of modern physics with a premise that all things and phenomena of the universe are explainable in physics frame of reference. The premise however is not all-encompassing. There are many things in the universe that are not understood! Several pertinent questions were raised in Chapter 1. Additionally - why is there the phenomenon of 'desire' in living-beings at their varied levels of evolution and why is the desire only in living-beings and so on? There may be reference frames other than the physics reference frame that we have not discovered yet and there may be reference frames that we do not have the ability to discover. There are many logical questions that belong to different frames of reference. Therefore, what if the *munn* is a real phenomenon outside the 'measurability' condition of modern physics?

The Five Principles of Modern Cosmology

A contemporary theory of the universe is only the theory of the 'observable universe' that the astronomers are able to observe with their most powerful telescopes. A theory that explains most of the observations from the observable part of the universe is deemed a good theory and is then generalized for the rest of the universe that might exist beyond the observable universe. Since the existence of conscious-beings is a cosmological reality, a cosmology theory is incomplete if this reality is not addressed in it because one cannot disregard the phenomenon that is responsible for excavating cosmological theories and asking intelligent questions.

To better organize the approach for developing knowledge of the universe, cosmologists have adopted five basic principles to guide them towards the ultimate brilliant discoveries from the cosmos. These principles are listed below.

(1) The principle of uniformity:
'*The laws of physics, verifiable in the laboratory, that apply to the observable objects, apply to all objects of the universe.*'

(2) The Cosmological Principle:
'*Except for local irregularities, the universe is homogeneous and isotropic* (uniform density in all directions)'.

One of the cornerstones of modern astrophysics is the cosmological principle. The basic premise of the Cosmology Principle is that the observers on Earth have no privileged view of the universe and that the laws of physics must be the same everywhere. Many observations back up this premise. For example, the universe looks more or less the same in every direction, having the same distribution of galaxies everywhere we look.

In recent years, however, some cosmologists have begun to suspect that the Cosmological Principle may be wrong. They point to the evidence from the study of Type 1 supernovas, which appear to be accelerating away from us, indicating that the universe is not just expanding but accelerating away. The curious thing is that this acceleration is not uniform in all directions. Instead, the universe seems to be expanding faster in some directions than others. But how good is this evidence? Is it possible that the preferred direction is a statistical mirage that will disappear with the right kind of data analysis?

Rong-Gen Cai and Zhong-Liang Tuo at the Key Laboratory of Frontiers in Theoretical Physics at the Chinese Academy of Sciences in Beijing have re-examined the data from 557 supernovas throughout the Universe and re-crunched the numbers. They confirm that the preferred axis is real. According to their calculations, the direction of greatest acceleration is in the constellation of Vulpecula in the Northern hemisphere. That is consistent with other analyses and also with other evidence such as other data showing a preferred axis in the cosmic microwave background radiation (CMB). This has caused the cosmologists to rethink about the validity of the Cosmological Principle itself. But it also raises exciting questions: why does the universe have a preferred axis and how should it be accounted for in the Big Bang model?

More will be said on this preferred axis of the universe in Chapter 9 on the Vedic Big Bang Scenario where mantras after mantras state that the universe is a spinning system, spinning around a *Skambha* of energy, the physical axis of the universe. A spinning universe is a unique Vedic idea where the *Skambh* is the pillar of energy going through the centre of the universe, serving as its axis of rotation. This is in the likeness of a bar in the centre of a galaxy around which the entire galaxy spins. The consequence of this Vedic idea is discussed in detail in Chapter 9 where it is also pointed out that the hypothetical Dark Energy, thought as the only possible expanding force of the universe in the modern Big Bang cosmology may not be a reality. The rotation of the universe right from the onset of the universe along with the property of gravitation can explain the expansion of the universe and the associated red-shift in the spectrum of the far away galaxies.

(3) The Anthropic Principle:

'The laws of nature and the values of the fundamental constants of physics are so critically just right that life is enabled on Earth and the intelligent conscious-beings are capable of observing the universe'.

(4) The Principle of Equivalence

'All bodies made up of matter or anti-matter have positive or zero mass and that the gravitational mass and inertial mass are equivalent'.

(5) The Mach's Principle:

'The inertia of every body is by the distribution of matter in the universe'.

These principles help in building a cosmology theory except that for most cosmologists the Anthropic Principle (AP) continues to be a sticky point. This is because a scientist has to explain why the probability of life occurring on the Earth is tied to the laws of Nature so much so that even minute changes in them will not enable life on the Earth – it is as though the Nature's laws were designed to have life on the Earth. Most scientists think that there is nothing special about having life on the Earth because life on the Earth is just another accidental phenomenon. Most biologists are completely satisfied with Darwin's theory of evolution of the species leading to higher intelligent beings, the humans.

The problem with regard to the AP that a cosmologist/physicist faces is the one created by cosmology/physics itself. Physics has yielded several universal constants related to the nature of matter and the laws of Nature, measured as well as calculated. The values of these universal constants are so critical that if they were slightly different, the life on Earth would not exist and therefore intelligent life on Earth would not exist in the first place to derive these constants. So, the problem is – why is the universe in such a critical balance? Under the AP they feel that they have an obligation to explain this phenomenon of criticality of physics constants leading to the existence of life. While the other four cosmology principles help set up the boundaries of the framework for developing cosmology theories, the AP is part of the query on which to keep an eye but is not effective in building the theories. Cosmologists have come up with the explanation for the

AP; however, to find an answer they have resorted to hypothesizing about other universes outside our own wherein there is no life.

Therefore there seems to be an apparent disconnect between the AP and the other cosmology principles. The Vedic perspective on this apparent disconnect is a subject of another book where the discussion leads to a positive connectivity between the two dissimilar realms, matter and consciousness; after all the AP arises due to the existence of the phenomenon of consciousness. The take-off point for the Vedic rishis' creation principle is in the simultaneity of both, the consciousness and the matter.

Because the forthcoming chapters are devoted to the subject of cosmology from a completely different time in human history and from a completely different discipline of thought (the Vedas), it is of benefit to the reader to first know the modern advances in cosmology. The foundation of all modern cosmology theories is the use of hypotheses, at times counter intuitive and unreal. The soundness of a theory is justified by the premise that they are mathematically valid and yield convincing explanations for what is being observed in the universe.

Presenting an in-depth knowledge of contemporary cosmology is beyond the scope of this book; however, it is essential that at least a qualitative overview of some of these cosmology hypotheses is presented here and the broad features are discussed. Wherever necessary the contrasting and relevant Vedic ideas are brought in to show that the latter are the outcome of highly disciplined minds, are scientific in their basic character and are sufficient to portray an interesting picture of cosmology. If nothing else, these ideas make one wonder as to what made those Vedic rishis to come up with cosmological principles so far back in human history. These ideas, devoid of myths or mythical characters, are at a qualitative level but they certainly raise the question – why are they similar to the ideas emanating from the thinkers of modern times while being unique, mysterious and distinctly dissimilar? Not just that, but these ancient ideas pose questions for the modern scientists, which are closely related to the present reality of cosmology.

Modern Cosmology Theories of the Universe

In 1927, Lemaître put forward the cosmology theory according to which the universe had a beginning with an enormous explosion at one dimensionless point. This theory popularly known as the Big Bang Theory is presently the most accepted cosmology theory. However, it is not a final theory yet because not everything about the assumptions in the theory and about the universe are explained by it and there are several problems with the theory that have been pointed out and discussed by other groups of cosmologists. In the following a broad outline of the Big Bang Theory is presented first and then some of its limitations are discussed that have led to uncompromising opposition by a good number of scientists. To strengthen the Big Bang Theory itself the Big Bang cosmologists have introduced several approaches and hypotheses. These include the so-called inflationary theories; vacuum fluctuation theories, time-independent theories, strings theories and so on. Physicists, with all these approaches, have been able to explain several observed phenomena of the well-established universe yet the true nature of the matter of the universe, right after t = 0 seconds of the big explosion, is still not fully understood.

The Time-entity in most of these cosmology theories is crucial. Even though this entity is not understood fully by the modern cosmologists, all things happen within it, thus making it a part of the cosmological development. A fair bit of discussion on the nature of the eternal entities, Time, Space and Nature, was included in Chapter 1. Some cosmologists have recently said that the Time is only an illusion of the mind. Others have said that it is real but began at t = 0 seconds of the Big Bang. A Vedic principle states that anything created in the universe cannot be a continuum. The illusion or a reality the Time is seen as a continuum. If quantized a cosmologist is unable to define its quantum. Pertaining Time there are three types of modern cosmologists: the traditional Big Bang cosmologists believing the birth of Time with the birth of Big Bang, the new cosmologists who work with time-independent field equations and those who consider the existence of Time before the t = 0 sec of Big Bang.

The Big Bang singularity being a non-descript entity, there is no history hidden in it which may be related to Nature's pregnancy period before the time-zero of the Big Bang. The metaphor of Nature's pregnancy is meant a preparation period such as prescribed in Vedic Big Bang when well-described primordial Golden-ovate spun in *paryaplava* (rotation while flotation) around an axis and developed *tapsah*, the heat energy, before the tremendous explosion broke it into 'uncountable pieces'. This period of Nature's pregnancy according to the Vedic references is one *dava-yuga* of 4.32 MY. May be someday the complexity of the cosmology theories of the universe will be as understandable as a reality in our daily lives, like the birth of a newly-born where its physical growth during pregnancy is comprehensible and finally the growth of the body is comprehensible. The Nature itself was that metaphoric egg of matter, waiting to burst opens its shell. Contrary to presenting a simple picture of the advent of the universe the modern cosmology is heading towards complexity. The account of this complexity is beyond the scope of this book. It is intuitive to think that if the increased complexity was a necessity to explain a theory, the theory may have had a wrong start!

The Contemporary Big Bang Hypothesis

If a force is applied to a stationary object the status of the object changes. It is displaced to occupy different space. This displacement relative to a reference point over a given time is seen as the motion in the object. The displacement relative to the reference point and body's new position after a given time are part of a frame. This reference frame is termed as the 'observational frame of reference' or the 'inertial frame of reference'. The branch of physics wherein the behaviour of an object is studied in an 'inertial frame of reference', under the influence of forces and the applicable laws of nature, is called mechanics. Science of mechanics has been developing since 'early modern' period, the period of Galileo Galilei (1564 –1642), Johannes Kepler (1571-1630) and Newton (1642-1727). It is referred to as the 'Classical Mechanics'. With his laws of motion because Newton contributed very significantly to the 'Classical Mechanics' it is also popularly known as Newtonian Mechanics.

In 1905 Albert Einstein with his Special Theory of Relativity revolutionized the way the state of an object was thought. He brought the Newtonian Mechanics into a new realm where two separate entities, the Space and the Time were brought together to be dealt as a single entity called the 'space-time continuum'. According to the Special Theory of Relativity the state of rest and the state of uniform motion, mentioned in Newton's first law, was not absolute, rather it was only relative. Therefore the Special Theory of Relativity is in fact an improvement over Newtonian Mechanics. The objects moving at very high speeds were to be treated with laws at one level higher than in Newtonian Mechanics. This gave birth to the new term 'Relativistic Mechanics'. That the speed of light was the fastest in the universe, as mentioned in the above text, is an important outcome of the Special Theory of Relativity. When the speed of an object is small compared to the speed of light there is no difference between the outcomes of Newtonian Mechanics and Relativistic Mechanics.

In addition to several other outcomes of the Theory of Relativity an important outcome was the equivalence of mass and energy. This means that matter can be converted into an equivalent amount of energy. Mathematically, the mass m and its equivalent energy E are related through the famous mass-energy equation of Einstein, $E = mc^2$ where c is the speed of light.

Einstein introduced the idea of curved space-time in 1916 and published his theory as the Theory of General Relativity which unifies the Special Relativity and Newton's Law of Universal Gravitation. Einstein published a set of ten equations known as Einstein's field equations which take into account the fundamental force of Nature - the gravitational force. He represented the gravitational force as a curved space caused by the presence of matter. It can be said without going into the details of its outcomes that the gravity now was seen differently than before and was just a property of the geometry of space and time. When the universe is thought as nothing but mass, energy, space and time, Einstein's field equations opened a window of opportunity for the physicists to apply them to study the behaviour of cosmos and try to solve the outstanding puzzle; the beginning of the universe.

Although the space-time relationship is applicable to the universe where the embedded heavenly objects are found in perpetual motion, Einstein first believed that the universe as a whole was a static system. Contrary to this belief, a Russian physicist Aleksander Friedman (1888-1925) using the solution of Einstein's field equations published in 1922 his theory of the expanding universe. Georges-Henri Lemaître (1894-1966), a Belgian Roman Catholic priest, a mathematician, an astronomer and a professor of physics, applied Einstein's General Theory of Relativity to cosmology. In 1927 Lemaître published the 'Hypothesis of the Primeval Atom' where he deduced that the universe began with the explosion of a primeval atom and as in an explosion the material started to fly outward. The matter of the universe contained in this dense primeval super-atom burst creating fireworks of galaxies spreading out in a growing sphere from the centre of the burst. He theorized that this explosion, which started at a single point, was the beginning of Time, implying no Time existed before the onset of this gigantic explosion.

The British astrophysicist Sir Fred Hoyle (1915-2001), who was in disagreement with the theory, sarcastically coined the term Big Bang for this big explosion and the name Big Bang stuck to Lemaître's 'Hypothesis of the Primeval Atom'. The astronomers and cosmologists in the recent past had determined differing ages of the universe and with more accurate measurements of distant galaxies they have arrived at a consensus, that the Big Bang took place about 13.8 BY (billion years) ago when it threw matter and energy out in all directions with tremendous speeds. The universe thus is an ever-expanding system since t = 0 sec of the Big Bang.

After Friedman died in 1925, Edwin Hubble in 1929 discovered from his observations of distant galaxies that their light exhibited Doppler shift towards the red part of the light spectrum that they emitted. This meant that the distant galaxies were flying away from each other at an increased speed and therefore the universe indeed was expanding and was far from being static as Einstein had once postulated.

The Doppler shift or Doppler Effect is the change in frequency (or wavelength) observed by an observer when the object producing it is moving toward or going away from the observer. An example of this is an observer standing by the train-track and listening to the whistle of the engine of a fast approaching train. If for example the whistle of the engine was tuned to the middle C note of a piano at the frequency of 261.6 Hertz (Hertz = cycles/sec), the observer will hear the whistle detuned from C toward C-sharp (increased in frequency) when the train was coming toward the observer. When the whistling engine of the train passes the observer, the observer hears the C note and as the train rushes away from him the note C drops toward the lower note B at the frequency of 246.94 Hz. The sound waves travel in the air at a speed of about 1,235 km/hour. The calculation suggests that for one to hear a shift of a complete note from C down to B, the engine has to travel away from the observer with a speed of about 73.4 km/hour. Conversely, if one measures the shift in frequency, one can estimate the speed of the train. The same principle is used in astronomy to estimate the speed of faraway galaxies using the observed shift in frequency of light that they emit (instead of sound in the case of an engine).

The subject of the sense-organ for sound frequencies is the ear and the subject for the sense-organ for light is the eye. As sound is emitted at different frequencies in the audio range, example being many octaves of piano keys, light is also emitted at a wide range of frequencies called a spectrum. A spectrum in the visible range exhibits colours that a human eye differentiates. Seven colours in the solar spectrum is an example. Of this spectrum, the human eye can see from the red colour (lower frequencies) to the blue (higher frequencies). Therefore, if a distant galaxy was observed with its spectrum shifted toward red, the obvious conclusion was that it was moving away from the observer. Even though the ultimate determination is achieved with a sense organ the accurate detection in the spectrum shift is carried out by the photo-electric spectrometer technique coupled with the telescope. Hubble discovered from his astronomical observations of faraway galaxies that their spectrum was shifted towards the red-end of the spectrum. He also observed that the farthest the galaxies the greater the redshift and therefore the faster they were receding away from the observer. Since then the ever-expanding universe became a permanent feature of the Big Bang Hypothesis. However, the source of the red-shift in the spectrum of galaxies is not free of discrepancies. Nevertheless, the Big Bang Hypothesis is the most accepted hypothesis for the origin of the universe.

Lemaître believed that the Big Bang was a unique event and that the universe would keep expanding with cosmos receding from each other as in the case of the fragments of an explosion and will not return to the original point of explosion. If the universe continues to expand it will attain extremely low density and become extremely cold and end up in a 'Big Chill' - a favorite term in cosmology for this scenario. The reason for the fragments of an ordinary explosion on the Earth not returning to the origin is that the origin does not attract them back. If it were not for the massive gravitational attraction of the Earth bringing fragments of an explosion down from where-ever they have reached, the moving fragments would follow Newton's first law and would continue to fly away from each other.

Contrary to the universe ending in Big Chill, Stephen Hawking made a prediction that the gravity of the universe itself would slow the expansion down and make the universe fall back toward its centre. If the universe starts to reverse under its own gravity, it will collapse back to where it started. The popular expression in cosmology for this fate of the universe is the 'Big Crunch'. The Big Crunch should ultimately turn it into the original space-time singularity. Then the natural question is what would happen to it at that point? What will prevent it from exploding again and create another universe? And what happens to the Time and the Space entities? Do they get crunched as well? There is no specific logic to disregard the argument that in a Big Crunch the universe will not go back into its original state of 'singularity' and there is no sound logic to suggest that it would not explode again and begin creating another universe. Therefore, intuitively one can surmise that the creation-dissolution is a cyclic process, just what the Vedic Cosmology Principle has suggested; albeit with differing steps and time epochs.

Some modern cosmologists have in fact started to believe that the universe would shrink to a point in space and another Big Bang would occur. This would make the process repeat endlessly. Philosophically this condition is termed as 'eternal'. The term in Vedic Philosophy for this cyclic behaviour of the universe is *srishti-pralaya-chakra* (creation-dissolution-cycle).

The Big Bang Hypothesis must provide answers to several related questions. If the Time has a beginning at t = 0 sec of Big Bang, it must have an end in the Big Crunch. Does Time die when the Big Crunch comes? What creates the Time and what causes it to age? What causes it to be born again after Big Crunch? For the Time to die in Big Crunch, the Time has to start reversing sometime in the life of the universe and end up going as low as Planck-time to come to its death at Big Crunch. Would it do that? Or instead of reversing, does it face a 'sudden death'. If the answer to the question is that the Time does not die in Big Crunch but continues increasing through the Big Crunch it would create a time-dilemma:

- *"that a new Time is created after each Big Crunch in the presence of the old Time still running, thus rendering infinite number of different Times running parallel".*

The answer – that the Time is what the field equation or a computer algorithm created at t = 0 sec in a given mathematical step, is not sufficient. If the Time is a real entity that participates in the birth of a real universe the answers to these questions have to come from outside the frame work of the cosmologist's field equations wherein there is no mechanism to create time with its initial ingredients.

As has been said before the universe according to modern Big Bang Hypothesis emerges at Time = 0 sec out of nothing when its density and temperature are infinitely high. The temperature is measured in degrees Kelvin (° K). The Kelvin Temperature scale begins at zero ° K which on Celcius scale is -273 ° C. When a substance is at 0° K (- 273 ° C) it is said to be at Absolute Zero Temperature. When it is said that the temperature of space-time singularity, before the Big Bang, was infinitely high it means that it is unimaginably high!

The enormous explosion breaks the Big Bang singularity and within an infinitesimal fraction of a second, between time t = 0 seconds and 10^{-43} seconds (1 second divided by the number with 43 zeros after 1), the universe expands to acquire a size of a ball with radius of 10^{-35} cm (centimeter) and its temperature drops to 10^{32} °K. To put it in common expression, the universe within 10-trillion-trillian-trillion-millionth of a second grows to the size of a dot with a radius of only about 100-trillion-trillian-billionth of a centimeter with the temperature dropping to about 100-trillion-trillian-million °K. Because the Time-entity and the Space-entity are also said to have been born in the Big Bang, they also grow from t = o second onwards.

This negligibly small time duration of about 5.4×10^{-44} sec (about 10^{-43}) is known as the Planck-time after Max Planck who is also credited to be the founder of Quantum Theory. The radius length of about 1.616×10^{-35} meters that the universe acquires at the end of Planck-time is known as the Planck-length. The Planck-time is the duration of time that light takes in vacuum to travel one Planck-length. Therefore it is assumed that the matter of the universe travelled at the speed of light. The state of matter within the Planck Era is such that the regular physics principles do not apply to it and therefore logically no information about its state can be extracted. Physicists define this state of matter as "complete symmetry". By that is meant that up to Planck-time there is no material diversity in the matter of the universe. The whole matter is said to be of one kind, whatever that is! From all standards of measurability and perceptibility the matter of the universe in that time duration is just an indefinable, invisible and immeasurable blob of matter.

The birth of the universe begins with the creation of the building blocks of the future universe. These building blocks are the fundamental particles of matter. This also means that the Nature or the matter of the universe is not a continuous entity. The entire matter of the universe seen in all its states, properties, shapes and colours is basically made up of tiny corpuscles. These fundamental particles are not of many types; rather they are countable on finger tips. It is their combinations in different proportions under different laws of Nature that creates such diversity in Nature from micro to macro level. The hypothesis for the birth of the universe has to keep an eye on the model that explains how the primary building blocks of matter interact with each other and under what laws they create the early universe. This model is the Standard Model in particle physics. The physicists have developed this model over a period of a century which explains how the fundamental particles behave with each other and what the relationship is between each and what forces govern their participation in the universe. A brief overview of the fundamental particles and the forces governing the matter of the universe is presented in the following.

Fundamental Particles

The fundamental particles are of two types, named quarks and leptons. There are six different particles in each type, grouped in three pairs each under the label of 'generation' - first second and third generations. This classification is shown in Table 2.0. The 'first generation' quarks and leptons are the lightest and most stable. The 'second generation' and 'third generation' particles are heavier and less stable. Second and third generation heavier particles quickly decay into smaller and the next most stable particles. It is the 'first generation' quark pair and lepton pair that form the commonly measurable and observable stable matter of the universe. The electron, muon and tau, all have an electric charge and the extremely light neutrinos are electrically neutral. Quarks are not found free in nature. They are all used up in forming the nuclei of atoms of elements of the universe.

The larger particles are formed with multiple quarks and they are called hadrons. Hadrons decay into stable hadrons and other particles. The stable hadrons are the nuclear particles named protons and neutrons. Two up-quarks (u) and one down-quark (d) form one proton which is a positively charged nuclear particle. In the same way two down-quarks and one up-quark form a neutron which is a charge neutral nuclear particle.

Table 2.0: Nomenclature of Quarks and Leptons.

Particle Type				
Quarks		**Generation**	**Leptons**	
1	Up-quark	First generation	1	electron
2	Down-quark		2	electron-neutrino
3	Charm-quark	Second generation	3	Muon
4	Strange-quark		4	muon-neutrino
5	Top-quark	Third generation	5	Tau
6	Bottom-quark		6	tau-neutrino

The smallest atom in Nature is the Hydrogen atom formed by one proton with an electron orbiting around it. The nuclei of elements heavier than Hydrogen have both, the protons and neutrons.

It was mentioned in Chapter 1 and will be discussed in detail in Chapter 5 that for the Vedic science and philosophy the *Prakriti* (matter of the universe) is inert and quantized (corpuscles) into a triplet of fundamental particles called *aapah* with three types within it, the *sata, raja* and *tama*, each with different property. According to the Vedic thought, all substances and forces in the universe, in physical and metaphysical realms are formed out of differing combinations of these three types of fundamental particles. It is mind-boggling that the ancient Vedic *rishis* should have found that matter at its fundamental level is quantized into three types of particles and not a single type. Is it a coincidence that a triplet of quarks must form a stable nuclear particle, proton and neutron in modern physics?

Four Fundamental Forces of Nature

What keeps the tiny quarks and electrons bound together to form the atoms of elements in Nature, what creates conglomerates of atoms and molecules to form great celestial objects, what keeps the celestial objects bound to each other as in the solar system and what ties together the solar systems (stars) in a galaxy to form galaxies? The answer to all these questions is – the forces. Not one or many forces but exactly four different fundamental forces that operate in their individual regimes. After the matter is created in the Big Bang and it goes through the Planck Era, a total of four types of fundamental forces in Nature are revealed that continue as a permanent feature of the universe. Owing to their strength, effective range, nature and properties, they are named as - 'Strong Force', 'Weak Force', 'Electro-Magnetic Force' and 'Gravitational Force'. A few salient features of these forces are mentioned here briefly.

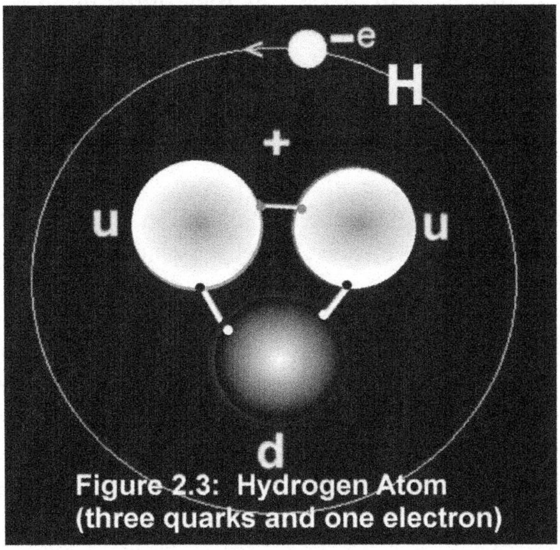

Figure 2.3: Hydrogen Atom (three quarks and one electron)

These forces are distinct, diverse from each other and effective in their respective realms. Not one is derived from the other. The 'Strong Force' and the 'Weak Force' are effective in very short ranges at sub-atomic levels; of the order of 10^{-13} cm. The 'Strong Force' owing to its property of strong attraction holds the nuclei of atoms together. If this force was not as strong as it is, the nuclear particles forming the nucleus of different atoms will fly away from each other putting an end to the diversity in matter. The 'Weak Force' is responsible for radioactive decay of elements. The 'Electro-Magnetic Force' creates all electric and magnetic effects. It is responsible for the attraction between the negative and positive electrical charges and the opposite magnetic poles and it is also responsible for the repulsion between similar charges and similar magnetic poles.

It manifests between pieces of matter that carry electrical charges. Fourth force, the 'Gravitational Force', causes attraction between matter and matter owing to the property of mass in matter. The 'Electro-Magnetic Force' and the 'Gravitational Force' operate up to the infinite range and the former is many times stronger than the latter. The 'Weak Force' is weaker than both, the 'Strong Force' and the 'Electro-Magnetic Force', but it is much stronger than the 'Gravitational Force'. It is the 'Gravitational Force, operating over the cosmological scales, that is most important in the space-time context of the functioning universe. It is this force that keeps the celestial objects bound to each other. These four forces are also the physical laws of Nature governing the matter of the universe. The absence of even one or even a slight change in any one of them will not enable the type of the universe we have or the universe may not even form.

None of these forces that create diversity in nature existed before or during the Planck-time of 0.539×10^{-43} seconds, from the $t = 0$ seconds of the universe, but the matter of the entire universe was created during this short time period. This state of new matter is not definable by physics. This is as good as saying that the state of matter at that point in time is Veda mantras' *Asata*-state of matter.

What Creates Forces in Nature (Matter)?

According to the modern physics / Quantum Physiscs the force between two fundamental particles is created by the exchange of packets of energy. These packets or corpuscles of energy are known as the force-carrier particles. The particle physics term for these discrete particles is "bosons". This exchange is depicted in the figure between the 'u' and 'd' quarks. Each force has its specific boson symbolically shown in Figure 2.3 with small dots on 'u' quarks originated from the 'd' quark and the dots on 'd' quark originated from the two 'u' quarks. The force carrier boson for the "Strong Force" is named as the "gluon" and for the "Weak Force" it is named as the "W and Z boson". The force carrier particle of the "Electromagnetic Force" is the "photon". The force carrier of the "Gravitational Force" is supposed to be a particle of gravity named "graviton", but it has not been discovered yet. Even without the discovery of a "graviton" the "Gravitational Force" is independently dealt with successfully in the Theory of General Relativity.

Physicists believe that all four fundamental forces are unified in the Planck Era into one single force. But unified into what is yet to be discovered by the physicists. In theory anything can exist but it may also be possible that some of these forces were not even born yet in the Planck Era! When this is put to a logical test it reveals a few things. Because there is no diversity in matter in the Planck Era the 'Gravitational Force' owing to mass-mass attraction could not have been born yet. Mass in a particle is said to have been imparted by the well-defined but not discovered yet Higgs particle in undefined matter of Planck Era, the 'Gravitational Force' could not have been there either. A brief commentary on Higgs particle is in the Vedic creation scenario in Chapter 8. Physicists in their effort in developing the Standard Model to unify all forces have been able to unify three of the four forces but the 'Gravitational Force' continues to be beyond the grasp of this unification. The basic question remains – when and why the Nature created these forces?

The Vedic Perspective on Force Transfer

To the best of this author's understanding the Veda has not explicitly defined the four forces. However, the Vedas do have a perspective on the transmission of two forces from one point to the other; the two forces being the electro-magnetic force (light) and the gravitational force. A brief discussion on this ancient perspective is presented below.

In modern physics these two forces are transmitted in the form of waves with their force-carrier particles respectively being photon and graviton. These forces do not need a medium for transmission. They transmit better in vacuum. The electro-magnetic force instead travels the fastest in vacuum with a maximum speed of $3x10^8$ meters per second which is the speed of light. Contrary to this the other forces of mechanical nature like the sound waves do not have a force-carrier particle and need a medium to transfer and do not travel in vacuum. The force of sound creates a compression wave in the medium which travels from one point to another. Thus we observe that some forces need a medium to transfer and the other forces transfer better without a medium. The question is –why is there non-uniformity in force transfer? Based on the fundamental definitions of the entities in Vedic Cosmology this author has understood that **all forces need a medium to transfer whether it is an electromagnetic force, a gravitational force or a mechanical force created by sound waves**. This point is elaborated upon in Chapter 9; here it would suffice to say that in the Vedic perspective there is no such thing as vacuum. The so-called vacuum of space is also filled with the undetectable all-pervading most fundamental particle-field of *Aapah* that serves as a medium for the electro-magnetic and the gravitational forces. All forces create compression waves in the *Aapah*-field to transfer just like the sound waves create compression waves in air to transfer.

Irrespective of how one looks at it the philosophy continues to follow in the proximity of physics to raise questions –

(1). *"How does one explain why the modern quantum mechanics arrived at three fundamental particles (quarks) forming a stable nuclear particle which is similar to the fundamental definition of matter in the Vedas- the triplet of sata-raja-tama?"*

(2). *"What properties or mechanisms in the Big Bang create four forces?"*

(3). *"Why are there four fundamental forces in Nature, why not just one or five? What led to this optimal balance in the count?"*

A fundamental force between the fundamental particles manifests as a property of matter when matter manifests from the undefined state in the Planck Era into a defined state later. The inert matter being the follower of the physical laws it does not have a property or a law unto itself to create the four physical laws and to impose them back on to itself. So, more questions arise:

(4). *"Does the fundamental particle create itself and the property of force within itself, or did the matter and force co-create fundamental partcle"? Which is the 'cause' and which is the 'effect'? Why do the fundamental forces not have their own evolutionary history leading to their most accurate present finalized state?"*

What all takes place in the tiny universe right after t = 0 sec to t = 10^{-43} sec in the Planck Era is still being understood in theoretical physics. The small Planck-time is an extremely small time, yet it is mathematically important and it is this time interval that determines the future of the universe in terms of the nature of matter itself and the final shape of the universe. This small time sector in Big Bang is still a sort of missing link not understood fully like the missing link in Darwinian evolution. In this margin of time the Strings Theory of the universe is also invoked. In the Strings approach the attempt is to bring together the Quantum Mechanics and General Relativity to develop Quantum Theory of Gravity which then may be able to tell the whole story of the creation of the universe from t = 0 seconds onwards.

We are familiar with the three dimensions in geometry to which the fourth dimension 'time' was added by Einstein in his Theory of Relativity. The Strings Theory works with a total of 11 dimensions and predicts multiverses - the existence of many universes other than ours. No one has seen these dimensions of Strings Theory and no one can prove or disprove them. Therefore some physicists have started to rethink the validity of this whole concept.

The Big Bang Hypothesis faced the limitation that the General Relativity could only be applied to the well-defined matter in the form of mass, energy and the gravitational force that is exerted by the mass in matter. Going back in time to find out how and when it all got triggered, the General Relativity extrapolates the Big Bang Hypothesis towards t = 0 sec and leaves it at the doorstep of Planck-regime because what worked in the Planck regime is the Quantum Mechanics and not the General Relativity. Also, a relationship between Quantum Mechanics and Gravitational Mechanics has not been established yet. The physicists in Quantum Mechanics and quantum gravity have been hard at work for decades to develop a Unified Field Theory to reconcile the two and expect to explain everything about the universe from t = 0 sec onwards revealing the fundamental nature of matter as well. A connotation for this expected unifying theory in literature is the "Theory of Every Thing".

Since there has not been success in unifying all forces the aspersions are being cast on the validity of the mathematical singularity of Big Bang itself. In the Big Bang singularity the temperature of the singularity is infinitely high and its mass (energy) is also infinitely great which violates general relativity and common sense that one derives from the observed universe. Temperature is the inherent nature of matter and energy. If there is temperature in the Big Bang singularity there must be matter in that singularity. Dealing in Quantum Mechanics one could create enough matter invoking vacuum fluctuations to create a universe. The problem in this process is that an equal amount of anti-matter is also created. Then the question is – what do you do with it? Create an anti-universe, and where to put it? For a general reader it would be mind-boggling, to say the least, and it also defies the cause-and-effect principle which prevails in the inheritance of the human mind that something is created out of something and nothingness cannot be the cause of something. A pigeon appearing out of nothing from a magician's hat even is owing to the reality that magician and the pigeon have preexistence.

In the literature on cosmology, the state of matter during Planck-time or the state at t = 0 sec has no specific name that would represent its nature or properties. It is generally referred to as matter and/or anti-matter, which spontaneously appear and disappear in empty space. Although the Veda mantra does not stipulate the time duration of an epoch or the mass of matter, the state of matter at that time, as mentioned earlier, is assigned a meaningful term, the *Asata*-state (RigVeda, 10/129/1). In the Vedas and Vedic Literature the descriptive adjective for this state of matter is *avyakta*, which in English is 'indescribable' or 'indefinable' or 'inexplicable'. From the modern physics one learns that the matter in its real existence during the creation phase exists in full-fledged diversity, with Hydrogen, oxygen, iron, gold, compounds and substances, which in the *Asata*-state of dissolution precipitate into the indefinable uniformly mixed three forms of inactive *aapah* particles, *sata*, *raja* and *tama* with complete absence of any dynamic activity. Therefore, as discussed in the Vedic Cosmology Scenario in Chapters 7, the Vedic universe begins with cold matter without any temperature in it. The *Asata*-state in Vedic cosmology transits into *Sata*-state at t = 0 sec of the Vedic Big Bang and remains in the *Sata*-state during the whole creation till the Big Crunch when the matter returns into the *Asata*-state again. The Big Crunch of the Vedic cosmology is however different from that postulated by the Big Bang Theory. This difference will be elaborated up on in Chapter 10, Volume 2).

Table 2.1a : Expanding universe t = 0 sec to 13.8 billion years.

Era (Epoch)	Duration from t= 0 sec	Size of Univ. (radius)	Temp.
Planck	**0 to 10^{-43} sec**	**10^{-35} m**	**10^{32} °K**
Cosmic Inflation	**10^{-36} to 10^{-32} sec**	**1 m**	**10^{27} °K**
Quark-Lepton	**10^{-32} sec to 10 sec**	**$4x10^{13}$ km**	**$5x10^{9}$ °K**
Radiation	**100 sec to 3000 Yrs**	**$5x10^{14}$ km**	**10^{9} °K to 60,000 °K**
Matter	**3000 Yrs onwards to present 13.8 BY**	**$4.4x10^{23}$km**	**2.7 °K**

The universe according to the Big Bang Hypothesis goes through different epochs of time that cosmologists have named on the basis of what went on in that epoch. Without going into the details of the epochs their salient features are listed in Table 2.1a. One finds in the Big Bang literature a considerable variation in the numbers associated with the epochs, however, the numbers in the Table 2.1a represent more or less a realistic picture of the progress in the Big Bang physics.

When the universe was only 10^{-36} sec old and less than a meter in radius, it was subjected to sudden exponential expansion by a factor of trillion-trillion-trillion. This mechanism is termed as the Cosmic Inflation. The particle physicist and cosmologist Alan Harvey Guth of Harvard University introduced in 1980 the idea of Cosmic Inflation in the Big Bang Hypothesis. There is no provable physics reason for this sudden Cosmic Inflation occurring only once in the universe's lifetime, cramped in a negligibly small time span between the Planck and the Quark-Lepton eras and violating some well-known physics rules. Such an expansion in such a short time-duration does not ever occur again in the created universe. It is not even close to a physical phenomenon that one can observe anywhere in the universe or create in a physics laboratory. The only reason for this imaginary exercise is that this shock-treatment to the universe makes the Big Bang Hypothesis viable for creating a universe and explaining the phenomena and features in its present state after 13.8 BY.

The Cosmic Inflation hypothesis is doomed if the time-window is not infinitely short and specific. It is a mathematical exercise to make our universe unique. Small variations in parameters would create universes different from ours. The specific set of critical parameters also makes the creation suitable for life to exist later when Earth-like celestial objects are born in the universe. With the other sets of parameters other universes can be created. It is these other uncountable combinations of sets of parameters creating uncountable other universes that some cosmologists think do exist outside our universe. Cosmologists insist that the process of inflation is not unique to our universe and they put it metaphorically as - our universe is like a bubble among many universes on the 'tree of bubbles' (the tree of universes).

One needs a cause for Cosmic Inflation to occur. The cause is said to be primarily the combination of two ingredients, the 'inflationary energy' and gravity. The hypothetical 'inflationary energy' has to be extremely dense such that its density during inflation does not change and the gravity should have the property of repulsion rather than attraction, which is contrary to its nature in the rest of the universe's life. It is this gravitational repulsion in the early matter of the universe, which under the high energy density in matter causes the space to expand suddenly. Additionally, the quantum fluctuations in the universe's matter within this short window in time are thought to be the cause for creating islands of matter that are the future galaxies in the later life of the universe. It is also during this time epoch that the mode of the universe is supposed to have been set to create a universe with the property of flatness. It is not that the Cosmic Inflation automatically yields a flat universe it is the arbitrary mathematical massaging of the Big Bang equations that results in a universe to match the one presently observed in astronomy. This drastic hypothetical step in Big Bang cosmology has been discussed at length in the literature on modern cosmology, however, some salient features of the 'inflation debate' are presented in the following under 'Deficiencies of Big Bang Hypothesis'.

The particles of matter, anti-matter and radiation appear after the inflation era. The universe starts with equal number of matter and anti-matter particles. After the Inflation ends and before 10^{-10} seconds in the life of the universe, the matter and antimatter annihilate each other giving off high energy gamma radiation and other matter and anti-matter particles. This results in a universe that is dominantly filled with radiation and matter. The Big Bang Hypothesis further suggests that for some reason more matter is created than antimatter. Quantum Physics explains this matter-antimatter disparity; however, it does not explain why the matter should have the dual nature of matter and antimatter.

CHAPTER 2: Present Understanding of the Universe – A Critical Scrutiny of the Contemporary Big Bang Hypothesis

When the universe is only a few meters in diameter, still within a tiny fraction of time, it is filled with a hot dense soup of quarks and leptons. Negatively charged electrons belong to the lepton category and are found free as well as in the bound state. Quarks are not found free in the universe because they all are quickly consumed into the next stage of particles called hedrons. There are two types of hedrons: baryons and mesons. A baryon is formed of three quarks. A meson is formed out of a quark and an anti-quark. It is the baryons that form the nucleus of an atom. This era of the formation of baryons is also known as the Baryogenesis Era when most fundamental forms of matter start to form which then become the main cause for the formation of the observable universe.

The temperature drops to 10^{10} °K about 1 second after the Big Bang and the homogeneous sphere of the primordial matter is about $4\text{x}10^{10}$ km across. At 100 sec when the temperature of the universe is about 10^9 °K (1 billion °K) the nucleo-synthesis begins forming Hydrogen and Helium atoms, ending up with about one-quarter Helium and three-quarters of Hydrogen by mass. The matter density at the end of this time period is equal to the radiation density. During the Radiation Era, the photons of light of different colours are produced including the microwave part of the spectrum. George Gamow (1904–1968), a well-known Russian-American theoretical physicist and a cosmologist, predicted from his studies of Big Bang nucleosynthesis and the stellar nucleo-synthesis that the Big Bang would produce sufficient radiation in the microwave region of the spectrum that should be observable now. Formation of neutral Hydrogen produces this radiation. This radiation is named as Cosmic Microwave Background Radiation (CMBR). This CMBR was created in all directions at every point in the universe about 379,000 years after the Big Bang. Since then the Hydrogen and Helium condensed into stars and galaxies while they were being accelerated away from each other owing to the initial momentum they acquired in the Big Bang explosion. Now they occupied the universe that was 46 light-years across ($4.35\text{x}10^{23}$ km); a light-year being the unit of distance which light covers in one year at the rate of 300,000 km/sec. Despite the expansion of the universe the CMBR was expected to retain its isotropic distribution throughout the universe's growth. This radiation was finally discovered by two American physicists from Bell Laboratories, Arno Allan Penzias (1933) and Robert Woodrow Wilson (1936). In 1976 they were awarded a Nobel Prize for their discovery. The discovery of this 'radio remnant' of the Big Bang provided support to the Big Bang Hypothesis and helped cosmologists to improve the Big Bang Theory.

The temperature of the universe drops to 3,500 °K when the Big Bang is about 300,000 years old. The atoms are formed, matter dominates in the universe and it becomes transparent. At the age of about 500,000 years, the matter of the universe cools further to about 3000 °K and is about 3,000,000 light-years across ($3\text{x}10^{19}$ km). After about 100 million years to 1 billion years the stars begin to form as part of the galaxies that are now counted to be in billions. According to NASA's recent observations with the Hubble Space Telescope, there are over 150 billion galaxies in the visible universe that may contain over $7\text{x}10^{23}$ (700 thousand-million-million-million) stars. Of these stars some turn into supernovae that explode. Observation of supernovae provides considerable information on star formation.

There are about 100 billion stars within our Milky Way Galaxy, the home of our solar system. Our solar system was formed about 5 BY ago and our Earth within it was formed about 4.6 billion years ago. The Canadian scientists of the University of McGill in Montreal in 2008 discovered the oldest rocks on the Earth that are at least 4.28 BY old and are considered to be the remnants of the formation of the first primordial crust on our planet. The first life on Earth according to modern biology appeared some 4 billion years ago when a living cell acquired the nature of a 'replicator' to produce one more of its kind. The process continued to evolve producing bigger conglomerates of such cells ultimately forming bodies of living beings. This is how modern biology, more or less, hypothesizes about the beginning of life on the Earth.

The matter of the universe now in the form of stars and galaxies is still flying out at great speeds. The observations of these cosmological objects provide an estimate of how long ago the Big Bang took place and how far the material of the universe has reached. The present age of the universe is about 13.8 BY. This age of the universe is based on what has been observed thus far with the upper limit of the sensitivity of the instruments in astronomy. The universe certainly is beyond that limit. The diameter of the universe that we see around us within the perception of our best observing instruments is at least 93 billion light-years wide. Thus, the present observable universe is about 8.8×10^{23} km across and continues to expand. The open literature on the Big Bang has pictorially delineated the process of Big Bang with differing emphases that a reader can benefit from.

Endowment of Big Bang Creation

In the present universe there are a large number of inherent forms of matter and energy and the laws of Nature to govern them. These different forms under the laws of Nature exhibit specific characteristics that are seen in the form of motion and change of state. Only the important ones with their broad definitions and descriptions are included here. As new theories and measurements become available the information on the subject of cosmology continues to change. A brief commentary on the matter, energy and shape of the universe is presented in the following.

Observable Matter

Observable matter is the part of the universe that interacts with electromagnetic radiation, latter in its visible part of the spectrum is termed as light. Atoms and molecules interact with radiation through absorption and reemission. The process of reemission also includes reflection of light that makes the atoms and molecules visible. The observable matter is termed as 'baryonic matter' because the fundamental particles forming the nuclei of atoms are named as baryons (proton, neutron). It was mentioned earlier that right after the onset of the Big Bang, the matter of the universe goes through transitions from undefined state to definable state when matter particles called quarks first appear. A neutron or a proton formed out of three quarks each weighs about 1.7×10^{-27} kg. The other kind of prevalent charged particles, also formed in the early stages of Big

Bang, are electrons. An electron is negatively charged and weighs only 9.1×10^{-31} kg (1,868th of a proton). For a general reader it should suffice to say that because the opposite charges attract each other, a proton captures an electron to form a nucleus of a stable atom of Hydrogen (chemical symbol H). Hydrogen atom is the first created lightest element in the universe (Figure 2.2). The captured electron does not just go and stick to a proton; rather it goes into an orbit around the proton under the influence of the electro-static force of a positively charged proton. This is something similar to the Earth, with a mass 330,000th of the Sun, orbits around the Sun under the Sun's gravitational attraction on the Earth. In a classical sense the electron is orbiting around the proton (nucleus) at a very high speed of over 2 million meters per second just like the Earth goes around the Sun, although considerably slower; only about 30,000 meters per second. The diameter of the Hydrogen atom including the orbit of its electron is 10^{-10} meters (ten-billionth of a meter). The diameter of a proton alone being 10^{-15} meters (one thousandth of a trillion meters) is therefore 100,000 times smaller than the Hydrogen atom itself.

The differing number of nuclear particles in a nucleus of an atom makes one atom different from the other and heavier than the other. The next heavier atom formed in nature is Helium (He) which has two protons and two neutrons in its nucleus with two electrons orbiting around it. The number of protons in an atom determines its atomic weight. After H and He with atomic weights of 1 and 2 respectively, heavier elements right up to radioactive Uranium with atomic weight of 92 are formed at a later stage of creation of the universe. About 75% of the mass of the universe is in Hydrogen, about less than 15% in He and the rest of the heavier atoms are proportionately less abundant in the universe.

A multi-atom unit of an element is known as the molecule of that element. The nature presents itself in the form of a large number of chemical compounds that are used by nature to reveal itself in the form of conglomerates much larger than single atoms, conglomerates ranging from the bodies of living-beings to the solar system, the galaxies and so on. It is the matter of the universe that reveals the universe itself to our eyes which receive light coming from the objects of the universe. The light is also a state of matter that reveals other forms and states of matter. Light is energy. Matter and energy go hand in hand in revealing the Nature. Matter is found in three major states; solid, liquid and gas, the latter not being observable owing to its molecules separated from each other. The air, a mixture of molecules of nitrogen, oxygen and other gases, is thus invisible to the eye.

The night-sky, lit up with the stars gives an overwhelming impression of abundance in nature. When the size of the observable universe is considered and the observable matter in it, in the form of free Hydrogen, Helium, other atoms and molecules and galaxies are accounted for, one feels that this resplendent universe is packed with matter. This however is a mere 'optical illusion'. The universe in fact is quite a rarified entity and fits the definition of a vacuum. The stars in individual galaxies and individual galaxies themselves are far apart from each other with large expanses of the void of space separating them.

If the total observable matter of the universe is accounted for and total volume of the observable universe is determined the matter density of the universe can be determined. The density of a substance is measured by finding how much substance is present in a unit volume occupied by that substance. For example, a volume of water 1 cm by 1 cm by 1 cm (1 cm^3) weighs 1 gram. Therefore, the density of water is 1 gram per centimeter-cube (1 g/cm^3). Cosmologists provide differing estimates for the overall density of matter in the observable universe; however, compared to the water density of 1 g/cm^3, an acceptable rough estimate for the density of the universe as a whole is only $3x10^{-31}$ g/cm^3 This suggests that the space of the universe with all that matter in it on an average is a vacuum. This value of the universe's density is 100 billionths below the attainable vacuum in the laboratory. This density of universe is equivalent to finding one Hydrogen atom in a 2 meter-cube of space. Of the total matter-energy balance of the universe the observable matter forms only 4%; the rest being in the form of energy. Although we see clumps of matter in the form of stars and galaxies, the universe on large scales, on an average, is considered to be homogeneous with uniform structure throughout and isotropic with uniform density in all directions, fulfilling the condition of the second Cosmological Principle listed earlier.

Dark Matter and Dark Matter Mystery

In addition to the observable matter, which is a reality of the universe, cosmologists have hypothesized the presence of another kind of matter that is not seen or measured directly. It is thought to have mass causing gravitational effect on the observable matter. It is popularly known as the 'missing matter' of the universe and cosmologists also address it as the 'Dark Matter'. In 1933, Fritz Zwicky of the California Institute of Technology studied the Coma cluster of galaxies and found that its total mass based on its total brightness was several orders of magnitude higher than the estimate based on the number of galaxies near its edge. This discrepancy between the estimated and observed mass could be explained if there was more mass there than observed. This led to the idea of the 'missing matter' or the 'Dark Matter' of the universe.

In 1975 Vera Rubin and Kent Ford from Carnegie Institute of Washington presented their discovery at the American Astronomical Society that most of the stars in spiral galaxies orbit around the galactic centre with about the same speed; contrary to the expectation that the outer stars should orbit slower in the same way as objects of our solar system where outer planets have longer orbital periods. This was thought as a confirmation for the presence of more matter in the galaxy than was observed because without it the outer stars would fly away from the galaxy. It was the invisible Dark Matter which was at least half of the total galactic mass and seen as a dark halo around the galaxies. The rotational and orbital speeds in galaxies and several other phenomena associated with the observable part of the universe are attributed to the gravitational influence of this hypothetical invisible matter. It forms about 22% of the total matter-energy equation of the universe.

What is Dark Matter made of? The answer to this question at present is speculative. Since its observed property is gravitation and it being non-interactive with the radiation, the most entertained hypotheses for its origin are those suggesting its makeup from non-baryonic particles (i.e. no protons, neutrons)

CHAPTER 2: Present Understanding of the Universe – A Critical Scrutiny of the Contemporary Big Bang Hypothesis

The Chapter 8 presents several mysterious Vedic mantras (AtharvaVeda, 10/8/7, 13) that state by asking a question first that the cosmos are formed out of half the mass of the universe then where is the other half? In these mantras the expression such as अर्धेन विश्वं भुवनं... (*ardhen vishvam bhuvanam…*), meaning – "*From the half is created the entire cosmos...*" and सप्तार्द्धगर्भा भुवनस्य रेतो... (*saptaarth-garbhaa bhuvanasya reto…*) meaning – *"Those seven born out of half the embryo, the primordial matter, are the first matter-seed of the universe..."* have considerable cosmology hidden in them that is revealed in Chapter 8. The answer comes in the other mantra (RigVeda, 1/164/36) that the other half is used up right in the beginning of the creation to create seven *Deva* energy fields that advance further the creation of the universe. These mantras are discussed in detail with word translation under the sub-heading of "Matter distribution between *Devas* and Observable Universe" in Chapter 8. Not just the matter of the universe but they assign a proportion of 50% to what is being observed as the observable universe with the other 50% being invisible.

In the context of cosmology, the invisible other half of the total mass of the universe in the Vedic mantras, is either the Dark Matter of contemporary cosmology or it is the regular matter beyond our observable limit of the universe. Nevertheless when a mantra assigns a hard number to the matter of the universe, one ought to take notice of it. It is also discussed in Chapter 9 that the third component of the *Aapah* triplet designated as the *tamah-aapah* (dark-*aapah*), which is the fundamental state of matter, in this author's thinking, is universally distributed and serves as the invisible and undetectable 'viscosity of the space' that moderates the speed of objects in space. This author also surmises that it is this ubiquitous *tamah-aapah* (also means dark), that sets the limit on the speed of light $3x10^8$ meter/sec. The Vedic perspective of this invisible *tamah-aapah* serving as the 'viscosity of Space is further discussed in Chapter 9.

The physicists accept that the Dark Matter exists and plays an important role exerting its gravitational force on the matter of the universe. They are hard at work to understand what Dark Matter is, how was it created, when was it created or as the Veda suggested - was it always there as the $1/3^{rd}$ of the "material-cause" of the universe called the *tama-aapah*? As discussed in Chapter 5 in Volume 1 and Chapter 9 of Volume 2 that in the Vedic physics the most fundamental particles are a trio of *sata, raja* and *tama*; the *tama* being the most inactive.

With regard to the present understanding of the Dark Matter, Prof. Dan Hooper of Fermilab, Chicago University, summarized in an article in the "New Scientist", Vol 244, No. 3256, 16 November 2019 as:

"…. *the stubborn illusiveness of dark matter has left many physicists and cosmologists surprised and confused. In droves, we are returning to our chalkboards, revisiting and revising our assumptions – and with bruised egos and a bit more humility, desperately attempting to find new ways to make sense of a very dark and hidden universe.*"

The Vedic ideas may provide some possibilities!

Energy

The universe is more energy than mass. Although there is equivalence between mass and energy through Einstein's equation ($E = mc^2$), the difference in mass and energy is that mass exhibits weight and energy is treated as a mass-less commodity. Most common energy is in the form of electro-magnetic radiation covering a wide spectrum from short wavelength radiation in the form of X-rays to long wavelength radiation in the microwave region. The radiational energy in the infrared region of the spectrum between visible and microwave is commonly known as the heat energy.

Optical telescopes are utilized to image and measure the stars and galaxies. The microwave radio telescopes are utilized to measure microwave radiation. The stars and galaxies are also imaged in this part of the spectrum. As mentioned before, a homogeneous permanent energy feature of the universe is the CMBR. This low level radiation uniformly coming from all direction is also measured with high sensitivity microwave radio telescopes.

Because a force relates to energy one could perceive gravitational energy corresponding to the gravitational force. Energy is also available as kinetic energy owing to the motion of the object. It can be in the form of potential energy stored in an object against a force.

A completely different source of energy is hidden in the vegetation world. This energy is the molecular bond energy that is consumable by living-beings. These different forms of energies are observable through the effects they create. They are measurable and harnessable for practical use. In the Big Bang cosmology there is yet another energy of mysterious nature that is said to be permeating through the space of the universe but is not harnessable for practical use like the other forms of energies. It is the Dark Energy, also addressed as the Vacuum Energy.

Dark Energy / Vacuum Energy

It is suggested in Big Bang Hypothesis that the gravitational attraction between the galaxies tending to pull galaxies inwards opposing the expansion of the universe, should ultimately slow the expansion of the universe down and finally bring all galaxies closer together. In this scenario the universe collapses into one mass, returning back to space-time singularity where it all began. The logic prevails that if the Big Bang explosion created the universe sending matter to far distances the reverse process of collapsing the universe under gravity should send matter to the condition it was in the Big Bang singularity. This means that all matter of the universe must come together and disappear into one dimension-less point of space where Space, a real entity holding the universe, must also disappear into nothing. However, what one observes presently is that the universe is expanding at an accelerated rate. The observations of a very distant supernova in 1998 by Hubble Space Telescope showed that when the supernovae had exploded when the universe was only about half its age, the universe was expanding at a much slower rate than now. Presently the universe is said to be expanding at an alarming rate with the farthest observed galaxies seen flying out with the highest acceleration. The astrophysicists express this unusual expansion as the 'cosmic acceleration'. This would happen when some additional force/energy was acting against the gravitational force due to the mass of the universe. For such energy to expand the universe in all places and in all directions, it has to permeate throughout the space of the universe. Not knowing what this energy is, the cosmologists have called it the Dark Energy.

A steady-state universe, as Einstein had first conceived, would not collapse on itself due to its own gravitation. He modified his field equations in General Relativity by introducing a constant called the 'cosmological constant', represented by the Greek letter lambda Λ. To oppose the action of Newton's gravitation constant G in his equation that represents gravitational force, Einstein introduced the positive constant Λ (with other necessary parameters) on the other side of the equal sign of the equation. Just as the matter contributes to the gravitational field which pulls matter inward, Georges Lemaître in 1934 related Einstein's cosmological constant Λ to a hypothetical energy named 'Vacuum Energy'. The tendency of this energy should be to push matter outward to counter the gravitational pull. The idea was that if energy exists in vacuum it must exert outward pressure to oppose the effect of gravity.

Einstein had proposed in 1917 that the vacuum of empty space is not a 'nothing'. It has properties of its own. His premise was that the cosmological constant in his equation, countering the gravitational attraction, represented the energy of the empty space and that this energy was constant and permeating throughout the universe. This energy was a property of space and it does not weaken when the universe expands because the more the universe expands, the more space is created and therefore more Dark Energy is available. Not much is known about the true nature of this mysterious energy as to what its structure is and how it interacts with matter and the other energy forms in the universe. It continues to be addressed as the Dark Energy. Also, because no new matter is created in the universe but more space is created owing to the expansion of the universe, more and more Dark Energy is created accelerating the expansion of the universe. One easy deduction from this scenario is that the universe is tending towards a Big Chill!

The question is – how does the space acquire this energy in the vacuum of the space? The existence of 'Vacuum Energy' has been observed through the manifestations of several phenomena. Owing to its presumed action this uniform background energy by definition must be all-pervading through the universe. It is said to exist in space with or without the presence of matter. Since there is equivalence in mass and energy, the Vacuum Energy is expected to exert its influence on objects of the universe through its contribution to the cosmological constant, the critical value of which would determine if the universe would expand or contract and at what rate.

The Vacuum Energy is presently thought of as a phenomenon in the realm of Quantum Mechanics, however, the mystery of the actual nature of Vacuum Energy and what creates or created it, is not understood well. The most accepted mechanism is said to be the appearance of 'virtual particles' (particle-antiparticle pair) in the vacuum of space which annihilate each other giving out a burst of radiation, resulting in increased outward pressure in the empty space causing the expansion of the universe. For Vacuum Energy to play the vital role in sustaining the expansion of the universe, it has to have its right proportion in the total mass/energy equation of the universe. Because the universe is said to be expanding at an accelerated rate than what it used to be, the Vacuum Energy must have increased. The astrophysicists think that as the universe expands and more and more vacuum in space is made available to increase the creation of the virtual particle-antiparticle pair, the pressure of the Vacuum Energy must increase.

One difficulty in accepting the hypothesis of the virtual particle-antiparticle annihilation being the source of the Vacuum Energy is the estimate of the energy itself. The error estimates by cosmologists are infinitely larger than acceptable. Therefore, the question as to how the vacuum of space gets energy remains outstanding.

The astronomical observations provide estimates of the energy density of the Dark Energy (Vacuum Energy) in terms of the mass density units to be $7\text{x}10^{-30}$ grams/cm^3. Although this is a negligibly small number it is a factor of higher than the mass density of the universe, $3\text{x}10^{-31}$ grams/cm^3, enough to cause accelerated expansion of the universe. In proportion to the mass of the universe the estimates show that the Dark Energy is about 68.3% of the total mass-energy equation of the universe. The remaining 26.8% is the mysterious Dark Matter and 4.9% is the regular matter.

A Vedic Perspective – No Dark Energy Needed

Astrophysicists have suggested several alternate theories for the origin/mechanism for the Dark Energy but none are considered to be fully convincing. It is often suggested now that Einstein's theory of gravity itself could be incorrect and needs a fresh look. The Vedic Creation Principle (Chapter 6) and the Vedic Creation Scenario (Chapter 9) together provide an alternative to the mysterious Dark Energy, not considered in modern cosmology yet. When it comes to cosmology, the Vedic mantras do not resort to suggesting hypothetical mechanisms or a divine intervention to explain a physical phenomenon in Nature, such as the expansion of the universe. Their premise always is that whatever has been created is real and no virtual phenomena are created by vacuum of space. The mantras do point out that the Space is stretching, the universe is expanding and the mechanism of expansion is part of the dynamics developed by the universe during its primordial stage when it was preparing for the Vedic Big Bang. In the Vedic principle for the creation of the universe, the entire mass of matter that was going to be the universe began as a conglomerate in ovate shape named the *Hiranyagarbha* (Golden-embryo) which remained in *paryaplav*, the rotation around an axis while in floatation, for a specified period of 4.32 million years.

The Vedic term for the universe's axis of rotation is *Skambha* (the pillar of energy) which supplies energy to the entire matter of the universe. The mantra references emphasize that the *Hirnyagarbha* during this period develops tremendous *tapasah*, the heat energy and the angular momentum, supplied by the *Skambha*. Then *Hiranyagarbha* explodes in a Big Bang creating uncountable clumps of matter that form galaxies and celestial objects within them. Each celestial object of the universe, all the galaxies as groups of stars and the entire universe as a whole inherits the angular momentum from the parent *Hiranyagarbha*. The all-pervading centrifugal force of the universe owing to its inherent angular momentum causes the galaxies to fly outward expanding the universe against the inward pointing centripetal force caused by the gravitational force owing to the mass in matter.

The *gurutva* (gravitation) in matter causing contraction and the centrifugal force causing expansion strikes a dynamic balance in the rotating and expanding universe. The rotation of the universe around its *Skambha* axis which is also said to be the source of energy for the rotating universe, like a bar at the centre of a galaxy, causes the flattening of the universe. The Veda mantras point out the fact that the universe is a flattened cart-wheel like system. Modern cosmology may have a different meaning of the flatness of the universe; however, on the basis of the observations it stipulates that the universe is flat; just like what the Veda stated. For the Big Bang theory to create a flat universe the cosmologists introduced required parameters in the Cosmic Inflation.

The Vedic scenario does not need a hypothetical Dark Energy and all the related hypotheses to go with it. The physical action of rotation and the accompanying centrifugal force are real phenomenon. To explore the Vedic possibility in modern cosmology one has to set up the momentum equations within the gravitational consideration and determine the rate of expansion of the universe. Once considered the Vedic expansion mechanism due to the rotation of the universe, as discussed in Chapter 9, the expansion of the rotating universe may not even be as much as suggested by the astrophysicists because the part of the observed red-shift in the spectrum of the faraway galaxies could just be an artifact of space-time geometry in a spinning universe. Also, because the Space-entity according to its Vedic concept acts only as a non-reactive provider of a home for all the substances of the universe and therefore serves as the container of the objects of the universe, it does not have the property of creating energy through virtual particles. Therefore, yes Einstein's theory needs a fresh look. A detailed discussion on this subject is included in Chapter 9.

Deficiencies in Big Bang Hypothesis

A large number of astrophysicists, cosmologists and astronomers adhere to the most acceptable Big Bang Hypothesis and carry out research to learn more about the universe. Many of them have used space-based Hubble Telescope and other satellite data to substantiate their previous findings and have reported new findings to support this hypothesis. There is however the other side to this coin. There have been a large number of dissenters of this hypothesis, included among them are well-known astrophysicists, cosmologists and astronomers. The list includes some of the well-known physics personalities such as Sir Fred Hoyle, the Nobel Prize winner Hannes Alfven, astronomers Geoffrey Burbidge, Halton Arp and some thirty other scientists who signed a joint dissent letter, published in a weekly British science magazine, the New Scientist. Over 500 more have signed-in since then (New Scientist, May 22, 2004). The list of dissenters includes scientists and engineers from famous academic institutions around the world. The interested readers may dig into this controversy by accessing considerable literature on web sites. Examples of such references are: "The Top 30 Problems with the Big Bang" by Tom Van Flandern of Meta Research; "The Big Bang Theory under Fire" by William C. Mitchell (Phys. Essays Vol.10, 2, June 1997) and more.

There are scientists who are hundred percent confident about the soundness of the Big Bang Hypothesis. Others having no better alternative carry on their research within its framework. Then there are some who have spent over two decades working with the Big Bang Hypothesis, Cosmic Inflation mechanism included, who now have started doubting its foundations and are looking for improvements in the Big Bang Theory itself or even looking for other alternatives. The most fundamental reason for this awakening in cosmology since Guth introduced the idea of Cosmic Inflation in 1980 is that it created more questions than it answered and it created more doubts than it alleviated. In an article (Scientific American, April, 2011), Paul J. Steinhardt, one of the theory's creators, the Director of Princeton Centre for Theoretical Science at Princeton University in USA said:

> "*Matching experiments trumps any theoretical argument. But the strange twist to this story is that the predictions of the early 1980s were based on a naïve understanding of how inflation actually works - a picture that has turned out to be dead wrong*".

To get around this naïve understanding the Big Bang cosmologists came up with a new picture of the Big Bang. This new picture included eternity of inflation in islands of space, each representing a different universe. This led to the Multiple Universe Hypothesis with inflation still a continuing theme in the Big Bang Hypothesis. If all such universes outside our own and non-provable use Cosmic Inflation, they have not been free from the *naïve inflation*. From the perspective of the Anthropic Principle it is beginning to look as if the cosmologists are going out of our universe to win an argument!

Continuing with Cosmic Inflation in a non-naïve way with more assumptions and with different mathematics seems to be the solution for graduating the naïve Big Bang to a new acceptable level. However, creating an infinite number of universes outside ours using eternally going inflation hardly changes the argument because if one can create a universe with inflation one can create many universes with inflation. So why go to another universe to prove that the Cosmic Inflation occurred in our universe? This does not prove anything! The article by Steinhardt has presented a good overview of the problems with inflationary Big Bang and has brought in the views in favour of and against the Big Bang Hypothesis. Steinhardt's article summarized quite eloquently under the sub-title "IN BRIEF" thus:

> "*Cosmic Inflation is so widely accepted that it is often taken as established fact. The idea is that the geometry and uniformity of the cosmos were established during an intense early growth spurt. But some of the theory's creators, including the author (Steinhardt), are having second thoughts. As the original theory has developed, cracks have appeared in its logical foundations. High improbable conditions*

> *are required to start inflation. Worse, inflation goes on eternally, producing infinitely many outcomes, so the theory makes no firm observational predictions. Scientists debate among (and within) themselves whether these troubles are teething pains or signs of a deeper rot. Various proposals are circulating for ways to fix inflation or replace it".*

The cracks in the theory or possibility of deeper rot in it felt after two decades are owing to the powerful human psychological phenomenon that pervades through conscious-beings. This phenomenon, briefly discussed in Chapter 1, is addressed in Vedic terminology as *sanshaya*, which is 'doubt' of a substance (entity). The brief discussion on *sanshaya* (doubt) in Chapter 1 was with reference to the *sutra* (2/2/17) formulated by the pre-Mahabharata Vedic philosopher Rishi Kanaad. Several *sutras* later Rishi Kanaad stated that '*vidyavidyatashcha sanshayah* (2/2/20)', which translates into – "*Sanshaya* (doubt pertaining a substance) *is created by both, the knowledge as well as the non-knowledge (ignorance) about the substance*". This means that complete knowledge of a substance uproots doubt about the substance and complete ignorance about a substance does not create any doubt about the substance. For example, a galaxy far away that has never been seen and is beyond astronomical observation cannot create doubt about it. In the present context the primary 'substance' is the universe itself and we have incomplete knowledge about it. Therefore, the universe ought to create *sanshaya* (doubt). The Big Bang Hypothesis not being able to create complete knowledge of the universe is not free of *sanshaya* (doubt). Therefore, the Big Bang Hypothesis is the source of knowledge as well as the source of doubt (ignorance) as stated by Steinhardt, cited above, "*...cracks have appeared in its logical foundations....*". The *sanshaya* is a discontinuity in the thread of knowledge. From Rishi Kanaad's *sutras* one can safely deduce that '*complete knowledge of the properties of the substance* (or of something), *completely neutralizes doubt about the substance'*. The big Bang Hypothesis is deep in *sanshaya.*

To remove newly created 'doubt' about the universe one has to first have a complete grasp of its ordinary properties and then extend the effort to gather its extraordinary properties such as how big it is, when was it born, is this a pure physical system or a combination of matter and consciousness. This obviously leads to developing of theories that should reveal its prodigious properties. The whole effort is to remove doubt about the universe by predicting and proving all its properties. In the process of un-knotting the thread of knowledge if more knots accrued, the likely-hood is that the process did not have a good start. A new approach altogether is becoming a necessity to reveal the universe's prodigious properties.

A few important features of the Big Bang Hypothesis are discussed in the following which are the cause for doubts and controversies.

Horizon Problem and Cosmic Inflation

It was mentioned above that the Cosmic Microwave Background Radiation (CMBR) is observed to be extremely uniform from all directions in the sky. This suggests that the universe is homogeneous in all directions and is also thermally uniform corresponding to a CMBR temperature at 2.735 °K (-270.265 °C). After the CMBR discovery, a question was raised in the early 1970s - why is the universe so homogeneous with a constant temperature in all its regions, when these regions are so far apart now that even light would not have reached from one to the other in its life history of 13.8 BY? To explain what this means, several similes have been used in the Big Bang literature; the heater in a room is one. It takes some time for the heat to transfer from one end of the room to the other and it takes even longer for the entire room to come to the same uniform temperature. When this occurs, we say that there is a causal connection between the heater, the air in the room and all sections of the room. But such a causal relationship is not expected in the case of the universe which is so large that heat (CMBR) could not have reached from one end to the other in the life of the universe 13.8 BY. However, the uniformity in observed CMBR exhibited a causal connection between all parts of the universe. That the entire universe must have acquired homogeneity so early in its life is also known as the 'Horizon Problem' in the Big Bang Theory. This led to the doubts about the validity of the Big Bang Theory as it stood at the time. The expected solution to this problem was suggested – that the homogeneity in the universe must have started right in the Big Bang when the universe was small. This necessitated some tweaking in the Big Bang Theory itself. This led Alan Harvey Guth in 1980 to the idea of the Cosmic Inflation, suggesting that the early universe right after Planck Time, from 10^{-43} sec till 10^{-32} sec was dominated by a 'False Vacuum'. The concept of 'False Vacuum' suggests that at extreme densities of matter, as in a singularity, the matter instead of having gravitational attraction starts to have strong repulsion.

In a recent article on the PBS website, entitled "An Eternity of Bubbles", Allan Guth has explained the mechanism of Cosmic Inflation in simpler terms. In this exponential swelling of the universe it doubled in size at every step of 10^{-37} second. It kept doubling at each step of this time till the repulsion due to False Vacuum lasted. It seems that it lasted only between 10^{-43} sec to 10^{-32} sec, increasing the size of the universe by a factor of about 10^{32} (32 zeros after one). The False Vacuum is said to be unstable and decayed fast 'converting its energy to a hot soup of ordinary particles'. After this little blip in time the hot Big Bang mechanism carried on with the universe expanding normally. Cosmic Inflation was not necessary right after Planck Era but if it was inserted the homogeneity of the universe with the same temperature everywhere was explained and also explained then was the same intensity of CMBR in all directions. This was the arbitrary rationale for introducing 'Cosmic Inflation'.

After introducing 'Cosmic Inflation' in the theory, one had to be imaginative enough to move towards proving other future phenomena in the functioning universe. For example how the future gigantic galaxies were to be formed was to be taken care of within Cosmic Inflation epoch. This is done by theoretically creating fluctuations in matter such that the centers of the galaxies are predetermined as the nodes in fluctuations where the future galaxies would collect matter from their surrounding space. The seeds of billions of galaxies were to be separately placed when the universe was not even a meter across. If the universe created homogeneous CMBR, 379,000 years after the start of Big Bang, when the universe was about trillion-trillion km across, the Big Bang cosmologists should have no reason to think that the universe was homogeneous right from the beginning.

The doubt about the Cosmic Inflation was – what was the source of this high False Vacuum that has an opposite nature to gravitational attraction or is it just the nature of matter right after the singularity breaks? We understand that two positive or two negative charges repel each other. But this is not said to be the case in the primordial state of matter in the Planck-time regime. We also know that the gravitation and strong nuclear forces exhibit forces of attraction between matter and matter. Then it is not clear what exactly the nature of this False Vacuum is. In other words, why does the primordial matter of the universe have a repulsive property, only momentarily, only once in its life when it is at an infinitely high density? The high density in the first place is due to the property of attraction. Physicists, keeping an eye on the reality of the universe, ought to shed some light on why a tiny volume of matter for only the trillion-trillion-trillionth of a second switches its nature of attraction to repulsion and then back to attraction? Is the False Vacuum for real?

In a symmetric explosion in space, like a fire-cracker exploding up in the sky, the fragments generally fly out occupying a spherical volume around the centre of the explosion. The astronomical observations of the universe reveal that the universe is not a spherical shell like a soap bubble; rather it has uniform matter density throughout. The cosmologists call such a universe as a 'flat' universe wherein there is no space-warping and inhomogeneity. To explain this observed situation, the Big Bang cosmologists insist that the Big Bang explosion was not a usual explosion. The Big Bang cosmologists attribute flatness of the universe to the mechanism of Cosmic Inflation wherein the expanding matter of the universe assumes a mode that yields a flat universe. In the Vedic scenario on the other hand, discussed briefly above and in detail in Chapter 9, the rotation of the universe around an axis, in addition to the expansion of the universe owing to the centrifugal force of rotation, flattens the entire geometry of the universe in its outside dimensions – a wheel-like shape. The flatness of the universe has a different definition in modern cosmology.

Although the Cosmic Inflation theory is unverifiable, it was further developed by other cosmologists and has been accepted as a virtual tool to explain the problems in the Big Bang Theory. The size of the universe at the end of the inflation was only a few meters across. Therefore the Big Bang cosmologists concluded that this was the time when the universe that we observe today was causally connected and acquired its homogeneity that persisted for the rest of its life. The inflation theory also suggested that the CMBR has had close connection with the whole matter of the universe from the time the CMBR was produced.

It is the Cosmic Inflation era from where the matter of the universe transits from the Quantum Physics regime where one set of physics rules applied to the Relativistic Physics regime where another set of physics rules apply with no relation between the two sets of laws. From what this author gathers from the open literature, presently there is no specific Quantum Physics mechanism to support Cosmic Inflation. The Relativistic Mechanics applies to the universe that had already come out of the Cosmic Inflation with all the seeds of all the important observed features of the later universe already planted in it. This means that, before it enters into the general physics regime, the universe's behaviour for the rest of its life is predetermined in the Quantum Physics regime. On the basis of Quantum Physics alone, with the Uncertainty Principle being its important pillar, it is not possible that the universe in that early state would be homogeneous with the same properties everywhere. The Quantum Physics should randomize rather than create a homogeneous universe. It is counter intuitive to think that a condition of homogeneity is a result of a turbulent explosion.

There is another twist to CMBR radiation. Some scientists think that the CMBR that is measured now may have another source and a simpler explanation for its origin. Before Penzias (1933) and Wilson (1936) discovered CMBR, Sir Arthur Eddington in 1926 had calculated the minimum temperature that a body in space attains when illuminated by just the starlight. His calculation yielded this temperature to be 2.8°K which is very close to the observed temperature 2.735 °K corresponding to the CMBR. It certainly raises questions about Cosmic Inflation being responsible for carrying out the process of homogenizing the early universe.

Multi-Universes – the Eternity of Bubbles

The False Vacuum is worked into the Big Bang Hypothesis to explain homogeneity of CMBR and therefore of the universe itself and it is also said that it is an unstable entity (see Guth's article, "An Eternity of Bubbles', on PBS website). The physics question therefore is - would the unstable nature of the False Vacuum not introduce instability in a homogeneous substance embedded in it, the substance being the early universe? On one hand the homogeneity in CMBR is extremely high with a measurement accuracy of 1/100,000 (one in one hundred thousand) and on the other hand it has come out of the unstable False Vacuum energy. This sounds counter-intuitive and does not seem commensurate with the physics laws that we observe in a functioning universe. In Guth's article the suggestion is that the False Vacuum energy never disappeared; rather its pieces randomly decayed to create new "bubble" universes at an ever-increasing rate. When the False Vacuum energy breaks into pieces to create new "bubbles", all bubbles being different they create all universes different from each other. Also there is no physics law or logic to create all universes homogeneous from a very unstable entity, the False Vacuum. If the False Vacuum continues to break into smaller pieces, what stops them to break into each other to create smaller universes within bigger universes? We do not see a smaller universe within our, expanding in our direction. Then no common physics logic should suggest that the bubble that created our universe has to be highly homogeneous within itself.

Our universe is said to be just one of the universes on this infinite tree of bubbles. Guth also says that if inflationary theory is right the universe is far larger and much older than was thought and that it includes not just one but infinity of Big Bangs creating an 'eternity of bubbles'. This has led to the Multiverse Cosmology wherein mathematical experiments are performed to create a universe that certainly would be different from ours because to create our universe only one critical-valued set of parameters was required. Ours therefore is a unique universe. The critical set of parameters, also called as the fundamental constants of physics, are so fine-tuned for our universe that without them being so, the universe would not produce life on Earth as well as in other places of our universe. This is also why our universe is unique. This uniqueness is the subject of the Anthropic Principle listed earlier among the cosmology principles and is taken up in another text. As far as the Multiverse Hypothesis is concerned, it remains an un-provable hypothesis that continues to engage in creating universes outside ours that cannot be proven by observations.

If the term 'bubbles' in Guth's metaphor of 'eternity of bubbles' is replaced by the term 'tortoises', the new metaphor would be 'eternity of tortoises'! With that said it sounds like the story of the 'little lady' in Hawking's book when she had a disagreement with the famous British scientist and emphasized that the Earth was supported on the back of a tortoise and that it was tortoises all the way, each supporting the one before it! The difference between Guth's metaphor of the 'eternity of bubbles' and the metaphor of the 'eternity of tortoises' is that one is wrapped in mathematical garb and the other is not and one crosses the boundaries of the universe and the other is limited to the Earth, the latter disprovable by observations and the former un-provable, yet both are based on counting the numeric up to infinity.

Flatness Problem

When the universe came to be, it could assume a likely shape: a sphere like a drop of water, or a closed curved surface like a soap bubble, or an open curved surface like the shape of a saddle, or flat like a sheet. Based on observations the cosmologists say that the universe is flat without any curvature in it. It is counter-intuitive to imagine that when the universe was only 10^{-35} sec old, confined in a point in the space having uniform mass/energy distribution in it, it would yield a universe without a warped up space. Massive celestial objects are supposed to massively warp the space around them! Not only that, the modern cosmology suggests that universe's history was decided in that short Quantum Physics time-regime. Nevertheless, the meaning of flatness of the universe is addressed in the following as the "Flatness Problem".

It was known for decades and has also been confirmed from the recent measurements of CMBR from COBE satellite and Hubble Telescope that the density of matter in the universe was very close to the density required to make the universe flat. What does this flatness mean? The flatness property of the universe for cosmologists/astronomers has a different meaning than commonly envisaged. One comes across a lot of confusion about the flatness of the universe. The flat universe does not mean that it is flat like a sheet with negligible thickness compared to its length and width. This flatness has nothing to do with the outside dimensions of the universe, rather it relates to the geometric properties exhibited inside the universe. The observable universe is spherical all around us and it does not look flat. In fact the entire universe could be spherical or ellipsoidal but it is still said to be flat. This property of flatness is observed inside the space of the universe.

To explain this, the astronomers take up the triangle-geometry. If a triangle is drawn on a flat plane like a table top, the sum of the three angles formed by the triangle add up to 180°. If a triangle is drawn on the surface of a sphere, a curved surface, the sum of three angles of the triangle is less than 180°. If a triangle is drawn on an open curved surface like a saddle, the sum of three angles of the triangle is greater than 180°. Another example to explain the flatness of the universe one draws two parallel lines on a surface. The parallel lines drawn on the flat surface will never meet even if extended to infinity. If parallel lines are drawn on a sphere such as the Earth, the lines running north-south, the parallel lines at the equator would begin to converge and meet at the pole. In case of a saddle shaped surface the two parallel lines will diverge.

Our universe is said to be flat because if a triangle is drawn in it in any plane formed by distant galaxies the angles of the triangle will add up to 180°. It is therefore also said that our universe obeys the Euclidean Geometry and this simple geometry could be used to deal with the universe. There is no space warping in such a universe.

The flatness is also related to the density of the universe. The balance in the density of the universe is so critical that if it were a billionth more than what it is, the universe would have collapsed under its own gravity within a fraction of its age and if its density was a billionth less, it would keep expanding. No doubt that the criticality in the universe's density is mind-boggling and it is equally mind-boggling as to how the colossal universe with extreme mass took birth in an explosion! The Big Bang explosion to a modern cosmologists is a special kind wherein all nice things have happened and if they did not the Cosmic Inflation could make them happen!

The Big Bang cosmologists surmise that the Cosmic Inflation flattened the universe when it was subjected to 'False Vacuum' when it was also homogenized, made isotropic and led to further expansion. The flat versus the highly curved universe problem is also being explained at the Quantum Physics level. The Big Bang algorithm is tweaked to take care of the flattening of the universe during the Cosmic Inflation lasting for a duration of only trillion-trillion-trillionth of a second.

To learn that the Vedic Cosmology also creates, naturally, a flat universe, is mind-boggling! But this flatness relates to outside shape of the universe. The flattening of the Vedic universe is owing to the centrifugal force generated by the inherent rotation of the entire universe discussed in Chapters 9 and 10. It should suffice here that the rotation of the universe flattens it in the shape of a wheel.

Magnetic Monopole Problem

Without elaborating upon this deficiency of the modern Big Bang Hypothesis the author summarizes that: among all kinds of fundamental particles created in the modern Big Bang mechanism, the particles called the 'magnetic monopoles' should also be created, but they are not observed anywhere in the universe.

Red-shift Problem

The American astronomer Halton Christian Arp is well-known for his '1966 Atlas of Peculiar Galaxies' and other publications that raise doubt on the validity of the Big Bang Theory; one of his important publications is the book entitled 'Seeing Red: Redshifts, Cosmology and Academic Science'. In an article he said: *'There is now a fashionable set of beliefs regarding the workings of the universe, greatly publicized as the Big Bang, which I believe is wildly incorrect'.* Among other things, Arp's dissention from the Big Bang Theory is based on the relationship between the amount of redshift in the spectrum and the recession speed of the galaxies. The higher redshift represented greater speed of recession of the galaxies and that they were also farther away from an observer on the Earth. Arp in his observations found that some quasar-pairs (quasi-stellar objects) exhibited high redshift that would put them at great distances from the observer on the Earth, but they in fact belonged to a nearby galaxy that showed low redshift of its own. Also, older quasars which are supposed to be the farthest astronomical objects must be moving away with the greatest velocities. The farthest quasars however did not fit in the highest redshift category because the older quasars exhibited lowered redshift. This leads to the conclusion that the speed of a cosmological object was not necessarily responsible for its redshift and conversely, a redshift did not necessarily represent the distance of the object. Also, the present general understanding from Big Bang that the farthest galaxies were moving away with the greatest speeds was not true either.

Cosmological objects, including the Sun, exhibit intrinsic redshift of their own, even when not moving away from the Earth. The intrinsic redshift in all illumined cosmological objects relates to their state and age. The other source of redshift is thought to be the Compton Scattering where a high energy light photon upon hitting a free electron or hitting an electron loosely bound in an atom, is scattered with its wavelength shifted towards red, just what the Doppler Effect does in the case of a moving source. Those who suggest that the observed redshift from far away galaxies is the accumulation of Compton Scattering in the light passing through the inter-galactic matter and through the predominantly present Hydrogen, argue that in the Big Bang Hypothesis the notion that the 'red-shift equals distance', is erroneous and needs a re-examination. If the universe is homogeneous and isotropic the red-shift must represent that too. But it seems there is observed asymmetry between the east and the west red-shifts. Most of the details related to the red-shift are an issue in the Big Bang. They are not included here for they relate to complex concepts specific to astronomy. The red-shift will however be discussed again in Chapter 9 in relation to the Vedic Cosmology with new mechanism for the creation of the universe. Part of the red-shift according to the Vedic scenario could just be an artifact of the dynamics of the rotating universe.

Dark Matter/Dark Energy Problem

What we observe and see as a part of the splendour of Nature in the form of stars and galaxies is only 4% while the rest of the 96% that we do not see is supposed to be the hypothetical Dark Matter and Dark Energy discussed above individually. Dark Matter and Dark Energy are said to be permeating through the universe and forming a huge proportion of it, however, no laboratory experiments have been able to harness either of them. For balancing the field equations to get desired results the Big Bang Theory hinges on these two unknown entities of the universe. They are said to be parameters for 'fine tuning' by the adherents of the Big Bang but are thought to be the 'fudge factors' by the dissenters. The problem is that the true nature and composition of either is not known.

Other Thoughts on Big Bang Hypothesis

In addition to the questions raised by other physicists on the Big Bang Hypothesis with the Cosmic Inflation being a sub-hypothesis within it and False Vacuum energy as a sub-sub-hypothesis to create Cosmic Inflation, this author has some संशय: (*sanshaya* = doubt) and the questions of his own.

(1). Big Bang Hypothesis at Cross-roads

The momentary False Vacuum with the property of repulsive gravitational field is said to be the sole cause for the explosion and Cosmic Inflation in the Big Bang. The 'naïve Cosmic Inflation' of the 1980s, as Steinhardt labeled it, created a single universe. The 'smart Cosmic Inflation' is said to continue creating multiple universes at other points in space outside our universe. In this never ending Cosmic Inflation, always present somewhere in space outside our universe or possibly even within it, the overall Big Crunch (the collapse of universe unto itself) scenario does not seem likely. Even if our universe is to go for a Big Crunch, the physics logic dictates that it will never take place within the framework of the Big Bang Hypothesis, even if the gravitation dominates the Dark Energy. It is the Cosmic Inflation that would not let it happen! The universe cannot come to an end. This is discussed below.

To understand this let us take an imaginary reverse ride on the universe moving towards the Big Crunch. To move towards a Big Crunch means the gravitational force will pull all galaxies closer together increasing the matter density of the universe. While increasing the matter density the universe will reach a point in time when its density is the same as it was at the end of Planck Era at 10^{-43} sec after the onset of Big Bang. Recall that this also was the time when the ghost of the False Vacuum manifested itself at the end of 10^{-43} sec. An unavoidable situation must arise at this point– the density of matter in this Big Crunch reaches back again to what it was when the universe was being created. Such a density value is reached before the matter even reaches that finishing line called space-time singularity.

At this point in the universe's life the unavoidable indeed is the apparition of the False Vacuum again. This False Vacuum must do again what it was hypothesized to do when the universe had entered from Planck Era to Inflation Era. The repulsive gravitation in matter under the False Vacuum sends the universe back on a new course of Cosmic Inflation leading to the creation of exactly the same universe that we live in now. Obviously the ghost of the False Vacuum would not let the universe go right back to the space-time singularity where it had started first. This is equivalent of saying that the ghost of 'repulsive gravitation' will never let the universe reach back to its original condition of space-time singularity. Therefore, this is also equivalent of saying that the matter density condition creating the False Vacuum, again becomes responsible for the Big Bang explosion.

Thus, if the False Vacuum is a reality, the true Big Crunch will never be reached. The consequence of this is that the universe will oscillate between the Inflation Era and its most expanded era. This is also a kind of birth-rebirth scenario for the universe. The big question to cosmologists is – if the Big Crunch is part of the modern Big Bang Hypothesis, it is meaningless to start the universe from the space-time singularity because that point is never reached, then why not begin the universe from the Cosmic Inflation and save 10^{-43} seconds from the universe's life? If the cyclic birth of the universe is discounted by the Big Bang cosmologists they have only one choice for the fate of the universe - the Big Chill. The universe in a Big Chill scenario can be created only once because in an eternity of Big Chills there are no more space-time singularities left. The consequence of this is that our universe is the last.

Hawking and others who professed that the gravitation of the universe ultimately will bring all matter of the universe to a Big Crunch have to consider the ghost of the False Vacuum which will never let the universe pass-by towards the dimensionless Big Bang singularity. The hurdle of the False Vacuum, if real, poses one other logical difficulty that the Time entity pre-existed at 10^{-43} sec. Then who is competent enough to proclaim that the Big Bang began at t = 0 sec in a singularity which can never be reached again till eternity? If the False Vacuum is a real phenomenon, the Big Bang singularity is only an illusion in the mathematical construct of cosmologists and if the Big Bang singularity is a real phenomenon then the False Vacuum is an illusion. The only way out of this 'cosmic dilemma' is to advocate the Big Chill.

So, now the question is – is it the Big Crunch or the Big Chill? One of the following two things should happen:

(a) If it is a Big Chill the universe is a onetime phenomenon and the Space-entity cannot again form a particle let alone a universe?

or

(b) If it is a Big Crunch then the universe is in a permanent mode of rebirth cycle oscillating between its full-fledged life and the False Vacuum in Inflation Era without ever reaching back to its space-time singularity.

The author summarizes this situation in the following deduction:

Deduction:

"*If the False Vacuum is a reality the space-time singularity is only a mathematical imaginary condition. Conversely, if the matter of the universe reaches the Big Crunch acquiring its true definition of space-time singularity, the False Vacuum is an imagination. If there are no convincing answers to these questions then the Big Bang Hypothesis itself is modern cosmology's big imagination.*"

There is a likely-hood of some explanation about the False Vacuum dilemma discussed above; however, there is recent thinking in cosmology advocating for a perennial cycle of birth and death. This is no different from the qualitative Vedic scenario with the difference that the universe in the Vedic cosmology takes as long a rest in the *pralaya* (dissolution phase) as long it is active in its *srishti* (creation phase). Several USA scientists at Princeton and Cambridge are working on a theory different from presently conceived in Big Bang, called the "Cyclic Universe Theory" (Steinhardt and Turok, May 5, 2006, Science) where the universe comes into existence with a bang not once but in endless cycles of death and reincarnation. Obviously the present cosmology is at the crossroads, not knowing which route to take and where the universe would end up, in a Big Chill or in a Big Crunch. To put it metaphorically, as it stands now, the universe at the hands of contemporary Big Bang Hypothesis is a confused entity!

The term False Vacuum itself is confusing because on one hand it is a vacuum but on the other hand it is not a real one. Etymology of the term 'false vacuum' conveys that 'it is a phenomenon giving the false impression of vacuum'. A vacuum by definition is devoid of matter and therefore it is supposed to be at low pressure or no pressure. The False Vacuum should then have high pressure. This is logically possible when there is matter in it yet exhibiting absence of matter to make it false. A Big Bang scenario is far from this where the early universe conveys the impression of matter at high pressure (opposite of vacuum) and does not convey the impression of a strong repulsion which is implicit in the term 'false vacuum'. Why introduce terms in science that increase confusion in the Big Bang Hypothesis?

(2). Matter Speeds in Cosmic Inflation and Quark Lepton Eras – A Puzzle

Others may have raised questions about the questionable speeds of matter in the Cosmic Inflation and Quark Lepton eras that are in violation of the Theory of Relativity. The Big Bang cosmologists may have even put forward some convincing explanations for it. However, this author wishes to point out an interesting puzzle.

The matter of the universe from the onset of the Big Bang at $t = 0$ seconds onwards, rushes outward expanding the universe from a point in space-time singularity to ever increasing dimensions. As presented in Table 2.1a, a Big Bang era (epoch) is assigned the time of its beginning and the time of its end. The size of the universe corresponding to each era is also listed. An average speed of the moving matter within an era can be calculated from the change in radius of the universe in that epoch divided by the time interval of that epoch.

The Table 2.1a is basically repeated here as Table 2.1b with small differences in parameters. The last column presents the calculated average speeds of matter in the five eras of the universe as the multiple of the velocity of light c $= 3\times10^{8}$ m/sec. The numerical factor with c when greater than 1.0 indicates that the average speed of matter in that era is greater than the velocity of light.

Table 2.1b: Average speeds of expansion in the Big Bang.				
Era (Epoch)	**Duration from t= 0 sec**	**Radius of the Universe**	**Speed of matter in different eras**	**Times the speed of light $c = 3x10^8$ /s**
Planck	**0 to $5.4 x10^{-44}$ s**	**$0.8 x 10^{-35}$ m**	**$1.482x10^8$ m/s**	**0.0 to 0.5 c**
Cosmic Inflation	**10^{-36} to 10^{-32} s**	**1 m**	**10^{32} m/s**	**$3.33 x 10^{23}$ c**
Quark-Lepton	**10^{-32} s to 10 s**	**$4x10^{13}$ km**	**$4 x10^{15}$ m/s**	**$1.333 x 10^7$c**
Radiation	**100 s to 3000 Yrs**	**$5x10^{14}$ km**	**$5.285 x10^6$ m/s**	**0.0176 c**
Matter	**3000 Y to 13.8 BY**	**$3.4x10^{23}$ km**	**$7.812 x 10^8$ m/s**	**2.604 c**

Notice that the three middle epochs exhibit speeds of matter higher than the speed of light with the Cosmic Inflation and Quark Lepton Eras registering speeds of matter ridiculously higher than the speed of light. This is a puzzle created by the Cosmic Inflation hypothesis. It is not the puzzle that inflation should have the highest speed because the inflation by definition is arbitrarily designed to accelerate the process, but what is being accelerated is the matter of the universe which should follow common physics. A process to have speeds higher than the speed of light is in violation of the Theory of Relativity and is therefore quite puzzling. The highest of all speeds being $3.33x10^{23}$ times the speed of light is in the Cosmic Inflation Era. With the value of the velocity of light c taken into account the highest speeds attained by matter in Cosmic Inflation is $1.0x10^{32}$ m/sec. This in common expression is one hundred-million-trillion-trillion meters per second, which for all practical purposes is an infinite speed. The question is – does the primordial matter of the early universe travel with such speeds in Cosmic Inflation Era?

Even in the beginning of the Radiation Era, when the universe is dominantly radiation, the speed of radiation (matter) is higher by a factor over 10 thousand than prescribed in the physics of the present universe. Do photons of the Radiation Era travel faster and then slowdown in another era? There is no known physics process that would bring all photons back to their speed called 'the speed of light'. The photons travel with the speed of light because by nature they are mass-less. What will make them even lighter to fly at billion-trillion times their normal speed? Why talk about the photons, even the matter of the early universe is spreading out at these incredible speeds.

Notice from the first row in Table 2.1b that the matter in the first era (Planck Era) moves by acquiring maximum speeds that are only about half the speed of light, which is reasonable. The matter in this regime may be of undefined nature but it is matter and not the radiation (light). The matter is to move with speeds less than the speed of light, which is alright in the Planck Era, but as soon as the Cosmic Inflation begins, the matter is shown to expand violating Relativistic Mechanics. The simple logic is that for the Cosmic Inflation to have speeds higher than the speed of light, the matter in that regime must have lost mass and become something subtler than a photon of light to travel at such higher speeds, as shown in the second row of Table 2.1b. There is no known mechanism in physics that would suggest that the conclusion of Einstein's Relativity could change under certain circumstances that would permit matter speeds higher than the speed of light. Any matter approaching the speed of light by the definition of General Relativity should exhibit infinite mass. It is during the Radiation and the Matter Eras (last two rows in Table 2.1b) when the speeds are shown to return to reasonable values. Therefore, from a normal physics perspective there is something inherently wrong with the Big Bang Hypothesis with Cosmic Inflation as its soul.

To explain a situation such as this the cosmologists could resort to more hypotheses. Would that make progress in finding the 'truth'? If the resolution of this speed discrepancy is based on more non-provable hypotheses, without provable physics principles, does it belong in the physics arena? It seems that the problem is not with the resolution of discrepancy but with the Big Bang Hypothesis itself. The Cosmic Inflation has no physics rationale to be considered in cosmology but it is there because it helps create a good-looking imitation of the universe. It is time to give a better start to the Big Bang Hypothesis, if that is the hypothesis to carry forward. The modern cosmology needs a fresh start.

From the articles on cosmology this author noted some explanations on these physics-defying and commonsense-defying speeds of matter, especially in Cosmic Inflation and Quark-Lepton Eras. One explanation is that it is not the matter of the universe that is travelling beyond the speed of light, it is the space-time continuum itself that is expanding at tremendous speeds and the matter is only taking a ride on it. Implicit in this premise is that no physics laws are violated because the matter itself is not moving with these ridiculous speeds. This raises several other questions. If the newly formed matter in the space-time singularity during the Planck Era follows the rules of Quantum Physics and exhibits only a third of the speed of light and later in the rest of the life of the universe the matter follows physics but if the Cosmic Inflation creates speeds of the order of 1×10^{30} m/sec, the Cosmic Inflation must lie outside the realm of physics? The speed of matter right up to the radiation Era is non-physics.

If one says that it is the space-time singularity that behaves that way and not the matter in it, then it is not sufficient to assert that the Space and Time are real entities but a cosmologist has to explain what is at the fine structure of these quantities that makes them behave differently and only once? What physics principle causes the space-time to expand the matter at these speeds? To put it differently - what is the known physics mechanism that stretches the space as well as time to expand the so-called space- time continuum at extreme speeds?

Energy is responsible for expanding a substance. For energy to expand a substance, the energy has to permeate the structure of the substance. Those who suggest that it is not the matter but the space-time that is expanding at such high speeds, have to accept that the expanding energy permeates through the structure of the space-time continuum. So, now the new question is - does the 'space' or 'time' or the space-time continuum have a structure? More hypotheses could be brought in to guesstimate the answers to this question; however, the fact is that physics does not seem to have an answer to this question. Therefore it does not sound logical to say that during Cosmic Inflation when the universe expands only the space-time expands and the matter only takes a ride on it.

Another problem with the argument that the space-time expands with unrealistic speeds and the matter only takes a ride on it relates to the interaction of the matter and the expanding energy, whatever that might be. At its most fundamental level even the matter is not a continuum but it is a corpuscularized entity because it is created particle by particle. Therefore, the primordial matter of the universe after the Planck Era has a structure for the expanding energy to permeate. There is no strong logic to suggest that the energy permeates the space-time structure and not the matter structure. Therefore if the space-time is expanding at ridiculous speeds the primordial matter embedded in space-time must also move at the same speeds. So, now the new question is- is the matter expanding the space-time that holds the matter or is the space-time expanding and giving a ride to the matter?

Another suggestion to resolve this speed dilemma is that because the expansion is so fast that the matter has no time to move and therefore there is no speed in matter and therefore there is no speed violation. If the 'space-time' has time to expand, the space also has time to expand. Then why does matter not have time to move? The example of two high-speed jet planes flying in opposite direction, separating from each other with twice the speed is generally presented. This example is not valid here because the speeds involved (Table 2.1b) are extremely high.

(3). Contemporary Big Bang in Ancient Logic

When the inherent logic in the Big Bang Hypothesis and its hypotheses within hypotheses is examined from its beginning in a dimensionless space where space and time are first born, not knowing out of what and with what mechanism and when the False Vacuum in the Cosmic Inflation is said to cause a big explosion, pushing matter at speeds transcending both, the imagination and physics, this author puts the label of हेत्वाभास: (*hetvaabhaasah*) on the Big Bang Hypothesis. Rishi Gautam's Nyaaya Sutra (Nyaaya Philosophy: the logic system) considered as the oldest of the six Vedic Darshans written in the far antiquity of the Vedic Civilization, anterior to the Mahabharata period near 3140 BCE, defines the term *hetvabhasah* as '*something that sounds logical but it is not*'.

Rishi Gautam's full *sutra* is - सिद्धान्तमभ्युपेत्य तद्विरोधी विरुद्ध: (1/2/6) (*Siddhaantam-abbhyupetya tadvirodhi viruddhah*) which translates into:

> "*Accepting one principle on the basis of another principle which is based on pseudo logic*". (Nyaya Darshan, 1/2/6)

(4). More Problems with the Contemporary Big Bang Hypothesis

Not detailed here, however, there are a number of other Big Bang deficiencies pointed out by physicists, astronomers, astrophysicists and cosmologists. These are accessible to a reader from the open science literature. To name just a few of these deficiencies: density discrepancy between the present and the infant universe; inconsistency in age and evolution of galaxies; too many assumptions in the Big Bang to create the right proportions of elements; some cosmological formations seem older than the Big Bang universe; no proof as yet of equal amount of matter and anti-matter being formed in the early universe when the present universe is dominated by matter only. There are more such concerns.

Other Modern Cosmology Theories

There are a number of cosmological theories under the general heading, 'Classical Cosmological Theories'. Only a brief discussion on them is within the scope of this book. Some of these theories fall under "Steady State Cosmology" where the universe is considered to be infinite and is in a perennial process of new matter creation. In a steady state scenario there is no beginning or end of the universe. These theories came forward as an alternative to the Big Bang Hypothesis. Galaxies in Steady State Cosmology are considered not to be moving away from each other and therefore the universe is not expanding. The basic premise in steady state cosmology is that the matter is continuously being created and being destroyed (disappears), sustaining the eternity of the universe.

Jayant Narlikar, an astrophysicist, the author of Hoyle-Narlikar gravity theory fame, supports the steady state cosmology and also implicitly suggests that the General Theory of relativity 'started on a wrong foot'. It implies that the Big Bang Hypothesis approached the problem of the origin of the universe in an unnecessarily complex manner using the space-time continuum which ends up into singularity where the physics breaks down with vanished geometric definition of space at infinite temperatures. Narlikar rewrote Einstein's field equations and in his theory (Annals of Physics, 107, 325, 1977), Newton's gravitation constant G is no more a constant, rather it decreases with time, the mass of a particle varies with time and is a function of the rest of the mass of the universe, the space is flat and time is uniform. It seems that Narlikar's theory explains intrinsic red-shift and the red-shift in general. It is also supported by observations and thus remains congruous with deductions of relativity.

The doubts and questions about this theory relate to the steady state of the universe itself and that the mass of a newly created particle is proportional to the rest of the mass of the universe. If the universe is infinite, its mass must also be infinite, then how does a newly formed particle somewhere in the universe sense as to how big it has to be born to maintain proportionality and to keep a balance? What is the mechanism of this instantaneous communication between all parts of the universe to keep the mass of the universe steady?

It could be said that under the Steady State Cosmology the universe is infinite in space and time and it continuously creates the feed that is being consumed by the perpetual machine of the universe. There are other contemporary theories of the universe that put a limit on the age of the universe, some accept an age much smaller than Big Bang's 13.8 BY and others go up to a trillion years.

Another cosmological theory is the "Cyclic Universe Theory'. It was put forward in 2002 by Paul Steinhardt of Princeton University and Neil Turok of Cambridge University (May 2006, Science) wherein the universe recycles with new matter and energy created about every trillion years when two sheet-like branes of Strings Theory collide along an extra dimension of space. In the proposed cyclic universe, the Dark Energy responsible for the universe's expansion and represented by the Greek letter lambda Λ, discussed earlier, is theorized to diminish with time and with every cycle of the universe's birth. This process is said to have led to the creation of a unique universe each time. The present value of Λ is said to have created our universe which is unique the way it is. The mechanism of this theory is different but the eternal cycle of the universe's reincarnation is similar to the thought in the Vedic Cosmology.

Author's Queries and Deductions from Modern Cosmology

An essential precondition for the Big Bang Theory is the pre-existence of all matter (Lemaitre's super-atom) at infinite temperature, at one point called the space-time singularity. On the other hand the particle physicists insist that the universe came out of the space-time singularity out of nothing. If not pre-existent, all matter of the universe in the Big Bang Hypothesis is said to have manifested at one dimensionless point in space-time singularity. A common understanding from the behaviour of the present universe is that the temperature is always exhibited by the presence of matter. Therefore, matter being the shelter of temperature, it is imperative that matter pre-existed in the singularity, or co-existed simultaneously. Without this requirement one cannot say that the singularity was at an infinitely high temperature before the Big Bang. If the Space was non-existent in space-time singularity but it was created then at what point was its start?

If one assumes that the Space is nothing but an effect of the presence of matter, then it is not a physical entity by itself because an effect cannot assume the status of an entity. Also, those saying that the Space and the Time are a part of the same continuum without their independent existence, must then agree to the logic that it would not be meaningful to call them two separate entities. Pertaining to the Space and the Time entities, several arguments based on the Vedic frame of reference were presented in Chapter 1. More unique insight into these entities is presented in Chapter 5.

One cannot form a space-time continuum without the Space and Time being two independent and different entities. On the other hand if two entities are merely the "effects" of two different 'causes', one cannot create a continuum with two different 'effects'. This author emphasizes the philosophical principle that two different 'effects' cannot form a continuum because one does not see that happening in the universe. This basic principle negates the idea of forming a space-time continuum. When the fine structure of the space-time continuum (if it exists at all) is considered, the logic in forming the space-time continuum itself does not stand.

The Time must precede the Space because otherwise in what would one create a Space for the Space to do the next thing, like its expansion, which is a time-dependent phenomenon? Since we are talking about creating infinitely small steps of the entities and if the Time is created first, the very first step of Space would lag behind the time-step. If the Space and Time are created in succession one will miss the other at least for a small time span before taking part in a continuum. The only way they form a space-time continuum is to string them in succession with time first and with a change in the nomenclature – 'time-space' continuum rather than 'space-time' continuum. Also before the step of Space is created, the step of Time is alone. Then who is to guarantee that the step of Space would not miss out the step of Time and render the space-time continuum to be only an imaginary process! If the two entities are quantized, the time quantum must precede because otherwise what time-stamp will the first quantum of Space have? Can anything be created without it being in Time? It does not seem likely. If somehow their first quanta are created simultaneously, to begin forming the space-time continuum, cosmologists must explain what physical process would create two entities of differing nature simultaneously. The cosmologists do not know this yet. It is not obvious from the Big Bang Theory as to when, before or after $t = 0$ sec the Space was made available and to what extent, for the big explosion to take place in. Therefore, the Big Bang should have either a mechanism to create both, the Space and Time, or there are two independent mechanisms for each with simultaneous onset to create them simultaneously. For the time being the modern cosmology has not reached a point where clear answers to these questions are available. Meanwhile a certain degree of confusion about Space and Time certainly persists in the Big Bang Hypothesis.

The Vedic Cosmology does not have to worry about when Time and Space were created because these entities in Vedic Cosmology are eternal entities.

Most of universe's history of 13.8 BY has been spent to form celestial bodies and galaxies and just keep expanding. The universe after spending about 9 BY created our solar system with the Earth. Anthropology suggests that the thinking humans with analytical ability emerged on Earth only a few million years ago. It is difficult to explain why humans have not thought so deeply and scientifically about the cosmos 1 million years ago or 5,000 years ago? Why did they study the cosmos only in the past few centuries? Why is it that a theory like the Big Bang with the concepts of Time, Space and Matter worked out only recently? Why did the intellectual potential required for discovering the Big Bang did not exist in humans, centuries earlier? This apparently defies the principle that governs the rate of the evolution of a biological system that ultimately led to the thinking human-beings. We will not take this up here and instead just state that there have been times before, far in the antiquity of Vedic culture, when the sages, thinkers and the *rishis* have thought deeply about the origin of cosmos and have left behind intriguing theories and principles some of which are worth looking into. They not only addressed the anthropic enquiry regarding our unique universe and the subject of cosmology, but they have also shed considerable light to shatter the myth that any serious progress in science-thought has taken place only during the past few centuries.

The chapters that follow are an attempt to bring out the science-thoughts that are ancient yet reflect universality. If the vocabulary pertaining to a specific subject exists in a society the subject itself must have been studied by that society. The existence of rich terminology in cosmology in the ancient Vedas indicates that the subject of cosmology was explored in the far antiquity of the Vedic Civilization.

Some Vedic Terms of mechanics used in Vedic Cosmology/Astronomy

To talk about the creation of the universe without giving some details of the process and giving credit for the creation to the Creator, is perhaps sufficient for a spiritualist but not for an analytical spiritualist, a philosopher or for a scientist whose emphasis might be to know more about the process than the processor. The theistic religions generally have credited the Creator for creating the universe and have said that the details and the knowledge of the process of creation are 'with the Creator'. Therefore, there was no necessity for developing cosmological terminology. Consequently, the terminology in cosmology in religious texts is found at a limited level. However, one would find that the Vedic terminology pertaining to cosmology, astronomy and physics in general encompasses the concept of creation of the universe down to the detailed nomenclature for the objects of the solar system and the related physical principles and phenomena. This should certainly tickle the minds of the intellectuals with theistic and atheistic orientation alike. The rich Vedic terminology not only relates to the Sun and the solar system, it also leads one to obvious and intriguing details about the Moon and the Earth, presented in Chapter 8. It does not stop at just saying that the Sun illuminated the Moon and that the Earth was round and in orbit around the Sun, the Vedic Cosmology in fact talks about the geology and geological substances and phenomena pertaining to the Earth. And all this was put together in the far antiquity of the human civilization. The *rishis* of the Vedas had a mind-grabbing story about the creation of the universe and they expressed their thoughts in rich Sanskrit terminology that is bound to attract the attention of intellectuals sooner or later.

If asked to narrate the story of the creation of the universe, a cosmologist of modern times will gladly start off with the Big Bang singularity and soon he would introduce a cornucopia of technical terminology necessary for him to proceed towards the end of the story. Modern cosmology is so involved that even with a rich terminology at hand a cosmologist (with some exceptions) could easily be labeled as 'bad story teller' because hidden behind each term is a complex concept that if explained, will put a listener to sleep. Term-rich modern cosmology, even if the terms were borrowed from Latin or Greek, is however well-equipped with terms to explain complex phenomena. If not from the Latin or Greek a modern cosmologist would create new terms when the need arises. The meaning conveyed by the term Big Bang is 'big sound' that no one heard! Yet that seemingly meaningless term becomes meaningful with the history of the universe hidden in it.

A natural question arises – is the Vedic cosmology term-rich enough to tell the story of the origin of the universe? The short answer to the question is yes. The validation to the answer requires listing and defining a large number of Sanskrit terms pertaining to the subject of cosmology, which is deferred to the chapters that follow. To give a flavour of the Vedic terms of mechanics that relate to cosmology, the author is presenting here just a few terms that reflect the richness of the Sanskrit language and the intellectual depth of the wisdom of the Vedic *rishis* who used those terms in their mantra expressions.

A Sanskrit term is a lens for its own meaning to create its instant real image. For example a frequently occurring term is *chakraprasaran.* It represents two actions: '*rotating around an axis and spreading radially, keeping its circular shape*'. Although the term has occurred in the context of Vedic cosmology to be discussed later, a simple example of *chakraprasaran* is the ripple effect created on the surface of still water when a pebble is thrown in it, where the spreading of the matter (water in this example) is in a circular fashion. The term has been used in the Veda mantras in the context of the matter of the universe spreading in space while spinning around an axis.

The other term close to it with the same prefix of *chakra* (circular) is *chakrampari* (RigVeda, 7/63/2) which stands for '*circular motion around an axis keeping the rotating object within a circumference*'. Its application, as described in this mantra, is for the Sun's control on the orbits of the objects of the solar system. Factually in that RigVeda mantra the usage of *chakram-pari* is in conjunction with three other most used terms of mechanics, समानं (*samaanam*) means **invariant** or **stable;** अविवृत्सन् *(avi-vritsan)* means **repetitive circular motion** and धूर्षु (*dhoorshu*) means the **axis** (of rotation). The subject matter of the mantra being the *Surya* (the Sun), the usage of these terms creates a picture so vivid that one can have no other meaning of it but - *"The Sun keeps all planets rotating around their stable axes in their stable orbits"*. This is not the only mantra teaching science and delineating what it teaches, a large number of them express similar thought explicitly. More of such mantras that form the basis for the Vedic Cosmology Principle are presented in this book.

One interesting observation for a linguist is the sound of the word *chakram-pari.* The pronunciation of the word *chakram* is not very far from the pronunciation of the term 'circum' in modern mechanics; also having the same meaning. The whole Vedic term *chakram-pari* is not too far in sound from 'circum-peri' leading to 'circum-perimeter' or 'circumference'. Not just that but *chakram-pari* also has the same meaning as the term circumference. Other examples of terms in mechanics are: *pari-kraman* which defines the '*step-wise motion on a circle*', while *paryaplavan* (Shatpath Brahmana, 11/1/6/2) is '*a large scale circular rotation around an axis of a system in floatation*', like the rotation of the solar system or a galaxy in space. The term *syandan* (Shatpath Brahmana, 3.9.2.1) is '*flow in togetherness*' (in reference to the flow of *aapah* the primordial matter of the universe), *urmim* is *vibrations or waves*, विततः (*vitatah* in RigVeda 10/129/5) is '*instantaneous expansion*'.

All thoughts and terms in the Veda mantras are to enhance the knowledge in all realms of the creation, the physical and the spiritual. When one comes across these specific terms of physics and mechanics in particular, one cannot help but emphasize that they are primarily to enhance the knowledge of the physical aspect of the creation. The sheer presence of such specific terms in the mantras, with specific physics meaning contained in them, tells us that there is physics in the Vedas. The terms in this small sample and more to come in the forthcoming chapters, relate to the early stages of the universe, the objects of cosmology and the solar system. These specific terms lead to the conclusion that the Vedas in the antiquity of civilization talked about cosmology, astronomy and the heavens. This is worth looking into from the science perspective. In the chapters that follow, the mantras and individual Vedic Sanskrit terms are analyzed and interpreted to build the story of the universe from the Vedic perspective. The next chapter is devoted to proving the historicity of Vedic references used in this book. All Vedic references were formulated in the distant antiquity of the Vedic Civilization, anterior of 5,000 years.

CHAPTER 3: The Historicity of Vedic References - The References on Cosmology Precede 5,000 years

The Vedic Reference Hypothesis

The reference material used for this book is in three basic categories: 'dated', 'partially-dated' and 'undated'. The dated material includes published books, records and articles that are available from publishers, libraries, journals and web-sites with authorship and dates in known history. An example of this category comprises the recent peer-reviewed published journal papers by a large team of Indian national scientists on the geological discovery of the ancient, now extinct mighty River Sarasvati that flowed from the Himalayan foot-hills to the *Paschim Sagar* (Arabian Sea) in the North-West India. Another example would be the historical records at the four different Aadi Shankaracharya Mathas (monasteries), written serially with calendar dates starting near 500 BCE till now. An important feature of these records is the congruity among them even though they were maintained independently in different locales of the Indian subcontinent, that were arduous to travel from one to the other. These records commenced from the laying of the foundation-stone by Aadi Shankaracharya himself at Dwarka Matha in Gujarat state on *kaartik badi trayodashi* in the year 2648 of Yudhishthir Era (YE). This exact date in the luni-solar calendar is the 13th of the dark lunar phase of the YE Calendar that was in use at the time. This could be converted to the day, month and year in Common Era (CE) or Christian calendar with accurate computer algorithms now available. However, a quick round off within the accuracy of a month yields that the above date falls in October 492 BCE.

The term 'recent' in historical accounts is generally used for events well-known and it relates to the age of a society. For a Western society the references of World War II may be considered recent whereas in a millennia-old Vedic society, the written records at Shankar Matha may be termed as recent. However, because the records at the Mathas are in calendar dates based on luni-solar calendars well-implemented and practiced in the society, with frequent entries of other important celebrations at these Mathas, they could be termed as the 'dated' material as well. These records have enabled this author to establish the historical accuracy of the Vedic references used in this book.

The 'partially-dated' references are those that are relatively ancient, that are referenced in authentic texts and are available as published books, manuscripts or records. Although some references may have the day, the month and the year of a calendar, the others may only record calendar year or the century. The ancient history books such as Rajtarangini written in Sanskrit by Kalhan in 1148 CE on the genealogy of the Kashmir kings going back to 5000 years from now, falls in that category. For the historians who are interested in data accurate to within a year, a book such as the Rajtarangini could be better than a 'partially-dated' reference. The age of the book cited in the published material in terms of centuries or millennia could be considered in the category of 'partially-dated'. Some of the references in this category may only have the name of the author and the year or a specific century assigned to his/her written text. Examples in this category are books and manuscripts referred to in the book of Satyarth Prakash written and published in 1881 CE by the great Sanskrit scholar of the Vedas, Swami Dayanand Sarasvati, mentioned a number of times in the previous chapters.

The third category of 'undated' references will generally include ancient writings for which it is difficult to assign a date or a period. These are the significant books that helped guide the values of the society, its education system, its governance, science, philosophy and religion. The identity of these books has remained as real as the existence of the people who have been reading them, learnt them by heart, hand wrote them in the form of manuscripts and books, and now are available in printed and published form. The ancient Vedic Literature beginning with the four Vedas and their Brahmanas (commentary books) right up to the six Vedic Darshans are included in this category. Of the six Vedic Darshans (philosophy books) the last two in fact fall in the category of 'partially-dated' reference material, because they relate to the Yudhishthir-calendar of the Aryans and are contemporaneous to two well-known milestones in the Aryan history: the onset of the Vedic natural calendar (*Srishti Samvat*) on February 18, 3102 BCE and the end of the Mahabharata War and the coronation of the emperor Yudhishthin in 3140 BCE.

References and the data presented in this chapter provide proof that the Vedic cosmological ideas presented in this book are indigenous to the Vedas, Smritis and Sutras and are based on independent thought of the Vedic *rishis* with no relation to the traditions and civilizations external to India such as the Judaic/Christian/Moslem traditions or the Egyptian, Babylonian and other ancient civilizations. Also these Vedic ideas, older than 5,000 years (3000 BCE), may have some remarkable similarities with the modern cosmological theories and principles, yet they are original and uninfluenced by modern cosmology, the latter taking off in the Western society during the past four to five centuries only. Even though the Vedic ideas are from the far antiquity of the Indian Civilization this author has arbitrarily placed a boundary near 5,000 years in the past, before which, or even closer to it, no other non-Vedic Civilizations generated ideas similar to those in the Vedas.

Though graspable only by the chosen few, the modern cosmological ideas and theories have developed systematically over the past century. When it comes to indigenous applied Indian Astronomy, it is based on observational and complex mathematical disciplines. This is contained in several well-known treatises in Sanskrit. When it comes to physics or cosmological principles laid down in Veda mantra formulations, they generally are not presented with pre-set mathematical equations to go with them. Nevertheless, the basic science thoughts in the mantras are similar in many aspects to those in modern physics and cosmology. The aim of this chapter is to establish that the two disciplines, modern cosmology and the Vedic Cosmology are separated by a big span of time and that the Vedic science-thought was not affected either by the Greek or by the early Judaic/Christian philosophy. The unidirectional flow of the Time entity dictates that the originality of the Vedic thought in the exalted minds of the *rishis* of antiquity was certainly not a beneficiary of the modern thought in physics and cosmology.

Not all ideas put forward in contemporary physics and cosmology were the final word of truth. Many deductions of physicists have weathered the times while many others have disappeared. Over a century alone a change in methodology has also brought about drastic change in cosmological ideas and scientific deductions. In contrast, a Veda mantra is an invariant entity. This invariance must have been arrived at after considerable mental exercise by a *rishi* or many *rishis*. Considering that the Veda mantras cite absolute numerical figures in billion years for the age of creation, the numerical time-division going down to milliseconds, the accuracy in the speed of light and so on, hints that such deductions by the Vedic *rishis* must have resulted from solid mathematical, observational and measurement methodologies. If not on the blackboard, or in the laboratory or with large aperture telescopes employed by contemporary scientists, the Vedic *rishis* must have had some real means to accumulate knowledge and give it finality in mantra formulations. Since there is no direct evidence of complex mathematical treatments in the available Vedic literature on cosmology it is also thought that it is lost in the antiquity.

Based on the 'dated', 'partially-dated' and 'undated' references used in this chapter and on the historical information gleaned from authentic writings, this author formulates the following **Vedic Reference Hypothesis**:

"The Vedic references on cosmology are ancient and precede 5,000 years"

This hypothesis is proven in the following pages. The aim is not to establish the absolute age of the Vedas and the other reference books in the Vedic Literature, rather it is to prove that the references from the Vedic Literature cited in this book are older than 5,000 years. To be specific, the Vedic references in this book are older than a milestone at February 18, 3102 BCE. This date in the ancient Vedic history, as mentioned before, has a direct relation to the social calendars of the Vedic Civilization and their relationship to the Vedic astronomy. The proof of the Reference Hypothesis is acquired by establishing the historicity of certain historical events in the ancient Indian history, as the Indians have recorded in their Sanskrit Literature and Sanskrit history writings, including the most recent state-of-the art genome study of the Indian populace. A comparison of this with the historicity theorized by Western Indologists over the past two centuries will also be taken up.

Proof of the Vedic Reference Hypothesis

The science of cosmology is the study of the history of the universe. The starting point of this history is also the starting point of the universe. If the starting point itself is misunderstood, there is a possibility that the universe is misunderstood or only partially understood. Nothing is final in the history of the universe yet. This is the situation even when exact science is used as an implement to excavate the history of the universe. More difficult from the exact sciences is the excavation of facts from the human history. Therefore, one has to depend on the best correlations and logical deductions to prove the Vedic Reference Hypothesis.

To understand the facts of the ancient history of a country or of a civilization one does not have the luxury of a cosmologist. To predict a universe, a cosmologist changes the parameters of the model, introduces new fudge-factors and carries out back-and-forth iterations to arrive at a final result. For one to understand the history of a country one has to depend on what existed there in the form of the written historical records, the literature, the monuments, archeological finds and other signs of development of its civilization. Compared to billions of years of history of the universe the history of the human civilization is negligibly small, yet the human history is riddled with contradictions and confusion. The best one can do is to logically analyze the information and draw a conclusion. The Vedic references utilized in this book have the same situation. The attempt by this author is not to resolve the historical miscalculations of some

Indologists alluded to in the previous chapters but to establish the positions of just a few important historical events in the Indian culture and civilization on a timeline segment of 5,200 years.

The citations of the Veda mantras in Darshan *sutras*, Smritis, Upanishads and the other Sanskrit Literature indicate that the references of the Vedas and the ancient Vedic Literature are anterior to the 5,200 year period. The reference to ancient texts is not just fleeting mention rather there is a strong historical correlation with the written Vedic Literature. A coordinate point in a history frame of reference needs three interdependent factors – a real geographical locale, real people and an actual time period. The specificity of these three factors together forms a specific milestone in the history of the place and the people. The importance of a milestone is reflected in the substantial contribution that a historical personality or the people at that time made to the society, be it in governance, or in the creation of literature or discoveries and inventions. A limited number of such milestones from the Vedic Civilization are identified and placed on a timeline with reference to the calendar in use.

Timelines A and B

The Indian history from two perspectives is represented by two timelines in Figure 3.1. The Timeline-A is based on the well-researched, reliably recorded historical information in the Vedic Literature. The Timeline-B is based on the

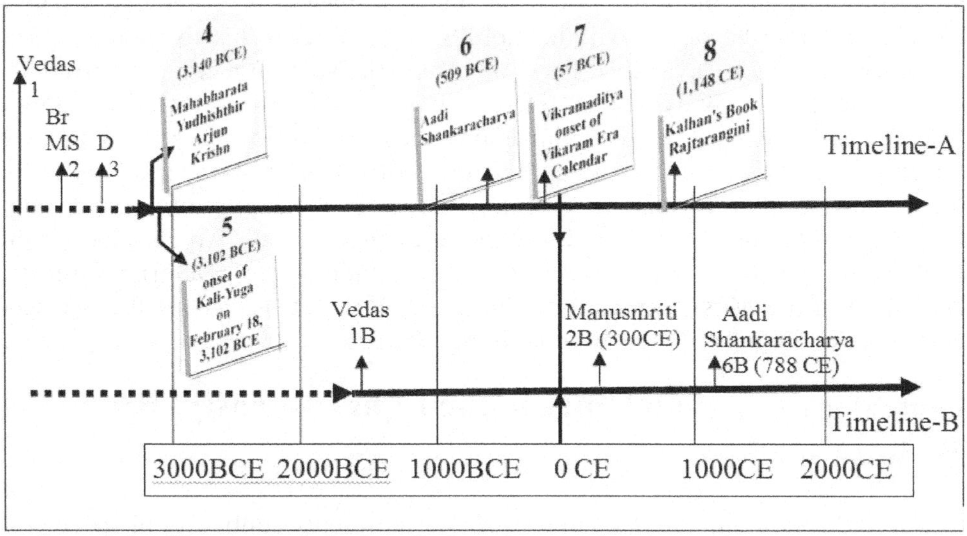

studies of the Western Indologists during the past few centuries.

Figure 3.1: Timeline-A for the historical events according to the Vedic Literature and the Timeline-B according to the Western Indologists. (Milestone # 4 at 3,140 BCE, #5 at 3,102 BCE, #6 at 509 BCE, #7 at 57 BCE and #8 at 1,148 CE)

Relevant to the text of this book the eight vertical arrows on the Timeline-A represent the eight major historical periods in the Indian history. The date in Common Era (CE) for the historical event is also enclosed in brackets at the marker, where available. Only three vertical arrows representing the three markers 1B, 2B and 6B are placed on the Timeline-B. The dotted line on the left end of the two timelines represents the distant past for which no specific historical events with dates are available from the Vedic Literature on Timeline-A or from the Indologists on the Timeline-B. Notice from the Timeline-A that the Marker-1 (vertical arrow) for the Vedas is placed left most on the dotted line. The Marker-2 on the dotted line of Timeline-A represents the Brahmana books (Br), Manusmriti (MS), Upanishads and numerous other literary works of Vedic Literature listed in Table 4.1 in Chapter 4. The letter D at Marker-3 of Timeline-A, also on the dotted line, represents four of the six Darshans (see Table 4.2).

Implicit in the Figure 3.1, as mentioned before, is that the Vedas are considered to have emerged in an unknown distant past of the Aryan Civilization long before the rest of the Vedic Literature. Marker-4 at 3140 BCE represents the events of the Mahabharata War, the epic book of Mahabharata and several contemporaneous great historical personalities; only three are mentioned on the Time-milestone-4.

The Mahabharata War, the book of Mahabharata, Yudhishthir and Yudhishthir-calendar were mentioned in Chapter 1. Emperor Yudhishthir at Time-milestone-4 is part of the long genealogy of the Lunar Dynasty at Hastinapur prior to Yudhishthir's coronation in 3140 BCE. Some more insight into the history at and around the Time-milestone-4 helps us strengthen the historicity of the Vedic references. Recorded genealogy is a vital part of the written history of a nation which cannot be ignored by those interested in the history of that nation. It is a window into the past and it certainly enables considerable insight into the Civilization. Several tables in the following present the recorded history available from the Sanskrit Literature, ranging from the ancient to the relatively recent times. An abridged genealogy of the Lunar Dynasty up to Emperor Yudhishthir is presented in the flowchart of Figure 3.2. Sometime after Yudhishthir's rule, the capital of Hastinapur was abandoned and the city of Indraprasth (modern Delhi) continued as the capital of the genealogy that followed Yudhishthir. A complete sequence of all Aryan kings from Yudhishthir down to the last Aryan king at Indraprasth (Delhi) up till the beginning of the Moslem rule is in Table 3.3. First let us look at the abridged genealogy of Yudhishthir's ancestors in Figure 3.2.

Genealogy of Aryan Kings in the Lunar Dynasty: Pre-Yudhishthir Era

A solid arrow in the chart of Figure 3.2 represents continuity in succession while a dashed arrow represents discontinuity that may measure up to many generations and centuries. The flow chart of Figure 3.2 begins with the block labeled Aryavart where the term '*aryavart*' means '*the place of the Aryans*'. As was mentioned before, Aryavart is India's original name before it came to be known as Bharatvarsh after the great King Bharat (fourth block from the top in Figure 3.2). Moslem invaders called it Hindustan (the place of Hindus), and the British called it India. After independence from the British in 1947 India's official name is back to Bharatvarsh. The second and the third blocks are not connected in a sequence because sometime in the antiquity of the Indian

Civilization the two well-known lines of genealogy branched off to rule from different regions of India.

The solar line started from the great King Manu of Manusmriti. A famous historical milestone in that dynasty was the rule of King Ram. The Solar Dynasty ruled from the ancient city of Ayodhya established by Manu himself. The Lunar-line started from the King Dushant who married Shakuntala the adopted daughter of Kanva Rishi, a prominent name in Vedic Literature. Included here is some anecdotal information about the kings in the rest of the boxes of the chart.

There were numerous kings in the Lunar-line between Dushant and Bharat. Centuries passed after King Bharat when a king in the Lunar-line ruled who was industrious and action oriented. Owing to his dedication and noble karma for the nation he was addressed with his virtuous name Kuru. The Lunar-line after King Kuru assumed another name, the 'Kuru-vansh', meaning the 'Kuru-genealogy'. Centuries later in the Kuru-dynasty was born a king named Shantanu. In the flow-chart of Figure 3.2 the last nine blocks are devoted to the successive five generations between King Shantanu and King Yudhishthir.

The name of King Shantanu's queen was Ganga who gave birth to a son named Vedvrat. Later in life, King Shantanu fell in love with a forest-dweller's daughter named Satyavati and approached the forest-dweller (fisherman) for his daughter's hand in marriage. But the fisherman put a severe condition on King Shantanu that he would permit his daughter to become Shantanu's queen only if her first born became the next ruler of Hastinapur. On the grounds of morality the king could not agree to this condition because the rightful ruler to the Hastinapur throne was the prince Vedvrat, already a young man. Unable to marry Satyavati but still deeply in love with her, the King Shantanu went into depression. When the son Vedvrat came to know about this he could not bear his father's suffering. He took a vow not to ever get married, not to claim the throne and instead to lead a celibate life. Because this was a sacrifice and determination of a very high order Vedvrat from there on was reverentially addressed as Bhishm - the honour bestowed on him for taking up an insurmountable challenge. By the time of the Mahabharata War, the old prince Bhishm came to be known as Bhishm-pitamaha, the Grandfather-Bhishm.

Shantanu married Satyavati who gave birth to two sons, Chitrangad and Vichitra Veerya. Chitrangad died in war and Vichitra Veerya became the king. Vichitra Veerya had two queens Ambika and Ambalika. He was a lustful person with loose morals. While still young he died of *kshaya-rog* (the term in Ayurvedic medicine for tuberculosis), leaving behind two sons Dhritrashtra and Pandu under the guidance of their uncle Bhishm.

Dhritrashtra being blind from birth, Bhishm seated Pandu on the throne of Hastinapur. Pandu married Kunti (birth name Pritha) the daughter of the King Kuntibhoj. Pandu had five sons: Yudhishthir, Arjun, Bhim, Nakul and Sahadev. By mistake Pandu killed a brahman pundit. Killing of living-beings is considered sinful but killing a brahman is considered to be a very serious sin; Pandu renounced the throne in repentance, took monkhood and went to the forest for meditation and self-realization. Pandu's first-born, Yudhishthir, was the heir-apparent of the Hastinapur throne but was too young to rule. Bhishm had no choice but to put the blind nephew Dhritrashtra on the throne of Hastinapur. This transition in the rule without change in generation is shown in the Figure 3.2 by the left-pointing arrow between the 11th and the 12th blocks shared by King Pandu and King Dhritrashtra respectively.

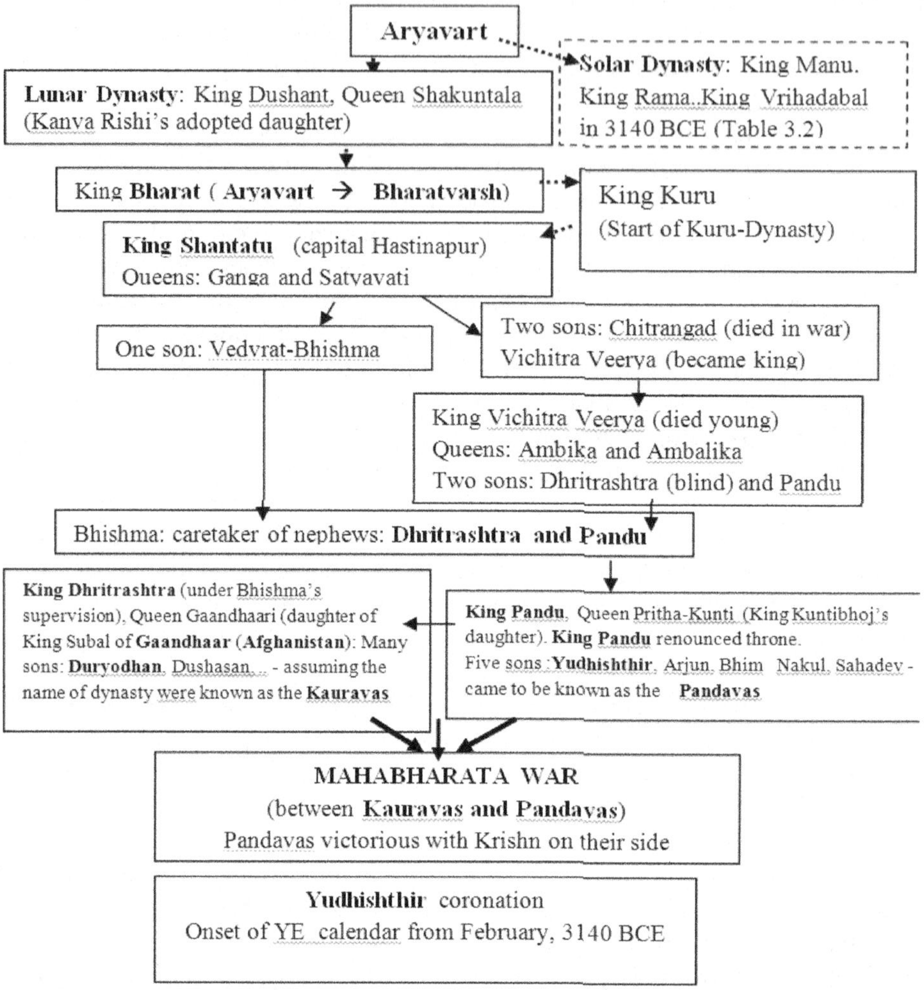

Figure 3.2: Abridged genealogy of the Kuru-dynasty within the Lunar- line.

Dhritrashtra was married to the princess of Gaandhaar (present name Kandahar in Afghanistan) which was ruled by King Subal. Dhritrashtra's queen being from Gaandhaar her historical name is Gaandhaari, meaning 'the one from Gaandhaar'. Dhritrashtra had many sons: Duryodhan, Dushasan … all younger to Yudhishthir, the oldest son of Paandu. Being the descendants of Kuru the sons of Dhritrashtra were called the Kauravas while the five brothers born of Paandu, also in the Kuru dynasty were addressed as Paandavas (born of Paandu).

The Paandava princes were known for their ideal character, ability and valor and the Kaurava princes were known for atrocities and misuse of power. The Kauravas felt threatened that at anytime or after their blind father died, the Pandavas, being the legitimate heirs to the throne of Hastinapur, would naturally assume the throne. This led to conflicts between the Kauravas and the Pandavas; initiated by the Kauravas. The Kauravas hatched numerous conspiracies to eliminate the Pandava princes and failed. To resolve the conflict, the grandfather Bhishm, the blind King Dhritrashtra, the military generals, the emperor's political advisors and diplomats decided to give the Pandavas the ownership of half of the Kuru kingdom.

The area of the kingdom given to the Pandavas was uninhabitable, covered with thick forest. The Pandavas with the help of Krishn and his brother Balram studied the lay of the land and after determining a suitable area for establishing the new capital for the new kingdom for the Pandavas, they requested the most respected scholarly Rishi Badrayan Vedvyas (Mahabharata, Aadi-parv, 34/34) to conduct the auspicious ground-breaking ceremony, to survey and measure the land and design the capital. The book of Mahabharata is arranged in sections called *parvs*, each parv having a number of chapters and each chapter has a number of shloka-couplets in it. The masterful strategic design of the city, its architecture and beauty are described in the chapter 34 of the Aadi-parv of Mahabharata in shlokas 35 through 49. The shloka 41 states that this 'heavenly city' at the shore of River Yamuna was named "Indraprasth", which meant - "the arrival of the most powerful king". Indraprasth was located in the area of modern Delhi, the present capital of India. Indraprasth continues to serve as the capital of the Lunar Dynasty after Yudhishthir.

The Kaurava princes being immoral and greedy wanted it all and continued conspiring against the Pandavas and unscrupulously managed to uproot them from their new capital at Indraprasth. The Pandavas prepared for war to take back what was rightfully theirs but the Kauravas did not budge. Yogiraj Krishn went to negotiate with the Kauravas and requested them to give Pandavas some land for them to at least settle down in their ancestral kingdom. Krishn in his meeting with the Kauravas advised them to avoid war which would always result in death and destruction. To keep all the land for themselves the oldest Kaurava Prince Duryodhan rejected Krishn's request and advice and also rejected the advice from his own father Dhritrashtra, mother Gaandhaari, Grandfather Bhishm and many scholarly people in the court. Duryodhan chose to go to war. While rejecting Krishn's request, the Kaurava Prince Duryodhan arrogantly said to him what is now a common expression among the Hindus when a similar situation arises - "*We will not give any land to the Pandavas - not even equal to the tip of a needle*". With Yogiraj Krishn's efforts wasted, the kings of entire Bharatvarsh began to envisage the conflict escalating into a war between the righteous and the evil. It ultimately led to the Mahabharata War. This war between the cousins became a war between "the good" and 'the evil'; the Pandavas being the good and the Kauravas being the evil.

Mahabharata War, primarily about the disagreement on the division of the kingdom, pulled in a large number of kings of Bharatvarsh, some taking sides with the Pandavas and others with the Kauravas, a scenario quite similar to the World War II of modern times. Krishn, who had moved his capital from Mathura to Dwarka in the present Gujarat State, vowed not to take direct part in the war, however, he chose the side of the righteous Pandavas and decided to be only a charioteer for Prince Arjun. After the rules of engagement for the war were finalized the war was inevitable. When Arjun arrived at the Battle-field of Kurukshetra in his war-chariot driven by Krishn, he realized that the unavoidable war loomed over the extended family and unmitigated destruction awaited. He expressed his innermost thoughts to Lord Krishn saying that this war was not worth fighting because his own cousins and dear ones would be killed. Krishn advised Arjun that it was no longer a war between cousins but that this conflict was a war between righteousness and evil and therefore he had no choice but to fight for righteousness. Krishn gave a sermon to Arjun about the importance of karma, duty to the society and the philosophy of life. This sermon at the battle-field is known as the Bhagvadgita. Arjun was convinced and became ready again to take part in the battle. The Kauravas were destroyed by the Pandavas in the Mahabharata War. Yudhishthir already known as the Dharmaraj (a righteous king, following dharma) was enthroned at Hastinapur. Yudhishthir's coronation was marked by the onset of a working calendar known as the Yudhishthir Samvat (Yudhishthir Era, YE) 38 years prior to the milestone at February 18, 3102 BCE which is the onset of the Kali-Yuga calendar within the Srishti-Calendar (Natural Aryan Calendar).

The present day Hastinapur town is in the state of Uttar Pradesh about 100 km north of Delhi. The Mahabharata War was fought on the plains of Kurukshetra, west of Hastinapur, west of River Yamuna, which is about 160 km north-west of Delhi. This war is also known as the Kurukshetra War.

Beginning in 3140 BCE Yudhishthir ruled for 36 years 8 months and 25 days and abdicated the throne to proceed to the Himalayas for meditation. From Yudhishthir onwards the Aryan kings ruled successively at Hastinapur and later at Indraprasth till the rule of the last Aryan King Yashpal was ended in 1192 CE by the Moslem invasion. Based on historical records found in Sanskrit Literature, Swami Dayanand Sarasvati has listed in the Satyarth Prakash the unbroken succession of Aryan Kings at Indraprasth. These data are presented in Table 3.3. From there on the change in the Mahabharata War rule at Indraprasth from Aryan rule to Moslem rule also brought the change in the name of Indraprasth to Dilli. The history of the Moslem rule at Dilli is well-known and it is also well-known that Dilli under the British rule changed again to Delhi. The present capital of independent India is situated in New Delhi.

The Marker-4 at 3140 BCE on the Timeline-A (Figure 3.1) also represents Lord Krishn and his pre-war sermon of Bhagvadgita to Arjun at the battle-field of Kurukshetra. This is also the period when the last two of the 6 Vedic Darshans (philosophies), the Meemansa Sutra by Rishi Gautam and the Vedanta Sutra by Rishi Vedvyas were written. Krishn, just before Mahabharata War had moved his capital from Mathura to Dwarka, establishing a new kingdom in the area now part of the Gujarat State. Dwarka is at the shore of the Paschim Saagar (ancient name of Arabian Sea). Sometime after Krishn's death near the onset of Kali - Yuga epoch in 3102 BCE, tectonic activity submerged the city of Dwarka in the sea. Under the national heritage program, the Indian archeologists and geologists of the Geological Survey of India have recently discovered and mapped the submerged ancient city of Dwarka. Dwarka continues to be a national heritage site and a pilgrim place for the Hindus. It is the reestablished Dwarka town at the shore of the sea where the brilliant monk Aadi Shankaracharya established an ashram (Matha) in October 492 BCE when he was only 17 years old. The Marker-6 on the Timeline-A in Figure 3.1 reflects Aadi Shankaracharya's historical significance in the Indian Civilization. The Matha that he established at Dwarka still continues to function as a seat of learning. In February 2014 it was a moving experience for this author to see some of Aadi Shankaracharya's personal items like his wooden foot-ware in the Shankaracharya library at the Dwarka Matha.

There is no dearth of recorded history in the ancient Sanskrit Literature. When one goes into the far antiquity of the Vedic Civilization beyond 3140 BCE (Timeline-A, Marker-4) one finds the history of the Solar Dynasty that started with King Manu who established the city of Ayodhya in the ancient region of Kosal. Rishi Valmiki authored Ramayana, the biography of King Ram, a descendent of King Manu. Rishi Valmiki in Ramayana has provided a substantial list of the names of kings in King Manu's genealogy. All this further tells us that the Aryans have been in India since the beginning of their Civilization in the far antiquity. The dotted block in Figure 3.2 pertains to the Solar Dynasty. It has been a tradition in India, which the Western scholars were not aware of, that the historians and pundits were appointed by the kings to record the history of their region. Therefore, in addition to describing the genealogy of kings there is a geographical description of the region that a king governed. A brief mention of the Solar Dynasty is made in Chapter 4 under Smriti and Ramayana. A brief genealogy of the kings in the Solar Dynasty of King Manu is presented here.

According to the ancient Indian tradition, a marriage is not only between the young couple but also between the two families. As a part of this tradition the members of the two sides are introduced to each other at the wedding ceremony and the genealogies of both sides are recited by the respective family priests. This tradition first cited by Rishi Valmiki in Ramayana, continues in traditional wedding ceremonies in modern day India. At the wedding of Prince Ram to the Princess Sita, the genealogies of the two kingdoms, Ikshvaku-kingdom (named after King Manu's son Ikshvaku) of Ayodhya and Janak-kingdom of Aryavart's Mithila region, were recited. The *Raj-purohit* (royal-priest) Vasishtha of the Emperor Dashrath (Ram's father) recited the genealogy from King Manu to King Dashrath and in the place of the *Raj-purohit* Shatanand of Mithila Kingdom, Sita's father himself, the learned King Janak known for his deep knowledge of the Vedas, recited Sita's side of the genealogy.

These genealogy-recitations in the court of King Janak are part of Valmiki-Ramayana. Rishi Valmiki has devoted 36 *shlokas* in the Baalkaandam chapter's 29th *sarg* (section) to cover 34 names of the kings in the Solar Dynasty from King Manu to King Dashrath. In the next *sarg* he has devoted 15 *shlokas* to cover 23 names of the kings in the Mithila-kingdom, beginning from the King Nimi to King Janak. In the Table 3.1 are listed the names of the kings in the Solar Dynasty only. King Vrihadbal, the last entry in the Table 3.1, was contemporaneous to the Mahabharata period in 3140 BCE. Rishi Valmiki's Ramayan and King Ram being well before the Mahabharata period, King Vrihadbal of course is not listed in Ramayana in the Solar Dynasty genealogy recital by the Royal Priest Vasishtha.

Even though the Puranic rituals prevail in the present Hindu society, the basic societal structure continues to function within the laws of Manusmriti and Hindus continue to consider King Manu to be the founder of the Vedic/Hindu society. Whether a Hindu is a ritualistic Puranic or a worshiper of the formless God, a Hindu believes that Manu's Manusmriti is very ancient and Valmiki Ramayan and Mahabharata respectively are historical accounts respectively of the Solar Dynasty and the Lunar Dynasty. Manu, Ram and Krishn for the Hindus have been historical figures, in that chronological order, separated by centuries and eons, who brought great transformation in the Aryan society. The books such as Manusmriti, Ramayana and Mahabharata are historical accounts in that chronology.

Table 3.1: Abridged Genealogy of Aryan Kings in the Solar Dynasty of King Manu
(Valmiki Ramayana, Baalkaandam, 29/1-36).

	King's name		King's name		King's name
1	Manu	13	Bharat	25	Agnivarna
2	Ikshvaku	14	Asit	26	Shighrag
3	Kukshi	15	Saagar	27	Maru
4	Vikukshi	16	Asmanj	28	Prashushruk
5	Baan	17	Anshumaan	29	Ambareesh
6	Anranya	18	Dileep	30	Nahush (Just and virtuous)
7	Prithu	19	Bhagirath	31	Yayati
8	Trishanku	20	Kukutsth	32	Naabhaag
9	Dhundhumaar	21	Raghu	33	Aj
10	Yuvnashva	22	Pravridha (Kalmashpaad)	34	Dashrath
11	Mandhaata	23	Shankhan	35	Ram (three step brothers: Lakshman, Bharat, Shatrughn)
12	Susandhi, Prasenjit	24	Sudarshan		....
					King Vrihadabal (in 3,140 BCE)

Western scholars accept that the Hindu society's basic rules are from Manusmriti, whenever it was written. Then there should be no difficulty in accepting Manusmriti's geographical map of the place of the Aryans/Hindus. When we look at Manu's geographical description of the place of the Aryans (to be discussed), we find no place else but India as the home country of the Aryans. It was intellectual dishonesty on the part of Max Muller et al. who disregarded Manu, his genealogy and his geographic description of the place that he ruled. On top of that, supporting the rootless Aryan Invasion Theory (AIT, discussed in the following) concocted behind closed doors in England by some ignoble ignorants is further proof of this intellectual dishonesty.

Accepting Manu's geographical map of Aryavart and his description of its inhabitants and those outside the borders of Aryavart render AIT defunct. Manu and Manusmriti are variously referred to in the book of Mahabharata and in the post-Veda ancient literature preceding the Mahabharata War and indicating that the Aryans of India are the native people of India from antiquity. Rishi Vedvyas the author of the book of Mahabharata is also the author of Vedanta Sutra and the interpreter of Rishi Patanjali's Yog Sutra and other books, all of which are a reality of the historical period around 3140 BCE. The genealogies of Figure 3.2 and Table 3.1 are not fictional.

This was also the period in the history of Aryavart when the *guru-shishya parampara* (guru-student Vedic tradition) prevailed with high literary achievements against the backdrop of the simplicity of Rishi Vedvyas's ashram. It was Rishi Vedvyas's first senior student, Rishi Jaimini, who completed his Meemansa Sutra before his guru completed Vedanta Sutra (see Table 4.2 in Chapter 4). The other three senior students of Rishi Vedvyas according to Vishnupuran (3/4/7-18) were Sumantu, Pail, Vaishampayan; the latter two continued teaching other students who are also named in the Vishnupuran (from "Vedant Darshan ka Itihaas", by Udayaveer Shastri). The completed book of Mahabharata in poetry was presented by Vaishampayan the third student of Rishi Vedvyas, to the Aryan King Janmejaya of Hastinapur (Delhi), the great-grandson of Arjun. All these scholarly people with their literary works that the Indians cherish today, cannot be termed fictitious and discarded for a fictitious AIT for which there is no evidence in the Aryan literature, history or the Puranas. The Mahabharata War was between the Aryan cousins. There never was an invasion of India by the Aryans of Asia Minor, as imagined by the proponent of AIT - Maxmuller et al. If the Aryans had come from some place outside of India, the literature that they created would have at least one mention somewhere to that effect. The Aryans known to be prolific in expressing their thoughts could not have had a memory lapse and forgotten their native place, Asia Minor or elsewhere.

Commonsense dictates that in an invasion of a country by a group of people only a limited number of invaders take part in the attack. When invaders have come to India, for example the Greek/Armenians with Alexander in the fourth century BCE or the Moslem invaders from 1192 CE onwards, only a limited number of men soldiers have come to invade and not the whole population of the invading country. A large part of the population of the invading country is left behind to carry on the business of that country. The invading militaries do not bring en masse their teachers, scholars and philosophers. Rather they are generally left behind where they are more useful to carry on with their duties in the areas of education and culture. There are no regions on the Earth other than India where the highly developed Sanskrit language, considerable Aryan culture and the ancient voluminous Sanskrit Literature exist. Neither in this vast Sanskrit Literature nor in the Indian folklore is any fleeting reference to be found of Aryan's coming from someplace else. On the other hand the overwhelming numbers of references that one finds in the ancient Vedic Literature indicate that India has been the place of the Aryans from as far back as antiquity. In the German culture and language there is considerable influence of Sanskrit and the Aryan culture. It is strange that Max Muller, himself a German, did not apply the same AIT to Germany first; that the Aryans invaded Germany as well!

The Aryans cremate their dead. This practice is not blindly followed by the Aryans because the Veda advised them to do so. It is because in their philosophic development they understood the meaning of death and the reality of matter and consciousness and the inherent separation between the two. This unique cremation practice is not found in other European enclaves. It is offensive for a foreigner to say that the Aryans of Aryavart were intruders in their own homeland. The Genome Project, discussed later in this chapter, will put an end to this insult that has prevailed for a century and a half.

Time-Milestone at February 18, 3102 BCE (Marker-5 in Figure 3.1)

The Srishti Samvat of the Aryans is a natural calendar that relates to astronomy rather than to the birth of a well-known person or to a coronation of a great king as in the case of the Yudhishthir Samvat, Vikram Samvat or the Christian Era. The term *srishti* means the 'creation of the universe' and *samvat* means the calendar year. The Srishti-calendar therefore represents the number of years elapsed after the creation begins functioning and the solar day becomes meaningful. In other words the Srishti Samvat represents the establishment of the solar system. More discussion on the Srishti Samvat is in Chapter 5. A sub-calendar within the Srishti Samvat represents the onset of a shorter cyclic epoch of a specific length known as a Yuga. As this author has understood it, the onset of the sub-calendar relates to the solar system and the astronomical position of the planets. The **Marker-5** on the Timeline-A (Figure 3.1) represents the onset of such an epoch, called Kali-Yuga. The duration of a Kali-Yuga is 432,000 Earth-years after which another epoch and sub-calendar within the main Srishti-calendar will commence with another Yuga name with a different duration. The Kali-Yuga begins in the 38th year of Yudhishthir Era (YE). The historical records after the onset of the Kali-Yuga calendar show that from there on the Vedic society has used the YE and the Kali-Yuga Era calendars in parallel. In the context of historicity, only limited astronomical details related to the calendars are presented here.

The onset of Kali-Yuga is stored in the time division of *din, muhurt, ghatka, kalaa, kaashtha* and *nimesh,* where *din* is a solar-day (day+night) and the *nimesh* is $8/45^{th}$ of a second. As Swami Dayanand Sarasvati had expressed - the Aryans have been tracking the calendars by adding the days and years to the onset of Kali-Yuga and therefore to the composite *Srishti Samvat,* in the same way as *the ledgers are kept by accountants*. For example the accurate onset of Kali-Yuga in terms of popular units was on **February 18, 3102 BCE at 2 hours 27 minutes and 30 seconds**. This serves as the base for all future calculations. Additionally the planetary positions near the *yog-taara* (identifying star) of the Vedic constellation named Revati are stored for the six planets - Saturn, Jupiter, Mars, Venus, Mercury, Moon (including both of the Moon's opposite nodes - northern (*Rahu*) and southern (*Ketu*)) and the Sun. These astronomical data are recorded with the accuracy of *raashi, ansh, kalaa* and *vikalaa,* which in the terminology of modern astronomy respectively are: the count of the zodiac sign, and the angle in degree, minute and second. There are 12 zodiac signs as in the Western astronomy each subtending an angle of 30 degrees out of the 360 degree circle. Of the 28 Vedic named constellations the Revati (AtharvVeda, 19/7/5) with its 32 stars appearing in the shape of a fish has its identifier star (*yog-taara*) same as the *Zeta Piscium* in modern astronomy.

Observed from the Earth, like all planets, the Sun is also observed moving through the constellations within the central bend of stars called the zodiac. The Sun's apparent path through the zodiac called the ecliptic is not parallel to the celestial equator of the celestial sphere (night-sky dome); rather it is tilted showing the Sun above it for half the year and below it for the other half. The star *Zeta Piscium* representing the Revati constellation falls right on the path of the Sun in the zodiac.

These astronomical data are based on the observations made by ancient *rishis*. The astronomical data with mathematical methodology for calculating astronomical details are successively passed on to the scholars of astronomy groomed in Vedic learning centers. The calendars are prepared based on accurate calculations with predictions of the dates and times of festivals and celebrations most of which are related to astronomy and Nature. In the present day India the same calendars and almanacs are also used by astrologers to determine the planetary positions for astrology.

It is well-known in the history of modern astronomy that the written copies of the astronomical data recorded by Indian astronomers reached Europe in the 18th century. When these data were put to the test by well-known astronomers of the time, Jean-Sylvain Bailly (1736-1793 CE) and John Playfair (1748–1819 CE), they found that the recorded planetary positions corresponding to the onset of Kali-Yuga on **February 18, 3102 BCE** were within the accuracy of 37′ (minutes of an arc) which is about 0.6 degree angle. Vedic astronomy is a subject by itself that cannot be explored fully in this book. Relevant information on this however is included in Chapter 11 in Volume 2.

Now, the question – why exactly is this date established for the onset of the period called Kali-Yuga with the ensuing sub-calendar within the natural calendar of the Aryans? More details on the time division of the Aryans, in terms of the *yuga* periods in million and billion years on the higher side of the time scale and the unit of *nimesh* on the lower side, up to the fraction of a second, are given in Chapter 5 under the sub-heading of '*Kaalah-Tatva*: the Time-Entity'. Here a brief discussion is presented on how the Aryans arrived at this milestone on the timeline of their history.

Kali-Yuga onset and the Planetary Conjunction

The Vedic cosmology/astronomy principle upholds that all planets of the solar system, at the beginning of the creation, start their orbits around the Sun at the same time, from the same direction in space from the Sun. The creation in this context is primarily the creation of the solar system when the planets including the Earth are in orbit around the Sun. A day for the Earth is defined by the Earth's spin around its axis and a year by its revolution around the Sun. Therefore the creation in this context is the onset of the solar system. In fact this Vedic principle is not too far off from the modern theory of the formation of the solar system itself where an explosion on the Sun throws out in space a blob of matter that cooled and coalesced into different planets. The physics also tells us that this blob of matter under the gravity of the spinning Sun carried with it enough angular momentum for planets to spin and orbit around their progenitor the Sun. Therefore, the Vedic premise that all planets began their orbital journey around the Sun from the same space and time is after all a piece of science and not religion.

The Vedic astronomy principle further stipulates that based on their period of revolution, all planets orbiting around the Sun, position themselves in different constellations at different times (Surya Siddhant, 1/25-27). When a planet starts from a constellation say, Revati, and arrives back at the same point in Revati, it completes one *bhagan* (orbit). Different planets over a long period of time complete many *bhagans*. Over a period of 4,320,000 years all planets are said to have completed different number of full *bhagans* such that they all are aligned pointing in Revati constellation at the same time. This astronomical phenomenon is well-known as the conjunction of planets. The period of 4,320,000 years when all planets completing the exact number of their *bhagans* (orbits) return back into the same constellation at the same time of the year, is a derivation indigenous to Vedic astronomy. This period of 4.32 million years is called a Maha-Yuga. A Kali-Yuga period is one tenth of a Mahayuga when there is a partial planetary conjunction. Over the centuries from time to time there are always partial planetary conjunctions. With sophisticated computer algorithms in modern astronomy one can now predict or do back-calculations for conjunctions.

What is mind-boggling here is the periodicity discovered and established by the Vedic rishis that could not have been possible, as the European astronomer Bailly had remarked, without actual accurate astronomical observations of all planets including the Sun in the background of previously well-defined constellations. Such a planetary conjunction also defines the onset of the present Kali-Yuga epoch that commenced on **February 18, 3102 BCE** at the milestone at Marker-5 on the Timeline-A in Figure 3.1.

It is these astronomical data that went to Europe from India and were examined by the famous astronomers, particularly by the French astronomer Bailly (1736-1793) who confirmed their accuracy and remarked– "*...The calculation of Brahmans is so exactly confirmed by our own calculations that nothing but actual observations could have given so correspondent a result...* " (in "The Theogony of the Hindus", by Count Bjorustjourna, Pub. John Mu, London, 1844, p.82; taken from "Aryon Ka Aadi Desh Aur Unki Sabhyata" by Swami Vidyanand Sarasvati, Published by International Aryan Foundation, Mumbai, 1989). More details about the time-division of the Aryans is in Chapter 5, however, it would suffice here to say that the onset of the present *Kali-Yuga* marks the beginning of the *Kali-Yuga* Calendar.

Other Historical Markers on the Timelines

A discussion on the period of the Vedas and Manusmriti from the perspective of Max Muller et al. is included in Chapter 4. Max Muller in particular had placed the Vedas as shown on Timeline-B (Figure 3.1) at Marker-1B (1500 BCE) and Manusmriti (MS) at Marker-2B (300 CE). The Marker-6 at 509 BCE on Timeline-A was also placed by Indologists at Marker-6B at 788 CE. In the analysis that follows, all Markers on the Timeline-B will be proven wrong and those on the Timeline-A will be proven correct.

The Marker-6 on Timeline-A represents the advent of Aadi Shankaracharya, the great thinker, philosopher and the proponent of *Adwaita-Vedanta* (the philosophy of non-dualism). Aadi Shankaracharya at the Marker-6 is an important milestone in Indian history. Recall that this is the time when young

Shankaracharya took it upon himself to preach against the atheistic and agnostic religious sects in the Vedic society at that time, Buddhism and Jainism included. He toured India from south to north and east to west, challenged Buddhist and Jain philosophers for public *shaastraarths* (debates), defeating them every time. The rule of engagement for these *shaastraarths* was that the defeated person will convert to the winner's religion. These *shaastraarths* therefore led to the conversion of prominent Buddhist and Jain philosophers, kings and general masses, back to Vedic religion (now known as Hinduism). During his travel he established four Mathas (monasteries) at four corners of India. They are since known as Shankaracharya's Mathas which continue to be highly revered ashrams and the seats of learning even now after two and a half millennia. In the Matha tradition, established by Aadi Shankaracharya himself, a learned and dedicated person is appointed as the figure-head of the Matha to succeed the previous figure-head. All figure-heads are addressed as "Shankaracharya", whatever their birth names may be.

Western Indologists in their studies of Sanskrit Literature and Indian history shrank the Timeline-A into the Timeline-B (Figure 3.1) by theorizing that the advent of Manusmriti at Marker-2 and the Mahabharata War and the other historical personalities at the Marker-4 were all a myth. They also said that the important historical figure Aadi Shankaracharya existed well after the beginning of CE. The Western scholars not being fully aware of the Matha tradition that all care-taker Acharyas from Aadi Shankar onwards, till today, are also addressed as Shankaracharya, they placed Aadi Shankaracharya's advent all over the place on the map after the beginning of the Common Era (CE). It was later narrowed down to somewhere in the 8th or mid-9th century CE and finally erroneously to 788 CE. This is shown by the Marker-6B on the Timeline-B.

Aadi Shankaracharya was a great person of national and historical importance. In the following it will be shown from the hard-core data at the Mathas that he was born in 509 BCE. Thus, looking at the difference between the Marker-6 on Timeline-A and the Marker-6B on Timeline-B one finds that even for this important historical event in Indian history the Indologists created a confusion of 1,300 years. Under such a situation if this author successfully establishes here that Shankaracharya's advent was at 509 BCE (on Timeline-A) and relates this date and other specific dates in Shankaracharya's life to Yudhishthir Era calendar, then Yudhishthir himself is proven to be a real historical figure and not a fictitious, mythical character as suggested by the Indologists.

The three calendars presently in use in India, two Hindu calendars and the third brought in by the British, are chronologically after the King Vikramaditya (57 BCE), Jesus Christ (0 CE) and the King Shalivahan (87 CE). These three were real people. The Indian history is as certain about the reality of the two Aryan kings as the rest of the world is about Jesus Christ. Yudhishthir-calendar was long in use in India, in which written records are available for the time period before the Kings Vikramaditya and Shalivahan. Yudhishthir-calendar continued for a long period along with these two later Aryan calendars. Under this situation, there is no sound logic for one to suggest that the King Yudhishthir was an imaginary figure.

The strong correlation between Marker-6 on Timeline-A at 509 BCE and the Marker-4 at 3140 BCE is shown in the following. One cannot separate King Yudhishthir from Yudhishthir Era calendar and one cannot separate King Yudhishthir from Mahabharata War and the contemporaneous existence of King Krishn, latter being for the Hindus what Jesus Christ is for the Christians. Based on the written historical records kept in Shankaracharya Mathas, the historicity of the King Yudhishthir, Yudhishthir-calendar and Mahabharata War is validated, thus validating Timeline-A.

Additionally the data recorded in yet another calendar, Kali-Yuga Samvat in parallel with the Yudhishthir calendar further validates all markers on the Timeline-A. The onset of Kali-Yuga Samvat cannot be disputed because it did not commence like another calendar. It did not commence at midnight or at sunrise at GMT (Greenwich Meridian Time), it commenced, as mentioned before, exactly at 2 hours 27 minutes and 30 seconds in GMT on February 18, 3102 BCE. This astronomy based calendar is as scientific as it can get. While some Shankar Mathas have recorded succession history of care-taker Acharyas since 509 BCE onwards in the Yudhishthir-calendar with its beginning in February in 3140 BCE, the other Mathas have independently recorded the succession history of the care-taker Acharyas in the Kali-Yuga calendar with its beginning 38 years into the Yudhishthir calendar.

History-Keeping in Aryan Calendars

History-keeping, on a short and long term basis has been ongoing in India using the social calendars as well as in the *Srishti-calendar* (Natural-calendar).

Table 3.2: Indian Calendars

Calendar	**In the Year 2018 CE**	**Comment**
Yudhishthir-Samvat (YE)	**5158 (YE)**	**started from the coronation of King Yudhishthir at vernal equinox in 3140 BCE**
Kali-Yuga Samvat (KE)	**A+5120 (KE)**	**Natural sub-calendar (within *Srishti-calendar*) started on Feb 18, 3102 BCE. A is the accumulated number of years from the beginning of Srishti Calendar.**
Vikramaditya Samvat (VE)	**2075 (VE)**	**started from the coronation of King Vikramaditya of Ujjain at vernal equinox in 57 BCE**
Shalivahan Samvat (SE)	**1937 (SE)**	**started from the coronation of King Shalivahan of Ujjain at vernal equinox in 78 CE**

Four Aryan calendars are summarized in Table 3.2. For example the Yudhishthir-Era (YE) in the Common Era year of 2018 CE was 5158 YE (3140 BCE+2018 CE = 5158 YE). The relationships between the other three Aryan Calendars and the Common Era are also shown in the Table 3.2.

The Vikram Samvat started from the time of the coronation of the great King Vikramaditya of Avantika (present name Ujjain) on March 22, 57 BCE. The onset of Vikram Samvat/Era (VE), 57 BCE, is shown by the Marker-7 on the Timeline-A (Figure 3.1). The Shalivahan Samvat/Era (SE), popularly known as Shaka Samvat, started 135 years after VE with the change in the dynasty at Ujjain. It began in the Common Era 78 CE. Although computer softwares are available now to prepare Hindu (Vedic) calendars with predictions of important dates and festivals, the Indian system based on Surya Siddhant (Sun Principle) is still followed for preparing the *panchangs* (five-limbed almanac/calendar) for Hindu festivals and celebrations in both the VE and SE calendars.

The five limbs of the *panchang* are: (1) *Tithi* (lunar phase: a 12° sector between the Moon and the Sun, total 30 sectors); (2) *Nakshatra* (one of the 27 constellations wherein the Moon is observed from the Earth: a named angular sector width of 13°:20′); (3) *Yog* (degree of conjunction between the Sun and the Moon: 1 Yog is equal to 13°:20′); (4) *Karana* (the angle of a half *Tithi*: 6° between the Sun and the Moon) and (5) *Vaar* (solar-day of the week). The ancient introduction of the five-limbed calendar into the Vedic society itself points to the fact that the Indian astronomy was highly developed from antiquity. Obviously the five-limbed Indian calendar is more detailed than the Common Era calendar. This is indicative of the level of civilization of the society. The ancient records of history and historical events recorded in VE and the two other Vedic calendars older than VE will further help prove the 'Vedic Reference Hypothesis'. It is also informative and instructive to look at some features of the Common Era (CE) or the Christian calendar, also known as the Gregorian calendar.

In the Vedic system there is a unique connection between astronomy and the names assigned to the months of the year. This uniqueness distinguishes the Vedic/Hindu calendar from the other non-Vedic calendars, for example the nomenclature of the months in the CE calendar has no relation to astronomy. The other striking difference between a Vedic calendar and the CE calendar is a shift of about two months in the start of the year. The year in the Vedic calendars begins in the month of March.

Christian Calendar Discrepancy

This author finds the Common Era/Christian Era (CE) calendar a mixed-bag of names and numerals for the months and provides an interesting insight into the development of the CE calendar and its possible ancient connection to the Vedic astronomy.

Not many people may have noticed that some of the successive names of the months in the CE calendar are based on numerals. For example, the months of September, October, November and December have the names of Sanskrit numerals. This becomes obvious when the common suffix of 'ber' is dropped from each of the names. Without the suffix the names of these four months are: Septem-, Octo-, Novem- and Decem- . The Sanskrit pronunciation for the numerals 7, 8, 9 and 10 respectively are: *saptam, ashtam, naval* and *dasham*, which sound very close to the prefixes of the names of these months in the Common Era calendar : Septem-, Octo-, Novem- and Decem-. This indicates a possible historical connection between Sanskrit and the names of these months. However, the fact is that these months are: 9th, 10th, 11th and 12th in the CE calendar and not 7th, 8th, 9th and 10th as the Sanskrit names of the numerals indicate. Therefore there is a two month shift in naming the months in the CE calendar with respect to the Vedic calendar. In addition to a two month shift in naming the months in the CE calendar there is about a two month shift in the beginning of the year as well. *Chaitra* is the 1st month of the Vedic calendar which begins around March 22 and ends in March. So there is a shift of about two and a half months between the two calendars.

Looking at the above mentioned dual discrepancy, this author formulates a deduction here as:

> *"Due to some historical reason the Christian Era (CE) calendar got shifted with respect to the Vedic calendar by two months without changing the names of the months."* (Shiv Raj Pal, 2006)

Is there an ancient connection between the Vedic calendar beginning in March and the CE calendar beginning in January? Yes there seems to be and to check this let us begin with the definition of the terms 'day' and 'night' in the two calendars. By definition the term 'day' corresponds to a bright period with sunlight and the term 'night' corresponds to the dark period with the absence of sunlight. The idiosyncrasy of the CE calendar is that the names of the seven days of a week are hyphenated by the word 'day' (e.g. Sun-day, Mon-day, ….), while the day in CE begins at midnight which is in the middle of the dark period. Logically, if the beginning of a day is at midnight it would be logical to call the days as Sun-night, Mon-night and so on.

Within the contiguity of the bright and dark periods in a day, a complete-day is defined to be of 24 hours duration. The complete-day in the CE calendar is made up of a day (with light) plus two different half-nights. For example, a Sunday is made up of 12 hours with half-night leftover from the Saturday, plus the daytime of Sunday, plus the half night of Sunday. The question is - why split the night at midnight and call it a day while it is still night? This does not seem logical and is far from being scientific. In the Vedic system on the other hand the day begins at sunrise when the day-light begins and the night starts at sunset, when the darkness begins; a very logical delineation. In the CE calendar therefore in addition to the shifted serial numbers of the months and the month-misnomers, now we also have to worry about the split of two nights around a complete-12 hour day. A most likely reason for this miscellany in the CE system is discussed below.

The day change in the West begins at midnight at the Greenwich zero longitude, in England. Is there a direct correlation between the sunrise in India and midnight at Greenwich meridian? Just as in Yudhishthir's time the ancient city of Hastinapur at the geographical coordinates (28.4° N, 77.2° E) was prominent, the city of Ujjain at the geographic coordinates at (23.2° N, 75.77° E) came into prominence with the Emperor Vikramaditya. Vikramaditya had established a highly scholarly environment. It is said that he sent out scholars to teach math, science and Vedic philosophy when requested by the kings in the Middle-Eastern regions. Ujjain's prominence continued with Brahmagupta (598-

668 CE) who headed the astronomical observatory at Ujjain and wrote in 628 CE the text on astronomy, entitled "Brahmasphuta-siddhanta". Down the line yet another astronomer Bhaskara (1114 - 1185 CE) headed the astronomy observatory at Ujjain and wrote the text entitled "Siddhantashiromani" in two parts, "Goladhyaya" that deals with "mathematics of spherical bodies" and "Grahaganita" which deals with the "mathematics of planetary orbits".

With this background in place let us check if there is a correlation between the prominent city of Ujjain and Greenwich. Presently, the geographic longitude of a place is determined with respect to the zero longitude at Greenwich. Of the 360° longitudes covering the entire globe, the coordinates of a place are expressed as 'degrees east of Greenwich' and 'degrees west of Greenwich'. The 360° of longitude are equivalent to 24 hours; 15° per hour. The time zone of a place is therefore determined in hours-east or hours-west, depending on what longitude in the east or west of Greenwich the place happens to be. Since the longitude of a place determines the time zone of that place, the time difference between two places relates to the longitudinal difference between those two places.

The longitude of Ujjain with respect to Greenwich being 75.77°E, the time difference between Greenwich and Ujjain is 5 hours and about 3 minutes. The first day of the first Vedic month of Chaitra about 14 days before the spring equinox, marks the beginning of the Vedic year on March 10th. Let us look at the time difference between the sunrise at Ujjain, say for March 10, 2012, and midnight at Greenwich for the same day. The night of March 10 being about 13 hours and 28 minutes long at Greenwich the difference between the midnight (zero hours at Greenwich) and sunrise at Greenwich, is close to 6:44 hours. The sunrise time on March 10, 2012 at Ujjain is 06:40 AM. This is the time when the first day of the Vedic calendar begins at sunrise in Ujjain and at the same time the first day of the Vedic calendar begins at midnight at Greenwich. This means that when it is sunrise in India and a new day begins in the Vedic calendar it is very close to midnight in Greenwich and the same new day begins there, even though it is still in the darkness of night.

Due to the latitudinal change in the solar position during the year, the length of night and day changes during the year at all places on the globe. In spite of this, there is a strong correlation between the beginning of a day at midnight in the West at Greenwich and the beginning of the day with the sunrise in India. As mentioned before, it is logical to have the beginning of the day with the sunrise; the Vedic definition is more meaningful than the Western definition of the day where the day begins in the darkness at midnight at Greenwich. However, it seems that the definition of the day in the West with its beginning at midnight at Greenwich, did not begin arbitrarily or in isolation, rather it began in order to synchronize it with the more meaningful Vedic definition. This synchronization with the Vedic calendar was at the cost of splitting the night; half night to go with one day and the other half to go with the next day. This author's attempt to resolve the *Christian Calendar Discrepancy* is as in the following statement:

"There is a high likelihood that the Common Era (CE) calendar was influenced by the Vedic calendar. The influence of Sanskrit in the names of the months and their count is obvious but with an error of two month-

shift. To keep a link with the commencement of the day with sunrise in the Vedic calendar the followers of the CE calendar began their day at midnight. The correlation between the two calendars indicates the possibility of migration of the scholarly between Aryavart and Europe when exchange of culture and language took place."

Defunct Aryan Hypothesis (AH) and Aryan Invasion Theory (AIT)

It is discussed at length in Chapter 4 that the Aryans have been proficient in record-keeping and recording their history at different levels of the society. Contrary to this some Indologists put forward a misconception that the Indians were not good in recording their history and created new timelines for the Indian history (e.g. Timeline-B in Figure 3.1). The Indologists did not stop at that, they concocted the Aryan Hypothesis (AH) that there were two races in India. The Indians with lighter skin were Aryan invaders who settled primarily in North India and those with darker skin were the native Dravidians mostly in South India. Complementing AH was the Aryan Invasion Theory (AIT) that the Aryans in India are invaders from Asia Minor.

The AH was hatched behind closed doors on April 9, 1866 CE in a meeting of the Royal Asiatic Society of London, presided over by Right Honorable Viscount Strongfold (ref. Vidyanand Sarasvati, 1984, 1989). In that meeting Edward Thomas put forward the hypothesis that the Aryans from Asia Minor invaded the Dravidian natives of India and consequently there were two separate races in India. The implication of the AH and AIT was the British justification to colonize India just as the Aryans and other invaders had colonized India. The expected result of AIT was the division created between the people of North and South India and between the fair-skinned and dark-skinned Indians. This division continues to this day.

The Oxford Professor Max Muller present in the same meeting, never having visited India once in his entire life, gave legitimacy to this imaginary invasion hypothesis by assigning a historical period to it. To this he added that the Aryan invaders destroyed the Indus Valley Civilization around 1500 BCE which he related to the recently excavated Mohenjo-Daro and Harappan Civilizations. The immediate disagreement to the AIT came from several British authors of that time who had lived in India, had studied Vedic culture and literature and those who had participated in the excavation of the ancient archaeological sites. In spite of these disagreements the AIT continued to thrive with the added weapon of the 'divide-and-conquer' strategy.

The famous archaeologist of that time Sir John Hubert Marshall (1876 - 1958 CE), was appointed by the British government as the Director General of the Archaeological Survey of India who served at this post from 1902 CE to 1928 CE. Archaeological excavations under his supervision discovered two ancient cities of the Indus Valley Civilization, Harappa and Mohenjo-Daro. Although not a single piece of armament or any sign of strife between the assumed invaders and the presumed defenders was found in these excavations, the antithetic discovery by Marshal did not deter the AIT proponents. Marshal expressed his disagreement with Max Muller's Aryan invasion date of 1500 BCE and said that if there was an Aryan Invasion it could be somewhere between 2700 BCE and 3200 BCE. Even these dates proposed by John Marshall were offered without any substantiating evidence. So the date of AIT swung between 1500 BCE and 3200 BCE. One reason why Max Muller remained adamant about his suggested date of the Aryan invasion being near 1500 BCE could be to save another of his baseless hypothesis that the writing of the RigVeda and the arrival of the Aryans were not older than around 1500 BCE. Such was the pathetic wisdom of these professors and self-appointed historians. But the cat was out of the bag. There are those who still consider the word of a Westerner to be carved-on-stone and therefore the AIT continues in spite of massive evidence against it piled on by written historical records, new archaeological finds, carbon-dating and finally from the recent genetic research under the national Genome Project proving beyond doubt that the inhabitants of India, from north to south, are indigenous to India. If nothing else, the new results from the state-of-the-art genetic research, discussed at the end of this chapter, have put the last nail in AIT's coffin.

To prove the Vedic Reference Hypothesis that the mantra references on Vedic cosmology are older than 3140 BCE it is imperative to prove that the milestones on Timeline-A are real. This makes it necessary to shed some light on the AIT's dictum.

When a negative correlation between the development of Vedic Literature and AIT was pointed out by the indigenous Sanskrit scholars, the proponents of AIT insisted that the invading Aryans first settled in North India and then developed the rich Sanskrit language and then started creating Veda mantras and then established the tradition of learning and reciting them as the time progressed through their Civilization and culture. When the mind-boggling complexity of the Sanskrit language, its highly developed grammar, vocabulary, its unparalleled poetic excellence and its vast literature stared in the face of the proponents of AIT, they found themselves on shaky ground because all this progress required a long and stable historical period. To achieve such cultural magnificence in a short time, the invading Aryans had to be successful military invaders as well as literary geniuses; that defies commonsense. To get around this dilemma AIT proponents proposed that the Aryans must have brought the Vedas with them. Throughout the vast Vedic Literature there is not one single mention of another region or place from where the Aryans of India came. On the contrary, the Vedic Literature is filled with geographic references of places on the Indian sub-continent where certain *guru-kulas* (centers of learning with resident guru and students) were situated which produced some renowned thinkers, writers and philosophers who created post-Vedic literary works. There are lineages of families and dynasties that indicate that the Aryans are residents of India only. Also in the Indian Sanskrit Literature one finds references indicating the travels of individuals from ancient India to other continents and accounts of some even returning home to narrate their experiences.

Marshall's date assignment for AIT near 3200 BCE, well beyond Max Muller's date around 1500 BCE does not lend validity to the imaginary AIT. Marshal's figure is close to the Indian historical event of the Mahabharata War, fought between the cousins in the same dynasty and not between the Aryans and Dravidians as proposed by AIT. The Mahabharata is very well recorded in the Sanskrit Literature; the validity of its period is being proved in the following pages. The documented historical records as well as the astronomical sky inscriptions are used for this proof.

The book of Mahabharata giving a comprehensive account of the conflict is in fact a book of history. Yet in the book of Mahabharata there is no hint of any Aryan invasion or of the Aryans coming from elsewhere into India. On the contrary there are a number of references in Mahabharata indicating a continued stable Aryan Civilization in different regions of the Indian sub-continent, from centuries before Mahabharata, with lists of genealogies of the Aryan kings who ruled there. The ancient book of Mahabharata was not written with a foresight that someday it would come handy to discount a 19th century wild Aryan Invasion Theory. The book of Mahabharata compiled by Rishi Vedvyas and his deciples and submitted to the King Janmejay of Delhi soon after the Mahabharata War that ended in 3140 BCE. The Mahabharata War was fought in the field of Kurukshetra near Hastinapur (near Delhi). Both warring sides in this war addressed themselves as Aryans and no one was called Dravidian. The fact remains that the inventers and supporters of AIT did not read the book of Mahabharata or deliberately remained ignorant of this ancient book on ancient Indian history.

The fact that Manu and his book Manusmriti are cited a number of times in the book of Mahabharata but Manu's Manusmriti has no mention of Mahabharata and other contemporaneous historical personalities like Krishn, Arjun, Yudhishthi and Rishi Vedvyas the author of the sixth Darshan, Vedanta Sutra. Manusmriti is older than Mahabharata. When Max Muller places Manusmriti after 300 BCE, he has to place Mahabharata after 300 BCE as well. In the ancient Indian history, the authors of the original book of Mahabharata, not counting the recent insertions, indisputably are Rishi Vedvyas and his students. Max Muller therefore puts Vedanta Sutra and his author Rishi Vedvyas's period later than 300 BCE.

The period of 300 BCE in Indian history is well-known where there is no sign of Vedvyas. Buddhism and its *sutras* existed well before 300 BCE which Max Muller had read and translated. Manusmriti, Mahabharata and the vast Buddhist literature do not have any indication of an Aryan invasion. Max Muller was perhaps not aware of the genealogy of the King Manu, that his son Ikshvaaku was the first King of Ayodhya after it was established by Manu himself. Muller was not even aware of Manu's Sun-dynasty at Ayodhya, listed in Valmiki Ramayana, nor was he familiar with the genealogy of kings in the Lunar Dynasty that ruled from Hastinapur and listed in the book of Mahabharata starting with King Shantanu. Neither Ayodhya nor Hastinapur are in Asia Minor or elsewhere in the world but they are certainly in the Aryavart, which is India.

India – The Aryavart According to King Manu

People in other countries and continents west of India addressed the residents of Aryavart (Bharatvarsh, now India) as the people 'across the Sindhu River (Indus River)' or just the Sindhu. The invaders addressed the residents of Aryavart as the Sindhus. But the language deficiency changed the sound of the letter 's' into the letter 'h' and the word Sindhu turned into Hindu. The term Sindhu-sthan (place of Sindhus) turned into Hindu-sthan. It finally became

Hindustan. The distortion of a distortion having no limits, Hindustan changed into Indis and India. A distorted version of the Sanskrit term *sthaan* for 'place' changed into *taan* or *tan* which is found as the suffix in the names of many countries in the north-west of India: Baluchistan, Tajikistan, Afghanistan, Kyrgyzstan, Uzbekistan, Kazakhstan and so on. This also shows that the Sanskrit language at one time had reached regions far from India.

Just as the debate on AIT continues the debate between some linguists on the origin of the word "Hindu" still continues. Some suggested that the deficiency in pronouncing the letter 's' was brought into India by the invading Aryans which they improved upon later. The Aryans had to be somewhere on the Earth then why is it so difficult to accept that they could be living right there in India. The fact is that from the very beginning, starting from the Vedas, the Sanskrit language has had a large number of consonants and vowels giving rise to unique sounds including the sound of 's'. The single letter symbols to represent differing sounds of the same consonant are part of the Sanskrit alphabet set. As a sample, let us take the Sanskrit consonant ल which is equivalent to the 12th consonant 'L' in English. Without presenting here all the variations of ल, the five variations of the sound of ल in Sanskrit are: लं, लँ, ल्, ऌ, ळ. The symbols added to ल are respectively: a dot on top, crescent moon with a dot on top, a slanted dash underneath, sideways u under it and a reformed ल into ळ. They all change the sound of the basic consonant ल. The exact sounds of these five variations of ल are difficult to express with English alphabets, however, the closest sounds corresponding to these five variations respectively are: *lum, lumg,* half pronounced *la*, *lri* and *lrd*. The Sanskrit language, being highly developed and sophisticated, is not limited in sounds that humans produce. Therefore, the confusion between the consonant *s* and *h* could never arise out of Sanskrit and is rooted in some under-developed languages of the regions beyond the Aryavart.

The Vedas themselves provide fully developed sounds and terminology of their language, Sanskrit. For the sounds of *h* and *s* appearing side by side one has to look at the first RigVeda mantra. The third word *purohitam* utilizes the consonant h, pronounced as in the word 'him' and right after that its fourth word *yajyasya* utilizes the consonant s, pronounced as in the word 'say'. Both sounds thus pronounced separately in the very first mantra of the Aryans' first book would enable the Aryans to correctly pronounce the consonants *h* and *s*. The fact remains that Hindustan is not the name of India it is Bharatvarsh which used to be Aryavart. The AIT's propagation through the British education system in India has created a deep division between north and south Indians, the latter believing that northerners were the invader Aryans and the southerners were the suppressed Dravidians. The old name of India, Aryavart, was discouraged and it is not in usage now. Bharatvarsh as the official name of India was however accepted by most Indians.

The first known name for India in the Vedic Literature is what Svaayambhuva Manu stated in his Manusmriti, the Brahmavart, meaning the place of Brahma the first known human to have preached all four Vedas to the society. The alternate meaning of the word Brahmavart is 'the place of learned people', the Brahamanas. Five *shlokas* from the ancient book of Manusmriti that relate to this topic are presented below. Manu has given a geographic definition of what now is Bharatvarsh. He has done this in three steps - starting with Brahmavart, a small region of the greater Bharatvarsh and then defining its Maddhya Desh (Central Region) and finally defining the whole nation that he addressed as Aryavart. In addition to the geographic description of Aryavart, Manu has also briefly presented its characteristics in terms of sociology and ecology.

Once Manu's definition of Aryavart is in place the reader will understand that this Aryavart has been inhabited by the Aryans from the very beginning of the Aryan Civilization, they are its indigenous people and have never been the invaders from Asia Minor or any place else. Manusmriti's chapter and *shloka*-numbering in this book are according to the 1966 publication entitled "Vishuddha Manusmriti" with accurate Hindi translation of *shlokas* and enlightening interpretive commentary by Professor Surendra Kumar. In this publication of Manusmriti, both subjects, the Srishti-Utpatti (Birth of the Universe) and Dharma (duties) are included in the first Chapter unlike in other publications where the two topics are separated in two successive chapters.

The three *shlokas* on the geographic description of Aryavart are presented first. King Manu's first step in defining the nation that he ruled, is in the following *shloka*,

सरस्वतीदृषद्वत्योर्देवनद्योर्यदंतरम् ।
तं देवनिर्मितं देशं ब्रह्मावर्तं प्रचक्षते ॥ (Manusmriti, 1/136)

Transliteration: "*Sarasvati-drishdvatyo-r-devanadyo-r-yadntaram.*
Tam devnirmitam desham brahmaavartam prachakshate."

Translation:

The region between Sarasvati and Drishdvati, the rivers in union with virtuous sages and brilliant scholars, inhabited by devas the virtuous people with noble character and brilliant qualities, is said to be the region of Brahmavart." (Manusmriti, 1/136).

It is well known in Sanskrit Literature that the River Sarasvati with its magnificent roar did exist and it is also stated that it disappeared with time, sometime in the antiquity of the Vedic Civilization. In the Sanskrit Literature the River Sarasvati is revered and is known for its shores being the abode of *rishis*, *munis*, sages and ascetics where they created brilliant pieces of literary work such as the Brahmanas, Upanishads, Smritis and Sutras that form the backbone of the Vedic religion and philosophy and continue to provide an unwavering foundation for present day Hinduism. Since a lot of knowledge has flowed from the shores of Sarasvati like the flow of its water, the ancient *rishis* named it as the River Sarasvati. Considerable discussion on the term '*sarasvati*' is in Chapter 4. The term *sarasvati* appearing in differing connotations in the ancient Vedic Literature relates to different subjects. However, in the context of the Manusmriti *shloka* given above, Sarasvati is one of the two rivers serving as a boundary of the Brahmavart region.

The River Sarasvati of Manu's *shloka* is now extinct. Since this is the river with a sempiternal relationship with Vedic Literature and the Vedic Civilization, people continued to wonder about this legendary river and wondered if it existed in reality. The Vedic literature has presented a picture so vivid that people continued with the ancient belief that it had its origin in the Himalayas and fell into the Paschim Samudra (Western Sea) which is now known as the Arabian Sea.

The names of both rivers in Manu's *shloka*, Sarasvati and Drishdvati, are not found among the existing rivers on the current map of India. Based on Vedic Literature, several authors have depicted the presumed trajectories of these rivers on the map of India. Professor Surendra Kumar in the Hindi interpretive translation of Manusmriti has presented a geographic map of India depicting Sarasvati and Drishdvati rivers. In the present context a sketch of Northern India is shown in Figure 3.3 with the main rivers, mountains and places labeled. The rivers Sarasvati and Drishdvati are shown in magenta-dashed lines. The dashed line represents a river that is currently not on the geographic map of India.

Manu defines the region between the two rivers as Brahmavart shown in orange. But if the two main rivers named by Manu are not on the present Indian map, how can we ascertain that Brahmavart region ever existed in India? That the Brahmavart was part of a bigger region within India becomes obvious from Manu's *shloka* that followed. Not including all the *shlokas* given in between, the relevant *shloka* defining the boundaries of the Maddhya Desh (the Central Country) is,

हिमवद्विन्ध्ययोर्मध्यं यत्प्राग्विनशनादपि ।
प्रत्यगेव प्रयागाच्च मध्यदेश: प्रकीर्तित: ॥ (Manusmriti,1/140)

Transliteration: "*Himvdvindhyayor-madhyam yatpraagvinshanaadpi |*
Pratyagev prayaagaachch madhyadeshah prakeertitah."

Translation:

"*The region between the Himalaya* (in the north) *and the Vindhya* (in the south), *and also from east of Vinshan Region* (devastated region) *to the west of Prayag is called the 'Madhya Desh'* (central region)". (Manusmriti,1/140).

Manu in the earlier shloka used the rivers to define the geographic region of Brahmavart. In this shloka he defines a larger region Maddhya Desh using the mountains and identifiable geographic regions. The north-south span of the Maddhya Desh (Central Region) is defined by Himalaya in the north and Vindhya Mountain range, just south of river Narmada (Figure 3.3). Himalaya (meaning: the abode of ice) of Manusmriti is still Himalaya and Vindhya Mountain range of Manusmriti is still Vindhya with the added adjective of '*achal*', meaning 'stable'. Its common name now is Vindhyachal. The Prayag in the east is still Prayag. Prayag, the eastern point of Manu's Maddhya Desh, is the city of Prayag in the present Indian state of Uttar Pradesh, situated at the confluence of the rivers Ganga and Yamuna shown by the blue lines. This defining coordinate for the ancient central region of the Bharatvarsh has been a holy city and the oldest city in India (probably oldest in the world). The Maddhya Desh is present Maddhya Pradesh, a state in the Central India, though much smaller than the Maddhya Desh of Manusmriti. This area is shown in the map in light pink with its southern boundary along the Vindyachal Mountains. With so much continuity in the nation-defining terminology in Manusmriti the rivers Sarasvati and Drishdvati that bounded the Brahmavart region must also have been a reality during King Manu's time. All these land-marks do exist with their original names in the present Indian geography, except for the Vinshan Region to the west of Maddhya Desh.

A remarkable geographic observation in Manu's shloka is the label of 'Vinshan' for a region which is the western region of the Maddhya Desh. On one hand Manu states that it is the region of the great river Sarasvati and on the other hand it is also labeled as the Vinshan region. The meaning of the word *vinshan* is the region that 'perished' or was 'destroyed'. Geographically the Great Thar Desert in the state of Rajasthan should represent the Vinshan Region of Manusmriti. The southern part of it extends into the State of Gujarat with its name as Rann of Kutch. The course of Sarasvati shown in the map of Figure 3.3 by a dash-line is based on other ancient references, which are now confirmed through scientific rediscovery of the extinct River Sarasvati. The Vinshan Region on the map is arbitrarily marked by this author along the assumed course of Sarasvati. Further discussion on the Vinshan Region of Manusmriti is taken up later in the following.

Notice that Manu started with a narrow region of Brahmavart, expanding to the central region of Maddhya Desh. Manu's third step extends to encompass an even bigger geographic region to define 'Aryavart'. Manu's next *shloka* (1/141) therefore is the nation defining statement,

आसमुद्रात्तु वै पूर्वादासमुद्रात्तु पश्चिमात् ।
तयोरेवांतरं गिर्योरार्यावर्तं विदुर्बुधाः ॥ (Manusmriti, 1/141)

Transliteration: "*Aasamudraattu vai poorvaadaasmudraattu pshchimaat |*
Tayorevaantaram giryoraaryaavartam vidurbudhaah ||"

Translation:

"*One that exists from Poorva-samudra* (East-Ocean) *to the Paschim-samudra* (West-Ocean) *and bound by the mountains is proclaimed by the learned as the Aryavart* (the place of Aryans)." (Manusmriti, 1/141).

With the two seas on either-side and the Himalaya in the north (in previous shloka) Manu does not leave any doubt as to which country is the Aryavart. Also, it is not that he is naming this region as Aryavart; he is clearly crediting the 'learned' for proclaiming thus defined land to be Aryavart. Implicit here is that as indicated in his shloka, this land was known as Aryavart before Manu. If one defines the central region of a country it is obvious that the country extends on either side of the central region. Therefore the Aryavart extends beyond Sarasvati in the west probably up to river Sindhu (Indus) and beyond Prayag in the east up to Burma, ancient name being Brahma Pradesh, which now is Myanmar. Similarly the Aryavart must extend beyond the southern boundary of the Maddhya Desh at Vindyachal Mountain range. This also makes it clear that Manu had a full grasp of the geography of Aryavart. Therefore it can safely be concluded that according to Aryan literature, the Manusmriti being the present context, Aryavart is the most ancient name of the country where Aryans lived.

The suffix 'vart' to the words Brahma and Arya forming respectively Brahmavart and Aryavart means the 'place of' or 'inhabited by'. This should suffice to close the debate on where the Aryans lived. Brahma in the Vedic Literature is known for preaching the four Vedas to the Aryans in the beginning of the Vedic Civilization. Therefore the two names of the country were proclaimed in Manusmriti, relating respectively to the Vedas and the inhabitants. The Veda has adorned with beautiful adjectives the term used for a person following the Veda. The multi-meaning term '*arya*' stands for '*ishvar putra*' meaning 'God's progeny'. Virtue-based derived meanings of the term arya lead to different closely related meanings; most common being 'person with noble character and qualities'. The word arya has no race connotation in its meaning. Manu in his Manusmriti has also described the nature of the people living outside the boundaries of Aryavart. Such a nation-defining description for a country is perhaps unique in ancient history and literature. To suggest that the Aryans lived elsewhere before they invaded Aryavart (India) is a theory based on ignorance.

The rivers Sarasvati and Drishdvati, shown by dash-lines on the map of India (Figure 3.3), are extinct and exist in the Vedic Literature and legends only. Swami Dayanand Sarasvati while writing Satyarth Prakash in the late eighteen-eighties interpreted Sarasvati and Drishdvati to be the rivers Sindhu (Indus) and Brahmputra. According to him Aryavart extend from the centre of Himalaya in the north to Rameshvaram in the south.

Prayag, a most ancient City

The city of Prayag mentioned in Manusmriti got its name from what it stood for. The meaning of the word '*prayag*' is - "the place where the '*yagjya*' is performed with perfection" (*yagjya* is the sacred fire-oblation). The people performing *yagjya* are called the *yagjyiya* people for they have the welfare of others in their thoughts and actions and are considered to be devoted to the Almighty Brahmn and are therefore called brahmanas. Prayag has been a center of Vedic culture from the time it came into existence and continues to be important till today. Historically, the city of Prayag has a strong relationship with Maharishi Bharadwaj, an often-mentioned figure in Vedic Literature. His ashram was in Prayag. He was contemporary to King Ram of Ayodhya and Rishi Valmiki who wrote Ram's biography which is famously known as Ramayana.

Valmiki Ramayan has devoted 22 *shlokas* in the 42nd *sarg* (section) of Ayodhya-Kandam (chapter) on the historical concourse of Prince Ram, his wife Princess Sita and his younger brother Prince Lakshman with Rishi Bharadwaj at his Prayag ashram. The Prayag ashram is a significant stop in the journey of the trio after they left the Kingdom of Ayodhya for life in the forest. They spent one night with Rishi Bharadwaj at his ashram. Ramayana *shlokas* on Rishi Bharadwaj welcoming these guests to his ashram indicate that this was not the first meeting between Ram and Rishi Bharadwaj.

They had considerable intellectual exchange before retiring for the night. Next morning Ram took permission of Rishi Bharadwaj before leaving the ashram to continue forest-living for 14 years, to fulfill the commitment made by the obedient son Ram to his father, Emperor Dashrath. This unfortunate situation arose because Dashrath was forced to keep a promise that he had made to his first queen. The queen demanded that her son Bharat, who was younger to Ram, be enthroned and Ram be exiled to the forest for 14 years. Rishi Bharadwaj, reciting Veda mantras and showering his blessings on Ram, Sita and Lakshman, saw them off.

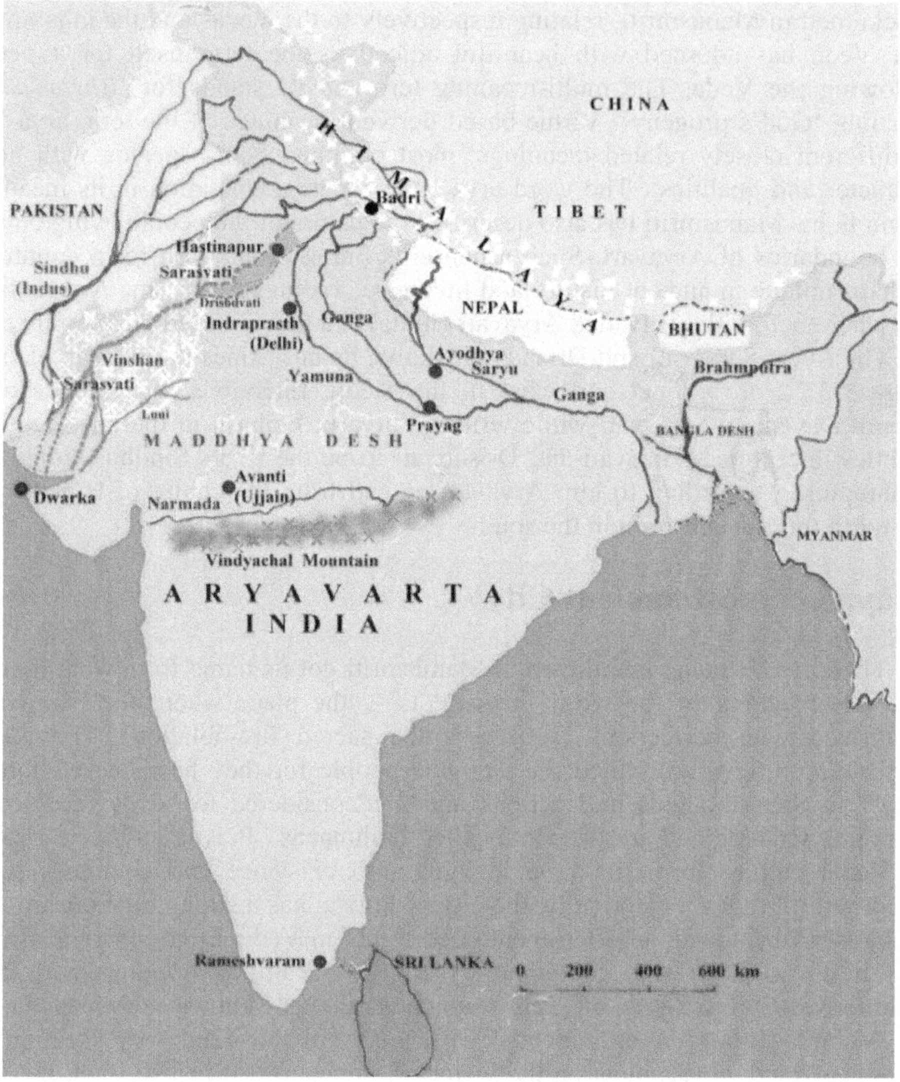

Figure 3.3: Map of Aryavart (India) with defining landmarks mentioned by Manu in his Manusmriti.

Ram, Sita and Lakshman crossed the River Yamuna at Prayag to visit the great poet Rishi Valmiki whose ashram was at Chitrakoot Mountain. It is this ashram where Valmiki was inspired to write Ram's biography (Ramayana) in shloka poetry in anushtup-chhand meter of 32 syllables. Prayag and Chitrakoot, having such incredible historical significance, continue to be highly revered pilgrim places for Hindus. Historical chronology has it that Ram being in Manu's dynasty, many generations after Manu, his contemporary sages Bharadwaj and

Valmiki are not mentioned in Manusmriti. On the other hand in Valmiki Ramayana there are numerous references to Manu and his dharma principles that he laid down in Manusmriti. Rishi Valmiki, citing Manu at places in the Ramayana has even reproduced Manusmriti shlokas with little variation.

Manusmriti's Prayag continues to be a real chapter in India's ancient and modern history. However, the Manusmriti's River Sarasvati is absent from India's present geographic landscape. This led some early Indologists to hypothesize, purely on speculative grounds, that the statements in Manusmriti were fictitious and that the book of Manusmriti is not necessarily authored by Manu and that it was put in poetic form sometime after 300-400 BCE. The well-known Indian history of this period discussed below contradicts these Indologists. This was the period when Buddhism peaked in India while Manusmriti's laws were already ingrained in the Aryan society.

River Sarasvati Corollary

To get India out of the claws of contemporary distortions in its history one has to make an additional effort to show that the nation-defining geographic features in Manu's *shlokas* in Manusmriti are real. The presence of the Himalaya Mountain range in the north, the Vindyachal Mountain range in the central region, the East Sea and the West Sea surrounding its land mass is indisputably accurate. The reference of the historical city of Prayag in Manusmriti was also substantiated by references in post-Manusmriti Valmiki Ramayana describing historical events and persons. The validity of the reference of the mighty River Sarasvati in Manusmriti is an added proof of the ancientness of this landmass. To shed light on this historical aspect of Aryavart, to advance the discussion further, let us begin with a corollary formulated by this author on River Sarasvati.

"*Manusmriti's River Sarasvati of nation-defining significance presently does not exist on the Indian sub-continent. If it is proven that the River Sarasvati was a reality at one point in the Aryan history with its existence in the region mentioned in Manusmriti, it will be self-proven that Manu, the author of Manusmriti, was a real historical person*".

This corollary is proven in the following.

Presently, a seasonal river called Ghaggar flows in the dry riverbed where the ancient course of Sarasvati is assumed to have existed. Ghaggar flows through the state of Haryana, continues in the desert of the western state of Rajasthan with its name changed to Hakra and it continues into Sindh region of Pakistan as Nara. It is Ghaggar's dry bed in Rajasthan where in 1893 CE an English engineer named C. F. Oldham (Ref: "Recent Research on the Sarasvati River", www.stephen-knapp.com) was riding his horse when he observed that this seasonal river had little water when flowed, yet its bed was 3 km wide. He thought that it could be the course of a bigger and wider river at some point in time and that it just could be the River Sarasvati that continued to exist in Hindu mythology! What Oldham envisaged was Sarasvati, but its discovery was to wait for a century, when India after gaining independence from the British, progressed in leaps-and-bounds in its scientific endeavours. Indian government initiated the "Sarasvati Project" to investigate the 'myth' of Sarasvati. The extraordinary feature of the "Sarasvati Project" was that the investigating teams were formed from multidisciplinary science disciplines with state-of-the-art equipment at their disposal. The researchers of India, many of whom are world-renowned scientists, discovered the extinct legendary River Sarasvati. The evidence is in a large number of science publications.

The Scientific Proof of Extinct Legendary River Sarasvati

The unraveling of the mystery of the invisible River Sarasvati started in the 1980s when simultaneously various scientific methodologies were applied. A coordinated team-work of Indian national scientists of international repute, utilizing most modern remote sensing systems and ground-truthing data revealed that the said enormous River Sarasvati existed in the distant past, in the same region. This coordinated team-work under the national project named "Sarasvati Project" included remote sensing scientists, geologists, geographers, archaeologists and climatologists. The long-term study involved satellite imagery from earth-observing satellites; LANDSAT of NASA (National Aeronautics and Space Administration, USA), Remote Sensing Satellites IRS-1C and IRS-P3 of ISRO (Indian Space Research Organization) and European Remote Sensing Satellites ERS-1 and ERS-2 of ESA (European Space Agency).

In addition to ground and satellite data a pool of palaeoseismic and palaeoclimatic records enabled the scientists to reconstruct the drainage evolution in north-western India. The term "palaeoseismic records" means the geological evidence of earthquakes from the prehistoric times and likewise the term palaeoclimatic records means the climatic records available from geological information from ancient times. On one front the LANDSAT imagery revealed the river valleys and their courses under the Thar Desert and the Rann of Kutch and on the other front the drilling and hydrological studies of the palaeo channels were used for the dating technique. It was discovered that the magnificent River Sarasvati of ancient Vedic Literature flowed where it was said to have flowed and that it had three major tributaries, Shatadru (Sutlej) originating from the Mount Kailash, Drishdvati and Yamuna from Shivalik Hills. These rivers formed the mighty Sarasvati and charted the course along which the present seasonal river Ghaggar flows. The satellite images show that at certain places the River Sarasvati was up to 8 km wide.

The Sarasvati Project was broadened to include research disciplines of other national research agencies of India, such as the NRSA (National Remote Sensing Agency), BARC (Bhabha Atomic Research Centre), ISRO, GSI (Geological Survey of India) and CWC (Central Water Commission). Based on the information from the published research, some of the important discoveries that relate to the River Sarasvati included the observation of the vegetation cover thriving on the loamy residual soil along the course of Ghaggar through the states of Punjab, Haryana and Rajasthan. In 1995, discovered in the data of IRS-1C, combined with RADAR imagery from ERS, were huge palaeo-channels under the Thar Desert of Rajasthan that were independent of any connection to Indus River. BARC found ground-water in the Rajasthan desert 50 to 60 meters underground with the water depth ranging from 15 meters up to 40 meter. It is estimated that such a huge quantity of ground-water is enough to sustain agriculture even during severe summer conditions.

After India's Pokhran underground nuclear test in 1998, BARC tested the underground water for radioactive tritium. In the process, they found at these Rajasthan desert sites the potable (drinkable) water of Himalayan glacier origin that was some 8,000 to 14,000 years old. They also found that this water was presently being recharged through aquifers from some locations in the north. The geological term aquifer represents a porous deposit of rock, like sandstone, that contains underground water available in wells. The presence of an 'Underground River' along the expected path of River Sarasvati was confirmed by data-streams from different research disciplines, leading to the conclusion that ISRO's findings of the River Sarasvati were based on scientific truth. The decrease of the levels of radio-active Carbon-14 along the downstream of the underground-river also establishes hydraulic continuity of the buried Sarasvati. Scientific studies suggest that the river dried up around 2000 BCE coinciding with the extinction of the Harappan Civilization.

The Cause for Sarasvati's Disappearance

The geological evidence suggests that the cause for Sarasvati's disappearance is the tectonic activity (earthquake) raising the plate near the Himalayan foothills, cutting the river off and diverting its tributaries; Yamuna to the east meeting Ganga and Satlaj to the west meeting Indus. The geological ground-faults near its origin and along its course are the geological evidence of the tectonic activity that took place in several phases. A large number of scientific papers on Sarasvati's existence and disappearance are the outcome of this multifaceted scientific study involving satellite observations of the region and ground measurements by the prominent scientists: Yashpal, B. Sahai, R. K. Sood and D. P. Agrawal in "Proceedings of Indian Academy of Sciences, Earth and Planetary Science", (Vol. 89, pages 317-331, 1980); K. S. Vaidiya in "Resonance", May, 1996; just to cite a few. They have deciphered the mystery of River Sarasvati and concluded that there were geological reasons and events for the disappearance of Sarasvati and Drishdvati, the nation-defining rivers of Manusmriti.

The well-known geological fact is that the formation of the Himalayas is the result of the plate-tectonic activity wherein the Indian continental plate is moving north pushing against the Tibetan plate. Recent geological studies have shown that the whole Shiwalik region from one end of the Himalaya to the other continues to be geologically active. The studies have discovered that owing to severe tectonic activity, the Shiwalik region was destroyed about 1.7 million years ago changing the course of rivers that were there earlier. This region has continued to be unstable, frequently changing the geology of the Himalayan foothills resulting in the change of river trajectories.

The geologists have found that due to the geological tectonic activity leading to earthquakes, large land areas have shifted, been destroyed and tilted, changing the course of the rivers several times in north-eastern India. It was one such or several such tectonic events in relatively recent geological times that diverted River Yamuna from the River Sarasvati and cut off Shatadru (Sutlej) near the town of Roper giving it a sharp turn and sending it west to the River Sindhu. The water supply to Drishdvati and Sarasvati was thus cut off.

Climatological change descended on the 'Vinshan Region' of Manusmriti which is now known as the Thar Desert of Rajasthan which according to new geological findings once was a lush region. When the region started to turn into an extended desert it engulfed whatever was left of Sarasvati, along with the settlements at its banks. After many scientific conferences and journal publications it has been established that the lost River Sarasvati is a scientific fact with a well-established course of the river. Mighty Sarasvati's run from its origin to the sea is estimated to be about 1600 km. Sarasvati's existence in India's glorious past is a historic fact.

Reconstructed Course of Sarasvati

From these new data gathered under the Sarasvati Project the prominent geologists have reconstructed the course of the extinct River Sarasvati. The same pool of data have also enabled reconstruction of the extinct and changed courses of other rivers in the same region. Based on these measurements and the follow-on geological correlations, a picture emerged that depicted the course of Sarasvati and its tributaries (see K. S. Vaidiya in the May, 1996 issue of the science publication, "Resonance"). This is shown in Figure 3.4.
In this figure a dotted line shows the ancient course of the river and a solid line represents the present course of the same river. Over several sections of the river-trajectories the ancient and present courses remained coincident. It is interesting to note that the River Sutlej with its Vedic name Shatadru, which now joins River Sindhu, used to turn near the town of Ropur to join Sarasvati.

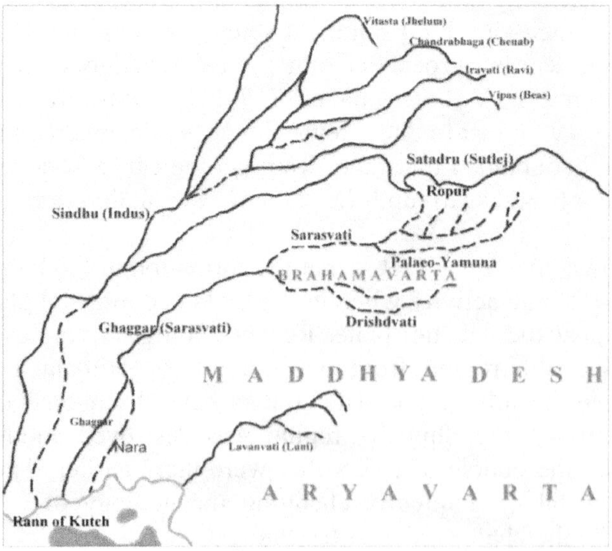

Figure 3.4: North-Western river system, past (dashed line) and present (continuous line) (Ref: K. S. Vaidiya, May, 1996, Resonance, pages: 19-28).

This grouping of the extinct ancient rivers presents the region between Sarasvati and Drishdvati that Manu addressed as the Brahmavart region, which stretched east-west and is a bit larger than as shown in the earlier Figure 3.3

The scientists have mapped about five additional tributaries of the extinct Sarasvati, including Yamuna (Palaeo-Yamuna) and the extinct River Drishdvati. The scientific data have thus yielded the final geographic picture with curves and twists in the course of River Sarasvati and its tributaries. The River Sarasvati with its source in the Shiwalik Hills at the foothills of the Himalayas in the present state of Himachal Pradesh, flowed through the areas of the states of Punjab, Haryana, Rajasthan and in the area called Kutch in Gujarat, immersing itself in the Paschim Sagar (Arabian Sea). Sarasvati flowed more or less parallel to the River Indus, its Vedic name being Sindhu.

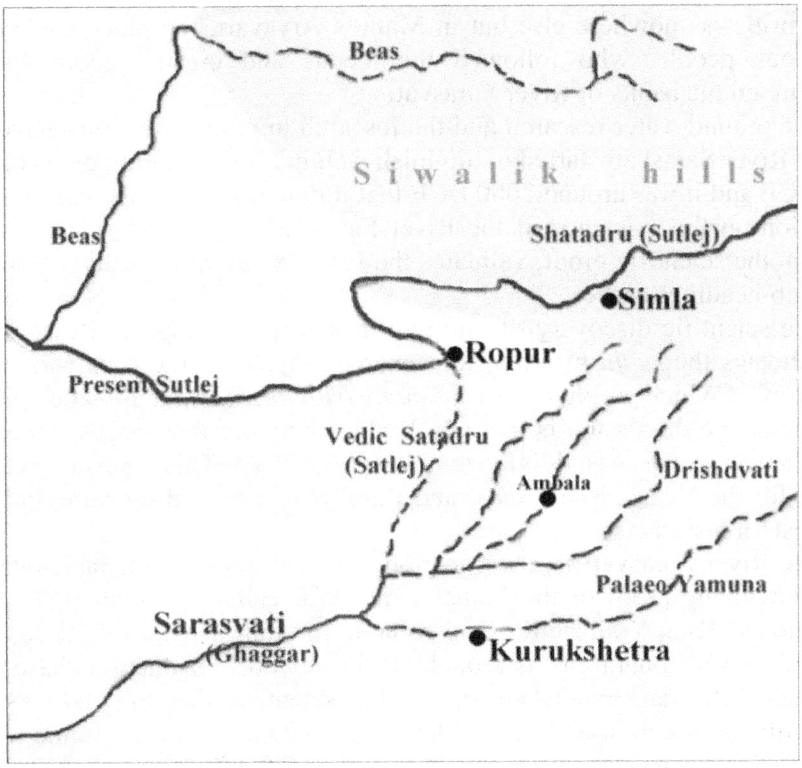

Figure 3.5: The Sarasvati river system based on Vaidiya, S. K. 1996, and Yashpal et al. (1980).

In defining the main course of Sarasvati there is no difference of opinion among the scientists of different disciplines that joined hands on the Sarasvati Project. There is no difference of opinion among them on the courses of two other rivers, Paleo-Yamuna and Drishdvati. The difference of opinion if any is on the position of Paleo-Yamuna and Drishdvati. The other group prefers to interchange the positions of these rivers from what is shown in Figure 3.4 to that shown in Figure 3.5.

Yashpal, B. Sahai, R. K. Sood and D. P. Agrawal (Proc. Ind. Acad. Sci., Earth Planet. Sci, Vol. 89, pages 317-331, 1980) with their extensive investigation from multi-facetted data have emphasized the grouping of tributaries as shown expanded in Figure 3.5.

For simplification of the diagram all details of other extinct rivers that have been discovered are not shown. The group of rivers that supplied water to Sarasvati originated from Shiwalik Hills shown in light green. The River Sutlej undisputedly was a big tributary of Sarasvati and flowed south from around Ropur along the dotted line to join Sarasvati before a cataclysmic event changed its course. Yamuna being a big river itself would add to Sarasvati's magnificence.

The discovery of the legendary rivers, Sarasvati and Drishdvati, based on the scientific methods and multi-disciplinary data tells us that King Manu did not imagine these rivers. These rivers were a reality during Manu's time. This would also lead us to the facts stated by Manu, that the River Sarasvati of his Manusmriti was nowhere else but in Manu's Aryavart, the place of Aryans, its indigenous people, who followed the Vedas and created enormous Vedic Literature on the banks of River Sarasvati.

The ground water research and the research in other disciplines have shown that the River Sarasvati started to diminish well before the period of Mahabharata 3140 BCE and it was around 2000 BCE that it practically disappeared. The above discussion on the existence of the River Sarasvati on the Indian sub-continent, based on the scientific proof, validates the River Sarasvati Corollary, formulated a few sub-headings earlier.

The scientific discovery of multiple tributaries feeding the River Sarasvati also validates the *rudhi* meaning for concrete objects of the term *sarasvati*, the etymology of which yields - *one with many tributaries*. An automatic conclusion from the above discussion is that any Vedic Literature that makes a mention of River Sarasvati is at least 4000 years old (2000 BCE). This takes us well before the age for the Vedas, Manusmriti and Ramayana assigned by some Indologists and Western historians.

The River Sarasvati is also mentioned in the book of Mahabharata. Since more convincing proof of the Mahabharata War ending in 3140 BCE with the coronation of King Yudhishthir is to come in the following, a reference to River Sarasvati in Mahabharata is as reliable as the scientific deductions based on the recent scientific data published by reputed scientists that the River Sarasvati existed till about a thousand years after Mahabharata. This is the same as saying that Sarasvati's waves ceased somewhere around 2,000 BCE.

Vedic Civilization at the Shores of River Sarasvati

It was the *rudhi* meaning of the term *sarasvati* that represented the mighty River Sarasvati. The River Sarasvati was named most appropriately because it represented an embodiment of 'dynamic action of perpetual movement of water with many tributaries'. The *yaugik* meaning of the term *sarasvati* that it is 'specific human intellect' (discussed in Chapter 4) was also fulfilled because its shores served as the abode of *rishis* and *munis* who developed knowledge at its banks, knowledge that is perpetuated through the Vedas and the Shastras of the Sanskrit Literature. Where there is knowledge, wisdom follows, which in turn leads to prosperity. There are all indications that the Ayravart at one time was a prosperous nation (see Valmiki Ramayana's Ayodhya Kaandam). Those references on prosperity in Aryavart are from the ancient references, however, the recent references related to archaeological finds along the shores of rediscovered extinct River Sarasvati exhibit a thread of continuity in Aryavart's prosperity.

Simultaneous with various scientific methodologies that were utilized in discovering the legendary River Sarasvati, the Indian archaeology made big strides in the past five decades. A large number of publications on Sarasvati re-discovery and archaeological sites on its shores are made available by Dr. S. Kalyanaraman who has set up a website, http:www.hindunet.org/saraswati, on "River Sarasvati and Civilization" enriched with 30,000 files on it. As he put in the summary of the colloquium held on January 5 and 6, 2003 (some details furnished below) he stated –

> "*The river (Sarasvati) is not a myth but is ground-truth and had drained in North west Bharat over a distance of 1,600 kms from Manasarovar, Mt. Kailash to Gujarat (Somnath, Prabhas Patan).*"

The rediscovered course of Sarasvati is the present course of the dry river Ghaggar. Out of over 2,600 archaeological sites discovered in the past five to six decades by the archaeologists, the sites that are termed as the Indus Valley Civilization or Harappan Civilization first discovered in 1922, about 2,000 of them are found on the banks of the River Sarasvati and are 3,000 BCE old. Several of these, too many to name here, are larger than Harappa and Mohenjo-Daro. Over a dozen other culturally vibrant sites on the banks of the River Sarasvati begin from Ropur in the north through Kaalibangan in Rajasthan reaching Dwarka with marine-riverine culture. A remarkable discovery among these sites is the prehistoric town of Kaalibangan in Rajasthan State that has furnished a treasure trove of information. It has exhibited that among the citizens of this town were priests, farmers, merchants and highly advanced artisans and craftsman. The discovered seals with inscriptions indicate that these people were educated. There were all signs of River Sarasvati once flowing in the same bed where now flows the seasonal river Ghaggar. Recent research articles indicate that the total number of discovered sites have gone up from 2,600 in 2003 CE to about 16,000 at the present time. It becomes obvious that the prosperity of these ancient towns was supported directly and indirectly by the 'vast and awesome' River Sarasvati. None of the sites along the River Sarasvati indicate signs of war between the locals and any invaders because the Aryans did not come from anywhere else. They were always here.

The above discussion on the extinct River Sarasvati based on its geographic information in the historic Manusmriti and its subsequent rediscovery based on multi-faceted modern scientific studies under the national Sarasvati Project of the Government of India should put an end to AIT and other groundless hypotheses on Vedas and Vedic Literature by some Western Indologists and their Indian followers. References of River Sarasvati validate our Timeline-A in Figure 3.1 thus establishing the antiquity of references on Vedic cosmology principles which form the basis for this book. Also established here is the fact that Manusmriti itself is older than Mahabharata and Ramayana. When one walks the path of history from Manu towards Mahabharata one finds references of River

Sarasvati in post-Veda and post-Manusmriti Sanskrit Literature. The River Sarasvati is crowned with many adjectives such as 'The Great River Sarasvati, 'most momentous', 'with ceaseless flow', 'a river with thunderous noise', 'vast and awesome', 'most holy River Sarasvati' and so on.

Why is Sarasvati important for this chapter?

The first reason is to establish the historicity of references, Manusmriti, Ramayan and Mahabharata. As mentioned above, Sarasvati for Manu was a nation-defining great river which after his time disappeared and which continued only in legends in post-Mahabharata Sanskrit Literature until it was rediscovered in recent times through scientific measurements and methods. The second reason is Sarasvati's association with Rishi Vedvyas who carried out astronomical observations from its shore, which he has mentioned in the text of Mahabharata. These astronomical observations have been examined by modern astronomers using modern-day computation of accurate equations in astronomy that take us back to 3000 BCE right to the time of the astronomer, Rishi Vedvyas.

Therefore, if the River Sarasvati existed even after the book on Mahabharata was written the Vedic references as old as and older than Manusmriti, used in this book, are older than 3000 BCE. This certainly frees us from the clutches of AIT.

Sarasvati and Astronomy in Mahabharata

The book of Mahabharata, right up to its last chapters, has numerous references of the River Sarasvati. The River Sarasvati's relation to the region of Kurukshetra, the capital city of Hastinapur and to Rishi Vedvyas is quite vividly depicted in the book of Mahabharata. An example from Mahabharata's Mausal-parv (Mausal-division), presented in the following is quite revealing.

Krishn being a just and powerful king fought several wars for just causes. His participation in such wars was on request from kings vulnerable in military to face attack from greedy and unjust kings. His participation was with the aim to establish law and order and *dharma*. After a while this great yogi chose to renounce war and violence. Along with his brother Balram, he had moved his capital, before Mahabharata War began, from Mathura to Dwarka at the sea-shore in the Gujarat Region. Krishn was now addressed with the new adjective of 'Ran-Chhod', meaning the one who renounced participation in wars. When Mahabharata War ended just before 3140 BCE, Yudhishthir was seated on the throne of Hastinapur and Krishn was already ruling peacefully from Dwarka. Krishn ruled peacefully from Dwarka for four decades.

Human nature being the way it is, strifes developed in the region of Dwarka. Towards the onset of Kali-Yuga epoch in 3102 BCE the region went into turmoil and civil war broke out between different groups and clans. This resulted in chaos and confusion with large numbers of people displaced from their farms and homes. Upon learning of the situation, the Emperor Yudhishthir of Hastinapur (Delhi) deputed his brother Prince Arjun to go to that region and help his best friend and charioteer in the Mahabharata War, none other than Krishn. Prince

Arjun intervened and helped to settle the strife-afflicted homeless people of the different clans. After that Arjun returned to the capital of Hastinapur with a stop at Rishi Vedvyas's ashram. A brief relevant description of the events according to the Mausal-parv (4/53-55) of Mahabharata goes thus:

> "*Arjun settled Saatyaki's son Yauyudhaani with all elderly and children, in the region at **Sarasvati**'s shore and made him the Governor. Then Arjun asked Vajra to be the ruler of Indraprasth. After settling the devastated people in anguish, Arjun returned. He reached **Rishi Dvaipayan Baadrayan Vedvyas**'s ashram with heavy heart and tears in his eyes. With utmost respect to the rishi, pronouncing his name he bowed at rishi's feet. Seated in great peace the rishi born to the mother Satyavati said to Arjun – 'Son, I welcome you, come and sit here.... and tell me why you are looking depressed and wearisome?' Arjun said – 'the one with lotus eyes, the one with most attractive dark complexion, the great Krishn has just died and also dead is his brother Balram... and I feel as if the ocean has dried up, the mountains are shaking and the sky is torn apart...'.*".

In addition to the human element in the narration there is also a part of the ancient history in it. The most relevant aspects of this description (abridged here) are the references to the River Sarasvati and to the Rishi Vedvyas, the author of the book of Mahabharata and the Vedanta Sutra. It shows the reality of the River Sarasvati and its usefulness for people to settle near it and prosper. It also shows that Rishi Vedvyas's ashram at the bank of Sarasvati and the city of Indraprasth at the shore of Yamuna River were places not too distant from each other. There are all indications that Sarasvati flowed just west of Kurukshetra, the battlefield of Mahabharata War (marked on Figure 3.5) and that River Sarasvati has direct relevance with the writer of the book of Mahabharata and the Vedanta Sutra, Rishi Vedvyas. This great *rishi*, a possible descendent from the region of the Badri-forest, with his full name being Badaraayana Krishn Dvaipaayana Vedvyas, was not only a philosopher who gave us the Vedanta Sutra, but he was also a great astronomer engaged in practical astronomy, observing planets and constellations from his ashram at Kurukshetra at the bank of River Sarasvati, about 300 km south from Badri (marked on Figure 3.3) and 160 km north from Indraprasth (modern Delhi).

To establish the historicity of the Kurukshetra War (Mahabharata War) a colloquium was organized by "Akhila Bharatiya Itihasa Sankalana Yojana" (All-India History Compilation Plan) under the auspices of the Mythic Society, Bangalore, India and Sir Babasaheb Umakant Keshav Apte Smarak Samithi Trust, on January 5 and 6, 2003. This conference was unique in that, in addition to historians and archaeologists, it was well attended by the scholars from across the world in science disciplines such as astronomy, geology and physics. The papers presented at this conference were related to Sarasvati references in the book of Mahabharata and the references related to astronomy in Mahabharata. Scholars have identified over 150 astronomical references in Mahabharata and pre-Mahabharata Vedic Literature of which 70 references are in the book of Mahabharata alone. These vivid astronomical records of celestial inscriptions are those that can be subjected to astronomical calculations and their dates can be ascertained by going back with the help of computer simulations. In the preamble to the colloquium there is a Mahabharata reference on Rishi Vedvyas, stating that –

"…day in and day out he* (Rishi Vedvyas) *is watching from the banks of the River Sarasvati the planetary positions on the skies."

His night-sky observations of over 70 planetary events are said to be like the astronomy records of celestial inscriptions.

Assuming that the orbits of planets have not changed significantly in the past 5,000 years, the modern astronomical calculations can take us, within acceptable accuracy, to 5,000 years ago to generate the astronomical map of the sky, predicting the planetary positions. An astronomer can shed more light on the subject as to what the sources of error are in determining celestial inscriptions 5,000 years in the past; however, a simple conclusion is that all variable parameters that constitute the orbit of a planet remain sources of error. We know that the time period of a planet for orbiting around the Sun is unchanged with time, but how fixed is it over 5,000 years is not known with 100% accuracy. Also, the period of a planet may be assumed to be invariable but its linear speed along its entire orbital path is variable, and affected by the gravitation of other planetary bodies in their respective orbits. The highest variation in planet's linear speed occurs when it is in the vicinity of the Sun. A planet's speed is accelerated from a given point on the orbit as it approaches the Sun and is decelerated from another point on the orbit on the far side of the Sun. In Vedic astronomy (in Surya Siddhant, Chapter 2) these positions/motions are defined by *sheeghr-uchch* (point from where motion increases), *mand-uchch* (point from where motion decreases). All planets are affected by this gravitational effect and it relates to their time periods and their orbits. Nevertheless, when all possible sources of error are accounted for the computer simulation would create a fairly accurate sky inscription. This is what was done using the latest computer algorithms by the astronomers who presented their results at this conference.

At the conference Dr. B. N. Narhari Achar, from the Department of Physics, University of Memphis (USA) gave a brief description of the various available planetary softwares and also reviewed the related work of other astronomers and astrophysicists. He critically examined the limitations and reliability of the simulations and the validations of Rishi Vedvyas' astronomical recordings in Mahabharata. He concluded that there was overwhelming astronomical evidence that the Mahabharata War took place about 3,000 BCE and the simulation of planetary events pointed to 3067 BCE, the latter figure being identical to the one derived by the astronomer Dr. Srinivasa Raghavan. The other number in this range was from P. V. Holey who based his comparative study of crucial events of planetary positions of astronomical dates in the reference frame of *nakshatra*, Julian and Gregorian systems. His conclusion was that the Mahabharata War began on the 13th day of November in 3143 BCE. The author of this book, based on the historical records found in Shankaracharya Mathas and genealogical records of the kings at Hastinapur recorded in Satyarth Prakash, which are discussed later in this chapter, placed the Mahabharata War on the Timeline-A in Figure 3.1 at the time-milestone Marker-4 at 3140 BCE. The differences between this author's time-milestone and the above numbers from the astronomers respectively are: - 140, -73, +3 years. These differences over a span of 5,000 years respectively are: 2.8%, 1.5% and negligible.

Regarding the existence of the River Sarasvati the book of Mahabharata itself states that the River Sarasvati was in the process of disappearing into the desert. At this conference there was a demonstration of modern technological tools such as Planetarium Software, *Panchang* Software compiled by a Japanese professor to produce the dates in Kali Yuga Era and relate them to Christian/Common Era. The overview of the conference is presented by Dr. S. Kalyanaraman.

Based on his astronomical analysis Dr. B. N. Narahari Achar presented the list of the following 11 dates that are consistent with the sky inscriptions observed by Rishi Vedvyas and laid down in the form of *shlokas* in the book of Mahabharata that Rishi Vedvyas and his students authored more than 3000 BCE. Based on what was mentioned earlier and the discussion that will follow the figure of 3000 BCE is close to the estimate 3067 BCE based on historical data of Table 3.3.

Dr. Narhari Achar's list includes the following:

1. Krishn's departure in Revati-nakshatra (Revati constellation) on September 26, 3067 BCE. (Revati the 26th Vedic constellation is ζ Piscium in Greek).
2. Krishn's arrival in Hastinapur in Bharani-nakshatra on September 28, 3067 BCE. (Bharani the 28th Vedic constellation is μ Arietis in Greek).
3. Solar eclipse on Jyeshtha amavasya (new-moon day of the Vedic 3rd month) on October 14, 3067 BCE.
4. Krittika full moon, lunar eclipse on September 29, 3067 BCE (Krittika is 7th Vedic month).
5. Mahabharata War starts on November 22, 3067 BCE with the planet Shani (Saturn) in Rohini-nakshatra and Brihaspati (Jupiter) in Revati-nakshatra. (Rohini the 2nd Vedic constellation is α Tauri Aldeboran in Greek and Revati the 26th Vedic constellation is ζ Piscium in Greek).
6. Winter solstice on January 13, 3066 BCE.
7. Bhishm's expiry on January 17, 3066 BCE.
8. A fierce comet in Pushya, October 3067 BCE. (Pushya the 6th Vedic constellation corresponds to the 10th Vedic month Pushya. Pushya constellation is δ Cancri in Greek).
9. Krishn's brother Balaram starts his pilgrimage along the shore of river Sarasvati on. Pushya-day on November 1, 3067 BCE.
10. Balaram returns from pilgrimage on Sravana-day on Dec. 12, 3067 BCE. (Sravana the 21st Vedic constellation corresponds to the 5th Vedic month Sravana.
    The Sravana constellation is α Aquilate Altair in Greek).
11. On the day Ghatotkachch was killed the Moon rose at 2 AM, December 8, 3067 BCE.

All these and the rest of the astronomical references in Mahabharata can specifically be pointed out in terms of Mahabharata's *shloka* number, chapter number and *parv* name. To relate just one, the most important of all from the above list is the 7th when Bhishm died. It is based on the two *shlokas*, 23rd and

25th of the 32nd chapter in Anushasan-parv. The two important observations related to this event are: the Sun has retreated north-ward and the corresponding date in luni-solar Vedic calendar given in these two *shlokas* transcribes to *shukla ashtami* (8th day of the bright-phase of the Moon) in the month of *Magha* (11th Vedic month) in *Rohini* constellation.

These of course are no ordinary observations of a religious nature that can be ignored. They are hard-core astronomical records that could be the oldest available and most reliable in ancient practical astronomy. There are some other rare astronomical events mentioned in Rishi Vedvyas's Mahabharata: The lunar-solar-lunar eclipse sequence occurring within a period of one month and one lunar-solar eclipse sequence occurring within just 13 *tithis* (lunar days).

Several different software packages were applied to simulate the sky 3000 years BCE and the results were presented and discussed at this conference. With due consideration to the variation in the results it was concluded that the Kurukshetra War (i.e. Mahabharata War) occurred around 3000 BCE and is a historical fact. The above conclusion on the historicity of the Mahabharata War that did take place about 5,000 years ago is based on the scientific methods presented at the above mentioned conference and the recorded ancient Vedic astronomy included. This leads to the proof that the references from the book of Mahabharata are at least as old as the book itself. The references for Vedic Cosmology drawn in this book from the four Vedas, Brahmana-books, Smriti and the six Sutras (Darshans) are in the antecedent of Mahabharata. Therefore, it can be said with confidence that this author's Vedic Reference Hypothesis, laid down right at the onset of this chapter that- *"The basic Vedic references on cosmology are ancient and precede 5,000 years"* is proven within the acceptable error. The time-milestone at Marker-4 on Timeline-A of Figure 3.1 which this author placed at the year 3140 BCE is also very close to the reality. The difference between the time of Marker-4 based on ancient Indian calendars and the conclusion based on astronomical calculations drawn at the said conference is only 140 years over a span of about 5,000 years. The error if any is only 2.8 %, which is quite acceptable in historical deductions and is even acceptable in most scientific methodologies. Therefore the Reference Hypothesis is fully validated beyond doubt.

This author feels that it would be of interest to a reader to read some anecdotal information about the legendary River Sarasvati. Also it is of interest to analyze a few more pronouncements of King Manu in his book Manusmriti.

Folklore and Legends about the River Sarasvati

The oldest reference to the term *sarasvati* is in the oldest book of Vedic Literature, the RigVeda (1/3/10-12). The term also appears in YajurVeda and AtharvaVeda. All terms in the Vedas are *yaugik*, signifying only the meaning of the root, indicating aspects of knowledge, not a physical objet. The etymology of the *yaugik* term *sarasvati*, which stands for 'intellect', is in Chapter 4. In rest of the Vedic Literature the term *sarasvati* appears generally in *rudhi* sense, the colloquial usage for concrete (physical) objects. In addition to the *rudhi* application in Manusmriti for the River Sarasvati, the Shatpath Brahmana (1/4/1/14), the commentary book on YajurVeda, also uses the term *sarasvati* for the same river. Using *yaugik* terms of the Vedas to name persons and objects of *rudhi* nature is a common practice in the Vedic religion. The mighty river with five tributaries being named Sarasvati was the most appropriate name for a river, the waves of which inspired *rishis* to create *chhandas* (vibrations) of knowledge that led to the composition of poetic *mantras* and the *sutras* of the Vedic Literature.

This Shatpath reference relates to the conversation that took place in the far antiquity of Aryavart '*at the shore of River Sarasvati*', between the King Videgh Mathav and his *purohit* (priest) Raahugan Gotam. It relates to *vaishvanar-agni* that takes on the form of *daavaanal-agni*. *Vaishvanar* and *daavaanal* are epithets of energy (*agni*), the former being the chemical energy and the latter being the fire that burns substances. An immediate example of the *vaishvanar-agni* is the chemical energy used in the digestion system in a living-being and the example of the *daavaanal-agni* is an ordinary wood-fire. So, a forest-fire is generally referred to as *daavaanal-agni*. There are a few sources of starting *daavaanal* fire in a forest, such as the friction in dry wood or lightning; however in this Shatpath Brahmana narration the chemical *vaishvanar* energy started the forest fire. This and subsequent Shatpath narrations give a glimpse into the ancient history of Aryavart. Word for word, English translation by this author, preserving the Shatpath style, goes thus:

> "The *king Videgh Mathav at that time was at the shore of the River Sarasvati. At that time the fire continued burning and moving towards the east. Gotam Raahugan and Videgh Mathav followed that burning fire. The fire dried all rivers in its way. One river named Sadaneera originates from the North-Mountain. The fire could not dry up this river. The brahmanas* (learned people) *had never crossed it before"*. (Shatpath, 1/4/1/14).
>
> "*But now many brahamans live east of this river. Earlier the land east of Sadaneera was uncultivable and was lying vacant. It was full of daldal (quick-sand) because the fire had not tasted it*". (Shatpath, 1/4/1/15).
>
> "*Now of course it is very fertile because the brahmans have performed yagjya on it and have given its taste to the agni (fire). It (*River Sadaneera) *has strong flow even during the onset of summer season. Vaishvanar-fire had not burnt it* (because the forest fire did not cross the River Sadaneera). *Therefore this region is very cold.* (Shatpath, 1/4/1/16).
>
> *"Videgh Mathav addressed the agni – where am I? The agni* (the energy of thought process) *replied – "Let there be your home on the east side of the river". Up till this day this river forms the boundary between the geographic regions of Kosal and Videh. People (in Videh region) are addressed as Videh because they are the descendents of Vedegh Mathav.* (Shatpath, 1/4/1/17).

There are several evident and mysterious aspects of the above Brahmana narrations. In their conversation the King Videgh Mathav and his *purohit* (priest) Raahugan Gotam differentiate between the two classes of energy and also, one energy initiating the other. They talk about severe forest fire that engulfs a large region and dries rivers in its swath. It is mysterious because severe forest fires do break out in modern times but none are able to dry up a river. The obvious observation in these narrations is the westerly wind system moving the forest fire from the River Sarasvati towards the River Sadaneera forming the boundary between the well-defined regions that exist on the Indian sub-continent even today. The literal meaning of the River Sadaneera is – 'the *river always with water*', as emphasized in the Shatpath narration as well.

It is discussed in Chapter 4 that the city of Ayodhya was established as the capital of Kosal kingdom by King Manu. The geographic regions of Kosal and Videh already had their names and therefore were part of Aryavart's ancient history. In the Shatpath narration the geographic region of Kosal begins from the west shore of the River Sadaneera. Ayodhya the capital of Kosal established by Manu is in fact situated at the west-shore of the river named Sarayu. Since Ayodhya has not changed its location it is possible that Sarayu is the changed name of the River Sadaneera. Since there is no mention of Manu or Ayoddhya in Shatpath narration the Videh kingdom may have been established by the King Videgh Mathav and his *purohit* Raahugan Gotam. This Gotam could be Rishi Gautam the author of Nyaya Darshan, the oldest of the six Vedic Sutras. Generations later in Manu's genealogy, the prince Ram of Kosal region had married Sita the princess of Videh Kingdom which was ruled at that time by her father who was known as Videh Janak. It can be deduced therefore that Videh Janak known in Vedic Literature as the king with high intellectual pursuits and having deep Vedic knowledge, was a descendent of King Videgh Mathav. The reference to philosophic dialogue (*samvaad*) between King Janak and the *rishis* is well-known in Vedic Literature.

The Shatpath narration between King Videgh Mathav and the *purohit* Raahugan Gotam makes it obvious that the River Sarasvati was a reality in the ancient past of Aryavart and that the geographic regions of Kosal and Videh were subsequently situated to the west of Sarasvati in the same order as narrated in the ancient book of Shatpath Brahmana and exist in the same order today. The difference however is that the major river flowing north-west of Ayodhya, between the regions of Kosal and Videh is River Sarayu (meaning- *dynamic like the wind*) and not Sadaneera (meaning - *river always with water*) named in Shatpath.

The Triveni Sangam

The River Sarasvati continues to be a legend for present-day Hindus of all strata. The most common legend of all is that the Invisible Sarasvati still flows and meets the famous rivers Ganga and Yamuna at their confluence near the city of Prayag. Although there is no sign of a third river at Prayag, the confluence of Ganga and Yamuna at Prayag has been known since time immemorial as the Triveni Sangam (confluence of three rivers). It is one of the important pilgrim places for the Hindus.

Let us now explore the possibilities of why the Ganga-Yamuna confluence is named Triveni Sangam (confluence of three rivers). In contexts of the discussion on River Sarasvati and the rediscovery of Sarasvati and Palaeo-Yamuna by the scientists, let us analyze first the term *yamuna*.

This author finds the meaning of the name Yamuna fascinating. The etymology of the term *yamuna* yields – '*one that moves in association with others and one that also moves separately'*. A major river generally stays its course year after year till a severe plate-tectonic event changes its course. Therefore, it is puzzling as to why the Aryans would give Yamuna a name that represents contrariety within itself? The question has no simple answer. However, in view of the new geological investigations and studies of land satellite imagery during 1980's and 1990's some light could be shed on this puzzle. As stated above, the multifaceted science investigations have achieved the mapping of the drainage system in North-West India, determining the past course of the rivers that still continue to flow, those that changed their course and those that are extinct now.

The analyses of ground and satellite data by eminent scientists show that at one time in the distant past, the River Yamuna, originating from its source named Yamunotri at 3,307 meters above sea level in the Himalaya, flowed south-west to drain into the now extinct River Sarasvati (Figures 3.4, 3.5). Therefore the Yamuna was Sarasvati's tributary and the Yamuna water flowed in Sarasvati. The ancient course of Yamuna between Yamunotri and the rediscovered Sarasvati was shown as Palaeo-Yamuna channel in the Figures 3.4 as well as in 3.5. Then a severe cataclysmic plate-tectonic event took place in the distant past changing the topography in the Shiwalik Hills, separating Yamuna from Sarasvati and turning Yamuna to assume its present south-east course of 1,384 km beginning at Yamunotri and ending at Triveni Sangam in Prayag where it offers all of its water to the River Ganga. Owing to the said cataclysm, the Aryans of the distant past who lost Yamuna from their region, suffered from the ensuing upheaval, were uprooted and must have followed Yamuna's new course looking for new settlements. Based on that observation and perhaps many centuries later the learned *rishis*, under the advanced Aryan system of knowledge-transfer, must have named it Yamuna. There is always a reason for an entity receiving a specific and meaningful name in Sanskrit. The contrariety in the meaning of the name Yamuna, '*one that moves in association with others and one that also moves separately'* is now obvious. Yamuna moves in association (with Sarasvati) and Yamuna also moves separately (from Sarasvati).

Similarly it is explicable as to why Prayag was named the Triveni Sangam, the *place of tri-river confluence*, the three rivers being Ganga, Yamuna and Sarasvati. Owing to geological reasons mentioned above when Yamuna broke off from Sarasvati, the water that would have become a part of Sarasvati was brought to Prayag by Yamuna. Later the entire River Sarasvati went extinct, however, the Prayag's adjective Triveni Sangam continued. The *rishis* must have travelled and studied the ecology of the region and surveyed the land before addressing Prayag as the Triveni Sangam. The invisible Sarasvati's water, now part of the visible Yamuna, flowed into River Ganga at Prayag, hence the name: the Triveni Sangam, the tri-river confluence.

Naming of a big river with a single name and characteristics reflected in that name cannot be given by the common folks living at the shores of that river. For example the river name Yamuna could only come from a learned person because the term '*yamuna*' conveys the characteristics of 'association' and 'dissociation'.

The river Yamuna in Yamuna because it has the association with the River Ganga and the dissociation with the River Sarasvati. Being astute thinkers the *rishis* encompassed geological history also in the nomenclature. Such finalization in the name of an entity of nature does not take place in a generation, in a century or even in many centuries. This is why the nomenclature of entities in Aryavart is remarkable and indicates the ancientness of the Vedic culture and its literature.

In the second *shloka* of 22nd *sarg* (section) of Baal Kaandam (first chapter) of Valmiki Ramayan, the River Ganga is referred to as '*tripathgaam*', meaning the '*one that has gone on three paths*'. The explanation behind this adjective for Ganga in Ramayan conveys that Ganga at *triveni sangam* has come together after following the three paths, the paths of Sarasvati, Yamuna and Ganga. This is no ordinary assertion.

The decisions made by the *rishis*, based on their long-term observations and the study in ecology and geology were respectfully accepted by the Aryans and firmly ingrained in Hindu psyche. Consequently, along the entire course of the major rivers, over thousands of years, the Ganga is still called Ganga, and Yamuna is called Yamuna even after numerous kings ruled from their shores. The kings also respected the decisions of the *rishis* and did not name and rename the great rivers after themselves as the rivers flowed through their respective kingdoms.

We will not take up here the three other nation-defining s*hlokas* of Manu (1/136, 1/140, 1/141) and the Vinshan Region metioned in Manusmriti, instead present a few extraordinary discoveries from Manusmriti.

In the Ramayana-Baal-Kaandam (3/1-3) Rishi Valmiki states that -*'At the banks of Saryu River the geographic region known as Kosal was full of prosperity and wealth with contented citizens. There the great king of people, Manu, had established the city of Ayodhya as his capital'.* Obviously there was no Ayodhya before Manu. As stated in Valmiki Ramayana (29/18), the first king after Manu, who ruled from Ayodhya was Manu's son Ikshvaku. It is likely that Manu came from another region of Aryavart to Kosal region and re-established his kingdom Ayodhya on the bank of the Saryu River; which could be the ancient Sadaneera.

The reference of one book in another book plays a significant role in determining relative historicity of the books. Rishi Valmiki in several shloka couplets of Ramayana encapsulated an interesting discussion between Ram and a ruler of a forest region in south India, which reflects this point. Ram, journeying through the forests of south India during his fourteen-year exile from Ayodhya, reached the Pampa Sarovar (Pampa Lake) where he met Bali the ruler of Kishkindha (360 km north-west from the city of Bangalore). In a dialogue between the two, Ram the follower of *dharma* (laws of moral and justice) unhesitatingly pointed out to Bali his unjust and immoral behaviour towards his younger brother Sugreeva by reciting two *shlokas* from Manusmriti on the subject of 'crime and punishment' and a king's duty in rendering the appropriate punishment. Before reciting the two specific *shlokas*, Ram addresses Bali, as in the Valmiki Ramayan *shloka* (Kishkindha Chapter, 13/20) - "(O Bali) *listen to what Manu has instructed in the two often cited shlokas wherein he established the rule of proper conduct. Most scholarly of righteous character accept it and I too follow it*". Then Ram recited to Bali the two *shlokas* of Manusmriti (8/318 and 8/316), in that order, 318th *shloka* before the 316th. These two *shlokas* of Manusmriti correspond to the two *shlokas* of Ramayana (Kishkindha, 13/21 and 13/22), with minor change in terminology. In the second line of Manu's *shloka* (8/316), Valmiki used his own poetic expression with a change of a few words, yet conveying the same meaning. This establishes the chronology between Manusmriti and Ramayana.

Ramayana also makes it obvious that even during Ram's time the city of Prayag, an important east point of the Central Region of the Aryavart, has been the confluence of the rivers Ganga and Yamuna. This places Ramayana after Manusmriti and conversely there is no reference of Ramayana or Mahabharata in Manusmriti.

The roar of the magnificent River Sarasvati is absent now. Any cataclysmic event that silenced its roar leading to its slow disappearance had to have occurred after the nation acquired its name Aryavart. In Manu's Aryavart are Prayag, Sarasvati, Drishdvati, Vindyachal, the region with the sea on the east and west. If the Aryans have come from outside India they must associate themselves with landmarks such as these. The fact is that no such place exists on the Earth elsewhere but in Aryavart (India). Therefore Manusmriti's Brahmavart is within Aryavart, the home of the Aryans the Bharatvarsh of today, which unfortunately is now addressed with mediocre and meaningless names, like Hindustan and India. On one hand we have Manusmriti's references, scientific research on River Sarasvati and the above logical discussion on the historicity of Aryavart and on the other hand we have the far-fetched Aryan Invasion Theory proposed by Western Indologists. Therefore, the AIT must be shunned by honest intellectuals because it is not based on an honest intellectual examination of history and instead was a byproduct of a judicious mixture of ulterior motives and ignorance. Other concocted hypotheses related to the Vedas, Vedic Literature, Vedic culture and history should also meet the same fate.

There are several other *shlokas* in the same chapter of Manusmriti that go beyond the geographic definition of Aryavart. These *shlokas* indicate the kind of people who lived in Aryavart and differentiate between the people living in Aryavart and those living in geographic locales outside Aryavart. The *shlokas* also give an insight into the knowledge pool of the people of Aryavart relative to not just its surrounding areas but the rest of the world.

A Remarkable Discovery from Manusmriti

Two *shlokas* from Manusmriti presented in the following also show that Aryavart has been the place of the Aryans from the very beginning, where they have indigenously developed their social structure and civilization. They developed knowledge not just in spirituality and philosophy but had wide knowledge of the sociology of the region well beyond the boundaries of Aryavart. Manusmriti, Mahabharata and other Vedic writings, including the Puranas variously tell us that the ancient Aryans of India had knowledge of the world and about the different people who inhabited it. Just two *shlokas* from Manusmriti are presented here to demonstrate the knowledge of the Aryans about the status of global ecology and education. All this awareness and the process of knowledge-gathering beyond the boundaries of Aryavart at such an early stage in the human history should send thinkers back to the blackboard to redefine human history.

Manusmriti's *shloka* (1/142) states:

कृष्णसारस्तु चरति मृगो यत्र स्वभावत: ।
स ज्ञेयो यज्ञियो देशो म्लेच्छदेशस्त्वत: पर: ॥ (Manusmriti, 1/142)

Transliteration: "*Krishnsaarastu charati mrigo yatr svabhaavatah* |
Sa gjeyo yagjiyo desho mlechchhadeshastvtah parah ||)

Shloka Translation:

"*And in the country where the most brilliant vigorous blackbuck grazes freely and fearlessly, inhabited by the yagjyiya people, know it to be that country (Aryavart) and beyond which are the countries inhabited by mlechchha"*. (Manusmriti, 1/142).

It may not be obvious immediately that this *shloka* deals with two different subject areas which are well-advanced in modern civilization: the science of world ecology and the world sociology. What has just been said may seem far removed from the two disciplines just quoted, however, the following discussion will show that the statements in the above *shloka* are very close to this author's assertion. This requires a brief interpretation of the *shloka* and this author hopes that the reader will value the validity and depth of the *shloka*. Let us discuss the second line of the *shloka* first.

Mlechchha-Countries of Manusmriti

There are two provisos in the second line of the *shloka* – that the Aryavart is inhabited by the *yagjyiya* people and the countries outside it are inhabited by *mlechchha* people. Therefore, it is self-evident that in this line of the *shloka* there is a comparison of the people of Aryavart with people living outside Aryavart.

The word *yagjya* (syllables g and j pronounced together) is a noun which is popularly written as *yajna* and wrongly translated by European scholars as 'sacrifice', the connotation of which has no place in Vedic philosophy and therefore is fallacious. The basic definition of the word *yajna* (*yagjya*) is: - '*any action requiring association of men or objects, and productive of beneficial results'*. Such an action is for the benefit of all living beings whereas the connotation in the term 'sacrifice' generally conveys the sacrifice of some being(s) for the benefit of others. Therefore a *yajna* (*yagjya*) is not equivalent to 'sacrifice'. There is no single English word that can justify the meaning of *yagjya*. Differently put, a *yagjya* is a sacred 'observance' of an act of non-violence. Thus by definition a *'yagjya'* is a noble deed and therefore is a *dharmic* action. Some examples of *yagjya* in a modern society are: unselfish acts of philanthropy, charity work, environment protection, saving the species and so on. The process associated with *yagjya* is a verb and addressed as *yagjya* as well. The participant or a devoted performer of *yagjya* is addressed as *yagjyiya*. This takes us back to the city named Prayag that Manu mentioned in *shloka* (1/140) and the one vividly described in Ramayana (Ayodhya, 40/1-22). Manu in his above shloka (1/142) is referring to the learned people with the adjective of yagjyiya who have the welfare of others in their mind and are devoted to the almighty Brahmn.

A colloquial connotation of the word *mlechchha* is 'impure and low', however, according to the translations and interpretations by Sanskrit scholars, including Professor Surendra, the word *mlechchha* means – '*one who is not educated and speaks in undeveloped language'*. Manu's implication in his *shloka* is that the high development in education and language uplifts a society leading to the development of philosophy and spirituality. The process of *yagjya* involves the participation of scholarly people having a deeper knowledge of the Veda mantras that are recited on the occasion of the act of *yagjya* for the collective benefit and upliftment of the society. In Manu's thinking the absence of *yagjya* in the society leads to selfishness and materialism. The context of his *shloka* indicates that Manu made a sociological comparison of the high development of the Sanskrit language in Aryavart and the relatively poorly developed language skills outside Aryavart.

The Blackbuck of Aryavart

In the above *shloka*, the expression – "*the most brilliant vigorous blackbuck grazes freely..."* in Aryavart, gives the impression that Manu in his nation-defining *shlokas* wanted to bring in some added poetic beauty to Aryavart by mentioning its pastures and the animals that grazed there. When one looks carefully at this *shloka*, one finds more in it than just that. This author observes that Manu in this *shloka* has only mentioned a specific species of blackbuck. So, there is an indication of the association of a specific species of deer and the people of Aryavart which is a relationship between the society and the ecology of Aryavart. Since Manu in this *shloka* draws a comparison between Aryavart and other nations exterior to it, Manu's *shloka* is a statement on global ecology. The mention of this species of deer in the *shloka* is the reason for the earlier heading "A Remarkable Discovery from Manusmriti". This deduction is supported by recent ecological observations discussed below.

Let us first look at the meanings of some terms in the first line of the *shloka*:

Krishnsaarastu = *Krishn* + *saarah* + *tu*
Krishn = root meaning is 'who attracts all', common meaning is 'black'
saarah = has several meanings: most brilliant, genuine, real, extract, strength, powerful, energetic; *tu* = and;
charati = moves, walks, roams, grazes;
mrigo= deer, buck; *yatr* = where, at the place; *svabhaavatah* = freely, naturally, without fear.

The emphasis in Manu's *shloka* is quite obvious that it is Aryavart where people with adjective of *yagjyiya* live with welfare in their thought and action enabling the coexistence of all living beings including one of nature's beauties - the *most brilliant vigorous blackbuck* that *grazes freely and fearlessly* in Aryavart's pastures. Implicit in the *shloka* is that in the other countries there was no blackbuck or if it did, it was extinct. That the blackbuck is not in other countries or if it is there it does not roam free, is a generalization which Manu can claim only when he has the ecological information about the region exterior to Aryavart. This is what can be easily deducted from this *shloka*.

In this *shloka tu* and *yatr* together mean '*and in the place where*' with the reference of *Aryavart* in the previous *shlokas*. The *krishn-sarah* or *krishn-mrig* of Manu's *shloka* is popularly known in India as the *krishn-mrig, kala hiran* or *krishn-jinka*. The English name for it is blackbuck (antilope cervicapra). Blackbuck still exists in India and is a protected animal under the Wildlife Protection Act of 1972. Killing it is a crime punishable by law. This species was near extinction but is now protected in its natural habitat, in sanctuaries and national parks of India. Hunting or killing a blackbuck is a un-*yagjyiya* act punishable with a jail term.

The pictures of the blackbuck in this text here are from the Mahavir Harina Vanasthali Park near Hyderabad City in the state of Telangana where the blackbuck is protected and declared to be the State Animal. The reference of the blackbuck in Manusmriti and its association with Aryavart is no ordinary statement by Manu. Here is why.

This author while learning about the blackbuck gathered that in recent times the scholars in geography and ecology have published a large number of papers on the blackbuck and its habitat. One striking fact emerged from these publications is that the blackbuck is **native to India only**. No scholar of ecology or geography today can state this fact without having global ecological knowledge. By the same token Manu could not have generalized that it is only in Aryavart where the blackbuck roamed freely in its natural habitat, unless he had global ecological knowledge on the species of animals. It is easy to gather global ecological information in modern times with the entire world connected through global travel and information technology. It is mind-boggling how Manu would have developed knowledge on global ecology, particularly on the blackbuck's habitat, that it lived only in Aryavart where *yagjyiya* people lived and where the blackbuck roamed freely.

Just as an ecologist of today would have to have knowledge of global ecology before he could deduce that the natural habitat of the blackbuck is only in India, so also, Manu had to have knowledge of global ecology before stating that same fact in his Manusmriti, millennia ago. The above *shloka* alone shows that he is not making a fleeting reference to some animal, any animal, but he specifically chose the blackbuck because he show-cased a unique ecological difference between Aryavart and elsewhere and also the sociological difference between the people of Aryavart and elsewhere. If there were blackbuck elsewhere and if there were people outside India with Sanskrit language and with *yagjyiya* discipline, he would be wrong to state what he has stated in the *shloka*. Manu within a single *shloka* differentiates Aryavart from other countries by connecting a specific species of animal to the Aryan people, exhibiting a remarkable universality in his knowledge. He could have mentioned other animals like an elephant, a horse or a cow but instead he chose an animal unique to Aryavart. The cow, for the people with *yagjyiya* discipline practising non-violence, is a holy animal and is considered to be the most beneficial animal to humans. But Manu in his *shloka* chose to cite the blackbuck with its exclusive existence in Aryavart.

Originally spread out over large tracts of open Gangetic plains (planes of River Ganga) of India, the blackbuck continues to be mainly in the States of Punjab, Haryana, Rajasthan and Gujarat which together are in the plains of the ancient River Sarasvati. The blackbuck is also found in other states of India and in a limited number in Nepal. In modern times, it was introduced for the first time into Texas (USA) from India and also into Argentina in the early 1900 CE. A thorough study of unpolluted Manusmriti would yield that a rich knowledge-base in varied disciplines has been the foundation for its creation.

The discovery by modern ecologists that the blackbuck has been native to India is a confirmation of what Manu laid down in his Smriti eons ago before Mahabharata. Manu could not have had this knowledge without him and or other scholarly observers of ecology and sociology traversing the globe! The startling fact on this comes from the five thousand year old *shloka* in the book of Mahabharata in its Shanti-Parv section, which indicates that Manu himself observed the ecology of the world and the people outside Aryavart before laying down the nation-defining *shlokas* in his Manusmriti.

In the book of Mahabharata the Grandfather Bhishm from his deathbed preached about the duties of a good king to the victorious Yudhishthir. He gave the example of the ideal ancient King Manu. Among several *shlokas* on this subject comes a *shloka* stating that:

ततो महीं परिययौ पर्जन्य इव वृष्टिमान्।

शमयन् सर्वतः पापान् स्वकर्मसु च योजयन्॥

(Mahabharata, Shanti-Parv, 18/14).

Transliteration: "*Tato maheem pariyayau parjanya iv Vrishtimaan.*
Shamyan sarvatah paapaan svakarmsu cha yojayan."

Translation:

"*Then, accepting the request of his subjects, the Emperor Manu, like a cloud with cooling effect, roamed the entire Earth preaching discipline, duty and good karma to people for planning a just society*". (Mahabharata, Shanti-Parv, 18/14).

The *shlokas* couplets in Manusmriti are no ordinary facets of literature, poetry or religious preaching. They actually lead one to uncover ancient facts. In addition to the significant observation of the blackbuck that this author has highlighted, Manusmriti also contains many mind-boggling facts related to science and society. An example of time-unit *nimesh* (1 *nimesh* = 8/45 second = 177.8 milliseconds) was presented in Chapter 1 and discussed in Chapter 5.

Manu's statement that "*the most brilliant vigorous blackbuck grazes freely...*" in Aryavart also sheds some light on the sociological makeup of the Aryans of Manu's time. The *yajna* being an act of nonviolence, the killing of animals at the time of *yajna* or sacrificing of animals in a *yajna* fire is inconceivable because doing so would be considered a bad karma and sinful. Not just for karma but for the purity and health of the physical body, Manu in the fifth chapter of Manusmriti has devoted a large number of *shlokas* on the subject of *bhaksh-aabhaksh* which means - '*edible – non-edible and what-not-to-eat*'. There he has made it abundantly clear that violence towards animals is an act of *a-dharma* (immoral and irreligious). Without discussing it in detail one *shloka* of Manusmriti is presented in the following that confers logic for practicing non-violence towards animals.

नाकृत्वा प्राणिनां हिंसा मांसमुत्पद्यते क्वचित् ।

न च प्राणिवधः स्वर्ग्यस्तस्मान्मांसं विवर्जयेत् ॥ (Manusmriti, 5/48)

Transliteration: "*Naakritva praaninaam hinsa maansmutpdyate kvachit* |
Na cha praanivadhah svargyastasmaanmaansam vivarjyet ||"

Translation:

"*Meat cannot be produced without killing animals and the act of killing being an unrighteous act it does not lead to salvation. Therefore meat-eating is completely forbidden*" (Manusmriti, 5/48).

Manu emphasized that meat eating was immoral because he simply followed the Vedas where the killing of animals and even being unkind to living beings is forbidden. To cite just one example here, a YajurVeda mantra (36/18) states –

दृते दृँ ह मा मित्रस्य मा चक्षुषा सर्वाणि भूतानि समीक्षन्ताम्

मित्रस्याऽहं चक्षुषा सर्वाणि भूतानि समीक्षे । मित्रस्य चक्षुषा समीक्षामहे ॥

(YajurVeda, 36/18)

Transliteration: "*Drite drim ha maa mitrasyaa maa chakshushaa sarvaani bhootani sameekshntaam, mitrasyaa-aham chakshusha sarvaani bhootaani sameekshe| Mitrasya chakshushaa samikshaamahe||*"

Mantra Translation

"O Illuminator of our minds let all living beings view me amicably and let my resolve be strengthened to view all living beings as friends and to love all living beings as I love myself." (YajurVeda, 36/18).

With that prayer offered to the Creator one cannot think of killing animals for consumption. This is why the *krishn-sarah mrig,* the brilliant blackbuck, grazed free in Aryavart.

Manusmriti has no mention of Aryans belonging to any other place than Aryavart (India) itself because any other place outside Aryavart belonged to *mlechchhas* with underdeveloped tradition and language. Manu throughout his Manusmriti has never advised to close Aryavart's doors to people from outside; rather he welcomes them with his advice for "all people of Earth to benefit from Aryavart". This is exactly what is stated in Manusmriti (1/139).

एतद्देशप्रसूतस्य सकाशादग्रजन्मन: ।

स्वं स्वं चरित्रं शिक्षेरन्पृथिव्यां सर्वमानवा ॥ (Manusmriti, 1/139)

Transliteration: "*Etaddeshprasootasya sakaashaadagrajanmanah | Svam svam charitram shiksheran-prithivyaam sarva maanavaa ||*"

Shloka Translation:

"For developing their individual character, knowledge and education all humans of the Earth should receive knowledge from the intellectuals and learned people (the brahmanas) born in this nation (Aryavart)*,"* (Manusmriti, 1/139).

Manu's emphasis in the above *shloka* is on पृथिव्यां सर्वमानवा (*prithivyaam sarva-maanava*) which translates into '*all people of the Earth*'. Manu welcomed people of other nations to come to Aryavart to gain knowledge. Manu makes no reference to Aryans and non-Aryans within Aryavart. All inhabitants of Aryavart according to Manu are Aryans.

Manu's statement that '*all humans of the Earth*' could learn from the Aryans of Aryavart indicates that the Vedic knowledge is indigenous to India and that all ideas in the Vedas are original to India and therefore the mantras and *sutras* on cosmology and other branches of science and philosophy presented in this book, are indigenous to ancient India. Manusmriti being older than Ramayana and Ramayana being older than Mahabharata, they all are in the past of the time-milestone at the Marker-5 at February 18, 3102 BCE on the Time-line A of Figure 3.1. This discussion based on Manusmriti puts an end to both, the Aryan Hypothesis and the Aryan Invasion Theory.

Astronomy and Age of the Vedas

Just as the sky inscriptions mentioned in the *shlokas* of Mahabharata led to the period of Mahabharata War and the planetary conjunction yielded the onset of Kali-Yuga period at the Marker-5 on Timeline-A (Figure 3.1) on **February 18, 3102 BCE** at 2 hours 27 minutes and 30 seconds, discussed earlier, many scholars have attempted to determine the historical period of the Vedas based on the statements in the Veda mantras. This is briefly discussed here.

Using the astronomy information in a mantra leading to when the mantra may have been written requires astronomy calculations. The fact is that there is considerable astronomy in the Veda mantras and throughout the ancient Sanskrit Literature to lead to meaningful deductions. An example from Valmiki Ramayana is the citation of sky inscription at Ram's birth (Valmiki Ramayana, Bal-Kandam, 4, 5). Ram was born in the month of Chaitra on 9th *tithi* (lunar-day) when five planets (Mercury/Sun, Venus, Mars, Jupiter, and Saturn) occupied their highest elevations in the Punarvasu Nakshatra (7th of 27) when the Moon rose together with Jupiter in the Kark Lagn (Cancer sign). This is quite a detailed inscription for an astronomer to perform a back calculation starting from the present positions of the planets and the Moon.

A Vedic reference on this relates to the "*Nakshatra Sukta*" (Constellation Hymn, 19/7/1-5) of the AtharvaVeda (fourth Veda) which has devoted five mantras for the sequential nomenclature of 28 constellations in the zodiac; 27 of these are presently used in the same sequence in the Indian astronomy as well as in the modern astronomy with Greek names. The same *sukta* (hymn) also instructs to use the Moon in astronomy because, as the *sukta* states, the Moon consistently scales these constellations. This was briefly mentioned in Chapter 1 and more details are in Chapter 11 of this book.

To follow this procedure the minimum condition from the Vedic perspective is that the chosen mantra for estimating the age of a Veda be interpreted correctly. The last thing one should do is to trust the translation of a neo-Sanskritist who relies on *rudhi* rather than *yaugik* meaning of the Vedic terms. The Vedic year's first month always begins near the vernal equinox on the first day after the full moon day. Therefore, to apply this knowledge of astronomy one needs a clue in a Veda mantra that is related to a constellation and the beginning of the Vedic year. Since the occurrence of vernal equinox keeps shifting on the celestial equator (a 360 degree circle) towards the next constellation at a rate of 1 degree for every 72 years, it will take 960 years for it to be located in the adjoining constellation. At this rate the vernal equinox will occur again in the same constellation completing a full circle of 360 degrees covering all 27 constellations in 25,920 years. The reason for this change is the wobble of the Earth's axis, also known as the precession of the equinox, which in Indian astronomy is "*ayan chalan*".

This methodology based on astronomy to ascertain the age of a mantra is then extended to finding the difference between the constellation mentioned in a mantra and the present constellation at the beginning of the Vedic year. The assumption in this method is that the mantra was written in the constellation mentioned in the mantra. This assumption is not scientific because within a mantra of the Constellation Hymn of the AtharvaVeda one can count at least 5 constellation names, which should take 960x5 years just to write one mantra. To write all mantras of the Constellation Hymn to include all 28 constellations, one needs 25,920 years. This does not make any logical sense. However, some very reasonable arguments have been made suggesting the age of the Vedas.

The eminent German Indologist Hermann Georg Jacobi (1850–1937) used the astronomical information in some Veda mantras to determine the true local time and based on the available inscriptions he suggested that the RigVedic hymns were composed as early as 4500 BCE. When he published his views in the Journal of the Royal Asiatic Society in 1908, the proponents of "Out of India Theory" welcomed them and the proponents of AIT suggesting Aryan invasion about 1500 BCE, were furious.

Bal Gangadhar Tilak (1856-1920) was a scholar of Indian history, Sanskrit, Hinduism, mathematics, law and astronomy. In his book titled 'Orion', he interprets a RigVeda mantra from 86th *sukta* of its 10th *mandala* (chapter) as having the vernal equinox in the constellation of Mrigsheersha (meaning Head of Deer). In his time the occurrence of the vernal equinox was in the constellation of Uttar Bhadrapadaas, six constellations ahead of Mrigsheersha. This is equivalent to 5,760 years which is about 4,000 BCE, which he estimated to be the age of RigVeda. This alone negates the AIT and Max Muller's age estimate of the Veda. Tilak's conclusion sets an arbitrary upper limit of about 6,000 years on the age of the Veda and the Vedic Civilization. But nothing stops one from extending the argument that the RigVeda was written in one of the previous cycles of vernal equinox yielding the age of RigVeda as 25,920 + 6,000 = about 32,000 years or 32,000 + 26,000 = 58,000 years and so on. The fact is that there is no clear indication of vernal equinox or even the said constellation of Mrigsheersha in the 86th sukta. Therefore, such scientific calculations are of value only when they are supported with the right interpretation of a mantra and reliable data or they do more harm than contribute to the progress of knowledge.

The other estimates and arguments for the age of the Vedas are remarkable for their degree of ancientness. Many text books and web sites have summarized the subject of the age determination of the Vedas (e. g. Akhandjyoti). A summary of other estimates without detail is presented below.

Avinash Chandra Das in his famous book, "Rigvedic India" took a different route to assign an age to the RigVeda mantra (10/136/5). He used the correlation between the geological data of the Indian subcontinent and the geographic information in a mantra. Knowing the derived time of occurrence of specific geological change in the sub-continent, he concluded that the mantra has to be older than 27,000 years to 50,000 years.

This author looked at Avinash Chandra Das's suggested RigVeda mantra and found no indication that it related to any geological event. The *devata* (subject) of the 136th *sukta* is *Keshi* (the Sun) and the mention of two seas in the 5th mantra does not relate to a geological event. The mantra only addresses the Sun – "*The Sun that is desired by the senses of those with cognitive function and who befriend the air for respiratory function, illuminates both oceans- the ocean water on Earth and the ocean of space (Antariksh)*". On the basis of the geological description of the continent and the astronomy, other authors have given higher ages of the Vedas. Pawagi in his book entitled "Vedic Fathers of Geology" has estimated that the Vedas are about 240,000 years old. Pundit Dinanath Shastry in his book "Veda Kal Ka Nirnaya" (in Hindi) determined the age of the Vedas to be anterior to 300,000 years. Jwala Prasad's analytical research in "Rigvedic Geology and The Land of Saptsindhu" pushes the Vedic age to beyond 500,000 years BCE. The website quoted Jwala Prasad's comment on his own finding – "This is staggering but what is the escape", implying that facts are facts.

The 1911 Noble Laureate Professor Maeterlink (1862-1949), citing the evidential findings of German Archeologist Hellad, has argued in his book entitled "The Great Secret" that the Vedas are at least 7,000,000 years old. Swami Dayanand Saraswati, the founder of the "Arya Samaj", based on the Vedic Yuga-calendar calculations stated that the Vedas are as old as the Vedic calendar. Based on Swami Dayanand Saraswati's estimate this author calculated the age of the Vedas (age of the Srishti Calendar) in 2018 CE to be 1,972,949,119 years.

To conclude, even if the age-determination by different methodologies used by scholarly people is not scientific and decisive, what emerges is that the Vedas are far more ancient than the usual definition of ancientness and the Vedic references for this book are certainly beyond the milestone at February 18, 3102 BCE.

Historical Records and Validation of Timeline-A

To understand a country's ancient history one has to depend on the archives of that country which can be in various forms, such as the country's literature, history books if they exist, coinage, stone inscriptions, discovered writings, historical objects and their carbon dating and so on. Without going into great details of what might be available in this realm for establishing the validity of the Timeline-A, let us examine the written historical records. Extremely valuable historical data to validate Timeline-A has come from two sources, Swami Dayanand Sarasvati (1825-1883CE) and Acharya Udayaveer Shastri (1894-1991 CE). Before discussing the validation effort let us briefly look at the background of these two scholars par-excellence.

Swami Dayanand Sarasvati's family gave him the name Mool Shankar. A few comments on Swami Dayanand Sarasvati have been included in the earlier text. A brief summary on his background will be of interest to some readers. Mool Shankar left home and took *sanyaas* (monk-hood) at the age of 22. After travelling and exploring the Himalayas and the forests of central India, the young yogi Mool Shankar reached Mathura with the desire to study the Vedas and other sacred books of the Vedic dharma under the tutelage of Swami Virjanand, a renowned Vedic scholar and an intellectual giant. The blind, 80-year-old, erudite yogi Virjanand had complete mastery over Vedic grammar and philology. Young Mool Shankar was fluent in Sanskrit and had learnt Vedas by heart before leaving home. The blind yogi Virjanand accepted Mool Shankar as his student.

Under the direct tutelage of Virjanand, in ancient Vedic tradition, Mool Shankar developed deep knowledge of Ashtadhyayi grammar, Mahabhasya, Nighantu, Nirukt, Vedanta Sutra, Manusmriti and numerous other texts of Vedic Literature. After completing his studies, his guru Swami Virjanand bestowed upon him the *sanyaas*-name and blessed him to go out and preach the message of the Vedas to the society and re-establish the Vedic dharma. From there on Mool Shankar was known as Swami Dayanand Sarasvati.

Swami Dayanand Sarasvati was a great thinker and a philosopher of the recent times and a profound scholar of the Vedas, Darshans and the other books of Vedic Literature. A prolific writer in Sanskrit and Hindi this celibate monk translated the Vedas with commentary and interpretation. He was a superb orator. He traveled through North India imparting the Vedic message to the masses and participated in numerous public debates with the scholarly pundits. These formal debates called '*shaastraarth*' were the traditional forums in the Vedic society perpetuating from the ancient times for the benefit of the masses so that they may listen and understand the differences in the views of the scholarly. The *shaastraarths* kept alive the critical thinking in the society and therefore were a source of social reform. Swami Dayanand Sarasvati observed that the Hindu society had gone very far from the ancient Vedic practices and instead was following anti-Vedic rituals, blind beliefs, worshipping reincarnations of various gods in different forms and supporting social ills. He decided to tackle the ills of the society by preaching and writing rather than retreating to the Himalayas for meditation. He was the cause for the new thinking about the Vedas and the ancient books of the Aryans and he gave birth to much needed social reforms in the Hindu society of the time.

He preached with references of the Veda mantras and regenerated an environment of critical thinking. In the first half of his super eminent book in the Hindi language entitled "Satyarth Prakash" he presented a fresh view of what an Aryan society used to be like and ought to be. In the second half of the book, he critically examined the prevalent practices of Hindu religion and the other non-Hindu sects and faiths as well. A meaning of the title of the book of Satyarth Prakash is -- 'Illumination of the True Meaning'. It was first published in 1881 CE and has 14 chapters in it. Reading the book of Satyarth Prakash, one perceives the emphasis on 'back to the Vedas' and one also gets a glimpse of the vastness of the ancient Vedic Literature that he read and has referred to in Satyarth Prakash. The contribution of this brilliant scholarly ascetic to Vedic Literature and to the social reforms in the Hindu society is immense. His knowledge of ancient Sanskrit Literature was deep. Just by beginning to read Satyarth Prakash one quickly gathers that Swami Dayanand had read a large number of ancient and rare books including ancient history books in Sanskrit Literature. The reference-rich writing is a fascinating feature of all the books that he wrote in a short period of time. In Satyarth Prakash alone he cited about 342 different ancient books from the Vedic Literature with about 1,450 different citations of mantras, *shlokas* and *sutras* that he interpreted and connected to the subject that he was dealing with.

Acharya Udayaveer Shastri was an Acharya in the true sense. The title of Acharya is bestowed on a person who has exhibited command on subjects like the Vedas and Shastras. He was a prolific writer. He wrote interpretive commentaries on all six Vedic Darshans and many more Vedic books including the history of the Vedanta Darshan. This author has benefitted immensely from his books written in the Hindi language. His research in Sanskrit Literature, adopted in this book, led to the establishing of the time-line in the Aryan history right up to the Mahabharata War that concluded with the coronation of King Yudhishthir in 3140 BCE.

In addition to the scholastic translation and interpretation of the Veda mantras and Darshan *sutras*, these two authors left behind a treasure-trove of historical information in their writings that has enriched this author's book in establishing the historicity of Vedic references.

First let us look at a part of the Indian history five centuries BCE.

(1). 300-400 BCE: Megasthenes on Krishn, Validation of Time-milestone-4 on Timeline-A

The Indian history of the fourth century BCE is quite well known from Indian historical accounts. This is the period when Alexander the Great (356-323 BCE) invaded India. After Alexander crossed the Indus River and entered the Punjab region of ancient India, then called Paurava (now in Pakistan), he faced Raja Puru and his strong army in the epic battle of 326 BCE on the waters of River Jhelam (Hydaspes in Greek). King Puru and his army fought so fiercely against the invading Macedonians that Alexander, who was well-known to have left complete devastation behind him after each victory and who even killed those who surrendered - men, women and children, was highly impressed by the bravery of the Porus (citizens of Paurava). This fight was so fierce that even in victory, Alexander's army lost courage to advance further beyond the River Ganges. Alexander had no choice but to enter into an alliance with the King Puru and appointed Puru as the satrap of his own kingdom.

East of the Paurava Kingdom, across the River Ganges was the Magadh Kingdom (present region of Bihar and Bengal) which was ruled by the Nand Dynasty. Alexander's army advanced as far as the River Beas and learnt in advance that on the other side of the River Ganga, about four miles wide and 200 yards deep, awaited two hundred thousand armed foot-soldiers, eighty thousand horsemen, eight thousand chariots, and six thousand fighting elephants. Alexander arriving at the shore of River Ganga ordered his army to cross the river. But seeing what awaited them on the other side, his army mutinied and Alexander the Great retreated and returned to Macedonia.

In 322 BCE Chandragupta Maurya overthrew the Nand Dynasty of Magadh, establishing the Maurya Empire with its kingdom at Pataliputra (present name Patna in Bihar State) that expanded well beyond Magadh. During the reign of Chandragupta Maurya his kingdom expanded in the Gangetic plains, westwards across central and western India on the heels of retreating Alexander the Great. He defeated all satraps that Alexander had left behind to rule on his behalf. By 316 BCE the Maurya Empire's rule extended from Assam in the east to Gaandhaar region in the west (present Kandahar in Afghanistan). It is the same

Gaandhaar or Kandahar in Afghanistan that was ruled by King Subal around 3200 BCE who had married his daughter Gaandhaari to the blind King Dhritrashtra of Hastinapur (see Kuru- dynasty in Figure 3.2). In addition to the east-west expansion Chandragupta's rule had also extended from Himalayas in north to the south India. Historians credit Kautilya Chanakya (350-283 BCE) the brilliant prime minister, for Chandragupta Maurya's grand success in establishing such a big empire.

The well-known historical figure of the Indian history, Kautilya Chanakya, is credited for authoring the books of Arthshastra (Treatise on Economics) and Nithishastra (Treatise on Public Administration and Military). Chanakya is not a fictitious author and these books are not fiction. They are considered to be marvels of Chanakya's brilliance and are available for one to read. He was a professor at Takshashila (present distorted name is Taxila), one of the oldest universities and the seat of learning for Vedic and Buddhist philosophies and many other subjects. It functioned over 1,200 years starting from the 7th century BCE and lasted till 5th century CE. Taxila is situated 55 km from Islamabad, the capital of Pakistan and is a UNESCO World Heritage Site. After Alexander's invasion and ensuing disruption at Taxila, Chanakya moved to Pataliputra and helped establish the vast Maurya Empire of Chandragupta, the dynasty that was to last from 322 BCE to 185 BCE. This period of Indian history is very well-known from both, the Indian and Western historical perspectives and is a highly reliable base of historical facts. The literary work during this relatively recent historical past of the ancient Indian Civilization does not mention that the ancient books such as Manusmriti, Ramayana and Mahabharata were written around Chanakya's time.

After Alexander's death, Seleucus-I (Seleucus Nicator), a Greek general from Alexander's army invaded the kingdom of Chandragupta Maurya. Chandragupta defeated Seleucus. As part of the treaty Seleucus married off his daughter to Chandragupta and also handed over some of the territories under his rule. For goodwill, Seleucus dispatched an ambassador named Megasthenes (350 BCE - 290 BCE), to the court of Sandrocottus (Greek for Chandragupta). Megasthenes was born in Asia Minor, was a traveler, a geographer and a Greek historian who lived in India for two or three years and travelled to important cultural and religious centers of India. He wrote a historically famous book titled "Indika", wherein he recorded his observations on the social system, judiciary and religious practices in Indian society. Although the exact dates of Megasthenes' embassy are not available the historians place Megasthenes' period sometime close to 288 BCE, the year Chandragupta Maurya died.

Max Muller stated in his own book, 'India - What Can it Teach Us?' that in India according to Megasthenes' account in Indika, *thefts were extremely rare and that they* (the people of India) *honored truth and virtue.* If this account of Megasthenes was credible then other accounts in Indika, even with his limited time and observation in India, have to be credible as well. While arbitrarily setting the age of Mahabharata and the Vedas, Max Muller certainly missed out Megasthenes' account of the city of Mathura and Krishn, who was the ruler there, before he moved his capital to Dwarka in the present Gujarat State.

Dr. Satyavrat Siddhantalankar, in the introduction of his Hindi translation of Bhagvadgita (1965 ed.), has discussed the connection of Megasthenes and an important piece of ancient Indian history. The following discussion is based on his reference.

Megasthenes' records in Indika describe a place called Methora and state that the people of this region worshipped Hiraiclese. The word analysis yields that just as Sandrocottus in Greek is Chandragupta in Sanskrit and the capital of the Emperor Chandragupta's empire Palibothra in Greek is Pataliputra in Sanskrit, Methora in Greek is Mathura in Sanskrit and Hiraiklese in Greek is Shree Krishn in Sanskrit. The transition of the word from one language to the other could be viewed as –

Shree → Siri → Sirai →Hirai (sound of the letter s changing to the sound of h) and Krishn → Klese. Therefore in Greek endings of the names where 'ese' is the suffix, Shree Krishn becomes HiraiKlese. Obviously, Megasthenes is referring to Shree Krishn. The *shree* adjective in Sanskrit is equivalent to and similar sounding English term Sir. One immediate conclusion here is that the Greek historian Megasthenes is talking about a historical person Hiraiklese of Methora, who is none other than Shree Krishn. Therefore around 300 BCE according to Megasthenes' account Lord Krishn was a real person - not a mythical figure. Important record in Megasthenes' Indika is the approximate period of Hiraiklese that 138 generations had elapsed between Hiraiklese (Shree Krishn) and Sandrocottus (Chandragupta Maurya).

Counting of generations by Megasthenes, between Krishn and Chandragupta, should tell intellectuals, historians and the rest that Krishn of Bhagvadgita and Mahabharata was a real historical person up to the time of Megasthenes. Historians usually use 20 years for a generation. This gives an average total of 138x20 = 2760 years between Lord Krishn and the Emperor Chandragupta Maurya. As mentioned above, Chandragupta's rule started in Pataliputra in 322 BCE. This takes us to Krishn's period of 2,760 + 322 = 3,082 BCE.

According to Sanskrit Literature, Shree Krishn died very close to the onset of Kali Yuga in 3102 BCE, the time-milestone Marker-5 on the Timeline-A. The two numbers, 3102 BCE and 3,082 with Megasthenes' account therefore are remarkably close. Megasthenes' records therefore help debunk unsound hypotheses of some Indologists. Krishn was an Arya of the ancient Aryan Civilization of India. Therefore, Megasthenes also puts a nail in AIT's coffin.

(2). The Case of Rajtarangini – A Book of Ancient History

About a millennium old book Rajtarangini (Raj-Tarangini), in Sanskrit, authored by Kalhan in 1148 CE, gives the genealogy and other historical description about the Kashmiri rulers. Based on the history in Rajtarangini, Acharya Udayaveer Shastri in his book "Vedanta Darshan Ka Itihas" (History of Vedanta Darshan, 1991) has offered valuable details and discussion. Based on Acharya Udayaveer Shastri's discussion on Rajtarangini's genealogy of the Kashmir kings, this author provides an analysis that takes the reader 5,000 years in the past. This author has supplemented the historical information given in Rajtarangini with the independent historical events recorded in the book of Mahabharata when Shree Krishn, Yudhishthir and Arjun were active historical personalities. To establish the context a limited historical account of Rajtarangini is presented below.

The first king in Rajtarangini's Kashmir dynasty is King Gonand. King Gonand according to Rajtarangini was present during the Mahabharata period. The Magadh Kingdom (present state of Bihar and Bengal together) at that time was ruled by King Jarasandh, an extremely ambitious and cruel king. King Gonand of Kashmir colluded with King Jarasandh and together they attacked Mathura from where Shree Krishn ruled. Let us tie the history of 3100 BCE together with that of the account in Megasthenes' book Indika presented above. It is the same Magadh kingdom of Jarasandh, ruled later in the 4th century BCE by King Nand when Alexander retreated from attacking the Magadh region. It is the same region which Chandragupta Maurya (Sandrocottus in Greek) took over from Nand in 322 BCE. Mathura is the same as Methora in Megasthenes' account in Indika where Shree Krishn ruled from. Shree Krishn is Megasthenes' HiraiKlese.

In the war imposed by King Gonand of Kashmir and King Jarasandh of Magadh on King Krishn of Mathura, King Gonand was killed by Krishn's step-brother Balram. Gonand's son Damodar was enthroned as the king of Kashmir but was revengeful towards Krishn for his father's death. During this period the Gaandhaar-royalty, which also was the maternal dynasty of the Kauravas of Hastinapur, hosted a *svayamvar* for the princess of Gaandhaar; *svayamvar* is the process of 'self-selection' of a life-partner by an unmarried girl from among the invited guest hopefuls. Among the many invitees was a prospective king named Vrishni. To add prestige to his presence in the gathering for the *svayamvar* at Gaandhaar, King Vrishni invited Krishn to accompany him to Gaandhaar. The young king Damodar of Kashmir came to know about Krishn traveling to Gaandhaar. He saw it as the opportunity to avenge his father Gonand's death. When King Vrishni's entourage, with Krishn accompanying him, reached the border of Kashmir, King Damodar attacked the entourage of Vrishni and Krishn. The young Damodar was killed by Krishn himself. Krishn, anticipating the ensuing disorder in a ruler-less Kashmir society, placed Damodar's pregnant young queen Yashomati on the Kashmir throne. She gave birth to a son who was named Gonand Dvitiya (Gonand II) after his grandfather.

According to Rajtarangini's description covered in 26 shloka couplets of its first chapter (1/57-82), Gonand II was the 4th king of Kashmir (after Gonand, Damodar and queen Yashomati). At that time the Mahabharata War loomed on the horizon and it polarized the kingdoms of India as they were forced to take sides. Kalhan, the author of Rajtarangini, writes that Gonand II being a child was not invited to take part in the great war of Mahabharata. Kalhan also writes (Rajtarangini, 1/83) that because all reliable references on the kings of Kashmir after Gonand II were destroyed, he had no information available on the next 35 kings of Kashmir to include in Rajtarangini. After acknowledging this gap in the genealogy of Kashmir kings the historian Kalhan lists the King Kush to be the 41st and continues on till Gonand III as the 55th king of Kashmir. Kalhan in Rajtarangini has given the description of 86 additional kings from king Gonand III up till the time of the writing of Rajtarangini in 1148 CE. Therefore from the first king Gonand up till Kalhan's writing of Rajtarangini, there have been 55+86 =141 kings in Kashmir.

No matter what logic one uses, one cannot brand Kalhan's Rajtarangini a fictitious book of history. Rajtarangini is a book of history of the kings of Kashmir. Kalhan has accounted for 141 kings of Kashmir up till 1148 CE. His description takes us to a point in history that is as real as reality. Mahabharata, the reality emerging out of Rajtarangini is to be faced by skeptic historians. To top it off Kalhan in his book relates Gonand, the first king of Kashmir and his son and grandson to persons, places and events – King Jarasandh, Magadh, Kashmir, Gaandhaar and the Mahabharata War. All this cannot be fictitious or mythological because Rajtarangini by Kalhan is a book of history accepted by historians as such. The independent historical account of Kashmir King Gonand inviting King Jarasandh of Magadh to takeover Krishn's Mathura ties in well with the historical account of Jarasandh.

The norm of the ancient kings of the Aryavart society had been that if a king was defeated by another king, the defeated king was to accept the rule of the winner. However, it was considered *a-dharmik* (*non-dharmic* = immoral) to kill women, children, farmers and Brahmins (knowledge givers and teachers) and it was *a-dharmik* to kill those who surrendered in battle. All kings who swore allegiance to an emperor were generally allowed to carry on as rulers of their respective kingdoms. The emperor, after defeating a king, would enthrone the heir-apparent of the same kingdom. An example was cited above from the Rajtarangini where Krishn, instead of grabbing the Kashmir kingdom for himself had instead enthroned the widowed queen.

Non-*dharmik* king, Jarasandh on the contrary attacked a large number of small kings in and around Magadh region, took over their kingdoms and took the defeated kings as prisoners in spite of their surrendering to him. Whereas the 1148 CE book Rajtarangini, written by Kalhan in Kashmir independently relates the activities of the cruel king Jarasandh along with the Kashmir king Gonand and their collusion against Krishn, the 3140 BCE book of Mahabharata gives considerable number of corroborative details about Jarasandh. His father's name was King Brihadrath and his son was Prince Sahadev. To declare his emperorship over a large region, Jarasandh had captured 86 kings and targeted to capture 14 more to consolidate his kingdom. The description states that all captured kings were kept in filthy prisons and his plan after capturing the remaining kings was to kill them all. Such a ruler who flagrantly disobeyed rules of dharma in the society was to be eliminated. Without going into the details of this part of the ancient history it should suffice here that Krishn, Arjun and

Arjun's warrior brother Bhim were sent by Yudhishthir to wipe out Jarasandh. Bhim killed Jarasandh and Krishn freed all the imprisoned kings who were perishing in Jarasandh's prisons. Krishn consoled the Prince Sahadev for his father's death and enthroned him to rule Magadh. The historical information of this period is corroborated well by two independent sources of recorded history, the book of Mahabharata around 3140 BCE and the Rajtarangini written in 1140 CE.

This section is concluded with a bit of math on Kalhan's data. The period, between Mahabharata War ending in 3140 BCE, as on milestone Marker-4 on the Timeline-A of Figure 3.1 and the undisputed year of 1148 CE for Kalhan's book Rajtarangini, yields time period of 4,288 years. The number of kings in Kashmir genealogy according to Kalhan is 141. This gives an average of about 30 years per king. Historians' thumb rule for averages for determining historical periods extending over 4,000 years should accept this average also. Therefore, the Mahabharata War between Kauravas and Pandavas, with Krishn as the charioteer of Arjun, and the other personalities mentioned in the book of Mahabharata are a proven part of the history at milestone Marker-4 on the Timeline-A in Figure 3.1.

(3). Ancient History from the Book of Satyarth Prakash

The remarkable book entitled "Satyarth Prakash", written by Swami Dayanand Sarasvati (1825-1883), was first published in 1881 CE in the Hindi Language. By the time Swami Dayanand wrote Satyarth Prakash, he had read a large number of ancient and rare books of Sanskrit Literature. Although his emphasis in Satyarth Prakash was not the history of the Vedic period or the history of the Indian culture and Civilization, one can still glean from his book considerable information on ancient Indian history. Throughout Satyarth Prakash and in particular towards the end of its 11th out of 14 chapters, it becomes obvious that this brilliant monk had even read ancient history books in Sanskrit Literature. Can there be a more reliable account of the history of a country than the historical records in the ancient indigenous literature of that country? No. Therefore one can reasonably put time-markers on a timeline running into the distant past of Vedic culture and history based on the historical information gleaned from Satyarth Prakash. For the sake of brevity the Timeline-A in Figure 3.1 has only a few time-markers, however, based on Satyarth Prakash one could populate the Timeline-A with many more important historical events and persons from the ancient Aryan history. Satyarth Prakash not only deals with the Vedas and Vedic philosophy but also critically examines the literary pollution created by some of the Indian pundits on the one hand and some of the Western Indologists and Sanskritists on the other hand that led to the erroneous Timeline-B of Figure 3.1. Satyarth Prakash provides strong evidence in support of several markers on the Timeline-A and takes us step by step to 3140 BCE lending yet another positive proof of the Reference Hypothesis (***"The basic Vedic references on cosmology are ancient and precede a 5,000 year period"***).

Recall that the time Milestone-6 and 7 on the Timeline-A relate respectively to Aadi Shankaracharya, the propagator of *Adwaita-Vedanta* philosophy and

King Vikramaditya of Avantika (Ujjain). Swami Dayanand Sarasvati in the 11th chapter of Satyarth Prakash gives considerable historical information about young Shankaracharya's contribution to the then ongoing debate on theism with the scholars of atheistic Buddhism and Jainism sects, including the fact that Jain opponents revengefully had poisoned Shankaracharya. Swami Dayanand writes that Shankaracharya lived about 2,200 years ago (as of 1880 CE) which is about 320 BCE. He also stated that a highly renowned king named Vikramaditya ruled from the city of Avantika (Ujjain) in Madhya Pradesh about 300 years after Shankaracharya. Swami Dayanand writes that the King Vikramaditya brought together warring kings and established peace in Aryavart. This Vikramaditya is the same king after whom the Vikram Era calendar started in 57 BCE that was discussed earlier. Vikramaditya is also known as a great king for his justice system and for his long life and rule.

Whereas the numbers of years mentioned in Satyarth Prakash are approximate round numbers, Swami Dayanand places Shankaracharya several centuries before Vikramaditya. This validates the historic order of Markers 6 and 7 on the Timeline-A. Based on the written records from Shankaracharya Mathas (monasteries) it is shown in the next section that Shankaracharya died in 477 BCE. One would therefore notice a discrepancy of about (477-320) 157 years between Swami Dayanand's surmise and Matha records on Shankaracharya. This discrepancy of about 157 years is in a historical account of 2,357 years before Swami Dayanand. To understand this discrepancy two points are to be kept in mind: Swami Dayanand Sarasvati used round figures for Shankaracharya's period with the main focus in that section of Satyarth Prakash was the *Adwaita-Vedanta* philosophy and not the settling of historical details of that period. According to Satyarth Prakash, Shankaracharya precedes at least about 320 years BCE. On the other hand the Western Indologists and a large number of their Indian followers have placed the Time-milestone-6 after 0 CE at 788 CE as shown on Timeline-B in Figure 3.1 creating a discrepancy of about 1,300 years. This artificially created myth certainly needs scrutiny. In the section on Shankaracharya, presented below, it is proven without a doubt that the Marker-6 for Shankaracharya on the Timeline-A in Figure 3.1 is where it is supposed to be, before the Marker-7 for the King Vikramaditya. For now the conclusion is that Swami Dayanand Sarasvati's fleeting reference of the period of Shankaracharya counteracts those who shorten the age of Indian history by 1300 years.

Genealogy of Delhi Kings on Timeline-A

Swami Dayanand Sarasvati, at the end of the 11th chapter of Satyarth Prakash, has provided records of the successive dynastic rule of the Aryan kings who ruled from Delhi starting from the Emperor Yudhishthir (3140 BCE) till the king Prithvi Raj Chauhan (1192 CE) when he was dethroned by the Muslim invader exterior to India. Swami Dayanand's records are from a history book written in 1725 CE that validate the markers on the Timeline-A. In Satyarth Prakash, Swami Dayanand provides full reference for the source of this historical information. Translated from Hindi to English by this author, Swami Dayanand states:

CHAPTER 3: The Historicity of Vedic References - The References on Cosmology Precede 5,000 Years

> *"...I have translated this material from the fortnightly publication from Nathdwara, Harishchandrika and Mohanchandrika, which are well-known in the kingdom of Udaipur Chittor Fort in the Mewar region of Rajputana. It will greatly benefit the country if the members of the Aryan society likewise researched the historical books and books of knowledge and published them. The respected editor* (of Harishchandrika and Mohanchandrika) *received an ancient book from his friend, which was written in Vikram Era 1782 (seventeen hundred eighty two) and published in the issues 19 and 20 of his two publications dated Maargsheersh, bright-lunar phase of Samvat 1939* (Vikram Era). *"*

In the above quotation the history book that provides the genealogy of the Delhi kings was written in 1782 VE (Vikram Era) which is 1725 CE). The lunar calendar date of publication issues 19 and 20 of the fortnightly is in the month of Margsheersh in the year 1939 VE which corresponds to October-November in 1882 CE. This author has reproduced below the genealogy of the Delhi kings in Table 3.3 as it is in Satyarth Prakash. This table is made up of sub-tables that represent transition of genealogy from one Aryan rule to another Aryan rule. In this table are interspersed historical notes by Swami Dayanand that he took down from the original publication. These notes primarily relate to the cause for the change of genealogy. These notes are original to the history book written in 1725 CE. A discussion on the writing process in ancient India is in Chapter 4. The writing with the black ink was carried out on several mediums such as the *bhoj-patra* (thin bark of birch tree), paper made of cotton or silk. The records and manuscripts were rewritten when they faded or got damaged. The history book from which the data came to Swami Dayanand must have been also rewritten several times over a time period over 4,000 years starting from Emperor Yudhishthir. Therefore the historical notes at the transition of dynasties and some anecdotal information about certain kings must have been copied by the successive writers, spanning over centuries. The overall impression therefore is that the *raj-purohits* (royal-priests), responsible for recording history for the kings, have recorded it with the intension of preserving the historical events. To test its authenticity this table is examined from several perspectives and some discrepancies with possible explanations are pointed out. The historical comments of the history book are in italicized quotes.

> *"At Indraprasth (Delhi) from emperor Yudhishthir to Yashpal have been 124 (one hundred twenty-four) kings who ruled over 4,157 years, 9 months and 14 days. They are presented here. There were 30 rulers in Yudhishthir genealogy who ruled for 1,770 years 11 months and 10 days. Their details are"*:

Table 3.3: Genealogy of Aryan kings of Indraprasth (Delhi). Number of years (Y), months (M) and days (D) for the rule of each king are shown in the last three columns.

No.	King	Y	M	D	No.	King	Y	M	D
1	Yudhishthir	36	8	25	16	Suchiratha	42	11	2
2	Parikshith	60	0	0	17	Shoor Sen II	58	10	8
3	Janmejaya	84	7	23	18	Parvata Sen	55	8	10
4	Ashwamedha	82	8	22	19	Medhavi	52	10	10
5	Rama II	88	2	8	20	Sonacheera	50	8	21
6	Khatra Mal	81	11	27	21	Bhima Deva	47	9	20
7	Chitta Ratha	75	3	18	22	Nrihari Deva	45	11	23
8	Dashta Shailya	75	10	24	23	Poorna Mal	44	8	7
9	Ugra Sen	78	7	21	24	Kardavi	44	10	8
10	Shoor Sen	78	7	21	25	Alamika	50	11	8
11	Bhuvanpati	69	5	5	26	Udaya Pal	38	9	0
12	Rana Jita	65	10	4	27	Durvana Mal	40	10	26
13	Rikshaka	65	7	4	28	Damata	32	0	0
14	Sukhadeva	62	0	24	29	Bhima Pal	58	5	8
15	Naraharideva	51	10	2	30	Khemaka	46	11	21

Historical note: '*King Khemaka was murdered by his prime-minister Vishrava whose 14 generations ruled for 500 years 3 months and 17 days, they are as following*'.

31	Vishrava	17	3	29	38	Kadruta	42	9	2
32	Pura Seni	42	8	21	39	Sajja	32	2	14
33	Veera Sen	52	10	7	40	Amar Chud	27	3	16
34	Anang Shayi	47	8	23	41	Amipal	22	11	25
35	Hai jit	35	9	17	42	Dasharatha	25	4	12
36	Param Seni	44	2	23	43	Veera Sala	31	8	11
37	Sukhpatala	30	2	21	44	Veerasal Sen	47	0	14

Historical note: '*King Veerasal Sen's prime-minister Veer Maha murdered him and became the king whose 16 generations ruled for 445 years 5 months and 3 days, they are as following*'.

45	Veer Maha	35	10	8	53	Tej Pal	28	11	10
46	Ajit Sinha	27	7	29	54	Manik Chand	37	7	21
47	Sarva Datta	28	3	10	55	Kama Seni	42	5	10
48	Bhuvanpati	15	4	10	56	Shatru Mardan	8	11	13
49	Veera Sen	21	2	13	57	Jeevan Loka	28	9	17
50	Mahipal	40	8	7	58	Hari Rao	26	10	29
51	Shatru Shala	23	4	3	59	Vira Sen II	35	2	20
52	Sangha Raja	17	2	10	60	Aditya Ketu	23	11	13

Historical note: *'Aditya Ketu was killed by the King Dhandhar of Prayag whose dynasty had 9 rulers ruling for 374 years 11 months 26 days. They are as following'.*

61	Dhandhar	42	7	24	66	Jeevan Raj	45	2	5
62	Marshi	41	2	29	67	Rudra Sen	47	4	28
63	Sanarach	50	10	19	68	Aarilak	52	10	8
64	Mahagadh a	30	3	8	69	Rajpal	36	0	0
65	Duranatha	28	5	28					

Historical note: *'King Rajpal was overthrown and killed by his General Mahanpal who ruled for 14 years 0 months 0 days. No more details about his rule'.*

70	Mahanpal	14	0	0

Historical note: *'King Mahanpal was attacked and killed by King Vikramaditya of Avantika* (Ujjain) *who ruled from Avantika for 93 years 0 months 0 days. No more details about his rule'.*

71	Vikramaditya	93	0	0

Historical note: *'King Vikramaditya was attacked and killed by King Shalivahan's noble Samudra Pal Yogi hailing from Paithan and who established rule* (over Indraprasth) *for 16 generations ruling 372 years 4 months and 27 days. They are listed in the following'.*

72	Samudra Pal	54	2	20	80	Amrit Pal	36	10	13
73	Chandra |Pal	36	5	4	81	Bali Pal	12	5	27
74	Sahoy Pal	11	4	11	82	Mahi Pal	13	8	4
75	Dev Pal	27	1	28	83	Hari Pal	14	8	4
76	Narsingh Pal	18	0	20	84*	Sees Pal	11	10	13
77	Sam Pal	27	1	17	85	Mahan Pal	17	10	19
78	Raghu Pal	22	3	25	86	Karam Pal	16	2	2
79	Gobind Pal	27	1	17	87	Vikram Pal	24	11	14

Historical note on item 84: '* *also Bhim Pal in some history books.'*

Historical note: *'King Vikram Pal of Indraprasth attacked Malukhchandra Vohara, the king of the Western India but was slain in a pitched battle in an open battle-field. Malukhchandra became the king of Indraprasth. His dynasty had 10 kings ruling for 191 years 1 month 16 days. They are listed in the following'.*

88	Malukh Chandra	54	2	10	93	Kalyan Chand	10	5	4
89	Vikram Chand	2	7	12	94	Bhim Chand	16	2	9
90*	Amin Chandr	10	0	3	95	Lov Chand	26	3	22
91	Ram Chand	13	11	8	96	Govind Chand	31	7	12
92	Hari Chand	14	9	24	97**	Queen Padmavati	1	0	0

Historical note on item 90*: *'His name elsewhere is also written as Manak chand.'*Item 97**: *'The Queen Padmavati was Govind Chand's Queen. She* *died heirless. Her nobles took a decision to put Hari Prem Vairagi on the throne and ruled themselves. His dynasty had 4 kings whose reign lasted for 50 years 0 months 21 days. Those details are' -*

98	Hari Prem	7	5	16	100	Gopal prem	15	7	28
99	Gobind Prem	20	2	8	101	Mahavahu	6	8	29

Historical Note: *'King Mahavahu renounced the throne, became a recluse and proceeded to the forest for meditation and self-realization. On hearing this, Adhi Sen the King of Bangal took over the throne of Indraprasth. His dynasty had 12 kings ruling for 151 years 11 months 2 days. His details are.'*

102	Adhi Sen	18	5	21	108	Kalyan Sen	4	8	21
103	Vilaval Sen	12	4	2	109	Hari Sen	12	0	25
104	Keshav Sen	15	7	12	110	Khem Sen	8	11	15
105	Madhav Sen	12	4	2	111	Narayan Sen	2	2	29
106	Mayur Sen	20	11	27	112	Lakshmi Sen	26	10	0
107	Bhim Sen	5	10	9	113	Damodar Sen	11	5	19

Historical Note: *'King Damodar Sen was very oppressive to his nobles. One of his nobles, Deep Singh won over the king's army and fought the king. The king was killed and was succeeded by Deep Singh whose dynasty had 6 kings ruling for 107 years 6 months 22 days. Their details are'-*

114	Deep Singh	17	1	26	117	Nar Singh	45	0	15
115	Raj Singh	14	5	0	118	Hari Singh	13	2	29
116	Ram Singh	9	8	11	119	Jivan Singh	8	0	1

Historical Note: ***'For some reason King Jivan Singh sent all his army to the north. On hearing this, Prithvi Raj Chauhan, King of Virat, attacked King Jivan Singh who fell in the battle and Prithvi Raj Chauhan became the King at Indraprasth who had 5 kings in his dynasty ruling for 86 years 0 months 20 days. They are described below'-***

120	Prithvi Raj	12	2	19	123	Uday Pal	11	7	3
121	Abhaya Pal	14	5	17	124	Yash Pal	36	4	27
122	Durjan Pal	11	4	14					

Swami Dayanand Sarasvati casts doubt on the last dynasty of Aryan kings (items 121 to 124) at Delhi and suggested to follow present historians, according to which the last Aryan King at Delhi was Prithvi Raj who ruled till 1249 Vikram Era (1192 CE) after which the Muslim rule began at Delhi.

Let us now analyze the data that Swami Dayanand Sarasvati presented from the old history book in Sanskrit. In spite of the apparent discrepancies of minor nature in the Table 3.3 the value of the data leads to the fact that these data render a positive proof of the Timeline-A and the historical markers on it.

Analysis of Indraprasth (Delhi) Genealogies

Assuming that the Aryan dynasties ended with Prithvi Raj Chauhan in 1192 CE, the sum total of Aryan rule from Yudhishthir to Prithvi Raj according to Satyarth Prakash data is 4,069 years, 11 months and 3 days, a period ruled by 120

Aryan kings. Rounding off this period between Yudhishthir and Prithvi Raj Chauhan gives 4,070 years. Added to it the period between 1192 CE and 2018 CE (826 years) gives 4,996 years; about five thousand years back in history.

An important correction is needed in this estimate of 4,070 years. The month in the Table 3.3 is assumed as a 30-day period. In a luni-solar Aryan calendar the 5 days are adjusted to keep all festivals synchronized, the month is counted as a 30-day period. Therefore 5 days a year remain unaccounted for in the Table 3.3. This results into a correction of 56.5 days in 4,070 years. With this correction the total period between Yudhishthir and Prithvi Raj is: 4,070 +56.5 = 4,126.5 = 4,127 years.

It is also obvious from the data of the Table 3.3 that a 30-day month has been used throughout the tables. Even though Yudhishthir ruled for 36 years 8 months and 25 days and renounced the throne to go to the Himalayas for meditation and self-realization, about half a dozen kings in his genealogy after him ruled on an average of about 80 years per king and the next half a dozen kings ruled an average of 70 years each. At the same time, a sudden change in rule by a change in the dynasty reduces the average period of rule by a king in the Delhi dynasties between Yudhishthir and Prithvi Raj Chauhan. The average ruling period in Delhi dynasty is about 4070/120 = 34 years. This is quite close to the 30.4 years estimated from the independent historical records for the stable Kashmir dynasty recorded in Rajtarangini the book written by Kalhan in 1148 CE; both rules running parallel in time at different locales.

These average ruling periods, derived from the years of rule and the number of kings in two different dynasties, are available from two independent history books written in two different locales at two different times. The history book cited by Swami Dayanand in Satyarth Prakash was written in 1725 CE for the Delhi rulers and the book of Rajtarangini discussed in the previous section was written by Kalhan in 1148 CE; two books written over half a millennium apart substantiate the historical fact that Mahabharata War, Yudhishthir, Krishn and other personalities were present near 3140 BCE

Megasthenes' records in Indika cited above had suggested 138 generations to Lord Krishn and not the actual number of kings. Counting generations and counting actual number of kings who ruled are two different things. For example the Queen Elizabeth II of England, coronated at the age of 25, has seen four generations during her single rule over 60 years by the year 2018 CE. The 138 generations between Hiraiklese (Shree Krishn) and Sandrocottus (Chandragupta Maurya) proportionately look much higher than both accounts, in Satyarth Prakash and in Rajtarangini. The difference is that the counts of these books provide years of rule by an actual king regardless of how many generations that a king ruled. The historians' average for the age of a generation is naturally much less than the average rule by the kings at Delhi or at Kashmir. Both approaches do lead to valuable historical information. While Megasthenes' generation count can lead to the approximate period in history, the data as in the Table 3.3 bring one closest to reality. It is therefore obvious that two history books provide more accurate historical information. Recording of years, months and days of rule by a king is as accurate as one can get. The count for Delhi dynasties based on year, month and day for the rule of each king is indeed most accurate of all.

The date for the start of Kali Yuga on February 18, 3102 BCE is accepted by most historians of Vedic tradition and some astronomers as well. If there is any disagreement on this, it is within a day or two because some scholars have considered the onset of Kali Yuga on February 20, 3102 BCE instead. However, the time elapsed from 3102 BCE at the onset of Kali Yuga to the last Aryan king at Delhi, Prithvi Raj Chauhan is: 3,102+1,192 = 4,294 years. This in terms of Kali Yuga Era is 4294 KE at Prithvi Raj Chauhan. The number of years estimated from the Table 3.3 between Yudhishthir and Prithvi Raj with correction is 4,127 years. The difference between the two calendar numbers (YE and KE) is therefore 4294 KE – 4127 YE = 167 years, KE starting before YE. This is not acceptable to the Indian historians of Vedic tradition because the difference between Yudhishthir's coronation (onset of Yudhishthir Era) and the onset of KE is 38 years after Yudhishthir. Therefore there is a discrepancy of 167+38 = 205 years. As mentioned before, the KE being the calendar based on astronomy has minimum error and is practically independent of human interference. This discrepancy could be attributed to the records in the history book which put Yudhishthir about 200 years after the onset of KE. Therefore the data of the table underestimates the total period of Aryan kings by about 200 years in about 4127 year-long period causing a deviation of about 4.5%.

The average rule by a king in Delhi dynasties being 34 years, this discrepancy is equal to about 3 to 4 rulers missed out in the history records of Table 3.3. To determine whether the Vikramaditya who ruled Delhi from Ujjain, included in the Delhi genealogy as 71st entry is one and the same king, the years from Yudhishthir to the beginning of Vikram Era (excluding the 71st entry) were estimated. This estimate is about 3,102 years. The estimate between KE onset in 3102 CE and 57 BCE is 3,045 years. Therefore the difference between the YE and KE is 3,102 - 3,045 = 57 years. Because the difference between YE and KE is exact 38 years the discrepancy is only 19 years over 3,000 years, which is about half a percent. Therefore, the King Vikramaditya is one and the same person. Also, most of the discrepancy in the history tables of 3 to 4 rulers is in the last one thousand years of the table. Over a period of five thousand years this is quite understandable. Considering that such a consistent and detailed data were even available, representing a five thousand year non-religious history, found somewhere in the world, by itself is remarkable.

The likely cause for such a discrepancy, small though it may be, could be in the process of recording history with the possibility of human error in the writing and rewriting the same records after long gaps to refresh the writing medium, the bhoj-patra or silk cloth. Nevertheless the listing of kings that Swami Dayanand included in his book Satyarth Prakash is from a book of history as opposed to the baseless contention of Max Muller that Indians did not write history. The nature of historical notes in between the change of dynasty tells us that the 1782 VE (1725 CE) book in Sanskrit is a credible history book.

One historical fact which becomes clear from the Table 3.3 is that the king at Serial No. 71 is Vikramaditya. Although Vikramaditya ruled from Ujjain, most kings of North India during his rule had accepted his emperorship including the ones at Delhi. Therefore the ancient historians must have credited him for the rule at Delhi as well and included him in the genealogy list of Delhi. This is obvious from the lone entry at the 71st row. When the years are added up the addition brings us to a period coincident with the rule of Vikramaditya of Ujjain in Central India. This Vikramaditya is the same long-lived, brilliant king known for his valor and justice. It is the same Vikramaditya of Ujjain in whose praise the Vikram Calendar of the Hindus (Table 3.2) was started in 57 BCE. He is known to have sent Sanskrit scholars and mathematicians to Middle East on request by the rulers there. Towards the end of his life King Shalivahan became powerful who sent his General Samudra Pal to overthrow Vikramaditya. King Shalivahan ruled from Avantika after Vikramaditya, over all regions including Indraprasth (Delhi). The dynasty at Indraprasth began with the rule of Shalivahan's general Samudra Pal; listed as the 72nd Aryan king under the Delhi dynasties in the Table 3.3. Twenty two years after Vikramaditya's death began another Hindu Calendar in King Shalivahan's name. Both calendars are still running in parallel; Shalivahan Calendar being more popular in South India.

The Table 3.3 has a few other discrepancies of editorial nature that if corrected would eliminate the difference between KE and YE. Thus the data in Table 3.3 ties several loose-ends and supports the Timeline-A.

The Reference Hypothesis has been already proven above in several different ways and now we present in the following section the final proof from the recorded history at Shankaracharya monasteries which is available for one to access at the present time. These records at the monasteries provide the connection between the YE, KE, VE and CE calendars telling us that the Vedic references are older than 5,000 years.

(4). Aadi Shankaracharya at Milestone-6 on Timeline-A

Aadi Shankaracharya and his monasteries were referred to in Chapter 1. This section presents very valuable written historical material related to Shankaracharya that proves the position of the time Marker-6 on the Timeline-A (Figure 3.1). At the same time it provides a proof for the position of the time Markers-4 and 5. This material also sets the ancient Indian historical record straight. Historical data and facts have been brought out by the great Sanskrit scholar and philosopher, the late Acharya Udayaveer Shastri (1894-1991CE), from his arduous research in ancient literature, particularly the literature preserved at Shankaracharya Mathas (monasteries). Acharya Udayaveer Shastri and his remarkable book "**Vedanta Darshan Ka Itihas**" (History of Vedanta Darshan) in Hindi language published in 1991 just before his death at the age of 97 have been the inspiration for this author to present these undisputable historical facts. The book's acronym, VDKI will be used in the following text. In VDKI he tied historical loose ends starting from the time of Rishi Vedvyas at the Marker-4 to the milestone at Aadi Shankaracharya at the Marker-6 on this author's Timeline-A, a period over three thousand years of Vedic History. To achieve this incredible feat he researched the literature well-guarded and preserved at Shankaracharya Mathas that Aadi Shankar himself established in the fifth century BCE. Acharya Udayaveer Shastri brings the reader slowly and carefully with cross-check of references from Aadi Shankaracharya up until contemporaneous care-taker *acharyas* of the Mathas. The Mathas are functioning well even to this day at the same locations where Aadi Shankaracharya had originally established them between 499 BCE to 482 BCE in different parts of India.

Acharya Udayaveer Shastri was a Sanskrit scholar, a great thinker, a Vedic philosopher with deep knowledge of Vedic literature. He was also a Vedic historian, a prolific Hindi writer with about a dozen books to his credit. He authored in Hindi language the interpretive translations of all six Vedic Darshans (Sutras). He was bestowed with high literary honors and titles like, Vidyabhaskar, Ved-Ratna, Nyaya-Vaisheshik-Sankhya-Yog Teerth, Vedantacharya and Shastrashevidhi. Acharya Udayaveer Shastri, having direct access to protected historical records at the present Shankaracharya Mathas uncovered valuable material that he included in his book VDKI. With his persistent effort and research in Sanskrit Literature at Shankaracharya Mathas, he brought out the historical facts that clear the fuzz of fog hovering over Aryan history of the past 5,000 years.

This author has greatly benefited from the knowledge-pool contained in Acharya Udayaveer Shastri's books on all six Vedic Darshans and his interpretation of *sutras* and Veda mantras. This author gratefully acknowledges the historical data given in VDKI that has been used here and analyzed to establish the validity of the Timeline-A in Figure 3.1 and all the time markers on it, thus validating the Reference Hypothesis that all references on Vedic Cosmology Principle used in this book predate 5,000 years, well before the beginning of substantive scientific activity in the West. Before presenting hard-core data a brief review of the background of Shankaracharya and the Mathas is presented.

The time Marker-6 at 509 BCE is the birth year of Shankaracharya. This great thinker and philosopher has been a towering personality in the historical and spiritual life of India. The word Shankaracharya is the euphonic combination of the words Shankar and *acharya* where the first word is the birth name and the second word is a literary title given to a highly learned person with deep knowledge of philosophy and who is a preacher and a teacher of that knowledge. The respectful title Acharya could also appear as a prefix to the name, something similar to the title of 'Doctor' or 'Professor' in contemporary practice. Shankar was an *acharya* in the true sense who taught and interpreted Vedanta Darshan, Upanishads and other ancient Vedic writings with particular emphasis on Vedanta Philosophy. His name in Sanskrit literature appears variously as Acharya Shankar, Jagadguru Shankaracharya, Aadi Shankara and Aadi Shankaracharya. The word *aadi* means 'the first'. The adjective Aadi is added to Shankaracharya to differentiate him from the other Shankaracharyas that followed him in his Adwaita-Vedanta order that he established. He travelled through India preaching Adwaita-Vedanta Philosophy and established the Vedanta order with four Mathas that he established. All caretaker *acharyas* in these monasteries are in fact addressed as Shankaracharya, whatever their birth name may be. This has caused some confusion in determining the time period of Aadi Shankaracharya. This is discussed in detail below.

He was a very brilliant soul, a genius of a man who learnd the important Vedic texts by heart at an early age and completed his education with his guru Govindpaad when he was just 9 years old. His educational achievement by this early age was equivalent to that of a regular student of 25 years or more. At the age of 16 he completed the interpretive translation and commentary of Rishi Vedvyas's Vedanta Sutra. Even today, with disagreements on philosophical principles of his Adwaita-Vedanta, all scholarly people agree that his interpretive translation of Rishi Vedvyas's Vedanta Sutra, is an unparalleled, heart-grabbing, literary creation with mastery of the Sanskrit language. With his interpretation of Rishi Vedvyas's Vedanta Sutra he introduced the philosophy of *A-dwaita* (non-duality) Vedanta, wherein everything in the universe is God. Therefore, at the root, it is a non-duality pantheism philosophy. Since his advent, his philosophy and the Shankaracharya Mathas have attracted scholarly people and have produced some well-known *acharyas*. It is his Adwaita Philosophy that gave birth to what is known today as the Vedanta Society. It is the same society that Swami Vivekananda started in Chicago when he visited North America, to address the Parliament of Religions, held at the Art Institute of Chicago in 1893.

As opposed to the traditional theistic Vedic religion and philosophy, there were two new non-Vedic religions prevalent in India during Shankaracharya's time, Buddhism and Jainism. God had no place in these religions. Shankaracharya decided to oppose these atheistic religions. He had barely crossed his teenage years but he got engaged heavily in many public *shaastraarths* (debates on philosophical and spiritual subjects) with Vedic scholars as well as with Buddhist and Jain philosophers. Shankaracharya with his brilliance impressed scholars, kings, rulers and masses and defeated Buddhist and Jain philosophers in the *shastrarths*. The immediate consequence of this was mass conversions from Buddhism and Jainism back to Vedic religion. Under the aegis of the kings and the rulers of his time he toured throughout India. In his short-lived life of 32 years, this brilliant young monk brought a historical transformation in the Indian society in re-establishing theistic philosophy.

The Shankaracharya Mathas continue to be the seats of learning and highly revered *ashrams*. The *ashram* at Kanchi-Kamkoti in South India, where Shankaracharya lived his last years, naturally assumed the status of the fifth Matha in his order. Now there are many Vedanta Society ashrams all over India and in many other countries. The historic tradition of appointing a figurehead caretaker *acharya* as a successor at a monastery was started by Shankaracharya himself which continues till today. The continuity of this tradition in some of the originally established five Mathas enables a historical connection to the antiquity of India over the span of the past two and a half millennia. The links in this chain of continuity are the *acharyas* who were successively appointed at the Mathas and the literature that was created there which gives insight into both, the philosophy and the historical records of administrative nature. The well-preserved and well-guarded written records at the Mathas that include the succession-order of caretaker *acharyas* from Aadi Shankar down to the present times happen to be of immediate interest to this author. It is this material that takes one from the present to the past over two and a half millennia ago, right to the date of birth of Aadi Shankar. It is all history. It is this historical material that has proven to be the most valuable evidence to resolve the confusion created by some of the Western scholars that is blindly accepted by quite a large section of scholarly Indians including the followers of the Adwaita Philosophy. These historical records will help us disprove the time Marker-6B on the Timeline-B.

Shankaracharya himself established the order of Adwaita Philosophy and the Shankar-Mathas to go with it. The well-established Mathas and their disciplined operation have carried the historical records forward with remarkable continuity.

Yet what one finds is that Shankaracharya's period is as much in historical confusion as any other period in Hindu religion. This is because those who have researched in this area have used a limited set of data without exploring a large amount of material available at these Mathas. Using scant information from Shringeri Matha and based on the book entitled- 'Vivad', the period erroneously assigned to Aadi Shankar is 684 CE to 716 CE. Also, based on one or two *shlokas* created by an unknown author, the birth of Shankaracharya is said to be in 787/788 CE. As shown below, none of these numbers make sense. The literature found at Shringeri Matha assigns Shankar's birth in 684 CE while the two *shlokas* assign it to be 788 CE with a difference of over 100 years. Obviously this cannot be correct. Acharya Udayaveer Shastri has set the record straight in his book VDKI.

Acharya Udayaveer Shastri in VDKI points out a well-known fact that the Shringeri Matha remained in a destroyed condition and was non-operational without caretaker *acharyas* for a long period of at least 700 to 800 years. Therefore any historical information from Shringeri Matha is not as reliable as from the other Mathas. Nevertheless, the literature from Shringeri Matha has been used by some scholars and some Western Indologists to assign the historical period for Shankaracharya and most historians have unquestioningly followed this and have been repeating the wrong information in their own writings. Contrary to this, the recorded data from the other Mathas, with remarkable continuity clearly delineates that Aadi Shankaracharya was born in 509 BCE. This means a discrepancy of 1300 years. The irony has been that many research scholars have worked in this subject area and wrote their dissertations, yet they were not able to find the birth place of Shankaracharya till the caretaker Shankaracharya of Shringeri, Acharya Nrisinhbharati (1879-1912) himself determined it from the records of the Mathas. He determined that Aadi Shankar's birth place was Kalady in the state of Kerala. Yet these Indologists and their followers insisted that Shankaracharya was born in the 8th or 9th century CE. For fear of losing face they defended their poorly thought out conclusions.

Considerable original literature was created at the Mathas over the centuries that needed to be studied and examined for shedding light on the history of Adwaita Vedanta, particularly about the disputed period of Aadi Shankaracharya himself. It was Acharya Udayaveer Shastri, no relation to the Mathas or to the Acharyas at the Mathas, who conducted deep research in the Matha literature and the preserved historical records in the five Mathas to set the record straight. Very carefully he established, step-by-step, the true details regarding the advent of Shankaracharya. Acharya Udayaveer in his VDKI has not only presented the historical aspect of Shankaracharya's period but has also presented voluminous historical information on the Sanskrit Literature, at Shankaracharya monasteries. Being personally close to three current Shankaracharyas of different monasteries, Acharya Udayaveer Shastri received from these monasteries very valuable research material, well-protected and preserved in the form of written records. Based on his book VDKI this author has gained considerable historical insight into the period from Aadi Shankaracharya to the present time. This author's limited attempt in this book relates:

- to establish the validity of the time Marker-6 on the Timeline-A

- to establish a relationship between the calendars of three eras, Yudhishthir (YE), Kali-Yuga (KE) and Vikramaditya (VE) that have been utilized for record-keeping at the monasteries
- to establish that the Marker-4 at 3140 BCE is historically valid

The Shankaracharya Mathas are held in high reverence by the Hindu society. The four Mathas originally established by Aadi Shankaracharya himself are:
(1) Jyotir-Matha at Badrinath (Uttaranchal Pradesh in North India)
(2) Sharada-Matha at Dwarka (Gujarat State in Western India)
(3) Shringeri-Matha at Shringeri (Karnataka State in South India)
(4) Govardhan-Matha at Puri (Orrissa State in Eastern India)

These Mathas are marked on the map of India in Figure 3.6 including Adi Shankar's birth place, Kalady in the Kerala State. With the help of King Sudhanva, Shankar in his short life-span covered north-south and east-west regions throughout India. The chronology of the Mathas that he established is in the same sequence as listed above, starting from Badrikashram (Badri Nath) in the Himalaya where he established the first Matha (marked by the number (1) on the map). To give the reader some idea of the distances that young Shankar covered in his short life, the direct distance between his birth-place Kalady near the southern tip of India to Badri in the Himalaya is about 2,250 km. The other direct distances are: between Mathas (1) and (2) about 1,450 km, between (2) and (3) it is about 1,000 km, between (3) and (4) it is about 1,200 km and between (4) and (5) it is about 900 km. Since the travel in 500 BCE was quite arduous, on foot or on horseback, over difficult terrain, young Shankar's travel itself was an unparalleled feat.

After establishing these four Mathas he went to live at an ashram in Kanchi-Kamkoti till he left his mortal body. Therefore the ashram at Kanchi has acquired the same status as the other four Mathas that Shankaracharya himself established. In addition to a Shankaracharya monastery being addressed as Matha it is also addressed as a Peetha, meaning 'the seat'. Shankaracharya formally appointed (ordained) an *acharya* at each Peetha from among his students and followers of high intellectual caliber who served as the caretakers of the assigned Peethas. Since then, it is the tradition at the Peethas that the caretaker *acharya* will choose an able successor when he decides to renounce the seat for a total monastic life for meditation and self-realization. Just as Shankaracharya's birth name was Shankar and a*charya* was the title of respect for a learned teacher at a high level, a caretaker *acharya* has a birth name but traditionally he is also addressed as a Shankaracharya, something similar to the practice at Vatican where all successive religious heads are called 'the pope'. To differentiate the first Shankaracharya from his successors the adjective Aadi (the first) is added to his name. He is therefore respectfully addressed as Aadi Shankaracharya.

The monasteries have kept written records in terms of dates and necessary details of the important events, particularly the date of appointment of an *acharya* to the Peetha and the date of his leaving. These details have been kept in a chronological order providing considerable historical information. The effort of this author presented in this section is mainly based on the information and data published by Acharya Udayaveer Shastri in VDKI.

The calendar chosen for the record-keeping in these monasteries is either Yudhishthir Era calendar (YE) or Kali-Yuga Era calendar (KE), transiting later into the presently used Vikramaditya (VE) calendar or Shalivahana (SE) calendar. Since the transition from YE and KE to VE in these records is without a break it is therefore a simple matter to establish that Yudhishthir was at the time Marker-4, at 3140 BCE. It is a simple matter because he was a real king and the YE calendar was launched at his coronation after Mahabharata War. When a calendar in the name of a king is introduced in a society, that king has to be a real historical person. The start of the YE and VE are real and provide historical validity for the life and reign of King Yudhishthir five millennia ago and for King Vikramaditya as a real king in the Indian history over two millennia ago.

In general, the process of carrying forward ancient records, involving calendar dates, would have some human element where errors get introduced inadvertently and in some cases some hearsay is erroneously added. Acharya Udayaveer Shastri has discussed this aspect of the history records as well and concluded that there is plenty of other research material at the same Mathas that would help resolve any discrepancies. Acharya Udayaveer Shastri's scholarship as a Vedic literary-historian makes it clear that, there is no discrepancy in determining the period of Aadi Shankar. The data on Aadi Shankar that is presented in tabular form in VDKI has high degree of consistency to draw valuable historical conclusions that help prove this author's Timeline-A.

Acharya Udayaveer Shastri in VDKI has presented tables of *vanshanumaatrika* of the *acharyas* who presided over the Matha administration. A *vanshanumaatrika* chronologically records the *acharya*'s name, year and period of his administration at the Matha. The term *vanshanumaatrika* is commonly used for 'family genealogy'; however, in this case it does not mean that an acharya had family relationship to the previous acharyas. The *vanshanumaatrika* in the context of the Shankar Matha represents only the 'succession order' of the acharyas at a Matha. Acharya Udayaveer Shastri got this information from the records at the Mathas including the succession order at Shringeri Matha where the records are not complete and has been the source of controversy for shortening of the historical period of Aadi Shankar, his birth in 509 BCE to 788 CE. Let us first discuss the valid data and then touch the controversial data.

Acharya Udayaveer Shastri has provided references of the large number of books and the written records that he used to present the succession order at the Mathas. These references are still available at the Mathas for anyone to study. Without going into all the details of his discussion and without presenting the uninterrupted sequence of all acharyas at a Matha, this author has presented in the following an abridged form of Acharya Udayaveer Shastri's tables. Those desiring to see the complete tables are advised to look into VDKI.

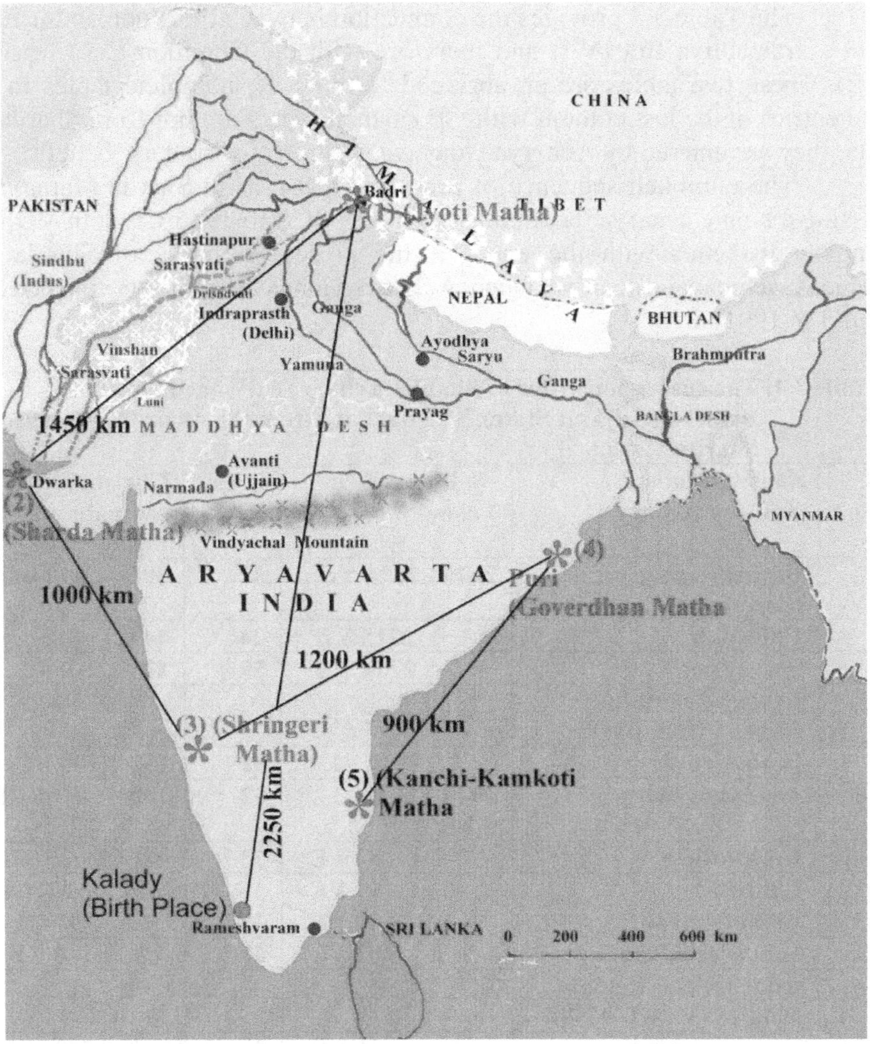

Figure 3.6: Map of India showing locations of Shankaracharya's Mathas.

In the following are two vanshanumaatrika (succession order) tables (Table 3.4 and 3.5) based on the records kept independently at the Sharada and Kanchi Mathas. These written records give sequentially the name of the caretaker acharya and the start and end of his period as a caretaker of the monastery.

Records at Sharda Matha

The record on vanshanumaatrika of acharyas at Sharada Matha in Gujarat State (Table 3.4) begins with the Acharya named Brahmswaroop. The name Brahmswaroop is the guru-given name to Sureshvaracharya; Aadi Shankar being his guru. It is indisputably accepted that Sureshvaracharya, a scholarly pundit was his first disciple. He was appointed as the first acharya of Sharada Matha by Aadi Shankar himself. He continued to be close to Aadi Shankar till Aadi Shankar died.

The Table 3.4 provides the connection between the Yudhishthir Era (YE) and Vikramaditya Era (VE) and therefore with the Common Era/Christian Era (CE). These two tables are an abridged form of the complete tables in VDKI. The entries in the last column with CE calendar years are not from Sharda Matha data, they are entered by Acharya Udayaveer Shastri for readers' benefit.

The unbroken sequence of presiding acharyas at Sharda Matha, without missing a single acharya, beginning from 491 BCE to 1945 CE is in VDKI. This compete list ends with the entry of the present Acharya of Sharda Matha, Abhinavsachidanand-teertha, when Acharya Udayaveer Shastri finished writing VDKI in 1991CE.

Table 3.4: The succession order of Shankaracharyas at Sharda Matha at Dwarka (Gujarat State). The data are from the records preserved at Sharda Matha.

S. No.	Sharda-Matha Name of Acharya	Indian Era From To	Pres. Yrs	Christian Era estimate From To
1	Brahmsvaroop (Sureshvar)	2649 - 2691 YE	42	491 - 449 BCE
2	Chinsukh	2715 YE	24	449 - 425
3	Sarvagyan	2774 YE	59	425 - 366
--	sequence continuous 4 to 8 (VDKI, p317)	---	---	--- ---
9	Brahmjyotsn	3040 YE	32	132 - 100 BCE
10*	Anandavirbhav	3040 YE to 9 VE	52	100 - 48 BCE
11	Kaladhitirth	82 VE	73	48BCE- 25 CE
12	Chidvilas	119 VE	37	25 - 62 CE
13	Vibhutyanand	154 VE	35	62 - 97 CE
14	Sphoortinilaypad	203 VE	49	97 -146 CE
--	sequence continuous 15 to 25 (VDKI, p318)	---	---	---
26	Narayanashram	836 VE	37	742 - 779 CE
27	Vaikunthashram	885 VE	49	779 – 828 CE
28	Vikramashram	911 VE	26	828 - 854 CE
--	sequence continuous 29 to 75 (VDKI, p318 - 320)	---	---	--- ---
76	Chandrashekharashra m	2002 VE	20	1925 - 1945 CE
77 **	Ahinavsachidanand-teertha	Present Acharya		1945 CE ---

Note: *change in Era from Yudhishthir to Vikram at serial number 10, **till VDKI published in 1991CE

Calendar Relationship in Sharda Matha Data

An important point to note from this table is the date of the entry at the 10th *acharya* when the Matha switched from Yudhishthir Era (YE) calendar to Vikram Era (VE) calendar. This right away provides the correlation between the two calendars. The fact that the 10th *acharya*, Anandavirbhav, was seated at the Sharda Matha for 52 years starting in 3040 YE and ending in 3040+52 = 3092 YE was recorded in the table as 9 VE. Therefore, the relationship between YE and VE is positively established that 3092 – 9 = 3083 Yudhishthir Era (YE) is 0 Vikram Era (VE) which is same as 57 BCE. Hence the YE began at 3083+57= 3140 BCE. This is what was laid down as the Marker-4 on the Timeline-A in Figure 3.1.

Records at Kanchi-Kamkoti Matha

Table 3.5: The succession order and Presiding Yrs. of Shankaracharyas at Kanchi-Kamkoti Matha, based on the records preserved at Kanchipuram in Tamil Nadu State, is given below. All serial numbers 1 to 68 are continuous. The table is abridged arbitrarily by this author.

S. No.	Kanchi-Matha Name of Acharya	Pres. Yrs	Indian Era (Last Year of Acharya in KE,VE)	Common Era (Last year)
1	Shree Shankaracharya	6	2626 KE	476 BCE
2	Sureshvaracharya	70	2696 KE	406 BCE
--	Continued...3 to 6	---	---	---
7	Anandgyan	69	3047 KE	55 BCE
8	Kaivalyanand	83	3130 KE	28 CE
9	Kripashankar (2)	41	3171 KE	69 CE
--	sequence continuous 10 to 14 (VDKI, p323)	---	---	---
15	Gangadhar (1)	12	3431 KE	329 CE
16	Ujjval Shankar (3)	38	3469 KE , 424 VE	367 CE
17	Sadashiv	8	3477 KE 432VE	375 CE
--	sequence continuous 18 to 35 (VDKI, p323-324)	---	---	---
36	Chitsukhanand	21	3860 KE , 815 VE	758 CE
37*	Vidyadhan	30	3830 KE , 845 VE	788 CE
38**	Abhinav-shankar (5)	52	3942 KE . 897 VE	840 CE
--	sequence continuous 39 to 65 (VDKI, p324-325)	---	---	---
66	Chandrashekher(6)	17	5010 KE, 1965 VE	1908 CE
67	Mahadev (8)	7 days	5010 KE, 1965 VE	1908 CE
68 ***	Chandrashekherendra Sarasvati	---	Present Acharya, 1965 VE --	1908 CE --

Notes: * Indologists theorized Aadi Shankaracharya's advent in 788 CE when Vidyadhan was the care-taker *acharya* at Kanchi Kamkoti Matha.
** The term 'shankar' in Abhinav-shankar's name was perhaps the reason why Western scholars thought that he was Aadi Shankaracharya, *** till VDKI was published in 1991CE

The record keeping in the Mathas were were not carried out in the Common Era (CE) dates as in the last column of Table 3.4. The last column entries in equivalent CE are by Acharya Udayaveer Shastri after he confirmed the correlation between the three calendars. To bring out the correlation between the three calendars, this author has presented only the necessary truncated entries from the unbroken continuous entries in VDKI. The Emperor Yudhishthir in 3140 BCE, Emperor Vikramaditya in 57 BCE and Jesus Christ at zero BCE are all real people of ancient history.

The first entry in Table 3.4 reveals that Aadi Shankaracharya's period was around 491 BCE when he appointed his first disciple Brahmsvaroop (Sureshvar) as the first caretaker Acharya of Sharda Matha.

Now let us look at yet another historical record, independently kept at Kanchi-Kamkoti Matha (labeled 5 on the map in Figure 3.6, starting from Shankaracharya up until the present acharya. The Kanchi Matha records are presented in Table 3.5. These data were recorded in a format different from the records at Sharda Matha (labeled 2 on the map, Figure 3.6). Notice that the entries are in number of years (third column) that an acharya presided over the Peetha of the Matha, followed by the column (fourth column) recording the year of the end of his administration.

Compared to the record at Sharda Matha in Table 3.4 the important feature to note from the fourth column of the Table 3.5 is the use of Kali-Yuga Era (KE) instead of Yudhishthir Era. Also from the 16th entry onwards, both, the KE and VE are entered till the end of the table. Plentiful historical information can therefore be extracted from these data in two different calendars. It becomes obvious from several of these differing features in the recording formats that the record-keeping at different Mathas was independent of each other. They ought to be independent because the straight geographic distance between the two Mathas is over 1,700 km and in those early days the mode of travel was on foot or on horse-back through thick forests of South India, making travel difficult between the Mathas

Kanchi Data Confirms Relationship between Calendars

Notice that in Kanchi Kamkoti Table 3.5 the calendar entries from the 16th *acharya* onwards in the 4th column are in two calendars, Kali-yuga Era (KE) and Vikram Era (VE). Picking any double entry in the fourth column yields a relationship in KE and VE. The calendar years at the 16th entry for the Acharya Ujjval Shankar (3) are 3469 KE and 424 VE. With VE being 57 years more than CE the corresponding CE to 424 VE is 367 CE. Therefore, in terms of CE the onset of Kali-Yuga Era is 3469-367 =3102 BCE. This is shown by the Marker-5 on the Timeline-A in Figure 3.1. The difference between the YE at 3140 BCE and KE at 3102 BCE being 38 years, is consistent with the genealogy of the kings at Delhi and Krishn's passing near the onset of Kali Yuga which is stated in the Mahabharata and the Puranas.

The bracketed numeral against an *acharya*'s name represents the number of repetitions of the word 'shankar' in the succession since the beginning of the Shankaracharya's order. The Table 3.5 for Kanchi has Aadi Shankaracharya as its first presiding *acharya* and Sureshvaracharya succeeding him. This is natural for the fact that Shankaracharya lived there for his last six years and his devoted first disciple Sureshvaracharya took charge of Kanchi, even though he was appointed as the first Acharya at Sharda Matha.

The very first *acharya* in the Kanchi succession order is Shree Shankaracharya; Shree being the adjective for respect and reverence in the same way as in Shree Krishn and Shree Ram. Aadi Shankar presided over the Peeta only for 6 years till 2626 KE. Shankaracharya's administration ended with his death in 2626 KE which is 476 BCE. If the age of Aadi Shankar is rounded off to 33 years and added to this figure one gets 509 BCE for Aadi Shankaracharya's birth year. The very first Acharya in the Table 3.5 is none other than Aadi Shankaracharya. This is consistent with the dates of events in Shankaracharya's life recorded at Sharda Matha, presented in Table 3.6. One other point to note from the Table 3.5 is that the second *acharya* after Aadi Shankaracharya is Aadi Shankar's first disciple Sureshvaracharya whose guru-given name is Brahmswaroop as the first caretaker Acharya of Sharada Matha. Sureshvaracharya's presence at Kanchi-Kamkoti Matha is natural because this most devoted disciple spent time with his guru during the period of the guru's ill-health and his last days. Sureshvaracharya-Brahmswaroop remained in charge of both Mathas with an overlap of about 28 years after which he appointed Chinsukhacharya at Sharda Matha; he himself remained at Kanchi-Kamkoti ashram.

While Sharda Matha completely switched the calendar from the YE to the VE in the first century CE, the Kanchi Matha continued entries in both, KE and VE, till the fifth century CE. The reason for the switch or to have both eras entered is that by that time the Vikram Era calendar was extensively in use in the society. This further suggests that one Matha did not automatically copy data from the other. Rather, the entries were made in different formats independently, at two different places, geographically separated with considerable distance.

The written records at these Mathas are not limited to recording the succession order of presiding Acharyas at the Mathas. They contain much more valuable information that should open the eyes of those historians who have been blindly following the word of the Western researchers. Without going into all the details that Acharya Udayaveer Shastri researched and presented in VDKI, this author has presented here several important dates related to Aadi Shankar's life to take the reader to the beginning of the sixth century BCE when Shankar was born and lived. Based on Sharda Matha data the important dates related to Aadi Shankar are summarized in Table 3.6. This author has added the bracketed explanation of the terms in the first column, the month/day/year in the second column and the closest CE year in the last column.

The dates for the events in Shankaracharya's life are all recorded in Yudhishthir Samvat (YE) where the name of the month, lunar-day with lunar phase of *shukla* (bright phase) or *badi* (dark phase) preced the year in YE. This record is remarkably detailed and complete. To keep the table short this author did not include other dates that are in VDKI that relate to Shankaracharya's meetings with the kings and other scholarly people.

Table 3.6: Based on the records at Sharda Matha in Dwarka (Gujarat State) the dates relate to Aadi Shankaracharya's life.

Events in Shankar's life	Date in Yudhishthir Samvat (YE) (month/day/year)	Closest CE
Janma (birth)	Baishakha, shukla panchami, 2631 YE (2/5/2631 YE)	509 BCE
Upnayan (thread ceremony)	Chaitra Shukla Navmi, 2636 YE. (1/9/2636 YE)	504 BCE
Sanyas (monk ceremony)	Kartik Shukla ekadashi, 2639 YE (8/11/2639 YE)	501BCE
Siksha (education)	till Falgun Shukla Dveetiya, 2640 YE (12/2/2640 YE)	500 BCE
Vedanta-Bhasya, other writings (Badri Ashram)	till Jeshtha Badi Amavashya, 2646YE (3/30/2646 YE)	494 BCE
Dehaavasaan (death)	Kartik Paurnamaasi, 2663 YE (8/15/2663 YE)	477 BCE

A well-known fact which all historians would accept is that Aadi Shankar lived only for 32 years. The record in Table 3.6 shows that he lived exactly for 32 years 6 months and 10 days. Another important feature of this record is that it is in Yudhishthir calendar, a calendar in use at that time. It also suggests that the onset of Yudhishthir calendar was 2,631 years before Aadi Shankar's birth, which in Common Era calendar is 509 BCE. The sum of the two numbers is 3140 BCE, a number at the Marker-4 on Timeline-A (Figure 3.1) that we have been talking about all along.

Table 3.7: Dates of construction and establishment of Shankaracharya Mathas. The serial numbers of the Mathas are shown on India's map in Figure 3.6.

	Matha Name	Place and State	Date Established	Date in CE
1	Jyotir-Matha	Badri- ashram[1] (Uttaranchal)	2641YE to 2645 YE	499 to 495 BCE
2	Sharda Matha	Dwarka[2] (Gujarat)	Kartik badi trayodashi, 2648 YE	October, 492 BCE
3	Shringeri Matha	Shringeri (Karnataka)	Falgun shukla navmi, 2648 YE	February, 492 BCE
4	Govardhan Matha	Puri (Orrissa)	Vaishakh shukla dashmi, 2655 YE	April 485 BCE
5	Kanchi-Kamkoti[3] Matha	Kanchi (Tamilnadu)	2620 KE[4]	482 BCE

Notes: [1]Badri-ashram, a pilgrim place, the possible place of Rishi Baadrayan Vedvyas the author of Vedanta Sutra and Mahabharata. [2]Dwarka is where Krishn established his second capital and ruled till his passing close to 3102 BCE. It is the legendary city that went under the sea and has been rediscovered recently. [3]After establishing the fourth Matha at Puri, Aadi Shankar went to live at a preexisting ashram/temple at Kanchi which also started to function as the fifth Matha.

On one hand there are records from Sharda Matha at Dwarka revealing that Shankar was born in 509 BCE (exact date of birth given in Table 3.6) and on the other hand there are some modern historians hypothesizing his advent in 788 CE, 1,300 years later.

If the Indian history hypothesized by these Western historians is accepted, which is also followed by many Indian historians, one has to follow the timeline similar to the Timeline-B (Figure 3.1) where Aadi Shankar is found at the Marker-6B at 788 CE. There is no provable solid argument to suggest that the Sharda Matha data are less reliable than the hypothesis put forward by these Indologists and modern historians.

The following discussion further strengthens the validity of the Timeline-A and further weakens the premise of the Timeline-B.

The next set of data from Sharda Matha in Table 3.7 gives information on the date of construction of the four Mathas that Aadi Shankar himself supervised. The foot-notes under the table are added for the reader's benefit.

Building a home or an important structure is an auspicious occasion in Indian tradition and culture. The building process has a ceremonial beginning and completion. The beginning act of laying the foundation stone is termed as *shilanyaas* and the first entry by the owner after the completion is termed as *grihapravesh*. Brought up in the Vedic tradition and being a great proponent of it too, Acharya Shankar would not begin and end the Matha construction any differently. Therefore one should not be surprised to find the construction year of each of the four Mathas on record. Note that the year of the beginning and the year of completion of the first Matha at Badri are recorded in YE. For the next three Mathas the date of completion (*grihapravesh*) was recorded in terms of the month and day with lunar-phase in Yudhishthir Samvat. This is no fiction; the life and practices at the time of Aadi Shankar indicate a high level of civilization with meticulous record-keeping and preserving of history. The date of initiation of the Matha at Kanchi-Kamkoti over 1,700 km away from Sharda Matha was independently recorded in the Kali Yuga calendar year. The calendar year there is 2620 KE which corresponds to 482 BCE. This is the ashram where Aadi Shankar spent his last six years.

One simple conclusion from this table is that all these five Mathas were established in the same order as accepted by most scholars and that is corroborated by the information of Shankaracharya's travels. Based on historical data from Kanchi Matha presented in the Table 3.7 it is obvious that Aadi Shankar presided over the Kanchi Matha himself for only 6 years. This takes us to 476 BCE and if the Sharda Matha data are assumed to be correct, the time span between his birth in 509 BCE and 476 BCE is about 33 years. This coincides with the undisputed fact that he lived for a period of 32 years, which is not very different from the fact that he lived for 32 years 6 months and 10 days as shown in Sharda Matha records (Table 3.6).

The Erroneous Shringeri Matha Data

It was mentioned above that the scant information from Shringeri Matha was the basis for the hypothesis that Aadi Shankar was born in the seventh century

CE. Also the two or three different views that place Shankar in the seventh and eighth centuries are contradictory to each other and are both without foundation.

Table 3.8: Succession order of acharyas at Shringeri Matha.

	Acharya Name	From (CE)	To (CE)	Presiding Period (yrs)
1	**Aadi Shankaracharya**	**684**	**716**	**32**
2	**Sureshvaracharya**	**712**	**773**	**60/61**
3*	**Nityabodhghanacharya**	**773**	**848**	**75**
4	**Gyanghanacharya**	**848**	**910**	**62**
5 to 33	**--- sequence continued ---**	**---**	**---**	**---**
34	**ChandrashekherBharti (3)**	**1912**	**1954**	**42**
35	**Abhinavvidyatirth**	**1954**	**--**	**Present *Acharya***

Note: *Modern scholars place Aadi Shankar in this period

The controversial Shringeri Matha data are presented in the Table 3.8 where the caretaker acharyas are listed with their administration periods. The list begins in 684 CE with Aadi Shankar and ends in 1954 CE with the present Acharya (when VDKI was completed). The total count of acharyas at Shringeri is 35 compared to 77 at Sharda Matha, 68 at Kanchi and 144 for Govardhan Matha. For the sake of brevity only the first four and the last two acharyas are shown in this table.

The records of Sharda Matha at Dwarka show the date of the establishment of Shringeri Matha in February 492 BCE (not 684 CE) and there is no date of the establishment within the accuracy of a month available in the Shringeri records. If the date of the first entry in this table is assumed to be the date of the establishment of Shringeri Matha it creates a discrepancy of 1,176 years. Except for the Shringeri Matha, all other *mathas* listed in this text are consistent within them.

A well-known historical fact is that Aadi Shankaracharya's first disciple was Sureshvaracharya. Therefore, it would seem logical at a first glance of the Table 3.8 that Aadi Shankar himself presided over the Shringeri Matha and then handed the charge over to Sureshvaracharya before Aadi Shankar died. It is also an undisputed fact that Aadi Shankar lived only for about 32 years: 32 years 6 months and 10 days to be exact according to the Table 3.6 based on the records at Sharda Matha. The Table 3.8 based on Shringeri dates suggests that Aadi Shankar served at Shringeri from 684 CE to 716 CE which is 32 years. Obviously, there is a problem with the record because Shringeri Matha was established by Aadi Shankar after he established the Mathas at Badrinath and Dwarka and his last Matha was at Puri and final residence was at Kanchi-Kamkoti and not Shringeri. Aadi Shankar was not born at Shringeri nor did he supervise Shringeri all his life from birth. By the time he started supervising the Matha building at Shringeri, he was about 17 years old. Therefore, at the most he could preside over Shringeri Peetha for 15 years during his remaining life and not 32 years. The numbers do not add up for those placing Aadi Shankar in the 8th century CE.

Without going into the details of Acharya Udayaveer Shastri's explanation of this discrepancy it is concluded here that the first two entries in the Table 3.8 are intentional and based on ignorance of Aadi Shankar's life and period. As discussed in VDKI (page 334) this could also be the result of a well-known ongoing rivalry between Shringeri and Kanchi-Kamkoti Mathas. It seems this rivalry continues to exist even today.

Based on VDKI (page 338) it is also well-known that the Shringeri Matha was in a destroyed condition and non-operational for at least a period of 700 to 800 years and remained without caretaker *acharyas* for 500 years. This discontinuity is also not shown in the Shringeri data. The first two erroneous entries at Shringeri perhaps are to lay claim to importance and status by having Aadi Shankaracharya and his first disciple Sureshvaracharya at the top of their succession order.

The profound research into Matha literature presented by Acharya Udayaveer Shastri in his book, 'Vedanta Darshan Ka Itihaas' (VDKI) provides direct benefit to the present author for proving the Reference Hypothesis. The Matha history provides unambiguous connection between four calendars used for recording the succession order of Matha Acharyas; the 4th calendar being the Shalivahan Era (SV) calendar. Since Vikram Era undisputedly began 57 years before CE, it was straightforward to establish a relationship between Yudhishthir Era (YE), Vikram Era (VE) and Common/Christian Era (CE), as shown in Table 3.4 and also to establish a relationship between Kali-yuga Era (KE), VE and CE as shown in Table 3.5. Therefore, the year of Yudhishthir's coronation at the end of Mahabharata War and the onset of Kali Yuga are accurately established to be 3140 BCE and 3102 BCE respectively. This further validates all the time-milestones on the Timeline-A.

The written records discussed in this chapter regarding Manusmriti, the discovery of the extinct River Sarasvati, Megasthenes' account in his book Indika, genealogies of Kashmir Aryan kings in the book Rajtarangini, the Satyarth Prakash account of the Aryan kings at Delhi and the historical records at the Shankaracharya Mathas all now confirm without a shred of doubt that Mahabharata is part of the Aryan history and the YE, KE, VE calendars have been as genuine as the CE calendar. The onset of these calendars respectively in 3140 BCE, 3102 BCE and 57 BCE is also a validation for the Reference Hypothesis that the Vedas are anterior to Mahabharata and therefore the Veda mantras in the Vedas are very ancient. The author is confident that the historical references provided here, followed by the discussion of the information and the logical conclusions drawn, all provide solid support for the Reference Hypothesis that the mantra references used to build the Vedic Cosmology Scenario in this book are older than 5,000 years, unaffected by modern science and philosophy.

Genetic Research under National Genome Projects

This final section of the chapter brings in references from the state-of-the-art comprehensive genetic research carried out on the Indian populace covering the north-south, east-west regions, all dialects, all castes ranging from brahmans to *shudra* including the *aadi-vasis* (tribal-people and forest dwellers), black and fair complexioned people and all those branded by Aryan Hypothesis (AH) as the

Aryans of the north and the Dravidians of the south. The National Genome Projects in collaboration with other international geneticists have discovered that all groups of Indians mentioned above, have within the acceptable scientific margin of error, the same DNA and that their DNA is different from all those who are exterior to India. These studies have also brought out the fact that the Indians with this common DNA have been on the Indian sub-continent well before the Holocene geological epoch exceeding 10,000 - 15,000 years in the past, going into the Pleistocene epoch extending from 12,000 years to 2.5 million years. The Vedic people of this sub-continent have been saying all along that the Vedas and the Vedic culture are extremely ancient. The Aryans of India did not come from elsewhere and the people of India are all Aryans. The Aryan Invasion Theory (AIT), a brain-child of Max Muller et al. has been debunked!

Indian phylogeny cannot be separated from Sanskrit Literature. The Sanskrit Literature is about the people of Aryavart, their culture, religion and philosophy, the people who addressed themselves as the '*arya*' (Aryan). The term *arya*, as explained before has no connotation of race or caste. The specific subject of 'traditional genealogy', a research discipline unto itself in the science arena employing multi-variate parameter analysis for a population is briefly discussed in the following.

When genetics, a branch in biology, is applied to 'traditional genealogy' it is called the 'genetic genealogy' where biological information contained in the nucleus of the cell of a living being is subjected to scientific scrutiny. This means testing genealogy by scientific techniques and methodology. Since the genetic research discussed in the following is important for defining the Indian populace in the context of different tribes, castes, languages and other genetic factors as well as in the context of the Indian phylogeny's connection to people outside India by way of agricultural migration or invasion as proposed in AIT, it was thought essential for the benefit of the reader that essential terms in genetics be defined first and then its application to Indian phylogeny be discussed.

The underlying question to be answered is, when and from where did the original founding population come to the Indian sub-continent, given the fact that the number of subsequent invaders compared to the continuing host population is negligibly smaller. The question relates to those inhabitants who developed Vedic philosophy and created the vast Sanskrit Literature 5,000 years ago and before. This question is answered on a scientific basis through the genetic studies and the answer, as mentioned above, is that these Indians have been on this continent well before the Holocene geological epoch exceeding 10,000 - 15,000 years.

Biological information in the cell of a living being is contained in its genome which is encoded in its DNA (DeoxyriboNucleic Acid). DNA is a specific assemblage of genes, which are discretely-defined units of molecular assemblage, which ultimately are formed out of carbon, oxygen, hydrogen, nitrogen and other elements in small proportions including some other trace elements. Thus the 'genetic genealogy' is part of the molecular biology. Within the mind-boggling complexity of a living cell itself there are certain molecular assemblages in it that are transferred from one generation to the next that help identify the biological connection between the generations. This is commonly called the inheritance of genes from parents. Among other genes that are transferred, it is the biological information contained in the Y-chromosome or Y-DNA that is of great interest in genetic research because it is passed exclusively from father to son establishing the paternal line from generation to generation. Mitochondrial-DNA or mtDNA on the other hand comes from the mother and therefore is used for tracing the maternal line. It is the mtDNA that has been studied to identify an ancient maternal lineage from which modern man evolved.

A most commonly used term in molecular biology is 'haploid'—having only a single set of homologous chromosomes. When the molecular evolution in species is studied, a haplogroup (HG) of similar haplotypes that share a common ancestor with the same DNA sequence variation is analyzed. To differentiate between haplogroups, they are identified by letters of the alphabet and numerals. In this nomenclature the different designations are assigned to Y-DNA and mtDNA.

The genetic research on human populations confined within different countries is relatively recent. Most important studies have been published within a decade or so. A review of these studies is beyond the scope of this book. Among other large number of publications, the reader is urged to read two important research papers on Indian genome studies that have provided extensive and up to date lists of references on the subject and published genetic studies on the Indian population, utilizing state-of-the-art technology and research tools. These studies have been conducted under the national genome programs of India in collaboration with a large number of national and international research institutes and universities. These studies have also included genetic data from the countries and regions outside India that were thought to have had some influence on the Indian genome. These two independent publications are: (superscript represents author's affiliation).

(A) '*A prehistory of Indian Y chromosomes: Evaluating demic diffusion scenario*'', by
S. Sahoo[1], A. Singh[1], G. Himabindu[1], J. Bererjee[1], T. Sitalaximi[1], S. Gaikwad[1], R. Trivedi[1], P. Endicott[2], T. Kivisild[3], M. Metspalu[3], R. Villems[3] and V. K. Kashyap[4]; (published in) *Proceedings of the National Academy of Sciences of the United States of America, Vol. 103, no. 4, 843-848, January 24, 2006.*
[1] National DNA Analysis Centre, Central Forensic Science Laboratory, Kolkata, India; [2] Department of Zoology, University of Oxford, Oxford, UK; [3]Estonia Biocentre, Tartu, Estonia; [4]National Institute of Biologicals, Noida, India.

(B) '*Polarity and Temporality of High-Resolution Y-Chromosome Distributions in India Identify both Indigenous and Exogenous Expansions and Reveal Minor Genetic Influence of Central Asian Pastoralists*', by
S. Sengupta[5], L.A. Zhivotovsky[6], R. King[7], S. Q. Mehdi[8], C. A. Edmonds[7], C-E. T. Chow[7], A. A. Lin[7], M. Mitra[9], S. K. Sil[10], A. Ramesh[11], M.V. Usha Rani[12], C. M. Thakur[13], L. L. Cavalli-Sforza[7], P. P. Majumder[5], and P. A. Underhill[7], (published in) *The American Journal of Human Genetics, Vol. 78, 202-221, February, 2006.*

[5]Human Genetics Unit, Indian Statistical Institute, Kolkata, India; [6]N.I. Vaiilov Institute of General Genetics, Russian Academy of Sciences, Moscow, Russia; [7]Department of

Genetics, Stanford University, Stanford, USA; [8]Biomedical and Genetic Engineering Division, Dr. A.Q. Khan Res. Labs., Islamabad, Pakistan; [9]School of Studies in Anthropology, Pandit Ravishankar Shukla University, Raipur, India; [10]University of Tripura, Tripura, India; [11]Department of Genetics, University of Madras, Chennai, India; [12]Department of Environmental Sciences, Bharathiar University, Coimbtore, India; [13]B. J. Wadia Hospital for Children, Mumbai, India.

It becomes clear from the above list of 27 international participants involving 13 different international institutions of high repute that the outcome of these independent studies is highly reliable and thus is of immense value. These studies have changed the superficial perception about the inhabitants of India; the reality is that they have lived on this continent since the pre-historic times. These papers have also reviewed the preceding genetic studies on Indian populations that a reader may wish to study.

Two studies with opposite conclusions were carried out before the above cited studies (A) and (B). Kivisild et al. (Americal Journal of Human Genetics, 72, 313-332, 2003) proposed that tribes and castes in India share considerable Pleistocene heritage (from 2.588 million years down to 12,000 years) with limited recent gene-flow between them. Opposite of this was the theory by Cordaux et al. (Curr. Biol., 14, 231-235, 2004), who concluded that the castes and tribes of India have independent origins. The two comprehensive studies (A) and (B) listed above have dealt with these two opposing deductions and have successfully tied the loose-ends. Sahoo et al., the authors of article (A) commented- "*Cordaux et al. defined, heuristically, a package of haplogroups* (J2, R1a, R2, and L) *to be associated with the migration of* IE *people*' (people of Indo-European language group) *and the introduction of the caste system to India, again from Central Asia.*" The term *heuristically* of course meant that the genetic sample taken by Cordaux et al. was inadequate and their conclusions were drawn without exhaustive effort.

To study genetic genealogy of a population one collects DNA samples that statistically represent that population. In spite of highly accurate DNA analysis of a sample using state-of-the-art instrumentation in molecular biology one ultimately resorts to statistical analysis of the biological outcome. If the sample is inadequate one comes up with analysis that leads to erroneous conclusions. According to the papers (A) and (B) such was the situation in the study by Cordaux et al. For the studies on genetic research on Indian population with a limited sample that preceded the authors of the two articles (A) and (B) listed above, the authors of the article (B) Sengupta et al, pointed out the limitations and deficiencies of the earlier limited samples thus –

> "*Although such studies have provided interesting insights into the genetic history of paternal lineages in India, there have been two limitations arising from a combination of ethnically ill-defined populations, limited geographic sampling, inadequate molecular resolution and inappropriate statistical methods. Another underappreciated factor in some studies is the innate complexity of caste origins, including the recognition that some castes have tribal origins. The present study* (i.e. the article (2) listed above) *addresses these deficiencies by using a large set of well-defined ethnic groups,*

wider geographic sampling, high molecular resolution to define HGs, comprehensive haplotyping with use of a large set of microsatellite markers..."

In addition to these short quotations drawn into this text the information in the two papers (A) and (B) gives exhaustive details on comprehensive data preparation, methodology, testing with most modern tools with high spectral resolution and on careful analysis. The conclusions based on these large mtDNA and Y-chromosomal studies with large number of haplogroups are dramatically opposite to the long-held view that the arrival of nomadic Central Asian pastoralists had shaped the genetic differentiation of India and had also contributed to the development of language, culture and introduced the caste system on this subcontinent. The studies instead have shown that on statistical basis, there is no difference in the genetic makeup of tribes and castes of India as had been surmised heuristically by Cordaux et al. and in several other earlier studies. The basic deduction of the studies is that, '*the age of accumulated microsatellite variation in the majority of Indian haplogroups exceed 10,000-15,000 years*'; leading to the fact that the Indians have been on the Indian sub-continent indigenously for an extremely long time. In terms of geological epochs, the Indians in India, at least since the 'pre-Holocene and Holocene-era', have shaped the distinctive South Asian Y-chromosome landscape and not the Indo-European expansions, attesting to the antiquity of Indian Civilization and its indigenous development.

The archaeological evidence based on findings in the state of Rajasthan that were related to the River Sarasvati and King Manu's *Vinshan Region*, the British Archaeological Reports, Oxford, UK (1982), radiometric dates for South Asian Archaeology (Univ. Pennsylvania, Philadelphia, USA) and many references in journal publications substantiate the fact that the settlement of India by modern humans using Middle Palaeolithic tools dates back to at least the Late Pleistocene period. As suggested by these and other authors the initial settlement of South Asia, between 40,000 and 70,000 years ago was most likely by the southern route from Africa because haplogroup M, the most frequent mtDNA component in India, is virtually absent in the Near East and South West Asia.

The authors of the article (A) analyzed Y-chromosomes from Indian haplogroups in the context of available data from West Asia, East Asia, Southeast Asia, Central Asia, the Near East and Ethiopia. While haplogroups R1, R2, L, O, H, J2 and C characterize more than 90% of the Y-chromosomal variation in all socio-linguistic groups of India, the total frequency of these haplogroups outside India is marginally low. Also the haplogroups E, I, G, J* and R1* dominantly present in the Near East among the Turks and Central Asia are rare or absent in India and haplogroups C3, D, N and O specific to Central Asia and Southeast Asian populations are virtually absent in India. The haplogroups that are found common between these regions and India are recent and not significant and that too exist particularly in Western India and the Indus region. To quote the conclusion from the paper (A):

"*It is not necessary, based on the current evidence, to look beyond South Asia for the origins of the paternal heritage of the majority of the Indians at the time of the onset of settled agriculture. The perennial concept of people, language and agriculture arriving to India together through the northwest corridor does not hold up to close scrutiny*".

The work on the Genome Projects in India has proven AIT to be a discard. It has proven beyond doubt that Aryans are people indigenous to India since at least 40,000 to 70,000 years. The Indians who insist to be Dravidians are in fact as much Aryans as the rest of the Indians. It is time to shed the myths of the Aryan Hypothesis and the Aryan Invasion Theory.

Dravidian group of Aryans

The Dravidian group of Aryans ought to understand that द्रविण or द्रविड (*dravin* or *dravid*) are the derivatives from the *yaugik* Vedic terms द्रविणोदः or द्रविणोदाः (*dravinodah* or *dravinodaah*), wherein the letter 'd' is pronounced soft as in the word 'the'. Both forms of the term द्रविणोदः (*dravinodah*) mean "*one who donates wealth generously for the benefit of others*". This term appears in several mantras of RigVeda in which four successive mantras are devoted to द्रविणोदाः (*dravinodaah*) *devata* in the *sukta* 37 of *mandal* (chapter) 2 of RigVeda (2/37/1- 4). When one studies the meaning of these four mantras one learns that the term *dravinoda* in these mantras is used as an adjective and relates to the people who are hardworking, prosperous, scholarly and generous at heart. In these mantras there is no suggestion of race or colour. Just as the Vedic term *arya* does not signify race of a person but means '*eeshvar putra*' (progeny of God) the term *dravid*, originated from the RigVeda, refers to the virtue of generosity in a person. The word *dravinod* has no other connotation and these mantras have no suggestion of belonging to the north or south of India or having a light or dark complexion. All Indians are Aryans.

Swami Dayanand Sarasvati, contemporaneous to Max Muller et al., read about their bogus Aryan Hypothesis and the Aryan Invasion Theory and pointed out their bankruptcy of thought. If Europe can achieve modernity in culture, science and technology within a millennia or two it is not difficult to conceive that the Aryan cultural heritage on the Indian sub-continent, flourishing over a minimum of 40,000 years would have developed Sanskrit, compiled the Vedas, formulated *sutras* and contributed to the vast Sanskrit Literature.

With all the evidence presented in this chapter it can be said with certainty that such spectacular literary activity took place in Aryavart's history, anterior to Mahabharata War close to the Time-Milestone-5 dated February 18, 3102 BCE.

CHAPTER 4: A Brief Review of Ancient Vedic Literature - the Source of References on Cosmology

> "*In the discussion of philosophical subjects, pre-conceived notions are the worst enemies to encounter. They not only prejudicially bias the mind, but also take away that truthfulness and honest integrity from the soul, which alone are compatible with the righteous pursuit and discernment of TRUTH. In the treatment of a question such as the estimation of the value of a system of philosophy or religion, extreme sobriety and impartiality of the mind are required. Nor is it to be supposed that a religious or a philosophical system can be at once mastered by a mere acquaintance with grammar and language. It is necessary that the mind should, by an adequate previous discipline, be raised to an exalted mental condition, before the recondite and invisible truths of Man and Nature can be comprehended by man. So is it with Vedic philosophy.*" (Pundit Guru Datta Vidyarthi in "The Terminology of the Vedas and European Scholars", 1888 CE).

These are the words of Pundit Guru Datta Vidyarthi (1864-1890), preserved in an 1888 CE publication in English, that he authored under the title of "The Terminology of the Vedas and European Scholars". Pundit Guru Datta Vidyarthi was a prolific writer in English with mastery of several languages of the world. Guru Datta's English translations of three Upanishads are splendid literary examples. For his linguistic and scholastic abilities, many accolades were bestowed upon him by the scholarly. Upon reading his essays, the English linguist Frederick Pincott (1836CE – 1896CE) wrote to him – "*There must be very few Indians capable of writing such idiomatic scholarly English; and it is great pleasure to me in communication with so thoughtful and studious a gentleman as yourself. My friend informs me that you are young and I am glad to hear that, for with your industry and attainments you may live to do much good to the cause of enlightenment of India...*"

Unfortunately, as said in his obituary in the journal Arya Patrika, "*A man, an uncommon man, a man extraordinary, a true, deep and profound Sanskrit scholar...dazzlingly brilliant...*", Pundit Guru Datta Vidyarthi died at the age of 26. Most of Pundit Guru Datta Vidyarthi's individual works in literature, numbering over two dozens, are compiled in a book titled – "Works of Pt. Guru Datta Vidyarthi", edited by Dr. Ram Prakash, Professor of Chemistry (Punjab University), published in 1998.

The author in this chapter presents a brief review of the Vedic Literature that has served as the resource for the references on Vedic Cosmology. The Vedic Cosmology is a part of the Vedic Philosophy to which Pundit Guru Datta Vidyarthi's statement in the above, "*...pre-conceived notions are the worst enemies...*" applies and to comprehend the Vedic Literature and to benefit from it "*...the mind should ... be raised to an exalted mental condition,...*"

Guru Datta had observed and analyzed the literary environment that persisted over one and a quarter centuries ago. Some of the prominent Western Indologists and Sanskritists of the 19th century, contemporaneous to Guru Datta, approached the Vedic Literature and Vedic Philosophy, wittingly and unwittingly, with pre-conceived notions and prejudice, which they blatantly expressed in their writings. This created a negative impression about the Vedas and the Vedic Literature.

There are several rationales for reviewing the ancient Vedic Literature in this chapter. It will benefit those students who developed high affinity for the Vedas and the Vedic Literature, inside and outside of India, and have been endeavouring to enhance their Vedic knowledge through their university curricula. The review will counter the *pre-conceived notions* of Indologists and Sanskritists that presently serve as the resource for the Vedic studies. In addition to providing a broad list of major literary works in the Vedic Literature this author points out what has enabled the preservation of the Vedic thought and the principles of science and philosophy since the unknown antiquity in the human civilization. This literature is the resource of the mantra and *sutra* references on Vedic Cosmology that are the basis for creating the Vedic scenario for the advent of the universe. The review will also enrich the knowledge base of those brought up in the Hindu tradition who have had reverence for the Vedas and the Vedic Literature but have had limited exposure to the ancient literary treasure.

Because the mantras and *sutras* drawn from the Vedas and the Vedic Literature form the basis for the Vedic principle of creation of the universe developed in Chapter 6 and the Vedic Big Bang Scenario developed in Chapters 7 to 10, it became imperative to address the *pre-conceived notions* and prejudices of the dominant Western Indologists and Sanskritists. It is the Western Indologists and Sanskritists who created false impressions and floated myths devaluing the Vedas and the Vedic Literature. For setting the record straight, this author has discussed briefly, with a touch of physics, the lattice of the Vedic Literature and the fundamentals of the Vedic methodology that created this unique literature with the built-in virtue of self-preservation and invariance, unaffected by non-Vedic influences.

Pundit Guru Datta was greatly inspired by his contemporary Swami Dayanand Sarasvati (1825-1883 CE), referred to several times in the previous chapters pertaining to the Vedic philosophy, the nature of Nature, the Space-entity and other contexts. Swami Dayanand Saraswati, a monk from his teenage years, was a great scholar of Sanskrit, Sanskrit grammar and the Vedas, a social reformer and a prolific writer in Sanskrit and Hindi languages. Based on Sanskrit grammar and the methodology for mantra translation, Swami Dayanand also pointed out with specific examples, the erroneous mantra translations by some prominent Western Indologists and Sanskritists. Nevertheless, the Indologists and Sanskritists continued to publish volumes on India's religion, culture, history and the Sanskrit Literature. The irony of the situation is that some of them had never even set foot on Indian soil. Unfortunately, the ill prepared Western Indologists and Sanskritists ignorantly misunderstood and intentionally misrepresented the rich treasure of knowledge from the ancient Vedic period of India. They left behind scathing commentaries dismissing the value of the incredible contributions of the Vedic *rishis* for furthering the human knowledge. Their negative opinion was widely accepted in the academic circles in the West and in British India.

Guru Datta's above quotation reflects his reaction to the mediocrity in interpretation of the Vedic thought and their biased judgment on Vedic philosophy and culture. In addition to the Upanishad translations in English, Guru Datta unhesitatingly countered the misconception of the Vedas by the Western Sanskritists and Indologists. He expressed this in his superb phraseology in the English language. Some examples of the shortcomings in the translations and interpretations of the Veda mantras by the Western Sanskritists will follow in this chapter.

The term Vedic Literature includes the four Vedas and the post-Veda Sanskrit Literature. It is vast and arranged in large number of categories. Some categories deal with specific Vedic subjects while the other categories deal with numerous other subjects. The creation of this literature by the Vedic *rishis* and sages took spans of centuries and millennia in the far antiquity of the Indian Aryan Civilization. It includes both, the science and philosophy, covering various subjects of intellectual diligence such as physics, mathematics, astronomy, cosmology, anatomy, medicine, music, philosophy, spirituality, language, grammar, logic, law, history, sociology, governance and various other areas.

The discussion in this chapter will include specifying the components of the Vedic Literature and it will also include a brief commentary on **the myths and misunderstandings created by the Western Indologists and Sanskritists**, namely that –

- the Vedas were composed about 1500 BC (F. Max Muller, in "India - What Can It Teach Us", (IV, VII), Oxford, December 1882)
- the 'Veda is decidedly not monotheistic...the Veda had worship of multitudes of gods and goddesses' (F. Max Muller, in "India - What Can It Teach Us", (VI and chapter VII on Vedic Deities).
- the Indians followed the oral tradition of teaching and learning and 'the writing was unknown in India before the third century BC' (Max Muller, in "India- What Can It Teach Us", (Veda and Vedanta).
- the Indians were not good in recording their history
- the Aryans of India are invaders from Asia Minor and the natives of India are Dravidians (Aryan Invasion Theory and Aryan Hypothesis)

All such misconceptions about the Vedas are debumked in this Chapter.

Western Sanskritists and Indologists, without carrying out adequate research in ancient Vedic history of India, irresponsibly said that the Indians were not good in writing their history. With that pretext, the Western Sanskritists took it upon themselves to set the historical dates for important events in the ancient Indian history and the dates for the advent of important books in Vedic

Literature, without providing proof. There have been a good number of these Western scholars, particularly in the late 18th and 19th century when modern writings on Indology began. A few well known from this period are cited in this text. Often referred to in this book is the prominent and dominant Indologist and Sanskritist, Professor Max Muller (1823-1900) of Oxford University in England. With the exception of several German Sanskritists and Indologists, a common factor among the domineering Sanskritists and Indologists from the British Isle, Max Muller included, was their superiority complex. Under that intoxication they had surmised that if no formalized philosophy and culture existed in Europe during half a millennia BCE, nothing much of value could possibly have existed elsewhere either.

In the above quotation, Pundit Guru Datta Vidyarthi expressed that *in the treatment of a question such as the estimation of the value of a system of philosophy or religion, extreme sobriety and impartiality of the mind are required,* which these scholars lacked. Those Europeans who thought contrary to such Western colleagues and instead praised the wisdom contained in the Vedas and the other Vedic texts were critically put down, more so by Max Muller himself than the others. Hence, this author chose Max Muller and his work on Vedic Literature as a contextual reference-sample to set the record straight.

With limited depth in Vedic philosophy and a poor grasp of the vast Vedic Literature in Sanskrit language, the Western scholars built theories on the meaning of Veda mantras, some out of ignorance and others with prejudice-laced ulterior motives. The fact remains that the ancient Vedic Literature is vast and is the product of the most sophisticated advanced Sanskrit grammar with uniquely rich Vedic terminology. To grasp all disciplines covered in the Vedas and the Vedic Literature, a student was generally required to study full-time in a *guru-kula*.

A *guru-kula* is a centre for learning (an ashram) where resident students become part of a guru's family, begin their education under the direct tutelage of the guru and leave the *guru-kula* after completing higher education. A student could return home after specializing in a specific discipline of Vedic education. However, to complete an extended syllabus on higher education, he/she was required to stay in the *guru-kula* for at least two decades away from home and under the direct guidance of a resident learned guru. In contrast with the solid grounding in the Vedas and the Darshans in *guru-kula* was the limited semester-tutelage in a modern university.

Guru Datta Vidyarthi noticed that in addition to having inadequate preparation in Sanskrit grammar the writings of Indologists were thronged with pre-conceived notions and prejudices. He also noticed that they had either ignorance of the preexisting well-defined Vedic methodology prescribed by the *rishis* (sages, philosophers, ascetics, thinkers) for understanding the mantras or they had complete disregard for it. The prescribed methodology for translating, interpreting and understanding a mantra has existed from ancient times, has never been a hidden secret and is readily available for those sincerely looking for it. However, to benefit from this ancient methodology one had to, before attempting to translate even a single mantra, bring oneself to a required level in Vedic learning. This straightforward methodology required thorough knowledge of the books of Brahmanas (first interpretive commentaries on Veda mantras),

Upanishads, Smritis, Darshans (philosophies), Nighantu (specific dictionary of Vedic terminology), Nirukt (Vedic analytical etymology) and Sanskrit *vyakaran* (grammar) which includes the 'science of word formation' and the 'science of sentence structure'. Swami Dayanand Saraswati in his translations of the RigVeda and the YajurVeda in the Hindi language had emphasized frequently that one should not even think of translating a Veda mantra without having pre-knowledge of these subjects. As mentioned before, he highly recommended one to study Nirukt, authored by Rishi Yask and Ashtaddhyaayi Grammar, written by the great grammarian Rishi Paanini. The literary scholars estimate that Rishi Yask lived before 1900 BCE and Rishi Paanini lived before 2707 BCE (ref. "Aryon ka Aadi Desh aur Unki Sabhyata", p.12, by Swami Vidyanand Sarasvati, 1989, "Vedant Darshan ka Itihaas", Ch. 3, by Acharya Udayaveer Shastri).

Max Muller was among several Western Sanskritists and Indologists who translated the RigVeda and numerous other Vedic and Buddhist texts from Sanskrit to English. The West considered him an authority in the Sanskrit language and his word on the Indian culture, language and history was considered final. All this he mustered up right in England without ever setting foot in India. If he had desired he could have personally observed and experienced the traditional Vedic teaching and learning process that had helped the *rishis* and philosophers to flourish in India. Several Vedic learning centers were still prevalent in British India.

Guru Datta Vidyarthi in his writings in English cited the English translations of the Veda mantras from Max Muller's RigVeda translation and pointed out the errors in the word-meanings and the misunderstandings of definitions and concepts of the Vedic terms (see "The Terminology of the Vedas and European Scholars", 1888 CE cited above). Many of these active Western scholars had read Guru Datta's writings, understood the problem, and developed a better understanding of the Vedic Literature. Unfortunately, several authors among the most known and respected Indologists and Sanskritists of that period (19th and 20th century) had already produced and published a large amount of literature that they could not retract without self-defamation and loss of high academic positions at their institutions.

To publish on Indology was an easy route for some in the race to 'publish for immortality' and hold high academic positions in their universities. They observed the prevailing religious practices of the Hindu society that were a derivative of the inferior education system adversely affected by centuries of the Muslim rule. This was the period when following the rituals blindly and worshipping multitudes of gods and goddesses with varied images and forms were the norm. Compounding the problem further were the readily available ritualistically slanted translations and commentaries of Vedic texts by the contemporary Indian pundits such as Raavana, Mahidhara (1589 CE), and Sayana (14th century CE).

After the revival of the Vedas by Swami Dayanand Sarasvati during the last part of the nineteenth century, with his strong message to the Hindus for shunning rituals and blind beliefs, there has been a gradual change among the Hindus of independent India. He travelled extensively covering a large part of India, preaching and pointing out the ills of the Hindu society, the ills of the misinterpretation of the Vedas by Puranik writers, including the deficiency in the translations and commentaries on the Vedas by Indian pundits. In his criticism, he did not spare the Western scholars, who were busy publishing misleading books on the Vedic wisdom. Citing Veda mantras, he emphasized that there is only one God in the Vedas and Vedic philosophy and that the Vedas are the books of knowledge unaffected by time and locale. There are no rituals in the Vedas and there is no social history in them. For the enriched Vedic history, one had to look into the rest of the Vedic Literature. Recognizing the urgency for the Hindu society to 'go back to the Vedas' he established a society in 1875 CE named Arya Samaj. In addition to creating an awareness of the Vedas in the Hindu society, an aim of the Arya Samaj has been to bring back the *guru-kula* education system and establish colleges with emphasis on the Vedas and the Vedic Literature in their curriculum. The Arya Samaj has since produced great scholars, philosophers and writers, including Pundit Guru Datta Vidyarthi.

The translation and interpretation of Vedic references presented in this book are based on the context and meaning of each word, without interpretive bias. The freshness in a Vedic thought is owing to the time-independent universality in Vedic thought and not owing to choosing the relevant English terminology. The reader will hopefully derive increased pleasure from this book and get intellectual benefit when he or she accepts that any principle with latent truth in it remains invariant with time, even when the language and the culture of a society go through changes. *So is it with Vedic philosophy* - to borrow the concluding sentence from Pundit Guru Datta Vidyarthi's quotation. Based on the standard that Guru Datta set, it is imperative for this author, as well as for anyone else, to make a sincere attempt for rising to an *exalted mental condition* to *comprehend* the meaning of the Veda mantras and then account effectively to share the knowledge with the eager and unbiased reader.

Generally, no one has difficulty in accepting that the Vedas and the Vedic Literature are very ancient. The difficulty however arises when a specific date or a period is assigned to the age of the Vedas and the Vedic Literature. With direct research in ancient Indian literature and history, the thinking about the ancientness of the Indian civilization and Vedic Literature is slowly changing for the better. However, the absurd theories of the Indologists are like cleavers, difficult to shake off once they cling. The European authors using mainly the Contemporarian Method, attempted to fix the dates and the succession period of the main Sanskrit texts of antiquity. The religious prejudice blatantly expressed by some Western Indologists did not help the situation. Guru Datta in his book, 'The Terminology of the Vedas and European Scholars' remarked,

> "*Various professors have fruitlessly tried to fix dates of these writings by searching in them, in most cases in vain, for any well-known historical facts.....our modern professors of Sanskrit should* (may) *have forgotten the invaluable maxim that nearer we approximate to the Literature of the period to which the Vedas belong for their interpretation, the greater would be our chances of the interpretation, being more probable and more correct*".

Some Western Indologists genuinely fascinated with the ancient wisdom of India devoted their entire life to understand it, yet they erroneously thought that the literature and the history of the Indian civilization could not be older than a thousand years BCE. Evident from the DNA studies of the Indian population presented towards the end of the Chapter 3 is that the same Indian gene pool has existed on the Indian sub-continent since the Holocene geological epoch exceeding 10,000 - 15,000 years in the past, going into the Pleistocene epoch extending from 12,000 years to 2.5 million years. Therefore the Vedas, Vedic Literature and their creativity ought to be extremely ancient. A glimpse of the vast Vedic Literature is in the following and some important Vedic terms are explained.

Vedic Trayi-vidya: Tri-knowledge

The Aryans have considered that the Vedas have been in existence from the beginning of their civilization for which a time-marker is difficult to assign. Most books of Sanskrit Literature contain, in one form or the other, the references of the Vedas but the references of other books are not in the Vedas. Therefore, it remains indeterminate as to how much development in language and grammar went on before the compilation of the Vedas in their final form. The Vedas are the source of the present Sanskrit grammar and the science of word formation. A remarkable feature of the Vedas, accepted by most scholarly, is that there has not been any change in them since they were first composed. No change of a mantra or a word in the mantra not even a syllable in a mantra has occurred since they were handed down first in some unknown ancient times in the Vedic Civilization. Therefore, the invariance of the final form of the four Vedas is a phenomenon in itself.

The Aryans of India, now addressed as the Hindus, continue to have high reverence for the Vedas despite the fact that the majority of the present-day Hindus are ritualistic and far removed from the core Vedic principles and Vedic philosophy. Nevertheless, they largely accept that the Vedas are the basis for the Hinduism and are the books of divine knowledge.

It is always said that the Vedas are the books of knowledge but what is this knowledge that the Aryans talk so much about? The starting point for the answer to this question is in the RigVeda, the first of the four Vedas. The RigVeda states that the Vedic knowledge is *tisrovachah* (RigVeda Samhita, 1/10/1/1), meaning that the Vedas contain ***trayi-vidya*** (*trai* or *tistro*: tri-knowledge). Shatpath-Brahmana (10/4/2/21-22) metaphorically stated –

"*The Creator saw that all forms of matter (bhootas), all forms of energies (all chhandas, stomas, all praanaas and devas) and the consciousness with virtue of amrit* (the property of non-dying, eternal) *are all included in trayi-vidyay. He desired that all forms of matter forming the universe, which is my body, be within trayi-vidyay.*"

This includes knowledge in three realms of creation - what is manifest, what is recondite and what is spiritual (or metaphysical). All Veda mantras with differing *chhandas* (metered poetry) reveal knowledge in one of these three realms. The proper Vedic terms for the mantras that belong to these three realms are said to be - *pratyaksh-krita*, *paroksh-krita* and *addhyaatmik*. They respectively mean - *one that is observable*, *one that is there but not seen* and *one that relates to spirituality*. In the Vedas and the Vedic philosophy, a number of correlated triplets within these realms of creation are also well recognized. A few examples of these triplets will make this point clear.

The knowledge of the universe that a conscious being desires and acquires comes from three different realms of creation. This knowledge-triplet in three realms of creation is *adhibhautik, adhidaivik* and *adhyaatmik*. Each term is a conjugation of two terms with the common prefix conjugate being *adhi* to each suffix conjugate *bhautik, daivik* and *aatmik* that represent three realms of creation. The closest translation of these full conjugations in the context of knowledge is – *adhibhautik: 'pertaining to the physical universe'*, *adhidaivik: 'pertaining to the visible and invisible forces and systems of nature that help sustain the created universe and affect life of living-beings'* and *adhyaatmik*: *'pertaining to the individual consciousness and the universal consciousness'*. A simplification of the three definitions could be - pertaining to the natural, pertaining to the super-natural and pertaining to the spiritual; covering all knowledge-realms in the created universe. The emphasis in this book is only on the two realms of knowledge – *adhibhautik* (pertaining to the physical universe) and *adhidaivik* (pertaining to the visible and invisible forces and systems of nature that help sustain the created universe). The latter includes the knowledge of Vedic Cosmology with major components of the inanimate creation being the physical forces, energy fields and matter on the one hand and the metaphysical forces operating to sustain the animate creation. Modern cosmology has attempted to address the *adhibhautik* aspect (pertaining to physics) of the creation of the universe. Since there is no getting away from the fact that life and consciousness are intertwined, the phenomena that relate to the conscious-beings, which also relate to the Anthropic Principle in modern cosmology, should be looked at from all three aspects of creation *adhibhautik, adhidaivik* and *adhyaatmik*. A unique perspective of the *adhyaatmik* consideration to the Anthropic Principle in modern cosmology is the subject for another book.

The next triplet associated with the Vedic knowledge is *gyan, karma* and *upaasana* that lead to salvation for eternal bliss - the ultimate aim of a conscious-being soul. These three respectively are to 'acquire knowledge (reducing ignorance)', to 'do right karma while in a physical existence' and to 'develop self-realization to uplift the soul'.

Each Veda mantra relates to knowledge in one, two or all three realms of *adhibhautik, adhidaivik* and *adhyaatmik*. This is the Vedic *trayi-vidya*, the triplet of knowledge. No Veda mantra is meaningless. It is the richness of the Vedic Sanskrit terminology in the inspired mind of a brilliant *rishi* that enables him/her to bring together the right words formulating a mantra in one of the unique Vedic meters. As mentioned earlier, the meaning of a Vedic Sanskrit term reflects the meaning of its root from which the other related words also originate. A single word on the other hand may have meanings that apply in three realms. The context determines what meaning should prevail.

According to the Vedas, three eternal entities coexist: *Paramatman-Prakriti-Aatman*, each component of the triplet having different functions, virtues and properties. They respectively are Creator-Nature-Souls. The inert *Prakriti* (Nature), the material-cause of the universe in corpuscular form at its fundamental level, is composed of triplets of fundamental particles. Serving as the building blocks of the universe (Nature), the three kinds of fundamental particles are *satah-rajah-tamah*; each component of the triplet has a different nature, different function and different properties.

The separation between the components of the Vedic triplet, Creator-Nature-Souls, raises a question - owing to their distinct properties if the three entities of the Vedic triplet are different from each other, where does the well-talked-about Hindu philosophy of non-duality, called the '*Advaita* Philosophy' fit in? As mentioned in Chapter 3 the *Advaita* Philosophy practiced today was introduced in 494 BCE by the 15-year-old extremely brilliant Aadi Shankaracharya. While writing the commentary on Rishi Vedvyas' Vedanta Sutra he introduced the *Advaita* Philosophy according to which everything in the universe is God (Creator), the visible matter of the universe is only an illusion and all conscious-being souls are small quanta of God. As opposed to this theistic pantheism of non-duality is the preexisting Vedic '*Trait* Philosophy' based on the eternal existence of three separate entities; Creator, Nature and Souls. The *Prakriti*-entity (Nature) in the Vedas and in the six Vedic Darshans (philosophies), including the last Sutra known as the Vedanta Sutra, is real, inert and an entity without illusion. It is merely an ingredient necessary for creating the matter and energy forming the universe. It is a separate entity from the Creator and the Souls. In addition, the souls are all separate individual quanta of consciousness, non-convertible into God or matter. The Creator for the Vedas is a conterminous unitary entity with the virtues of formlessness, all-pervading consciousness, non-convertible to matter, non-breakable into souls and so on.

Based on his theistic *Advaita* Philosophy of non-duality Aadi Shankaracharya challenged the philosophers of atheistic Buddhism for public debates and defeated them every time. Crisscrossing India for about ten years, Aadi Shankaracharya preached against the then prevalent Buddhism, practically wiping it out from the Indian society. Both, the *Advaita* Philosophy of Shankaracharya and the *Trait* Philosophy of the Vedic religion continues in the Indian society.

The rishis, ascetics and philosophers in the Vedic discipline pondered over the Veda mantras and created vast post-Veda literature in Sanskrit. They extracted the hidden knowledge in the mantras and presented it in simpler expression for the benefit of most in the Vedic society. This literary work in Sanskrit falls under the general heading of Vedic Literature because, the knowledge base in it is the Vedas and the subjects expounded in it are extensions and interpretations of otherwise concise Veda mantra formulations. The Vedic Literature is grouped under different literary sub-headings.

Without providing the names of all the books in a category, the broad categories of Vedic Literature are listed in the Table 4.1. This includes the works of ancient rishis yet this list is by no means complete. A tremendous amount of Vedic Literature exists in the form of translations and interpretive commentaries on the Vedas by the *rishis, acharyas* (authority on Vedic subject) and pundits, which are also an integral part of the Vedic Literature.

Even though the emphasis of this author in building the story of the Vedic Cosmology has been to refer to the Veda mantras, numerous relevant references from these categories are also drawn. Since some of the categories overlapped, the listed categories are in broad chronological order only.

All these books of the vast Vedic Literature are in the Sanskrit language. Because this author has referred to numerous books from these categories, a brief background of each category is presented below.

Table 4.1: Categories of Vedic Literature. Numbers in brackets represent the number of books in the same category written by different authors.

	Category Name (No. of Books)	Name of Book	Comments
1	Vedas (4)	(1)RigVeda, (2)YajurVeda (3)SaamVeda, (4)AtharvaVeda	All four referred to in this book
2	Brahmanas (4)	(1)Etreya, (2)Shatpath, (3)Saam (Tandya), (4) Gopatha	Other additional Brahmanas not listed here. (1) and (2) referred in this book
3	Upvedas (4)	(1)AyurVeda, (2)DhanurVeda, (3)GandharvaVeda, (4)ArthaVeda	One upveda for each Veda respectively
4	Smritis (28)	Manusmriti + 27 other, too many to name here	Manu-Smriti referenced extensively in this book
5	Ramayan (1)	Ramayan	Biography of Ram by Rishi Valmiki: History, philosophy, referred in this book
6	Upanishads (10) and Aranyakas	Too many to name here	Written by different *rishis,* some references used
7	Vedangas (6)	(1) Shiksha,(2) Kalpa, (3)Vyakaran, (4) Nirukt,(5)Chhanda, (6) Jyotisha	Many sutras on each. The (4) referred extensively in this book
8	Darshans (6)	Details in Table 4.2	Referred frequently in this book
9	Mahabharata (Bhagvadgita)	Book of Mahabharata, Bhagvadgita part of Mahabharata	By Vedvyas and his disciples,history, philosophy, referred in this book
10	[Puranas (18)]	Too many to name here	History, mythology

CHAPTER 4: A Brief Review of Ancient Vedic Literature - the Source of References on Cosmology.

The Vedas

The names of the four Vedas, mentioned in the first chapter, are Rik, Yajuh, Saam and Atharva. Following the conjugation rule in Sanskrit grammar called *sandhi*, these four terms conjugate with the suffix term '*veda*' yielding the names Rigveda, Yajurveda, Saamveda and Atharvaveda. The spelling and the sound of the individual words under the euphonic fusion of words according to the Sanskrit *sanddhi* rule are primarily preserved with minor change in spelling and sound at the fusion-point. This author has capitalized the letter 'v' in the term *veda* throughout this book for their names to read: RigVeda, YajurVeda, SaamVeda and AtharvaVeda.

Although there is strong connectivity between the preceding and succeeding Vedas, each Veda emphasizes a specific realm of knowledge. Thus, the RigVeda is for *gjyaan* where the term *gjyaan* stands for knowledge. YajurVeda is for *karma* where the *karma* stands for action, SaamVeda is for *upaasna,* where *upaasna* means '*that which brings one closer to the Creator*' and the AtharvaVeda is for *vigjaan* (*vi-gjyan = specific-knowledge*) which means science and applied knowledge.

The Meaning of the term "veda"

To know the meaning of the term *veda* itself one has to know about how the word formation takes place in the Sanskrit language. Without going into the details of this ancient science of word formation, a limited background is presented here. A reader may find this interesting and even enlightening. This author however deems it essential to present it here because throughout the framework of this book the mantra translations would often require the analysis of Vedic terms, particularly the analysis of word formation in Vedic science and cosmology.

A word in Sanskrit language originates from its *dhatu*. A *dhatu* is an independent meaningful entity by itself with properties. A *dhatu* generally is a smallest meaningful word from which originate many words which are related in their meanings. A *dhatu* can be as small as a single letter of the alphabet. Alternately expressed, a *dhatu* is a generative for words originating from it and all these words reflect the basic properties of the *dhatu*. All *dhatus* in Sanskrit language are generative of all the words that exist in this language. Thus, all individual *dhatus* together are the 'powerhouse' of word formation. There is no arbitrary or ad hoc creation of a word in Sanskrit.

A literary meaning of the term *dhatu* is also 'root'. Just as the single main root of a tree generates many branches in the tree, all having a dependent relationship with the same root, a single *dhatu* (root of words) generates many words. In Sanskrit, there are 1967 basic *dhatus* to form all words. In addition, the multiple *dhatus* forming single words further enrich the Sanskrit vocabulary. When a single word is formed from more than one *dhatu*, the word reflects the properties of each *dhatu* in it. Therefore, the foundation for word formation in Sanskrit is built on a methodology that is scientific and therefore is invariant and non-degenerative with time. Unlike many languages of the present societies, Sanskrit is more than a mere medium for expressing the meaning of the learnt words. The development of the process of word formation in Sanskrit from a generative *dhatu* (root) so early in the human history is by itself an incredible invention by the ancient Vedic *rishis*. When one examines a word in Sanskrit one goes to its *dhatu* to determine its meaning.

After creating a large number of words, a single *dhatu* in itself continues to be an un-enervating fountain of words. This is why a primary letter (syllable) of the Sanskrit script is termed as *akshara*, meaning '*the one that does not decay*', out of which the *dhatus* are formed. All words formed out of a *dhatu* are all linked to the parent *dhatu* and to each other through branching chains. With this brief background on word formation in Sanskrit let us now bring in the definition of the term *veda*.

The *dhatu* named *'vid gjyane'* in association with the *ghang pratyaya* forms the word '*veda*'. The *'vid gjyane' dhatu* is for creating words that relate to knowledge and the *ghang pratyaya* specifies the meanings such as 'an entity' or 'an action' or 'an enabler'. Therefore, *pratyaya* is that which enables the identity of the item for which it stands. This *dhatu* and this *pratyaya* together yield meanings of the term '*veda'* such as the 'knowledge', the 'substance of knowledge' and the 'instrument of knowledge'. After the Mahabharata War ended in 3140 BCE, the Sanskrit language started to decline. To put the declining Sanskrit language on the right track the great Sanskrit grammarian Rishi Paanini wrote the book of Sanskrit grammar, "Ashtadhyayi Sutra", with full analysis of *dhatu* and word formation. Estimates based on recent research in Vedic Literature and history suggests that Rishi Paanini lived a few centuries after Mahabharata. According to a prominent contemporary scholar of Sanskrit, Swami Vidyanand Sarasvati, the period of Rishi Paanini is estimated to be 2707 BCE. Rishi Paanini in his Ashtadhyayi (meaning – 'Eight-Chaptered Book') has related the *vid dhatu* to *satta, laabh* and *vichaarne,* which respectively are the reality of creation (existence), enlightenment and knowledge. These in conjunction with the above *pratyayas* determine the word *veda* to encompass knowledge of the Creator, knowledge of the creation, knowledge of the created substances. The last one is the enlightenment that one gets by understanding the nature of Nature and the nature of Nature's byproducts. When simply put, the implicit meaning of the term *veda*, derived from the above deeper analysis, is '*revelation of knowledge*'.

When one steps back for a moment and thinks about the effect of the above word-analysis on one's mind, it quickly becomes obvious that the formation of the word *veda* and its meaning creates an abstract picture that is comprehensible by the mind only (no other knowledge senses needed). There is nothing concrete in this abstract picture that could activate some of the five knowledge senses. There was nothing in this exposé for the eyes to see or the other senses to hear, smell, taste or feel. All five knowledge senses in this exercise remained completely passive and inoperative. When the meaning of the term conveys an abstract picture, comprehensible by the mind only, such a meaning is said to be *yaugik*, a term introduced earlier in Chapter 1. When one moves down on the word generative chain of a *dhatu* from the *yaugik* meaning of a word that is closest to its *dhatu* to other words on the same chain with ordinary meaning representing a tangible substance or an object, such a meaning is said to be *rudhi* (ordinary) meaning. Therefore, a *rudhi* meaning (the ordinary meaning) of the word *veda* translates to mean '*the book of knowledge*'. However, its *dhatu* being '*vid gjyane*' it still relates to knowledge. A book is a substantive object that one is able to see, feel, smell and even taste; thus tickling the knowledge senses.

With this quick exposé on how the words in Sanskrit language are formed and what they mean where and when, the reader certainly would appreciate that the Sanskrit language trains all mental faculties of a human being and exploits his/her full knowledge-spectrum. The reader would also understand why there is emphasis in Vedic studies for one to recognize and differentiate the application of the word with respect to a context. The context generally is either *yaugik* with extraordinary meaning closest to its *dhatu* or *rudhi* meaning relating to a concrete object. All words and terms in Veda mantras represent *yaugik* meaning and not the *rudhi* meaning. This fact escaped the Western Indologists and Sanskritists who translated the Vedas with *rudhi* meanings of the Vedic terms and were oblivious to the significance of context when interpreting Veda mantras.

With reference to the term *pratyaya*, mentioned above, the three items, 'an entity' or 'an action' or 'an enabler' that relate to the root of the word *veda,* have strong association with the subjects of the four Vedas. The reader by now may have gathered that there is no room for the obvious superficial meaning for terms in the Vedic Sanskrit. To glean and understand the meaning hidden in a Veda mantra, one has to transform oneself mentally to rise from an ordinary frame of mind to a higher frame - from the ordinary *rudhi* level to the *yaugik*, where a higher level of abstract thinking is necessary.

Yaugik (root-based) versus *Rudhi* (colloquial) Translation of Mantras

In the context of the mantra translation and interpretation of the Vedic terms, the usage of the two important terms introduced here, *yaugik* and *rudhi*, will frequently occur throughout this book. A brief elaboration on these two terms is in order.

In the Table 4.1, the Nirukt authored by Rishi Yask is listed under the Vedangas (item 7). Nirukt Sutra's importance was briefly emphasized in the first chapter as well. Nirukt renders the etymology of the Vedic terms and methodology for the Veda mantra translation and interpretation. While the study of Nirukt is essential for understanding the methodology for mantra interpretation and analytical etymology of Vedic terms, Nighantu, the ancient dictionary of specific Vedic terms is essential for developing the knowledge of Vedic terminology itself. Nirukt is a minimum prerequisite for understanding the hidden meaning in a mantra. The scholarly in Sanskrit Literature estimate that Rishi Yask's book of Nirukt that deals with the analysis of specific terms in specific mantras was written sometime about two millennia BCE.

Rishi Yask in Nirukt's first chapter has effectively emphasized that the terms used in the Vedas are *yaugik* (possessing derived meaning) as opposed to *rudhi* (terms having conventional or concrete meaning). When Pundit Guru Datta Vidyarthi expressed his concern about the consistent usage of *rudhi* meanings of the Vedic terms by the European scholars while translating the Vedas into English and other European languages, he defined these terms aptly in "The Terminology of the Vedas and European Scholars (1888 CE)".

Yaugik: *A term that has a derivative meaning, that is, one that only signifies the meaning of the term's root together with the modifications affected by the affixes. A yaugik word is purely connotative.*

Rudhi: *A term that is the name of a concrete object, where the connotation of the word gives no clue to the object denoted by the word. It means a word of arbitrary significance.*

These two definitions state that a *yaugik* meaning is the fundamental meaning that is closest to the root representing the property of the root and a *rudhi* meaning is the ordinary meaning that conveys the name and definition of a concrete object and is usually in common usage. Since the society functions with the usage of words under both definitions, one has to apply the right meaning in the right context.

Unfortunately, most translations of the Veda mantras by Indian pundits with Puranik emphasis and those produced by the Indologists and Sanskritists of the 18th and 19th century, are based on the *rudhi* meaning of Vedic terms. This has resulted in the translations of the Veda mantras that exhibit little knowledge content in them and are mostly nonsensical. This has also resulted in the misunderstanding about the Vedas, demeaned the high level of intellect of the *rishis* and therefore misled even the intellectuals who had no opportunity to study the Vedas.

A Veda mantra is a compact twine of knowledge, language and grammar. For this compact formulation the great philosopher Rishi Kanaad stated in his Vaisheshik Darshan (Vaisheshik Philosophy, in item 8 of the Table 4.1 and item 2 in the Table 4.2 to follow), that:

बुद्धिपूर्वा वाक्यकृतिर्वेदे । "*Buddhi poorva vaakyakritirvede.*"

(Vaisheshik, 6/1/1)

Sutra Translation:

"*The statements in the Veda are sound intellectual creation*".

(Vaisheshik, 6/1/1)

A relevant part from a statement in Mundak Upanishad (item 6 in Table 4.1) cited below emphasized that the four Vedas are the source of all *aparaa*-knowledge that conscious-beings acquire from creation,

तत्रापरा ऋग्वेदो यजुर्वेदः सामवेदोऽथर्ववेदः .....(Mundak Upanishad 1/5)

"*Tatraaparaa rigvedo yajurvedah saamvedo-atharvavedah....*"

Translation

"*Of the two kinds of knowledge (paraa and aparaa), the four Vedas – RigVeda, YajurVeda, SaamVeda, AtharvaVeda and their six Vedangas are the source of all aparaa-knowledge....*" (Mundak Upanishad 1/5).

The *paraa*-knowledge pertains to the Creator. It is acquired by the yoga practice and not just by the study of the Vedas and the Vedangas. The nature of the Vedic poetry is beautifully expressed in the 32nd mantra in the 8th section of AtharvaVeda's 10th chapter.

CHAPTER 4: A Brief Review of Ancient Vedic Literature - the Source of References on Cosmology.

अन्ति सन्तं नजहात्यन्ति सन्तं नपश्यति ।
देवस्य पश्य काव्यं न ममार न जीर्यति ॥(AtharvaVeda,10/8/32)

Transliteration: "*Anti santam na jahaatyanti santam na pashyati.*
Devasya pashya kaavyam na mamaar na jeeryati."

Mantra Translation:

"*Creator's ubiquity is perpetual and yet He is unobservable. O man! Observe the poetry (Veda) of such a Deva and have revelation of its nature. It neither withers nor does it enervate.*" (AtharvaVeda, 10/8/32).

To experience the invariance of the thought and the meaning in a Vedic mantra one has to have a deep knowledge of Vedic Sanskrit or have at least its *yaugik* translation in the right context, the context that is always prescribed by the *rishi* who had the revelation of the mantra. The methodology of understanding a mantra is as scientific as the methodologies evolved in modern science for understanding the result of the study of a physical phenomenon.

One always needs a language to express a thought. To express a phenomenon or a study one evolves a methodology and then ultimately resorts to a language and terminology relevant to the phenomenon or the study. Depending on the development of relevant terminology in that language, the final deduction could be verbose and descriptive or it could just be a concise definition or a formulation. For example, the terms 'spin' and 'orbit' are used to represent the Earth's 'repeated rotation around its axis' and its 'repeated revolution around the Sun'. To express a thought or a deduction in appropriate terminology in a compact formulation is the ultimate desire of an intellectual. A Veda mantra is such a formulation.

Let us take a simple example of the graphical representation of the growth of a tree with time. It required developing a methodology first which included the accurate measurement of the tree-height at regular time intervals, preparation of a data table with columns of two variables (tree-height and time) and then the plotting of a graph to delineate the growth of the tree. The final step in the methodology is to derive from all data points on the graph a mathematical relationship between the two variables. This mathematical relationship is the mantra for the growth of the tree. Recorded at different times each data point is independent of all other data points on the graph, yet when viewed in a relationship, which is the graph, each data point relates to all the other points on the graph.

In the same way, all the terms that formulate a Veda mantra relate to all the other terms in that mantra. The difference between the two formulations however is that while the formulation and ensuing conclusion on the tree growth will not change significantly if one data point was taken out. In the case of a Vedic mantra, not even a syllable could be taken out, because doing so would disturb its meter and alter its meaning as well. Therefore, a mantra formulation is in the framework of a methodology that in some ways is more scientific than the theoretical science methodology with unreal hypotheses. Sanskrit, the scientific language, takes the methodology in the direction of perfection. Just as arbitrariness in science methodology is a no-no, arbitrariness of word formation, as in some other languages, is absent in Sanskrit. Let us discuss a bit how terminology becomes part of an expression and then revisit the meaning of the term *veda* from a different perspective.

The theme of this book is cosmology, which is science. This book has drawn references from the Vedas where the thought process revealed in the form of mantras should indicate to us whether it is scientific or not. No one disagrees with the facts that the Sanskrit language is ancient, its structure complex, with a scientific method for new word formation, developed in India and is a marvelous invention of high order by the *rishis* in India's far antiquity. In the modern era when scientists create new science terminology, at times it is arbitrary and sacrifices logic or departs from scientific reasoning. Alternatively, words are borrowed from the Greek and Latin lexicons. Consequently the scientific terminology right now is a mixed bag of terms that had originated in Latin, Greek, other European languages, English and new English. The result is that a single scientific term may not necessarily convey the intended meaning. It may become necessary to fall back on the field of verbose sentences to understand the term's true meaning. There is nothing wrong in that, however, to express a full sentence by a few meaningful words certainly shows the advancement from simple to sophistication. Brevity is natural to science because mathematics is one of its tools. Nevertheless, assigning scientific nomenclature to substances or to a phenomena in modern science has to invoke the prevailing language of the society where a root (or roots) to create an appropriate word may or may not be available. Then forming a term out of already existing words is the appropriate second choice. An example of this in modern cosmology is the term 'black hole', coined in 1967 by the famous American physicist John Wheeler, who also coined the term 'space-time'. The term 'black hole' has come very close to what it is intended for, however, not quite completely.

A black hole is a most dense astronomical object, to the extent that nothing, not even light, can escape its extremely high gravitational attraction. Around a black hole is a mathematically defined surface called the 'event horizon' (coined by the physicist Wolfgang Rindler) such that the matter and the radiation (light) going past a black hole are completely pulled in by it without reflecting. It is black because no light or any form of radiation is emitted by it and no light falling on it is reflected from it (it seems there are departures to this conclusion). It is a hole because things fall in a hole and do not return. Hence the name 'black hole'!

The analysis of the term would indicate that the term 'black hole' could just be a misnomer. A hole in a substance, by definition, is the space devoid of that substance. The reverse is the situation in a black hole where, owing to its tremendous gravitational attraction, it carries highly packed matter everywhere in it than in its surroundings. There are no holes in a black hole and if there were any, they will fill first by the matter falling into them. A black hole in fact creates a hole in the space around and outside it, which is not an actual part of it. Therefore, this scientific name of "black hole" for a solid spherical cosmological object with extremely high matter density in it is a misnomer and strictly speaking should have been something else! Not familiar with the terms in cosmology, a person coming across the term 'black hole' has to read at least a paragraph to understand what it is. In the process, he also has to understand what 'space-time' and 'event horizon' are. Such a salmagundi is practically non-existent in Sanskrit nomenclature. I am sure other Sanskrit scholars could come up with a more fitting term from the Sanskrit nomenclature for a black hole, however, this author can think of a quick meaningful term for it. Just call it "*bhootbhakshak*"! This term is a conjugate of two terms, *bhoot* and *bhakshak*, which respectively mean – 'matter devourer'. Matter and radiation being *bhoot* in Sanskrit, the *bhootbakshak* is not a misnomer. To reflect the property of black hole being devoid of light as well, one could even call it, '*tamah bhootbhakshak*! It will become obvious from the forthcoming mantra references that the Vedic Sanskrit is extremely rich in scientific terminology. This richness is without borrowing a single term from other languages and without any other language influencing it and permits amazing brevity of expression.

The simple logic therefore should prevail that if scientific terminology existed in the Vedas it could not have been purposeless – there must be science in the Veda mantras. Yes, there is, plenty of it! It is incomprehensible to think that the *rishis* of the Veda were able to invent a language with a structure that is logical and scientific with rich science terminology but did not have any scientific ideas in them! As stated above, the subject matter of the AtharvaVeda is *vigjyan*; the common meaning of it being 'science', there must be some science in it! This book is a limited effort in extracting that science out of the mantras from the four Vedas and from the Vedic Literature created subsequently in the ancient times of the Vedic civilization. However, to evince further the scientific nature of the Sanskrit language, a few more words about *dhatu* are in order here.

Sanskrit: A Language with the Scientific Structure

Linguists and philologists have presumably written at length in different languages on the logical and complex structure of the Sanskrit language. This author briefly presents in the following some features of Sanskrit from a physicist's perspective. The purpose of this exercise is to delineate the similarity between the organization of the Sanskrit language by the Vedic *rishis* on the one hand and the development of methodologies to organize inventions and discoveries by contemporary scientists on the other; both methodologies aimed at the enhancement and dissemination of knowledge. As mentioned above, the term *dhatu* in the context of language stands for the 'root of a word'; all words generated from a *dhatu* must reflect that *dhatu's* properties. Language being an expression of the abstract, the term *dhatu* stands for a non-subsistent entity, which is the source of words and compound words that are also non-subsistent in nature. When one moves from the arena of Sanskrit language to the Vedic science arena of *rasaayan* (chemistry), the term *dhatu* also stands as a generic term for subsistent elements – hydrogen, oxygen, iron, copper and all the rest of the elements because out of an element are created a large number of chemical compounds. Thus, just as a *dhatu* (element) is the root-cause of all its chemical compounds, a *dhatu* (in language) is the source of words that originate out of it following the rules of Sanskrit grammar.

Just as the meanings of all Sanskrit words originating from a specific *dhatu* reflect the basic meaning of that *dhatu*, the physical and chemical properties of chemical compounds originating from a specific element are owing to that element (*dhatu*). The interesting similarity between Sanskrit and modern science is that there is similarity in fashioning the terminology in both. A basic difference between modern science and Sanskrit is that while an individual element in modern science is recognized by a specific term (name) which differentiates it from the rest, a *dhatu* (element) in Sanskrit may have a number of synonyms as well. The synonyms represent different properties of the same entity. Aryans from ancient times have had terms of many elements (*dhatus*) in physics/chemistry and many other inorganic and organic substances that are mentioned in the Veda mantras. There is even instruction by the Vedas to find these substances by excavating the mountains, hills and the ground in general. In one YajurVeda mantra (18/13) alone there are several *dhatus* (elements of chemistry) listed and also listed are other substances namely – stone, *mrittika* (clay), sand, wood, vegetation, precious elements/substances, blue-sapphire and other crystals, iron, gold, lead, tin and so on. The mantra instructs to excavate these useful substances for mutual benefit.

As an example, the Vedic terminology for some substantive *dhatus* (elements) taken from the AtharvaVeda (11/3/7, 8) and presented here in the table. Their corresponding common names in English are furnished in the second column. These elements are well-known as the '*dhatus* of *rasaayana*' (elements of chemistry). The usage of the same term *dhatu* in both arenas – for a substance in science and for word formation in language, is scientific. An element is the starting point for all chemical compounds produced by that element. Likewise, in Sanskrit a *dhatu* is the starting seed for a word in the language.

A science student in the modern day chemistry class learns that there are altogether 98 naturally occurring elements in Nature, beginning with the lightest first element Hydrogen (H) with atomic number 1 to Californium (Cf) with atomic number 98. The well-known radioactive Uranium (U) has the atomic number 92. About 20 more beyond the 98 elements have been artificially synthesized in the laboratory.

Sanskrit	English
Ayaha, loham	iron
Taamram, lohitam	copper
Trapu	tin
Seesam	lead
rasah, paarad	mercury
Rajatam	silver
hiranyam, haritam, suvarn	gold

To recognize one from the other, each element is assigned a symbol formed out of one or two first letters in its common name with the first letter capitalized. This symbol is the formula for that element, as C is for Carbon and Cu is for Copper. All elements are arranged in a tabular form called the 'periodic table'. Russian chemistry professor Dmitri Ivanovich Mendeleev is credited for organizing the elements in a periodic table first in 1869 CE. He placed all the elements in the ascending order of their atomic numbers, leaving gaps for atomic numbers for elements that had not been discovered by then. By observing the periodic trend and similarity in their properties, he placed them in rows and columns. It is interesting to note that a German chemist Julius Lothar Meyer also independently published his periodic table in 1870 CE and he also arranged elements in the same manner as Mendeleev did. Both, Mendeleev and Meyer coming up independently about the same time in two different locales with the same optimal arrangement of elements, which still stands today, is a reality. Two minds arriving at a single finality may be rare but is also a reality in Nature.

The minds of *rishis* eons ago arriving at a finality about discovering the *dhatus* (elements) of Sanskrit language is not a coincidence either, it is also a reality. Just as all the 98 elements of the periodic table form all chemical compounds in the universe and create the universe itself, in Sanskrit language the 1967 basic *dhatus* in Sanskrit language, extracted from the Vedas, create all primary words in the Sanskrit language. The development of Sanskrit with a *dhatu*-system and the development of methodology in modern science with an arrangement of elements in a periodic table are two scientific phenomena with the commonality in the method that took place independently, eons apart. The Sanskrit term *dhatu* having the same meaning as the term 'element' is not a coincidence or a chance factor. It had to happen because that was the highest level of development that the two scientific systems were to reach. There is a striking parallelism between the structures of both, the modern science methodology and the methodology in the Sanskrit language. A simple example in the following should demonstrate this commonality between modern science and Sanskrit.

Two *dhatus* (elements), C (Carbon) and H (Hydrogen) are chosen here to form several organic compounds. The following list includes just a few from a large number of organic compounds formed by Carbon and Hydrogen atoms with one or two C atoms and two and more H atoms in their formula. The chemical name of the compound is in the bracket.

CH_4 (Methane), C_2H_2 (Ethyne), C_2H_4 (Ethene or Ethylene), C_2H_6 (Ethane), …

Notice that the names of these compounds have common sounds with only minor variations. In the same manner if one chooses a *dhatu* from the Sanskrit language, one could represent all the words originating from that *dhatu* in a manner quite similar to the above example from chemistry. Our basic subject being 'knowledge' let us take the example of a root (*dhatu*) that generates a large number of words that relate to 'knowledge'. The root reflects its basic meaning and exhibits its presence in all the words that it generates; something similar to C in all organic compounds listed above. This particular entity in our example is the single syllable alphabet of Sanskrit script, ज्ञ, pronounced as *gjya* (each English letter g, j, y pronounced together). It generates a large number of Sanskrit words; only a limited number of short words generated from ज्ञ (*gjya*) are listed in the following. The English meaning of the word is also given.

ज्ञ (*gjya*) = knowing, ज्ञः (*gjyah*) = an intellectual person, ज्ञा (*gjyaa)* = to learn, to know, to perceive, .., ज्ञप्ति (*gj*yapti) = understanding, .., ज्ञान (*gjyaan)* =

knowledge, ज्ञात *(gjyaat)* = known, comprehended, understood,.., ज्ञातृ *(gjyaatri)*= a known person, an acquaintance, ज्ञात्रा *(gjyaatraa)* = one who knows, ज्ञाता *(gjyaataa)= one who has full knowledge of,..,* प्रज्ञा *(pragyja)* = intellect, सर्वज्ञ *(sarvagjya)* = omniscient, all-knowing ….,

The list of words, originating from ज्ञ *(gjya)*, goes on. This abridged list of words from one single letter ज्ञ is astounding!

From the example of the letter ज्ञ *(gjya)* creating meaningful Sanskrit nomenclature with a large number of words, one observes that there does not appear to be room in the methodology for much further improvement. We are led to accept the fact that the thought process of the Vedic *rishis* was scientific, final and remarkable. Their thought discipline for organizing their discovered knowledge for the maximum benefit to the society through developing the language in prose and poetry was excellent. It is amazing that all this took place in the distant past of the Vedic Civilization.

To keep words compact is another attribute of Sanskrit. Notice that in the above example a single letter ज्ञ *(gjya)* needed four English alphabets to write it: g, j, y, a. In addition, as listed above, the abridged list of words originating from ज्ञ shows that up until about a dozen words from it the words did not exceed two Sanskrit alphabets. Even when a word originating from ज्ञ becomes long, it continues to show the presence of ज्ञ in it and continues to project its basic meaning related to 'knowledge' in all words formed by ज्ञ. As soon as a word is pronounced, the sequential phonetics of the utilized alphabets in the word tells the meaning of that word. A total of 63 alphabets in Sanskrit script with their specific sounds create an immense field of expression. Most of the alphabets in Sanskrit are single syllable letters. Those interested in a glimpse into the development of these Sanskrit alphabets, that took place in the distant past, should look into the 'Varnochharan Shiksha' written by Swami Dayanand Sarasvati, cited a number of times. The meaning of the title of his book "Varnochharan Shiksha" is - "Education of Letter Pronunciation". Thus, the antiquity should not be seen as the cause for devaluation of the wisdom of the *rishis* who invented Sanskrit and wrote mantras in it in compact formulations with metered poetry, the poetry that is unique in itself. The beauty of the 'knowledge' is that it is ever youthful.

The word represents the name of an object, a phenomenon or a process. Therefore, the name must reveal the properties of whatever it stands for. Just as a *dhatu* generates numerous words, an object owing to its numerous properties would have numerous Sanskrit names as well, each name originating from the same or a different *dhatu*. The Sun and the Earth are good examples for this. This author has collected about 20 synonyms of the Sun and 33 for the Earth, tabulated in Chapter 11 in Volume 2.

Although a mantra will express exactly what it intends to convey, the right meaning of the mantra can be extracted only when the right context of the mantra is kept in mind. Additionally, the right linguistic tools are to be used to extract the right meaning from a mantra. Any adamancy on the part of a scholar to translate a mantra without the full grasp of the Sanskrit methodology would not be short of literary subversion. Professor Max Muller in this author's opinion is responsible for the greater part of this literary subversion. With time, this subversion led to the disrespect for Vedic thought. This disrespect started primarily in the 19th century when Sanskritists and Indologists insolently took it upon themselves to teach Indians how to write their own history and how to interpret the Vedic scriptures. Swami Dayanand Sarasvati objected to this. He strongly emphasized that one should go closer in time to the Vedas, as far as possible, to look for the meaning of even a single word in a mantra and study its usage in other mantras to understand the context and the application of its meaning. Then one could look for its interpretation in the Brahmanas, Smritis, Upanishads and Yask Rishi's Nirukt and Nighantu rather than looking into the recent interpretations of Mahidhara (1589 CE) and Sayana (14th CE).

It seems that Max Muller et al. did not come across the *nyaaya* (Vedic Logic System) phraseology of *'sthool rundhati-nyaaya'*, which means - "*As an object is approached closer and closer its true knowledge becomes closer to certitude*".

First thing that comes to mind on reading the principle of *'sthool rundhati-nyaaya'* is that the principle may apply only to *sthool* objects where the term *sthool* stands for an observable solid substance having form with dimensions. However, the principle is general and is equally applicable to the *sthool* as well as to the abstract; the latter could be a formless thought contained in a literary work. To understand better what a *rishi* in a Veda mantra might have meant and intended one should go as close as possible to the time of the formulation of the mantra. The best way to achieve this with a high degree of success is to look into the interpretation and commentary of the mantra in the ancient Vedic texts, as old as possible. The Brahmana books and Nirukt are these old resources.

Pundit Guru Datta Vidyarthi in his writings in the late 19th century had expressed it as the maxim for the then modern professors of Sanskrit and the Indologists. He put it in English write-up to alert the modern professors and Indologists of his time:

> *"The Nirukt even lays down explicit rules on the terminology of the Vedas which are as yet, quite unheeded by the modern professors. Niruktkara* (creator of Nirukt, Rishi Yask), *in the very beginning of his book, forcibly inculcates that the terms used in the Vedas are yaugik (possessing derived meaning) as contrasted with rudhi (terms having conventional, arbitrary or concrete meaning)".* (Guru Datta Vidyarthi in "The Terminology of the Vedas and European Scholars", 1888 CE)

The Puranik scholars like Sayanacharya, Mahidhara and Ravana carried out translations with *rudhi* meanings of Vedic terms and therefore portrayed the Vedas as having human history, mythology and worship of multiple gods and goddesses, while the Vedas have none of these but have only pure knowledge. Most Western Indologists and Sanskritists such as H. H. Wilson (1786-1860), Max Muller (1823-1900), Monier-Williams (1819-1899), Macdonell (1854–1930) and several more also produced in English the *rudhi* translations of the Veda mantras. The scholars who carried out research in the Vedic Literature generally follow these authors and produced dissertations with erroneous conclusions.

Yaugik to Rudhi Sequence

Being a physicist and a beneficiary of the valuable Vedic Literature, this author intends to give the reader an exposé of the dimensions that exist in the Sanskrit language. This is being done in a physicist's perspective rather than in a linguist's perspective. An understanding of these dimensions will help reveal the true meaning of the terms in the Veda mantras, specially the terms of physics and cosmology that have frequented the mantras and the *sutras*. This author has taken utmost care to follow the Vedic methodology to formulate the Vedic Cosmology Principle in Chapter 6 and to develop the Vedic Cosmology Scenario in Chapters 7 to 10. To exemplify that a single object may have a large number of synonyms and a single term (word) can yield a number of meanings, the author has chosen commonly used Vedic terms representing different meanings in different contexts, ranging from the *yaugik* meaning signifying its root, to the *rudhi* meaning signifying the name of a concrete object. This author has chosen two *yaugik* Vedic terms that exhibit how their status leads to their *rudhi* residuum. These terms are *sarasvati* and *Ayodhya.* These terms are of great significance pertaining to the history of the Vedic Civilization and the geography of the nation of Aryavart (India's ancient name) discussed in Chapter 3. Let us first take up the term *sarasvati*.

Analysis of the term '*sarasvati*'

The term *sarasvati* has already appeared in Chapters 1 and 3 and in this text a number of times in Swami Dayanand Sarasvati's name. This term in a name represents an honor given to a monk with high intellect who has mastery over the language and grammar and who devotes his time in learning, teaching and the dissemination of Vedic knowledge. Therefore, an immediate impression of the term *sarasvati* is that it relates to knowledge. The knowledge is something that is real in existence and abstract in nature. The term *sarasvati* has appeared numerous times in the Vedas wherein all terms are *yaugik*, representing the root meaning of the term. The term *sarasvati* in *rudhi* usage appears in the post-Veda ancient Sanskrit Literature for the name of a mighty river. However, Sarasvati being a river, a physical entity, the meaning of the term '*sarasvati*' here is *rudhi*. All Vedic terms being *yaugik,* the term *sarasvati* in the Vedas must mean something else.

The great Sanskrit grammarian Swami Dayanand Sarasvati has analyzed the term *sarasvati* and emphasized that the term *sarasvati* in the Vedas carries the *yaugik* meaning.

Sarasvati in the Puranik (mythological) sense is the goddess of learning, creativity and wisdom. Pursuit of knowledge being the theme here the term '*sarasvati*' still conveys the *yaugik* meaning. However, when the Puranik mythology depicts the image of the goddess Sarasvati as a female with four arms enthroned on a white swan, the meaning of the term *sarasvati* is *rudhi* and is exterior to the Vedas. Whether the *yaugik* meaning applied to the formless God in the Veda or the *rudhi* meaning applied to the Goddess with form in Puranik mythology one thing is certain that the word '*sarasvati*' relates to the knowledge and spirituality.

The *yaugik* Vedic term *sarasvati*, according to the interpretation given in the ancient book of Nirukt, is for God. With regard to the word-formation from the root (*dhatu*), the term *sarasvati* is derived from the root *sri gatau* (the dynamic action) to yield the term *saras* and suffixes *vat* and *i* added to it (*saras-vat-i*). While defining the virtue-based 100 names of God, Swami Dayanand Sarasvati in the first chapter of his voluminous book Satyarth Prakash, has included *sarasvati* as one of God's names. He defines it thus:

सरो विविधं ज्ञानं विद्यते यस्यां चित्तौ सा सरस्वती ।

"*Saro vividham gjyaanam vidyate yasyaam chittau sa sarasvati.*"

This translates into :

"*God's name is sarasvati because God possesses all diverse knowledge with details of words, meanings, word-associations and usage that form speech in humans*".

In the Vedic thought, all knowledge emanates from God, it is received by the minds of living-beings and transformed into speech for dissemination.

When one moves from its metaphysical meaning to its physical meaning, the term *sarasvati* stands for *vani* or speech or the vibrations that convey knowledge. In Nirukt (2/7/1, Nighantu, 11/22), the term *sarasvati* is listed as one among the 57 synonyms of *vani* (speech). Therefore, *sarasvati* in *yaugik* meaning is primarily for 'word' or for 'speech' through which knowledge is transmitted. The term *sarasvati* appears early in the RigVeda in three mantras of the third *sukta* of the first *mandala* (RigVeda, 1/3/10-12). The 12th mantra from that set is presented here:

महो अर्णः सरस्वती प्रचेतयति केतुना ।

धियो विश्वा वि राजति ॥ (RigVeda, 1/3/12)

Transliteration: "*Maho arnah sarasvati prachetayati ketuna.*
Dhiyo vishvaa vi raajati."

Based on Nirukt's etymological definition of the term *sarasvati* meaning 'speech', Swami Dayanand Sarasvati has provided interpretive commentary on this mantra and its translation in the Hindi language. This author's English translation of the same is given below. Before he furnished the mantra translation, the added comment made by Swami Dayanand is that this mantra has *vachkopmey-luptopma-alankara*, which means that the poetry of this mantra has a specific literary adornment in a decorative simile. The English translation of his Hindi translation of the mantra is:

> "*Just as the ocean's greatness is enhanced by the wind creating waves on its surface and by the Sun illuminating its vastness, revealing its precious contents and its usefulness, the Space (carrier of vibrations) and the Ved-vaani (voice,vibrations, knowledge) together illuminate the ocean of Vedic vocabulary. This is then transmitted through the voice (sarasvati) of the scholarly, enabling the elevation of the intellect in ordinary people.*" (RigVeda, 1/3/12)

Swami Dayanand, while translating the three RigVeda mantras (1/3/10-12) in his RigVeda translation, commented on the translations of the same mantras by the 14th century pundit, Acharya Sayana and the Indologist Professor H. H. Wilson (1786-1860), a contemporary of Swami Dayanand. Swami Dayanand with his detailed translation pointed out that there is no Vedic basis in these mantras for Acharya Sayana and Professor Wilson to imagine a female figure of goddess Sarasvati having a mortal body with four arms and riding on a swan or of a physical river Sarasvati. This nineteenth century monk Swami Dayanand a great scholar of Sanskrit and the Vedas, was straightforward in his conclusion - "*those who are inspired to translate or interpret a book of knowledge without knowledge are no different from those suffering from visual impairment*". (see RigVeda Bhasya-interpretation of the mantra (1/3/12) by Swami Dayanand Sarasvati).

The *sukta* (hymn) 61 of Chapter 6 of RigVeda (6/61/1-14) has 14 mantras in it whose *devata* (subject) is *sarasvati*. All these mantras relate to the human intellect and the human voice. The human voice, among 13 *karans* (vital functions of the mind) of the Vedic philosophy, is one of the instruments of the human mind that the mind uses to convey a thought by creating sound vibrations. This *sukta* may have been the basis for the ritualistic worship of the goddess *Sarasvati* revered as being able to bless her worshipper with a beautiful voice for singing. The *yaugik* meaning of the term *sarasvati* thus applies primarily to God, knowledge and the voice through which God-given knowledge disseminates.

The *yaugik* meaning of the word *sarasvati* also appears in the RigVeda mantra (9/5/8) where it is not related to a river or to a goddess with four arms. There are three *yaugik* terms used for 'knowledge' in this mantra. Because the knowledge is the 'cause' and the speech its 'effect', the three Vedic terms for knowledge are naturally related to each other. These terms are - *bharati, sarasvati* and *ila,* which stand for three distinct qualities of the human intellect-entity, the Sanskrit term for which is *Buddhi* (see Nirukt also). As discussed earlier, a single Vedic term may stand for entities with similar virtues, in the spiritual realm the word *sarasvati* stands for the Omniscient universal spirit and it stands for the *buddhi*-element (intellect) in the physical realm. At a *rudhi* level, it stands for the River Sarasvati, a goddess with image or just any female with that given name.

The term *bharati* represents '*bibharttiti bhartastsyeyam bharati*' – '*the intellect for developing knowledge in the subject of Bhramn (God)*', the term *sarasvati* represents – '*intellect for developing diverse knowledge*' and the term *ila* is for – '*applied intellect that is revered and praised by all*'. This mantra creates a visualization in the human mind that the three 'rivers of intellect', originating from the Brahmn (the Creator) flow on the plane of the *Buddhi*-

element (Intellect-element) of a living-being. Metaphorically, these three types of intellect are tributaries of the single river called 'knowledge', which in the Vedic thought is considered to be emanating from the creator Brahmn.

Two *suktas* of RigVeda (7/95/1-6 and 7/96/1-6), with six mantras in each, are important where according to Swami Dayanand's interpretation based on Nirukt and Nighantu, the *sarasvati* of the mantras is not a physical River Sarasvati. These mantras have 'decorative simile' to describe the meaning of the word *sarasvati* and when *yaugik* and *yoga-rudhi* meanings of the word, as emphasized by Nirukt are considered, its meaning is - सरो ज्ञानं विद्यते यस्या असौ सरस्वती । (*saro gjyanam vidyate yasya asau sarasvati*). This translates into - '*The one who possesses knowledge is called sarasvati*', as in the name of the great scholars like Swami Dayanand Sarasvati.

The answer to the question – how did the extinct mighty river get its name Sarasvati – lies in the *sukta* 75 of RigVeda's chapter 10 wherein the Veda teaches people how to name their rivers (RigVeda, 10/75/1-9). The *sukta* has provided 37 river-names with specific characteristics found in rivers. The ancient *rishis* learned from the Veda that if the river has the characteristic of सारांसि संति यस्या सा सरस्वती (*saaraansi santi yasya sa sarasvati*) which means '*the one which is formed by many tributaries*' and represents the *sri gatau dhatu* for tremendous motion (flow) call that river Sarasvati. Thus, whereas the *yaugik* meaning of the word *sarasvati* relates to knowledge, which purifies the inner conscience of a being, the *rudhi* meaning of the word *sarasvati* is for a dynamic river (Sarasvati) which enables the purification of the physical exterior of a being. The scriptures have emphasized that much Vedic knowledge has been developed on the shores of the mighty River Sarasvati.

In the perspective of the spiritual realm, the knowledge flows from the Omniscient Brahmn, reaches the sea of living-beings who are enriched by the tributaries of constantly flowing knowledge from all directions of the creation, thus enhancing the magnificence of the river of knowledge. The benefits of the physical river Sarasvati also have been enormous for the Aryans of the antiquity, making it a part of the river-system, which by definition would be a *devata*. A *devata* is the one who possesses brilliant properties, be it a conscious being or an inert entity of nature (a river), selflessly providing benefit for others in the *yajna* (selfless act) of sustaining life in the creation. Any other interpretation of *sarasvati* would be due to the confusion between a simile and a reality.

When viewed in a localized perspective the flow of knowledge into a conscious being is from the five realms of knowledge-senses culminating into a single Intellect-entity of the person. This is like the five tributaries forming a stupendous river. Therefore, the *rishis* of the ancient times found an immense river on the Indian continent with five tributaries and named it Sarasvati. The naming of the river Sarasvati (*rudhi* usage) is thus an after-effect of the mantras enabling appropriate vocabulary. The *yaugik* usage of the term *sarasvati* in the mantras however has much deeper meaning.

The aim of this discussion was to point out the methodology for extracting the meaning hidden in the mantra by using the *yaugik* meaning of the terms forming the mantra. The mantra always has a deeper meaning and purpose, to enlighten a person, rather than communicating some ordinary matters. Whereas the *yaugik* meaning of the terms is necessary in a mantra, one finds the application of *yauga-rudhi* and *rudhi* words for the names of substances. Just as the water in a river flows from a higher level towards the lower level, the meaning of the words flows from the high *yaugik* level to *yauga-rudhi* to *rudhi* level. Just as the reversing of the flow of a river can bring disaster, bringing the lower level *rudhi* meanings of words used in ordinary life into Veda mantra interpretation, would bring undesirable results!

Analysis of the term Ayodhya

Now let us take up the *yaugik* meaning of the term '***ayodhya***' that appears in AtharvaVeda (10/2/31) and the *rudhi* meaning associated with a concrete object/entity in daily life.

It has been customary in most civilizations to choose a name from the vocabulary already present in the scriptures that they follow. In the distant past of the Vedic Civilization, the great King Manu of Aryavart (ancient name of India) established a city for his kingdom and named it Ayodhya, the meaning of the term *ayodhya* being '*non-conquerable*', certainly a decorous name for a capital. Manu's city of Ayodhya continues to exist even today as one of the oldest cities, in the Uttar Pradesh state of India. The term *ayodhya* in the AtharvaVeda mantra (10/2/31) relates to the living human body and not a city. The meaning of the mantra is – "*...this human body is unconquerable (ayodhya), a perfect dwelling for the living functions and the place of the soul illuminated by Brahmn* (Creator)*"*. The reference to *ayodhya* in this mantra is for the remarkable human body well protected against outside attacks from diseases. Based on the term *ayodhya* in this mantra some scholars concluded that the AtharvaVeda was written after King Manu established the city of Ayodhya. Since Manu's Manusmriti is based on the Vedic knowledge and is suffused with the references of the Veda mantras, it would be right to say that Manu chose a meaningful pre-existing term from a mantra to name his city Ayodhya. Thus, the usage of the term *ayodhya* in the Veda mantra is *yaugik* and its usage in the name of a city, a concrete entity, is *rudhi*.

The ancient *rishis* have endorsed the fact that the Vedas are the source of all terminology for the human speech that is to be both, learnt and applied. The 'word' is the source of knowledge and it sustains knowledge. In the Vedic tradition, throughout the Vedic Literature down to Nirukt and Nighantu, it has been emphasized that the purpose of the Vedas providing rich terminology is to help the humans to develop knowledge and to apply that knowledge for successful functioning in their lives and society.

The Veda has provided rich terminology in a large number of areas of human endeavour. The terminology relates to the external world, the human body and the healthy functioning of its external as well as internal organs. As an example, one mantra of the AtharvaVeda (2/33/3) names several important internal body organs and functions: *hridaya* (heart), *klomn* (lungs), *haleekshn* (pancreas), *pleennh* (spleen), *yakn or yakrit* (liver). Knowing that the two human lungs are dissimilar in size (left lung being smaller) the *rishis* did not use the general term *klomn* for both. The Veda and the other *Ayurvedic* treatises address them with the separate names, the left addressed as *haleekshn* and the right as *Klom*.

CHAPTER 4: A Brief Review of Ancient Vedic Literature - the Source of References on Cosmology.

The term *hridaya* (*hri* +*da* + *ya*) for 'heart' is analyzed elsewhere in this book showing that the term *hridaya* is formed out of three different *dhatus* that create three actions that manifest heart's complete function. The first syllable ***hri*** is for ***harati*** to take, the syllable ***da*** is for ***dadati*** to give and ***ya*** for ***yati*** to keep going. It is the blood that it takes in, gives out and keeps pumping. Based on the Veda mantras, the comprehensive Vedic science of medicine has been developed and preserved in the treatises under the subject of *Ayurveda* (Science of Aging) which in the Vedic Literature is included under the Upveda (item 3 in Table 4.1). Providing details on *Ayurveda* science is beyond the scope of this book. For more insight in this area, the reader is advised to look into the AtharvaVeda. The Veda mantras furnish a large number of terms for the organs and functions of the human body. The Veda also mentions over a hundred medicinal plants for specific cures.

Examples of the enriched terminology in the Vedas for the benefit of the society, particularly in the area of physical sciences, were mentioned in the Chapter 1 citing the AtharvaVeda's (19/7/1-5) *Nakshatra Sukta* giving 28 meaningful terms to name the constellations in a sequence that is still in use in the practical astronomy of today.

The rivers have played an important role in the society. The rivers and river-systems flow selflessly for the benefit of the society and help sustain societies. Therefore, the Vedas have put the rivers under the designation of a *devata*, namely the *Naddyah Devata.* The fact that the River Ganga is Ganga and the River Yamuna is Yamuna from their respective sources at Gangotri and Yamunotri in the high mountain regions of the Himalayas all the way until the end goes to show that the Aryans had a very good final consensus in the way they named their rivers. The basis for a consensus was already provided in the RigVeda mantras (10/75/1-9) where the naming of the rivers according to their features and characteristics was advised and to name a river the Aryans just had to look that river template up. Since the Vedas were acceptable to all Aryans as the divine inspiration, following the wise advice of the *rishis* was an easy choice to build consensus in naming the rivers.

If the invading Aryans came to India around 1500 BCE, as advocated by Max Muller et al., could they have reached a consensus to name all rivers of the Indian sub-continent and enter 37 river names in RigVeda (10/75/1-9) under the subject (hymn) of *Naddyah Devata*? The *yaugik* terms '*sarasvati*' and '*ayodhya*' in fact pre-existed in the scriptures to represent different contexts from where they were adopted in *rudhi* usage to name two different concrete objects, a river and a city respectively. In the RigVeda's template of 37 river names the river Sindhu means a river '*with high velocity*', the Gange (Ganga) means a river '*going fast*', the Yamuna means a river '*moving in association as well as separately*', Sarasvati means '*with abundance of water and dynamics*' and Drishdvati means '*a river with abundance of stones*'.

The absence of name duplication of rivers and the absence of multiple names of the same river in different regions of the Indian subcontinent indicates that the Vedic methodology of naming the rivers was successful and must have been the result of a concerted effort using the river-name-template provided by Rishi Sindhu-kshitpraiya-medh, of the *Naddyah Devata* sukta of RigVeda. The basis for the implementation of the river-name-template throughout the subcontinent must have been with the sociological approach to the methodology in ecology, with the organized geographic survey of the entire country. Even after such incredible achievements under the conditions of presumably primitive communication system if some Indologists labeled the Aryans of India as 'shepherds' and the Vedas as the 'songs of shepherds', they only proved their mediocrity in scholarship. The singularity in river nomenclature for the rivers of Aryavart indicates that the river names were sorted out in the distant past. This leads to the significant conclusion that the Aryan Civilization is very ancient. The fact that the river Ganga was called Ganga by the time the book of Mahabharata was written over five thousand years ago and the Ganga was Ganga in the book of Ramayana written eons prior to Mahabharata, leads one to conclude that the Aryan Civilization and the Vedas have to be extremely ancient.

Organizational Structure of the Books of Veda

All mantras in the Vedas are metered poetic phrases grouped in *sukta*s (sections) with each *sukta* containing a number of mantras in it. A number of *sukta*s form a *mandal* or *addhyaaya* (chapter). A *devata* is associated with each *sukta* where the *devata* of the *sukta* is *sukta*'s subject matter. Each mantra has a self-sufficient independent thought in it on the subject (*devata*) of the *sukta*. Yet this subject does not relate to human history or events in the human civilization. Mantra by definition is the one that has knowledge hidden in it to be extracted with proper linguistic tools. A mantra is a unique poetic couplet in a specific meter. In mantra poetry, the meter of a mantra is not determined by the number of words used or by achieving the rhyming between two lines, as is usually in contemporary poetry. The meter of a mantra is determined strictly by the total number of alphabets in all of the words used in the whole mantra. This is the uniqueness of the mantra poetry.

It is relatively easy to compose poetry with two lines in a couplet such that the recitation of each line takes the same time regardless of how many words or syllables were used in each line. It is also relatively easy to write both lines ending with rhyming as well as having equal timing when recited. In a sensible poetic couplet, certain number of words is used but it is not the poet's worry as to how many total number of alphabets are consumed in all the words used. It is however difficult to have a couplet with the desired number of alphabets in all the words in the couplet. The word-count is not essential in mantra poetry as long as all words are *yaugik* and meaningful. It is the total number of alphabets used in all the words that is specific. It is not just one or a few mantras composed with this stringent literary condition but all Veda mantras are composed that way. The mantras in an entire *sukta* may be in the same meter or have mantras in several different specific meters, yet all follow the desired rule. Even highly scholarly Sanskrit pundits of today are unable to create poetry with the restriction of each line of the mantra couplet having the desired number of alphabets. If someone claims to have created a meaningful mantra by following this rule or has created several mantras on a given subject one has to realize that the uniqueness of the Vedic poetry is not limited to a few mantras or a few *suktas*; rather this is on a very large scale. It may be of interest to the reader that the RigVeda alone has 10,521 mantras in it with a total of 153,792 words utilized in them. All mantras are grouped into 1,028 *suktas* with 209 *devatas* (subjects) credited to 354 *rishis*, among whom are 21 *rishikas* (female *rishis*).

The term used in English translation for a *sukta* is 'a Vedic-hymn'. Each *sukta* is a collection of mantras with the same theme. It has a descriptive heading that includes *sukta's rishi,* its *devata, chhanda* type (meter of the mantra), *swara* (musical note) for the mantra to be recited in a monotone as well as several musical notes are marked on the words through the mantra as a vertical bar on top of the word and a horizontal bar under the word. These additional notes are to sing the mantra in a simple melody. As an example, the most recited *gayatri* mantra is presented below with singing notes marked on and under the words by a bar.

In the context of a mantra or a *sukta*, the word *rishi* stands for a sage who is a highly uplifted soul to whom the mantras were revealed in his or her mind. The word *devata* assigned to a *sukta* represents the subject of that *sukta*. Since the Veda mantras deal with a large number of subjects, there are a large number of *devatas* assigned to the *suktas* (hymns). This was misunderstood by Max Muller et al. to deduct that the Vedas worshiped multitudes of gods and goddesses.

The Veda mantras are basically in seven types of *chhandas* (meters). The *yaugik* meaning of the term *chhanda* is 'vibrations', the implicit meaning of which is - 'poetry with periodicity'. Therefore, the word *chhanda* in the present context means the 'meter' of the poetry. The first type of *chhanda* has 24 alphabets of the Sanskrit script. To explain the intricacies of the Vedic poetry the author chose the most-recited *gayatri-mantra* from the RigVeda (3/62/10), written below in one line with musical notes shown. This *gayatri-chhanda* has 10 individual words in it formed out of 24 alphabets of Sanskrit script. The mantra as recited with word conjugation is presented as follows:

तत्सवितुर्वरेण्यं भर्गो देवस्य धीमहि । धियो यो नः प्रचोदयात् ॥

(Mantra with word-conjugation)

Transliteration:
"*Tatsavitur varenyam bhargo devasya dheemahi | Dhiyo yo nah prachodaraat ||*"

The significance of the often recited *gayatri* mantra for the Aryans is akin to the often recited Lord's Prayer for the Christians. The mantra translation reads thus:

> "*O all-illume Creator of the universe, you are our protector and the purifier of our intellects. Invoking you in our intellects we meditate upon you to direct our intellects on the righteous path*" (RigVeda 3/62/10)

When word-conjugation is taken out and the singing notes (vertical and horizontal bars at the words) are dropped, the *gayatri-mantra* can be written with individual words in the same sequence as they appear in the mantra:

तत् सवितुः वरेण्यम् भर्गः देवस्य धीमहि । धियः यः नः प्रचोदयात् ॥

(Mantra with word separation: 10 words).

The single vertical long bar after two-thirds of the mantra and two vertical bars at the end of the mantra are equivalent to a comma and a period in a sentence in the English language.

After splitting the individual words into their individual alphabets, one can arrange them sequentially as shown below with each letter/alphabet marked by a numeral,

त^{1} त्स^{2} वि^{3} तु^{4} र्व^{5} रे^{6} ण्य^{7} म्भ^{8} र्गो^{9} दे^{10} व^{11} स्य^{12} धी^{13} म^{14} हि^{15} |

धि^{16} यो^{17} यो^{18} नः^{19} प्र^{20} चो^{21} द^{22} या^{23} त्^{24} || (24 letters/syllables)

Compared to the *gayatri-chhanda,* the next *chhanda-meter* called *Ushnik* has 28 alphabets in it. The successive *chhanda-meters* progressively increase in length with a step of four alphabets, going right up to 48 alphabets in a mantra, forming seven types of basic *chhandas*. The names of these *chhanda*-types are given first in two mantras of RigVeda (10/130/4, 5) under the subject heading of *bhavavrittam* (creation of the universe) and then all seven names of the seven types of *chhandas* are listed successively in a single mantra in AtharvaVeda (19/3/21/1). The names of the seven types of *chhandas* sequentially are Gayatri (24), Ushnik (28), Anushtup (32), Brihati (36), Pankti (40), Trishtup (44) and Jagati (48) and the number in the bracket represents the number of alphabets forming the *chhanda* (couplet). This is the concept of 'meter' in the Vedic poetry. It is unique and difficult to emulate. To create a meaningful poetic couplet using meaningful words that yield specific number of alphabets makes it unique and difficult.

So that we stay strictly in the realm of the *yaugik* interpretation of the mantras and do not go down the path of mistranslation of the Vedic terms in cosmology and physics, it is imperative to develop a little more insight into the framework of Vedic terminology and the historical aspect of Sanskrit and the Vedic Literature.

The Sanskrit vocabulary that exists within the four Vedas is vast. It led to the formalized grammar rules that were later laid down by the ancient *rishis*, which are still followed in the usage of Sanskrit today. As mentioned earlier, it was Rishi Paanini, the author of Ashtadhyayi Grammar Sutra, who formalized the Sanskrit grammar. It may seem odd that the Vedic vocabulary should exist first in the Vedas and then the rules of Sanskrit grammar were organized. This indeed is the situation because the Vedas happen to be the first books of the Sanskrit language and there is no reference or evidence found in the ancient books in Sanskrit Literature that anything else existed before the Vedas. The post-Veda *rishis* and the sages delved deep into the Veda mantras, discovered the roots (*dhatus*) of the words and later laid down the foundation of the Sanskrit grammar.

Now, coming back to the word *devata*, Rishi Yask in his ancient book of Nirukt (1/7) states that *devata* is a general term applied to those substances and entities whose attributes are invoked in a mantra. A *devata* is a real entity, not a god or a goddess of some kind. When a *devata*, a conscious being or an inert entity is invoked in a mantra or in a collection of mantras of a *sukta*, the purpose is to remind the attributes of the *devata* in the minds of the reciter and the listener. The context being very important in Sanskrit the term *devata* in differing contexts stands for different meanings as well.

A *devata* in a social context would include all those conscious-beings who manifest extraordinary qualities and selflessness who have a positive influence in the lives of others. Example of such a *devata* is mother, father and guru. The virtuous souls that have inspired and transformed societies are also termed as *devata*. The Creator in the spiritual context is addressed as *Maha-deva* the *great-deva* or *devata* of all *devatas*. Exterior to the Vedas, the deities in a Hindu temple are also termed as *devatas*. Human nature is to go for simpler practices and beliefs. The Vedas instead arriving at the concept of God that is conterminous, unitary, all-pervading, formless spirit is at a higher level of thinking. The fact that the Vedas, in most certain terms, propound monotheism in them was hard for Max Muller to digest. In his defense he stated, "*A large number of Vedic hymns are childish in the extreme, tedious, low and commonplace*" ('Chips from a German Workshop', ed.1866, p.27; taken from the introduction to 'Aryon Ka Aadi Desh aur Unki Sabhyata' by Swami Vidyanand Saraswati, publ. 1989 CE). One gathers from Max Muller's translations of the mantras that before embarking upon translating the Vedas, he did not read even the most fundamental guide prescribed for reading, understanding and interpreting the Veda mantras - the Nirukt by Yask Rishi.

The Vedas after being subjected to centuries of sustained literary onslaught have remained an invariant entity, available even today for close examination by those who seek the truth.

The Branches of Vedic Literature

Brahmana Books

The first books of Vedic Literature after the four Vedas, are entitled Brahmanas (Item 2 in Table 4.1). As Swami Dayanand Sarasvati explained, the term "*brahman*" for these ancient books is because they provide interpretation of Veda mantras the ultimate source of which is Brahmn the Creator. The term *brahmana* for a person signifies a learned and a scholarly person. A *brahmana* in the prevailing caste-system of present-day India is the one born in the *brahmana* family (upper caste), learned or not. The word *brahmana* is also spelt as *brahman*. To differentiate the brahmana book from brahmana person this author has used the term Brahmana with the capital B for the book and lower case brahmana for the person.

Whereas a Veda mantra conveys knowledge of a subject in a compact word formulation, a corresponding Brahmana being a commentary book on the Veda narrates it with emphasis and elaboration in simpler expression. The Brahmanas were written in prose and not in metered *shloka* or mantra poetry; nevertheless, their text is highly organized, as in a modern textbook, with proper chapters, sections, subsections and the statement numbers. In addition to providing interpretation of the mantras, the Brahmana books also advise how to recite the mantras and what mantras to use for specific occasions in the spiritual, the intellectual and the social realms. These books have laid down the foundation of the procedures and methodology in the society for charitable acts, celebrations and in general the implementation of mantras in daily life. The Brahmana books use examples, metaphors and similes to simplify mantra's message for the benefit of the common people and collectively the society as a whole.

Primarily there is one Brahmana on each Veda. The Brahmanas respectively for Rig, Yaju, Saam and Atharva Vedas are: (1) Etreya, (2) Shatpath, (3) Saam (Tandya) and (4) Gopath. The additional Brahmana books are Kaushitaki (or Saankhyayan) on RigVeda; Taitteeriya on Krishn-YajurVeda and Jaiminiya, Saam-vidhan; Aarsheya, Mantra, Devataddyaya, Vansh and Samhitopnishad are all on SaamVeda.

The Brahmana books being in prose were vulnerable for insertions by those with ulterior motives to pollute the Vedic Literature, especially after mythological practices became prevalent in the Hindu society. The non-Vedic sects with the emphasis on atheism added to this pollution. The society in general had high reverence for the *rishis* and the literary work that they created. Their literary work was carried forward as a part of the curriculum of the guru-*kulas* where the resident students learnt it, recited it and even preserved it in the form of hand-written manuscripts. This process helped carry forward the Vedic text accurately without even a slight change of a syllable. After the Mahabharata War when the social norms and education gradually declined, the ritualism and blind beliefs took hold in the society. It was this dark period when those pundits who wished to grab power and wealth inserted self-serving text and phrases in the Brahmana books. According to Swami Dayanand Sarasvati, such insertions remained unchecked from about a thousand BCE until about 500 ACE. Therefore, while studying a Brahmana book, one uses caution and discrimination to separate the pollution of insertion that usually is glaringly unfitting in the thought process of the original Brahmana phrases.

The term for an insertion in Vedic Literature is '*prakshipt*'. Therefore while reading a book of Brahmana, Manusmriti, Mahabharata or even Bhagvadgita one should look into the scholarly works published in recent times identifying the *prakshipt* parts. The difficulty however continues that on the one hand, the sincere scholars of the Vedas continue to discuss and point out the '*prakshipt*' text in the Brahmanas and in the other books of the post-Veda Sanskrit Literature and on the other hand, the printers continue to print the texts with *prakshipt* text in them. There is an increasing awareness now of the problem and identification of the inserted literary pollution.

Several references from the Brahmana texts are included in this book, particularly those related to the scientific thoughts. Since the most inserted *prakshipt* parts in Brahmana relate to the social context, the Brahmana references pertaining to science thought are generally found to be untouched by the *prakshipt* text (literary pollution).

Upvedas

Upvedas (Item 3 in Table 4.1) (pronounced as *oopveda*) are the treatises on specific academic subjects derived from each Veda. The AyurVeda is derived from the RigVeda and the AtharvaVeda. The AyurVeda deals with the 'science of aging' or the 'science of living', which extends into the field of Ayurvedic Medicine; prominently practised among the current medical systems in India.

The DhanurVeda, the science of archery is under the military. The GandharvaVeda teaches the art of music, other fine arts, and the ArthaVeda deals with the subject of economics and governance.

Smritis

Smritis (Item 4, Table 4.1) are the books authored by learned individuals at the level of a *rishi.* Of the 28 Smritis the oldest and the most known was authored by Svaayambhuva Manu, a great king of Aryavart. The King Manu and the *shlokas* from his Manusmriti were discussed in Chapter 1 in relation to the time-division and time conversion factors under the sub-heading of 'Age of the Universe'. Manusmriti was extensively referred to in Chapter 3 in relation to geographic definition of Aryavart and the nation-defining mighty river Sarasvati. It is one of the most valued and revered Smritis of the Aryans. Numerous *shlokas* from Manusmriti covering different subjects are still to come in this book.

Manu was a brilliant scholarly king of the ancient Aryavart Empire who established the present Aryan civilization. Rishi Valmiki's epic poetry Ramayan (Item 5 in Table 4.1), written well after Manusmriti, is a biography of King Ram in Manu's dynasty. Manu in his Smriti has called his nation, Aryavart. The term Aryavart means - '*the place of the Aryans*'. It was discussed in Chapter 3 how Aryavart's name changed from Aryavart to Bharat-varsh (now official name) to Hindustan and to India.

A large part of the ancient Vedic literary work is in poetry. A poetic couplet in the Vedas is addressed as a mantra and a poetic couplet exterior to Vedas is known as a *shloka.* The knowledge content in a mantra by definition is hidden, requiring extraction, while it is quite obvious and straightforward in a *shloka.* The composition of Manusmriti *shlokas* is in a specific meter called *anushtup chhanda* (*anushtup* meter) with a total of 32 alphabets; 16 alphabets in each line. Whereas the four Vedas are in seven basic different *chhandas* mentioned earlier, the entire Manusmriti is in *anushtup chhanda* only. For the easiness of *anushtup* meter the Valmiki Ramyana, Mahabharata, (Bhagvadgita within it) and numerous other books in Vedic Literature are also in *anushtup* couplets.

The preamble *shlokas* of Manusmriti state that the learned Vedic scholars of Manu's time, the *rishis* and the sages, came to Manu to hear his thoughts on various aspects of life. The book of Manusmriti is the compilation of his sermons to the visiting *rishis*. Manusmriti's first chapter starts with the Vedic principle for the creation of the universe, which closely relates to the Creation Principle in RigVeda's 129th sukta of its 10th chapter. Five of several *shlokas* from the first chapter of Manusmriti dealing with the creation of the universe are included in the Vedic Big Bang Scenario developed in Chapters 6 to 10 in Volume 2 of this book.

Manusmriti's other chapters contain advisement on the knowledge in all aspects of social life. Manu was a follower of the high ideals prescribed for a king by the Vedas. He was a great king and formulated the "Code of Conduct" for the individual and the society as a whole. Therefore, the book of Manusmriti is also known as 'Manu's Ordinance'. The code prescribed by Manu includes the duty of the judiciary under the king and the duty of the individual towards the society. The kings and rulers established their judiciary and the political framework based on Manu's Ordinance. The present Hindu society is based on Manu's code and therefore the Hindus revere Manu as the 'Law-Giver' of the society.

Based on the nature and the ability of a person (not caste), Manu grouped the society into four categories. These categories together are popularly known as Manu's *varna-system*. The four categories of this *varna-system* with their well-defined responsibilities towards the society are ***brahman*** (responsible for learning, teaching and dissemination of knowledge...), ***kshatriya*** (responsible for the protection of society, administration, ruling...), ***vaishya*** (responsible for welfare and growth of society, business, economics...) and ***shudra*** (responsible for providing direct labour for the service to the society...). This division according to him was fluid. The *varna* of a person was determined by a person's ability, nature and orientation. A person born in a ***shudra*** home (service-oriented) could transit into a ***brahman*** *varna* or into any other *varna* and vice-versa. Many *shlokas* make this fact very clear. The system rigidified into the caste-system much later during the past three millennia, otherwise the meaning of the term *varna* is to choose, which gives the freedom to the person for what he/she wants to be.

Manu's thoughts on the *varna*-system stem out of the Veda mantras in which the human body itself, naturally, has these four functions that analogously lead to the four *varnas* in the society in general. Manu derived the *varna*-system from the Vedas where functioning, be it of an individual living body or the society as a whole, depends on four natural divisions and of equal importance. Numerous references within Manusmriti and in the Vedic Literature show that the *varna*-system was never meant to be a rigid system like it is in the present Hindu society. It has turned into a caste-system with higher-caste and lower-caste connotations with little fluidity of movement between the four groups. The *varnas* relate to the human nature and not to the social strata in which a person is born. Without bringing in all the relevant references from the Vedas and other ancient Vedic Literature just two *shlokas* that define the *varna*-system's identity are presented below. These show the true intent of the *rishis* on which Manu based his Code. A *shloka* from Upanishad (item 6 in Table 4.1) states thus:

जन्मना जायते शूद्र: संस्काराद् द्विज उच्यते ।
वेदाभ्यासाद् भवेद् विप्रो ब्रह्म जनाति ब्राह्मण: ॥ (Vajrasuchi Upanishad)

Transliteration: "*Janmna jaayate shoodrah sanskaaraad dvij uchchyate|*
Vedabhayaasaad bhaved vipro brahm janaati braahmanah ||"

Translation (from Hindi text in Satyarth Prakash):

> "*All are **shudra** at birth. By sanskaar (cultivation of refined character) the person is called a **dvij** (reborn in the same life), the person becomes a **vipr** by studying the Vedas and the person becomes a **brahman** by developing knowledge of Brahm (God)*". (Vajrasuchi Upanishad)

Long before this Upanishad *shloka*, Manu's Smriti stated that the basic nature of a person tends to put him/her in a specific *varna*. The relevant *shloka* from Manusmriti is:

शूद्रो ब्राह्मणतामेति ब्राह्मणश्चैति शूद्रताम् ।
क्षत्रियाज्जातमेवं तु विद्याद्वैश्यात्तथैव च ॥ (Manusmriti, 10/65)

Transliteration: "*Shoodro braahmantaameti braahmanashchaiti shoodrataam. Kshatriyaajjaatmevam tu vidyaadvaishyat tathaiv cha.*"

Translation:

"Based on the nature and karma of the person *a* ***shudra*** *becomes a* ***brahman*** *and a* ***brahman*** *becomes a* ***shudra****. This applies also to a* ***kshatriya*** *and a* ***vaishya***" (Manusmriti, 10/65).

After Manusmriti, even Krishn in Bhagvadgita (18/41) 5,000 years ago upheld Manu's fluid *varna*-system and emphasizes that the *varna* determination should be based on the *nature of the person* (and not on the circumstances of his family).

One observes that the open Western society of today has moved along the lines of Manu's *varna*-system, according to which a person chooses a profession in which the person has developed capabilities. It may sound far-fetched; however, this author observes that the Western society also has the same four divisions with fluidity for people to move from one profession to another. For example, the scholarly professors, scientists, teachers and intellectuals are *brahmans* (teaching and learning). The president, prime minister, members of parliament, generals, soldiers and police are *kshatriyas* (administration and protection). The bankers, business persons, farmers are *vaishyas* (business, growing crops) and workers, servers, employees of corporations are *shudras* (service-industry). There is no fifth category because if there were one, the Veda and Manu would have included it.

Manu also provides guidance for different realms of a society, so much so that his Manusmriti has served as the foundation for the constitution of the society. Just to mention a few areas, it provides the rules for the division of property, for marriage and choosing the right partner, for the judiciary, for the duties of each of the four phases of human life-span of 100 years and so on. On preventing the inbreeding in the society, recall Manu's advice (3/5) cited in Chapter 3:

"*For the marriage of a boy the girl should be at least six generations removed from his mother's side and should be completely outside of his father's genealogy*".

Generated in the far antiquity of India this statement relates to the science of genetics and could lead to further genetic research. The statement goes beyond mere observation of the effect of inbreeding; it assigns definite numbers to the phenomenon. Considerable light from the advanced genetic research of today could be shed on the possible significance of the solid numbers prescribed by Manu for the genealogical distances to be maintained between the blood-relatives of a bridegroom and bride-to-be. Why was a boy to avoid marrying a girl within six generations on his mother's side and why to completely avoid marrying a girl from father's genealogy is worth studying. Manu also forbade parents arranging the marriage of their children without the children's consent. He listed eight classes of marriage (Manusmriti, 21/3) of which the *svayam-var* (choosing one's own partner) was the best.

There is no doubt that there are insertions in Manusmriti as well. These *prakshipt shlokas*, lead to misconceptions about Manusmriti's period of writing, its authorship and its reality. Max Muller et al. have gone as far as saying that Manusmriti was a recent development (a few centuries BCE). Since the Indian history during 4-5 centuries BCE is very well known when there was no king named Manu, Max Muller said that Manu, known as the first king of Aryavart (India), must be an imaginary person. This requires a bit of an exposition.

When the Vedic practices declined over the past two to three thousand years and mythological books (Puranas) of the Hindus were compiled, the caste system rigidified. Some brahman priests with the ulterior motive of promoting their own sect and registering control on the society inserted a large number of *shlokas* in the ancient books including Manusmriti. Some of the inserted *shlokas* diametrically opposed the general thought pattern of Manu. Owing to this literary pollution found in later compilations of the manuscripts of Manusmriti, the authenticity of Manusmriti itself was questioned.

After Swami Dayanand Sarasvati's awakening of the Hindu society in the late nineteenth century and after the Indian independence from the British Raj, scholars carried out considerable research on ancient Vedic Literature, including Manusmriti. These scholars realized the extent of the harm caused by the inserted *shlokas* as well as the harm caused by the distortion in Indian history created by Sanskritists and Indologists of the West. Their research effort identified the inserted *shlokas* and helped the publication of corrected versions of important books, Manusmriti being one. Among others, the effort by Professor (Dr) Surendra Kumar is remarkable. Manusmriti's citation in this book is according to the *shloka* numbering in 'The Vishuddha Manusmriti', an interpretive translation of Manusmriti in Hindi by Professor Surendra Kumar, published by Arsh Sahitya Prachar Trust (January, 1996). The possibility of insertion by literary polluters is minimal to nil in the section pertaining to cosmology in Manusmriti's chapter 1. This author has drawn some deductions on the amazing scientific observations found in Manusmriti. A detailed discussion on this and the historicity of Vedic references is in Chapter 3.

Ramayan

Rishi Valmiki is the author of Ramayan (item 5 in Table 4.1). Valmiki was a contemporary of the great and ideal king of Aryavart, King Ram (also written as Rama). King Ram was a descendant in King Manu's genealogy. This genealogy is also known as the Solar-Dynasty. Ramayan is Ram's biography written in Sanskrit poetry in *anushtup* meter. Ram's virtues being extraordinary, the author Valmiki through Ram's biography has provided a beautiful exposition of the desired social ideals, principles, duties and morality. Ramayan is also a window providing a glimpse of the life of the people and the level of civilization during

Ram-rajya, a common phrase in the present Hindu society for good governance. Ram's governance was so ideal that all governments in India, past and present, with the exception of the Muslim and the British rules, strove to emulate Ram-raj. Owing to the specific dialect of the author of Ramayan, the meter of the couplets and the *shaili* (grammatical aphorism) of the Sanskrit usage, an astute scholar of Vedic Literature could easily spot an insertion in the Ramayan.

An abridged genealogy is recited in Ramayan's 29th section of its first Chapter covering the kings from Manu until Ram's father, Emperor Dasharath. Rishi Valmiki in this list of kings of the Solar-dynasty has named 34 kings with some anecdotal information on some of the virtuous kings in that line. Ayodhya remained the capital of the Solar-dynasty since Manu established it. Rishi Valmiki describes the metropolis of Ayodhya at the time of Ram's father, Emperor Dashrath. Rishi Valmiki states in Ramayan's Baal-Kaandam (3/1-3),

> "*The geographic region known as Kosal was full of prosperity and wealth with contented citizens. The great king of people, Manu, had established the metropolis of Ayodhya as his capital at the bank of Saryu River. This world-famous metropolis (at the time of Ram) with beautiful wide roads was 12 yojan* (150 km) *long and 3 yojan* (36 km) *wide*". (Valmiki-Ramayan, Baal-Kaandam, 3/1-3).

The Ramayan also states that after Manu established Ayodhya the first king to rule Kosal region was his son Ikshvaku. In Manu's genealogy there was also a great king named Raghu. The same Solar-dynasty ruled from Ayodhya from Raghu to Rama and until Vrihadbal the king of Ayodhya in 3140 BCE. Ram's dynasty is also addressed as Raghu-vansh (Raghu-genealogy). Among other poetries that the great Sanskrit poet Kalidas wrote during the reign of King Vikramaditya; his poetry entitled "Raghu-Vansh", in praise of this dynasty, is considered to be very beautiful and an excellent piece of literary work. Kalidas wrote Raghu-Vansh sometime near 50 BCE when he was a court-poet of the King Vikramaditya who ruled from Ujjain in central India. Vikramaditya is the king after whom the Vikram-*Samvat* (Calendar) continues in the present Hindu society of India. The onset of Vikram-Calendar is with Vikramaditya's coronation in the month of March, in 57 BCE. The Luni-Solar Vikram-Calendar is prepared based on the indigenous astronomical calculations and printed for the usage of the Hindu society. It includes all the Hindu festivals, the days for fasting, the dates of astronomical conjunction of planets and constellations and the accurate predictions of the solar and lunar eclipses. The Vikram-Calendar's scientific astronomical information has become the bonanza for the non-scientific fatalistic astrology who use the astronomical positions of the planets and the stars for fortune telling; a money making racket in the name of Vedic science.

The Solar-dynasty continued after Ram with his two sons Lav and Kush. Innumerable generations passed in the continuity of this dynasty at Ayodhya. A Puranik account names the King Vrihadabal in Ram's dynasty ruling from Ayodhya during the Mahabharata War (discussed in Chapter 3). Just as in the World War II, most countries of the world had little choice but to join in the war against Germany, so also was the situation in the Mahabharata War. In this Kurukshetra War the king Vrihadabal of Ayodhya chose to fight on the side the righteous Pandavas. The Pandavas were victorious and Yudhishthir was coronated in 3140 BCE.

Upanishads

The word *upanishad* (item 6 in Table 4.1) is a derivative from three words: *up* (pronounced as *oop*), *ni, seedati*. When put together the meaning of the word *upanishad* is *'one that enables proximity to Brahman the Creator'*. Therefore, Upanishad is a *'medium for attaining eternal bliss'*. There are ten Upanishads that are considered ancient. They are Eesh, Ken, Kath, Prashna, Mundak, Maandukya, Etereya, Taitteeriya, Chhaandogya and Brihadaaranyak. Aadi Shankaracharya (509-477 BCE) in the fifth century BCE wrote commentaries on them. Swami Dayanand Sarasvati declared these ten as the Vedic Upanishads. Upanishads are the Vedic interpretive books that expound philosophical thoughts in the Veda mantras and express them in simpler statements that sometimes are in question-answer form - questions by students and answers given by their *rishi* guru. These Upanishads are non-sectarian and have universality in thoughts and principles. The books of Aranyakas are also in the same category as the Upanishads.

While writing a recent interpretive commentary on six Upanishads in the Hindi language, Swami Darshananand Sarasvati mentioned in his book 'Upanishad Prakash' that in addition to these 10 Vedic Upanishads there are at least about 240 other non-Vedic sectarian Upanishads that have cropped up since about 500 BCE. Authors belonging to various sects within ritualistic Hinduism and outside of it wrote these additional Upanishads. These so-called Upanishads lack universality of thought and are self-serving. There is even an Upanishad called *Allopnishad* written by a Muslim with Islam as its theme. An Upanishad is to convey universality of thought. If not, it is another book of a religious sect. Therefore, a student of Vedic Literature has to be cautious and knowledgeable to distinguish between the Vedic and non-Vedic Sanskrit Literature.

Vedangas

The Vedangas (item 7 in Table 4.1) contain six distinct branches of the Vedic Literature that a student of the Vedas is required to master. The six subjects under Vedangas lead a student of the Vedic studies to become a master of (1) the science of morals, (2) the science of dharma (3) the science of language, (4) the science of etymology, (5) the science of poetry and (6) the applied science from geology to astronomy. Researching these and commenting on them is beyond the scope of this book. During the Vedic studies a student learns about the practical methodologies such as writing and recording mantras and manuscripts, grounding in grammar (study of Ashtaddhyayi), Nirukt (e.g. Yask Rishi's Nirukt) for etymology and interpretation of Vedic terminology and the exposition to the knowledge in Vedic poetry for word meaning and accent. Through the six Vedic Darshans (philosophies), the student is educated in the logic system, yoga practice, the nature of matter and substances, the observational and mathematical astronomy, and many other subjects. To complete the Vedic syllabus that covers these branches of Vedangas, a student would devote his entire student career at a *guru-kula* (guru-family). There have been many well-known *guru-kulas* in ancient India and there are those established in recent times owing to the inspiration of Swami Dayanand Sarasvati. These education centres are dedicated to teaching Vedic Literature and ancient cultural traditions of the Aryans.

A student who has taken a crash course in Sanskrit, without a solid grounding in Sanskrit grammar and etymology cannot do justice when interpreting Vedic Literature. A solid grounding in Vedic learning is achieved by increased contact with the guru, which was the practice in the ancient *guru-kula* system.

Darshans

The six Darshans (item 8, Table 4.1) are listed in the Table 4.2. All six Darshans separately deal with six differing knowledge realms of Vedic philosophy. They are also popularly known as the Shastras or the Sutras. These six Darshans are addressed as Vedic Darshans to differentiate them from the Jain and Buddhist Darshans.

Some people thought that since the Darshans propound six different schools of philosophy, they were contradictory to each other. This is not the case because they all have their foundations in the Vedas where there are no contradictions. All six Darshans are in fact complementary to each other covering six different realms of the creation with a fair bit of overlap. The Darshans are not in poetic form; they are in the form of short statements called the sutras. A sutra is an aphorism with the formulation of a single thought and principle. It is therefore a formula. The ordinary meaning of the word sutra is ‘thread’. All sutras of a Darshan are connected together to form the philosophy of that Darshan. In other words, all principles of the philosophy expounded in that Darshan are threaded together. There are frequent references of the Darshans in this book. A Darshan expounds its subject matter completely from the beginning to the end and therefore the short comments under the ‘Subjects’ column of Table 4.2 provide only a glimpse into the contextual background of this book.

The six Darshans are placed in chronological sequence in Table 4.2 that is based on the scholarly discussions found in the Vedic Literature. The traditional Indian Vedic scholars have placed them in a sequence based on historical accounts in Sanskrit Literature and on their references found in other Vedic books.

Some Vedic scholars may not agree with this chronology; however, most would accept that the first four Darshans are anterior to the Mahabharata and the last two successively written by the two *rishis* who definitely were contemporaneous to Mahabharata. All scholars agree that the last two are in the same chronological order as in Table 4.2. Without going into the details of all the reasons, a brief commentary on why they are placed in the chronological order of the Table 4.2 is given below.

Table 4.2: Six Vedic Darshans			
	Darshan	**Rishi-Author**	**Subjects**
1	Nyaaya	Gautam	Examination by logic and proof. Cause-Effect in Creation. Logic and Capitulation, Logical fallacies....
2	Vaisheshik	Kanaad	Physical nature of the Universe (molecules, atoms,..), Time dependence of Action, nature of cause and effect.....
3	Saankhya	Kapil	Deals with Purush (Creator, Paramatma), Atma (Soul) and Prakriti (Nature/matter), Creation of the Universe, its primordial byproducts, functions and their relationship to living beings, Sukshm-Shareer (minute body) and connection between the matter, soul and consciousness....
4	Yoga	Patanjali	Treaties on Yoga (very popular in the West)
5	Poorv-Meemansa	Jaimini	Importance of karma (Jaimini was Rishi Vedvyas' student)
6	Uttar-Meemansa	Vedvyas	Brahm (The Creator). Other names of this Darshan are: Brahm-Sutra, Vedanta-Sutra and Shaareerik-Sutra

Nyaaya Darshan (item 1 in Table 4.2) deals with the origin of logic as the cognitive ability of human beings. Its author Rishi Gautam in his very first sutra (aphorism) lists 15 essentials for acquiring true knowledge. This list starts with the word *pramaan*, the meaning of which is, 'means of getting true knowledge' and does not start with the word *atha* in its first *sutra* as the other five Darshans do. The word *atha* stands for 'now let us discuss' or 'now let us begin'. The first *sutra* of other five Darshans begins with the word *atha*. The beginning of a Darshan with the word *atha* indicates acknowledging and respecting the previous Darshans. Scholars suggest that being the first Darshan the Nyaaya Darshan does not begin with the word *atha*. Therefore, the Nyaaya Darshan is thought to be the first among the six Darshans.

One finds in the Vedic Literature several interesting correlations to support the chronology of the next five Darshans. For example Rishi Kapil, the author of Saankhya Darshan (third in the list), also wrote the *bhaasya* (interpretive commentary) on the Vaisheshik Darshan (2nd in the table) authored by Rishi Kanaad. Therefore, chronologically Vaisheshik precedes Saankhya Philosophy. Rishi Vedvyas, the author of Vedanta Sutra (6th in the table) and the epic book of Mahabharata, is also the author of the *bhaasya* on the Yoga Sutra of Rishi Patanjali (4th in Table 4.2). Therefore, the Yoga Sutra must precede Vedanta Sutra. Rishi Vedvyas also authored the *bhaasya* on the Poorva-Meemansa written by his own student Jaimini, his contemporary. This makes it easy to place the Vedanta Darshan in the last place in the chronology of the Darshans. The prefixes *poorv* and *uttar* to the word Meemansa of the last two Darshans respectively mean 'first' and 'later' indicating that chronologically Jaimini the student completed his Darshan, the Poorva-Meemansa, before his guru Rishi Vedvyas completed his, the Uttar-Meemansa.

The literal meaning of the term '*meemansa*' is 'thesis' or 'dissertation'. Historically, Rishi Vedvyas's Uttar-Meemansa is also addressed as the Vedanta Darshan and therefore the Poorva-Meemansa of Jaimini is popularly addressed as just the Meemansa Darshan without the *poorv* prefix to it. The Vedanta Sutra also has a not so common name Brahm-Meemansa. The given name 'Vedanta' to it also signifies the 'end of the sequence of Vedic philosophy'. Another equally popular name for the Vedanta Sutra is Brahm Sutra because the subject matter of this Darshan is Brahmn, the God-entity. Swami Dayanand Sarasvati in his Satyarth Prakash (1881 CE publication) has referred to Brahm Sutra as Shaareerik Sutra as well. Thus, this last Darshan has five different names.

This bit of insight into the history of the last two Darshans also leads to the fact that both, Rishi Jaimini and Rishi Vedvyas and their literary works in the form of the two Darshans along with the book of Mahabharata, were contemporaneous. Based on the historical records discussed in Chapter 3, the historical period for the last two Darshans is expected to be just before 3140 BCE, the year of Yudhishthir's coronation and onset of Yudhishthir calendar.

Mahabharata

Mahabharata (item 9 in Table 4.1) is the epic book containing the account of the Mahabharata War fought between the Kauravas and the Pandavas of the Lunar Dynasty. The book of Mahabharata gives a detailed description of the Mahabharata War. It was written by Rishi Vedvyas the author of the Brahm Sutra and his disciples. Compiling it and putting it in poetic form was a joint undertaking by Rishi Vedvyas and his disciples. The Mahabharata war ended about four decades before the milestone in the Vedic civilization at **February 18, 3102 BCE,** when the Vedic Creation Calendar's Kali-Yuga epoch began. As discussed in Chapter 3 this date has a direct relationship to astronomy, making it the onset of the Kali-yuga Samvat (calendar). This calendar is called the Srishti-Samvat as well where the word *srishti* means the 'creation of the universe' and *samvat* means the calendar. Because of its stability the Srishti-Samvat has been more important in the Vedic society than other practical calendars commencing after kings who have come and gone. The Aryans have kept track of this *srishti* calendar since the beginning of the Aryan civilization. Mahabharata is an event, a great war within two factions of a long running Lunar-Dynasty. This dynasty ruled from the city of Hastinapur. The present day Hastinapur town is in the state of Uttar Pradesh about 100 km north of Delhi, the present capital of India. Hastinapur is located close to the River Ganga (Ganges). The Mahabharata War was fought on the plains of Kurukshetra, west of Hastinapur, west of River Yamuna, which is about 160 km north-west of Delhi. Therefore this war is also known as the War of Kurukshetra. The city of Indraprasth at the shore of the River Yamuna was newly established as the capital just before the Mahabharata War.

The war ended with the coronation of the Aryan king Yudhishthir, 38 years prior to the milestone in February of 3102 BCE. This history is well-known to the scholarly Aryans of India. Based on the astronomical data for the night-sky inscriptions available from Rishi Vedvyas's astronomical observations cited in the Mahabharata, it was established in Chapter 3 that the Mahabharata War was as real as World Wars I and II.

Rishi Vedvyas is variously referred to in the post-Mahabharat Sanskrit Literature. Vedvyas is a title and not his real name. The literal meaning of the term *vedvyas* is '*the diameter of the wheel of the Vedas*' with the connotation of the '*pivot of the Vedas*'. He acquired this name of honour because he was a great authority on the Vedas. His birth-name and the family name was Krishn Dvaipaayan where *dvaipaayan* means '*whose home was on an island*'; a descendant from an island. He was also addressed as Baadrayan, meaning '*whose home is in Badri*' probably because his immediate residence (ashram) was in the geographical area of Badri-vana. The Badri-vana is the Badri-forest situated along the foothills of the Himalayas. The Badrivana is known as such owing to the abundance of the *badri* fruit trees in this forest. The common name for *badri* fruit is *ber* (Indian jujube). Indigenous to India, the *badri* tree belongs to the Ziziphus mauritiana tree-classification. When fully pronounced, Vedvyas's name is Badaraayan Krishn Dvaipaayan Vedvyas. The references in the Book of Mahabharata and in other books in Sanskrit Literature indicate that he had a large ashram and *guru-kula* (residence school of higher learning) at the foothills of the Himalayas with many advanced level resident students. It is also mentioned in Mahabharata that he had an astronomical observatory on the shore of the River Sarasvati, which originated from the Shivalik Hills of the Himalayas. A place called Badrinaath in the Himalayan state of Uttaranchal is a likely place where his ashram once stood. It has been a highly revered pilgrim place for the Hindus over millennia.

The Himalayan pilgrim place Badrinaath has been of great spiritual significance that attracted Aadi Shankaracharya to travel from Kalady in Kerala state in Southern India to Badrinaath in the Uttaranchal in the Himalayas where he lived for about seven years before he went to establish the Dwarka Ashram in Gujarat. Badrinaath, also addressed as Badrikashram that Aadi Shankaracharya chose to establish his first Matha (monastery) in 2,600 years after Mahabharata and named it Jyotirmatha. The detailed historical account on Shankaracharya and the dates when he established the other four Mathas, during his travels is in Chapter 3.

Rishi Badaraayan Vedvyas belonged to a genealogy of learned people. Four generations of his family are known. His guru was his father Parashar and Parashar's guru was his own father Shakti. Rishi Vedvyas's genealogy ceased after his learned son Shukdev decided to devote his celibate life in learning and meditation. He renounced the world to lead a monastic life in search of self-realization. Thus there are no further descendants in Rishi Vedvyas' genealogy.

Rishi Vedvyas had four senior students: Pail, Vaishampayan, Jaimini (author of Meemansa Darshan) and Sumantu. After writing his account of Mahabharata, Vedvyas read it to his disciple Vaishampayan. After that in the Vedic philosophic tradition it must have gone through a critical examination and discussion phase leading to the finalization process where the highly valued outcome of the discussion was also included in the book of Mahabharata.

According to Vedic historical accounts, the Emperor Yudhishthir ruled for 36 years, 8 months and 25 days and renounced the throne to proceed to the

Himalayas for meditation and self-realization with Draupadi and his other four brothers, Arjun, Bhim, Nakul and Sahadev. Arjun's son Abhimannyu died in the Mahabharata War when Abhimannyu's princess-wife Uttara was pregnant. When Uttara gave birth to Arjun's grand-son, Yogiraj Krishn named him Parikshith. When Yudhishthir renounced the throne of Indraprasth, Parikshith's coronation took place when he was 35 years old. Parikshith ruled for 60 years, 0 months and 0 days when Janmejay followed on the throne in succession at Indraprasth. Rishi Vedvyas's student Vaishampayan recited and presented the book of Mahabharata to the King Janmejay on the occasion of the parliamentary session named '*sarp-satra*'. On the basis of some ancient references Swami Dayanand Sarasvati wrote in his book Satyarth Prakash that when the book of Mahabharata was first completed, it had a total of 10,000 *shlokas* (poetic couplets) of which 4,400 were composed by Rishi Vedvyas himself and the remaining 5,600 were composed by his four students.

The book of Mahabharata is in 18 divisions called the *Parvs* with each *Parv* identified with its subject heading. There are a number of chapters within each *Parv* with each chapter having a number of *shloka*-couplets. The book of Bhagvadgita, a well-known book in the West, is the sermon given by Yogiraj Krishn to Prince Arjuna at the battle field of Kurukshetra and it is a part of the book of Mahabharata. Bhagvadgita has 18 chapters in it which are the 23rd to the 40th chapters of the 6th *Parv* of Mahabharata, titled as the *Bhishm-Parv*. Yogiraj Krishn's birth name is the same as Vedvyas's birth name. However, both were different persons but contemporaneous great personalities of the Vedic civilization.

Krishn in Bhagvadgita, a part of the book of Mahabharata, has cited references of the Darshans. For example the Saankhya Darshan has been referred to 5 to 6 times in the *shlokas* (2/39, 3/3, 5/4-5, 13/25, 18/13). In addition, the *shloka* (6/35) in Bhagvadgita has in it the Yog Darshan's *sutra* (1/12) and so on. Therefore, the last two Darshan's must have been written just before the Mahabharata War began. Mahabharata has several references of Manusmriti, placing Manusmriti anterior to the Milestone at the onset of the Kali-Yuga calendar. Since all Darshanic principles in the six Vedic Darshans have their origin in the Vedas, all Vedas are before the Milestone.

The Vedic civilization started to decline several centuries after the Mahabharata War. This was also a time of turmoil and destabilization of the settlements along the shores of the mighty River Sarasvati. Due to cataclysmic events in the geologically sensitive region at the foot-hills of the Himalayas, the flow of the river was disrupted leading to the ultimate extinction of the River Sarasvati. As discussed in Chapter 3, these facts are now confirmed by new technology, satellite mapping and geological research. The excavations along the course of the rediscovered extinct River Sarasvati have revealed thousands of these settlements where Vedic civilization had once flourished. During the dark period after the Mahabharata War, the well-established *guru-kulas*, like Rishi Vedvyas', declined and perhaps disappeared with the high level of language and script also vanishing from the common masses. About a thousand years BCE the caste superiority and caste rigidity took root in the Vedic society, when some people in the priestly brahman community controlled education, keeping the knowledge of Sanskrit within the brahman community. For the financial gain and power some members of the priestly community propagated blind beliefs and as a consequence the ritualistic Hinduism gained the foot-hold. They gave their own erroneous interpretation to the Vedic texts and forbade women and the so-called lower castes from reading and studying the Vedas. To fulfil their own ulterior motives, some of them created self-serving *shlokas* and inserted them in the book of Mahabharata and Manusmriti. The book of Mahabharata, according to ancient references, as discussed by Swami Dayanand Sarasvati in his Satyarth Prakash, originally had 10,000 *shlokas*, which now has about 100,000 *shlokas*.

As a result of this literary pollution, one finds many inserted *shlokas* contradicting the original *shlokas* written by Rishi Vedvyas and his students. Confusion is imminent when inserted *shlokas* present un-Vedic opposing philosophical views and ritualistic practices. Most scholars accept the fact that the present Mahabharata is a corrupted version of the original and is artificially enlarged. This has led some Indologists to believe that the entire Mahabharata was written about 300 to 400 BCE and that it may altogether be a fictitious story. Fortunately the research by Vedic scholars of the Vedas, Vedic Literature and the recorded historical documents, has established the reality of the Mahabharata War and the book of Mahabharata.

Puranas

Puranas (item 10 in Table 4.1) are the religious books of Hindus with history and mythology in them. They are a set of 18 books which are a part of the Sanskrit Literature because they are written in Sanskrit and also because they have been written about half a millennia BCE through a few centuries CE, which by some standards is considered to be ancient. They are the source of ritualistic practices in Hinduism where there is prevalence of mythology, idol worship, worship of multitude of gods and goddesses in differing forms, belief in reincarnations of God in the form of a human being, ritualistic worship of gurus and many other such beliefs. The Vedic exhortation being diametrically opposite to this, the Puranas could be included in the Sanskrit Literature but could not be included in the Vedic Literature. This is why the title Puranas in Table 4.1 is in brackets.

Based on his deep knowledge of Sanskrit Literature and Hindu history, Swami Dayanand Sarasvati has included considerable historical information in the 11th chapter of Satyarth Prakash (1881 CE). From a history book written in 1725 CE he furnished the tables of successive genealogy records of 124 Aryan kings who ruled from Hastinapur (Delhi) over a period of 5,157 years starting near 3,140 BCE. The details on this are in Chapter 3.

He states that King Bhoj ruled from Avantika (Ujjain) about 500 years after the great king Vikramaditya. Regarding the Puranas, he writes that the Vaishnav-Hindu sect began 150 years after King Bhoj of Ujjain (about 600 CE). The founder of the Vaishnavait sect was Shathkopa. Munivahan and Yavanacharya (of Greek origin) later propagated this sect and popularized it by the well-known Ramanuja. The Vishnu-Purana was also written during that time for the propagation of the Vaishnavait-Hindu sect.

Likewise, the Shiv-Purana was written for the Shaivait Hindus. Many other Puranas also sprang up during this period of the Indian history. To give legitimacy, to gain authenticity and respect and to promote their sects the authors of these Puranas wrote them under the names of the great *rishis* of the past. Rishi Vedvyas, the author of the Vedanta Sutra and the book of Mahabharata, became their target. As a result, those following the Puranas believe that Rishi Vedvyas was also the author of all the 18 Puranas. This is also the time when some pundits who followed the Puranik Hinduism and other pundits who followed other non-Hindu sects, composed and inserted texts in the ancient books such as Ramayan, Mahabharata, the Smritis and several other post Vedic books.

Swami Dayanand Sarasvati wrote that during the rule of King Bhoj a group of pundits created Markandeya-Purana and Shiv-Purana in the name of Rishi Vedvyas, who in fact had lived over three and a half thousand years before King Bhoj. For bringing the great Rishi Vedvyas' name into disrepute and creating pollution in the revered Vedic Literature, King Bhoj punished the writer pundits and had their hands cutoff. Thereafter he issued a directive throughout his kingdom that anyone creating any poetry should do so in their own name and not in the name of the revered ancient *rishis* and the sages. Swami Dayanand Sarasvati also cites that the learned King Bhoj wrote a history book entitled 'Sanjeevani', which is one more example of the Indian tradition of recording history. Swami Dayanand Sarasvati cites a statement by King Bhoj from the book of 'Sanjeevani':-

> "*... at the time of King Vikramaditya* (57 BCE), *Mahabharata book contained 20,000 shlokas, at the time of my father there were 25,000 shlokas and in my half lifetime alone the book of Mahabharata contains 30,000 shlokas. If it continues to grow like this it will turn into a camel-load and if people created books like Puranas in the name of the rishis and the sages, the people of Aryavart will be entangled in the net of doubts and will become devoid of Vedic dharma and become bhrasht* (defiled)*"*.

The insertions of *shlokas* in the book of Mahabharata continued clandestinely in spite of King Bhoj's strong directive. As the King had predicted, the doubt and confusion persists in the interpretations of the scriptures. The underground process of writing of the Puranas continued after King Bhoj and their number increased to 18.

All Hindus accept that the Vedas are their most ancient books and there are no insertions in them by post-Veda writers and Puranik pundits. There are several reasons why the Vedas have not been sullied by insertions. The Aryans revere Vedas as the revelation by God in the minds of the *rishis* and therefore do not need further improvement. In addition, the unique Vedic poetry is difficult to emulate even by the scholarly. The Veda mantras being in unique seven different metered *chhandas* also made it difficult for a polluter to insert his dross. An important factor for preserving the integrity of the Vedas is the Vedic tradition of the scholars learning the Vedas by heart, mantra by mantra, in the same sequence and reciting them without even a slight variation of a syllable. For fear of a quick detection of the insertion, the Vedas remained untouched by the literary polluters.

In his effort to revive the Vedas and the Vedic practices and to point out the negative influence of the prevailing books of Puranas on the Hindu society, Swami Dayanand Sarasvati, born and brought up in a priestly brahman family himself, said that the true meaning of the word Purana was lost and misused by the brahman caste. To emphasize this in the 11th chapter of his book Satyarth Prakash he cited references from the Vedas and the Brahmana books. A reference from there is as follows –

ब्राह्मणानीतिहासान् पुराणानि कल्पान् गाथा नाराशंसीरिति ॥

Transliteration: (*Brahmanaaneetihaasaan puraanaani kalpaan gaatha naaraashanseeriti.*)

Translation:

"*The Brahmana-books* (Etreya, Shatpath, Saam and Gopath) *are addressed with five names: itihaas, purana, kalp, gaathaa and naaraashansi.*"

To explain with mastery the five Vedic terms in the above-cited reference by giving examples from the Brahmana books Swami Dayanand Sarasvati, the scholar par-excellence, dives deep into the ocean of Vedic Literature and brings out the pearls of literary definitions that otherwise would have remained hidden. He writes (in Hindi):

" *...the philosophical dialogue between the King Janak and Rishi Yaagyavalkya is **itihaas** (history); the description and elaboration on the creation of the universe and the other phenomenon of Nature is **purana** (account of, a reckoning); the process of defining, elaborating and manifestation of the ingenuity of Vedic terminology is **kalp**; description of a phenomenon of importance with examples and similes is **gaathaa** (i.e. rendition with proof and example) and the word **naaraashansi** stands for describing the inspirational and approved karma of people* ".

Some of the 18 Puranas do contain valuable history of the society in the form of the genealogy of kings and information on events in the society. Whereas the Puranik history of the society relates to the real persons of the Vedic society, the Puranas have also included commentary on the history of the creation of the universe. This history at times has some correlation to the Veda mantras and the statements in the Brahmana books and at other times, it contains absurd imagination on the subject of the creation of the universe. They are far from the depth of thought in the Brahmana books, which cover various aspects of the society as well as the aspects of the creation of the universe.

The 18 Puranas themselves are huge volumes of Sanskrit poetry and there are an additional set of 30 other Puranas termed as Up-Puranas (sub-Puranas) that also come in the category of Puranas, with a total *shloka* count of over 4 million (4,329,000 *shlokas*). Therefore, even for a diligent and eager student of Sanskrit Literature, who has chosen a Purana to be the starting point, the Puranik Literature alone will be quite overwhelming before entering the realm of the Vedas and the Sutras. As said before, the Sanskrit Literature is vast; it is an ocean of ancient literary work.

Western Indologists and the Western professors of Sanskrit of 18th and 19th century, including Professor Max Muller who translated RigVeda in English followed the writings and the interpretation of the Puranik scholars.

Contemporaneous to this new Vedic European literature in English and in other European languages, were Swami Dayanand Sarasvati's interpretive translations of the RigVeda and the YajurVeda in the Hindi language. His translation of course followed the methodology prescribed by the ancient *rishis*. At places in his translation, he has commented on the specific deficiencies in the translations of Indian authors like Sayana and Mahidhara as well as on the mantra mistranslations by the European Sanskritists, Max Muller included, where he made it abundantly clear that those authors lacked competency in the grasp of Sanskrit grammar and the language itself.

The references related to cosmology and the developments of other thoughts in this book are primarily drawn from the ancient texts listed in the Table 4.1. It was established in Chapter 3 that the main Vedic references on cosmology, used by this author, precede 3200 BCE and the translations of the mantras are based on *yaugik* meanings and ancient methodology prescribed in Nirukt and emphasized by Swami Dayanand Sarasvati.

Oral and Writing Traditions of the Aryans

The process of learning for all humans begins orally. It is debatable as to how soon after the conception of a child the process of oral learning begins. New references on this topic could be found in modern science publications, however, an ancient reference is found in the book of Mahabharata (item 9 in Table 4.1), written over 5,000 years ago. The process of oral teaching and learning according to this reference becomes operational for the child soon after conception. Prince Arjun and his queen Subhadra (sister of Krishn) talked purposefully about specific strategies to their brilliant son Abhimanyu when he was in his mother's womb. Mahabharata states that when Abhimanyu was only a teenager he implemented the complex *chakravyuh* war-strategy successfully without pre-training in it. It is believed that the process of learning through oral vibrations in the mother's voice begins when the child is still in mother's womb.

The knowledge senses according to the Vedic philosophy are seeded in a living being at the moment of conception and they develop from there on and become active right after the birth. It is also known that a large number of words and sounds are absorbed and stored in the child's brain before the child learns to speak, read and write. The oral process of teaching and learning is thus in place from the very beginning of the child's life which ultimately develops into a language for communication. The oral process could continue indefinitely into adulthood if the written script was not taught or nonexistent. The minimum requirement for oral learning is the preexistence of nomenclature of objects, phenomena and processes in the person's environment and some grammar to communicate the words in a given meaningful sequence. The larger the vocabulary in a language of communication, the higher is the possibility of social development and civility in a person and in the society as a whole.

Some civilizations have carried out their business in an oral tradition while others have invented scripts to carry forward the process of learning and communication. The process of passing knowledge from one generation to the next is better served when the writing tradition is in place along with the oral tradition. The society that emphasizes both traditions is better off than a society managing with the oral tradition alone or the society that has a high emphasis on the writing tradition but little value for oral learning.

There are misconceptions about the tradition of teaching and learning in Vedic Civilization of ancient India. Based on guesswork, hearsay and without citing a reference Max Muller who never set foot on the Indian soil fielded several myths about the Vedic Civilization. His myths about the Aryan Hypothesis (AH) and Aryan Invasion Theory (AIT) were debunked in Chapter 3. He emphatically fielded another myth that the Aryans had only the oral tradition of teaching and learning and the script did not exist in India until 300 BCE. He also emphasized that because there was no script the Indians did not record their history and were not good at it. Max Muller took it upon himself to write India's ancient history and compressed last 5,000 years into 2,000 years bringing most historical events from BCE to CE and declaring the rest as fictional.

Some Indians have high reverence for Professor Max Muller but other proud Indians are deeply saddened by the irreversible damage he has done to the Indian culture, literature and the Vedic Civilization. His translations of Sanskrit Literature, the RigVeda in particular, have remained like an irreversible chemical reaction where the participating molecules lose their identity. In his lecture series at Cambridge, entitled 'India - What Can It Teach Us?' (publ. 1882 CE) he writes in its VII chapter:

> "*Here then we have the first fact, viz, that writing, even for monumental purposes, was unknown in India before the third century BC.*"

He further writes:

> '*...the art of writing may have reached India before the time of Alexander's conquest (325 BC), its employment for literary purposes cannot date from a much earlier time.*"

Then after a few paragraphs, he wonders:

> "*Here then we are brought face to face with a most startling fact. Writing was unknown in India before the fourth century before Christ, and yet we are asked to believe that the Vedic literature (in its present form) ... goes back to at least a thousand years before our era.*
>
> *Now the Rigveda alone which contains a collection of ten books of hymns addressed to various deities consists of 1017 poems, 10,580 verses, and about 153,826 words. How were these poems composed – for they are composed in very perfect meter – and how, after having been composed, were they handed down from 1500 before Christ to 1500 after Christ, the time to which most of our best Sanskrit MSS belong?*"

There are a number of errors in his count. The total number of poems (*suktas* or hymns) in RigVeda is 1028, not 1017, and the total number of verses (mantras) is 10,521 and not 10,580. Max Muller's above-mentioned guesstimates within two pages of the same chapter of his book exhibit other errors as well. First, he stated without literary or historical analysis that writing was unknown in India before the third century BC. Next, he said that it was unknown until the fourth century BC. Finally, he wondered how, without the written manuscript, over a 1000 poems composed in perfect meter, could have been handed down from 1500 BC to 1500 AD. Max Muller's guesstimates reveal his inadequate historical analysis and inadequate study of the Sanskrit Literature.

India was under the British rule at the time of the influx of the Western Indologists into India; Max Muller remained in England. The British had devised a cunning strategy of 'divide and rule' India. For this, they implemented an education system that discouraged Indians from learning about their glorious past. Additionally, this system included newly created texts with distortions in the ancient Indian history. The British history reveals that Lord Macaulay in 1834 CE commissioned Max Muller to translate (mistranslate) the RigVeda into English for a payment of 4 shillings per page.

How then does a foreigner justify rewriting the history of another country? He has to come up with a solid reason for doing so. A solid reason could be that the natives were uneducated and needed his input. This however was not the case with the Aryan natives of India. The Aryans of the past had created vast Sanskrit Literature that the present Aryans had inherited in the manuscript form, in a highly developed script in a scientific language built on an admirable grammatical structure and philosophical foundation. Then if the foreigner still insisted on rewriting the history of the country, he had to create a myth that – '*the natives were educated but not in a habit of writing history*'. This myth worked for Max Muller who supplemented it by saying that the natives of India had an oral tradition of teaching and learning but did not have a script; therefore, they were not good at recording their history. To borrow an expression from his own quotation cited above, Max Muller came *face to face with a most startling fact*. This fact is: if Indians were able to hand down accurately the 'complex and vast poetry' of the Vedas in the written form to the later generations, at least from 1500 BC to 1500 AD as he said, then why were they not capable of writing-recording and handing down their ancient history? Obviously, he ignored the simple logic.

After creating the myth that "*writing was unknown in India before the third century BC*", he needed an explanation as to how the massive Sanskrit Literature could have flourished without a script. So he created another myth that "*...the art of writing may have reached India...*" presumably from somewhere outside of India, around the time of Alexander's invasion (325 BC). If his doubts about the historicity of the Sanskrit Literature were genuine, he could have learned from the Sanskrit Literature itself that writing was prevalent in the Vedic society along with the oral tradition.

Let us bring in some ancient references from the Vedic Literature to show that the writing process existed in India from ancient times. These references are the book of Nirukt by Rishi Yask and the book of Ashtaddhyayi-Grammar by Rishi Paanini. As Acharya Udayaveer Shastri has analyzed this point in "Vedant Darshan Ka Itihas" and some other contemporary Hindi writers have also pointed out that while interpreting Vedic terms in the Nirukt Sutra, Yask Rishi frequently uses the term – *uprishtaad vyakhyasaamah,* meaning '*we will establish this further in what follows*'. Such expressions suggest that there will be further discussion to establish the point being made in the written text of the book of Nirukt. With reference to the definition of terms and expressions, the book of Nirukt has also referred to about 14 other *acharyas* (scholars) either contemporary to Rishi Yask or before him. His references to these 14 Acharyas indicate the existence of literary works in textbook form prior to him. Rishi Yask's period by Sanskrit scholars is estimated to be anterior of 1900 BCE.

Rishi Paanini has used the term *lipi* in his book on Sanskrit grammar. The term *lipi* means 'script'. Rishi Paanini would use the word *lipi* only when the script existed in his time. Even if one wishes to discount Rishi Paanini's period, estimated to be before 2707 BCE (ref. Vidyanand Sarasvati, 1989), one fact remains that Rishi Paanini lived before Rishi Yask's period and certainly before the advent of Buddhism. It is also well known that Buddhism began before the time of Aadi Shankaracharya and was at its peak in India during his time. Based on deep literary research in ancient documents guarded at the Shankaracharya monasteries in India, Acharya Udayaveer Shastri (mentioned above) has established the dates of birth, death and other events in Shankaracharya's short life. This places Shankaracharya's birth on *Baishakha, shukla panchami*, 2631 in Yudhishthir Calendar, which is the 5th day of the 2nd month in the Vedic calendar. This translates to the month of April in 509 BCE. He lived exactly for 32 years 6 months and 10 days. This proves that Indians were very good at keeping historical records. Thus, the Indians have had *lipi* (script) at least in the 6th century BCE.

In his thorough analysis of the Vedic Literature and history, Pundit Yudhishthir Meemansak, a prominent Sanskrit scholar and a prolific writer of Vedic Literature, has provided in the first part of his highly acclaimed book 'Sanskrit Vyakaran-shastra Ka Itihaas' (History of Sanskrit Grammar), a proof of the fact that Rishi Paanini wrote his Ashtaddhyayi-grammar about 400 years after Mahabharata War. As mentioned before, and proven in Chapter 3 the Mahabharata War ended in 3140 BCE with the coronation of the King Yudhishthir. This takes us close to 2900 BCE. Therefore, *lipi* (script) in India existed well before the speculative dates of Max Muller and all those Sanskritists who accept his guesstimates of the time-lines of the Vedic Literature.

This author will now take the reader even further into the antiquity of the Indian civilization, well before Rishi Paanini's time, deep into the past of the Indian civilization anterior to the Mahabharata War, to show that the written script has been a part of the Vedic Civilization from its very beginning. The book of RigVeda itself has instructions for having the practice of writing and record-keeping in the society, which can be possible only when a script is in place. In the following are several references that clearly indicate the existence of script in India for writing and record keeping. Some historical records are also furnished here to show that the ancient Indian society recognized the importance of preserving and propagating its developed knowledge base for the benefit of the future generations. This is very similar to the way in which knowledge is preserved and disseminated in the modern societies.

Max Muller and the other Sanskritists and Indologists missed reading the following set of successive RigVeda mantras, which issue a directive to the judiciary of the society for record- keeping.

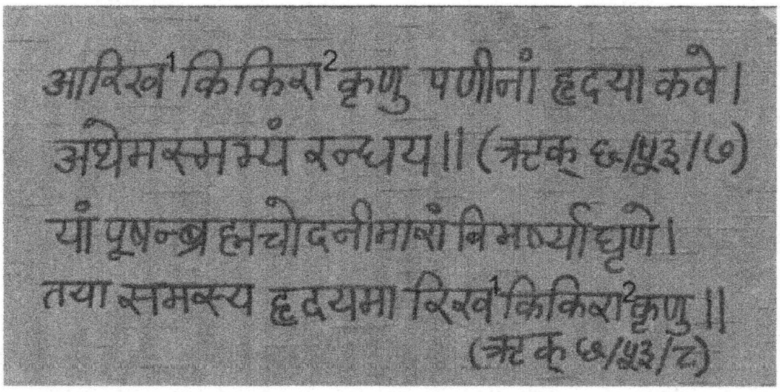

आरिख[1] किकिरा[2] कृणु पणीनां हृदया कवे।
अथेमस्मभ्यं रन्धय॥ (ऋक् ६/५३/७)
यां पूषन्ब्रह्मचोदनीमारां बिभर्ष्याघृणे।
तया समस्य हृदयमा रिख[1] किकिरा[2] कृणु॥
(ऋक् ६/५३/८)

Transliterations: "*Aa rikh*[1] *kikira*[2] *krinu paneenaam hridayaa kave. Athemasmabhyam randhaya.*)" (RigVeda, 6/53/7).

"*Yaam pŏoshanbrahmchodnimaaraam bibharsha ghrine. Taya samasy hridayama rikh kikira krinu.*" (RigVeda, 6/53/8)

Mantra translations:

"O learned people, prepare detailed written records covering all facts on unsocial and criminal persons and punish them severely to change their heart and thus provide comfort to the rest" (RigVeda, 6/53/7).

"O the learned one, propagator and upholder of justice, use your justice to separate the truth from untruth like a saw separates wood pieces. Prepare written record covering all facts on those who provide proof and who inspire for knowledge and wealth" (RigVeda, 6/53/8).

The directive is for both aspects of the judiciary - the execution of the clear-cut judgment and the writing of the detailed judgment. The Veda did not consider it sufficient for a judge to pass a judgment orally and move on to the next case. It instructed the judiciary to keep detailed records of all facts from both sides - the plaintiff's side and the defendant's side. The writing of a judgment by the judiciary is possible only with the preexistence of a script. In addition to the Veda taking for granted that the people already have invented a functioning script, an added beauty in the mantra is the simile of the 'wood and saw'. The Veda instructs a judge to pass a clear judgment just like a saw cuts wood into two distinct pieces, truth on one side and false-hood on the other. Thus, the Aryans, the followers of the Vedas, had read the RigVeda (script) had worked to iron out wrinkles in the laws governing their Vedic society and had developed a constitution to follow. The book of Manusmriti is such a constitution.

There are two important words that are repeated in two different contexts in the above two mantras; in one context the emphasis is on the criminal, the unsocial accused (7th mantra) and in the other context the emphasis is on the plaintiff (8th mantra). These words are आ रिख[1] and किकिरा[2] (*aa rikh*[1], *kikira*[2]). The words आ रिख[1] (*aa rikh*[1]) mean '*write in detail covering all facts*' and किकिरा[2] (*kikira*[2]) means '*written papers or written record or a report*' (all these meanings apply). These two mantras therefore summarize the whole process of a judgment by a judge where the record-keeping of a case is a must. This process of maintaining a record of a case in a judicial process, true in the present day society as well, is not possible without script.

The Vedas for the Aryans are the word of God. The expression – 'word of God', for Aryans has a different meaning than in other religions. God in the Vedic concept does not create sound for pronouncing words. This word is not like a deep voice calling from the sky. In the Vedic thought, all knowledge emanates from the all-pervading formless Conscious-being, the Creator, and each word is the result of the vibrations in that knowledge plane. These vibrations inspired the elevated minds of the *rishis* and sages in their meditation who gave these vibrations the sound (*shabd*) by pronouncing and subsequently giving them shape by writing in script. The Vedas are the collection of that divine knowledge put forth in the form of unique and perfect mantra poetry.

Therefore, it is natural for an Aryan to follow the instruction in a Veda mantra. As defined in the Veda itself the term *arya* means *eeshvar-putra*, which means the 'progeny of God' and therefore the term *arya* stands for '*a noble person*'. A progeny first begins learning to speak from its mother and is always eager to learn more. The Creator in the Vedas is therefore also addressed as a mother. The Aryans read and believe that all instructions in the Veda are for the benefit of all humanity and the rest of the creation and therefore an *arya* is to follow the written instruction prescribed in the Veda since the ancient times. To say that the Indians did not develop script until three centuries BC defies logic.

This author feels that inventing a script in which to express a language is not as difficult as developing the language itself. What is even more difficult is to formalize the grammar of that language. Then if the grammar of Sanskrit is extremely complex and scientific, it indicates a high level of development in the society. The script (*lipi*) in this author's thinking is of tertiary importance after the creation of a word (*shabd*) in the mind and pronouncing it through the voice (*sarasvati*). If the language is in place, the script would naturally follow.

The name for the Sanskrit script is Devnagari. The curvaceous letters stand in a unique style. It has been an independent script and has existed from the time of the Vedas to the present time. All the mantras in this text are in the same Devnagari script. To demonstrate the ancient practice of writing mantras on an ancient writing medium, this author hand-wrote the above two mantras on *bhoj-patra*. A discussion on *bhoj-patra* (bark of the *bhoj* tree) and other ancient writing media is in the following.

RigVeda states unequivocally that a hand is to be used for writing. Not just that, but the mantra also instructs the reader to prepare a system by which the writing process can be achieved much faster, a thousand times faster than hand.

उप[1] त्वा सातये नरो[2] विप्रासो यंति[3] धीतिभि:[4] ।
उपाक्षरा[5] सहस्त्रिणी[6] ॥ **(RigVeda, 7/15/9)**

Transliteration:

"Ŏop[1] tva saataye naro[2] vipraaso yanti[3] dheetibhih[4].
ŏopaakshara[5] sahashrinee[6]"

Mantra translation:

> "*Just as the learned persons with diligent effort produce akshara*[5] *(letters of script) using fingers of the hand*[4]*, O (man*[2]*) desirous of knowledge, prepare a well thought out system (instrument*[3]*) in collaboration with intellectuals to produce them thousands*[6] *of times faster than a hand*" (RigVeda, 7/15/9).

In this mantra, the word *naro*[2] *(naraha)* stands for human beings and *dheetibhihi*[4] according to the Nirukt (5/2/5/41) means - '*by the use of human hand'*. In this Nirukt reference, there is another RigVeda mantra with its Samhita reference number (5/1/23/1) being taken up by Rishi Yask for interpretation wherein the instruction is to create fire by a mechanical process. The common factor in the two mantras is the use of the human fingers. The instruction in the mantra under the Nirukt interpretation is to use the fingers to produce fire as - '*use fingers to rotate arani fast to create fire by attrition (friction)*'. Rishi Yask in his Nirukt (>1900 BCE) has given the definition of the Vedic term *arani*. An *arani* is an object '*that creates aran* (motion) *in the hidden agni* (fire) *by association of hand'*. Even in the present day Indian society the term *arani* is in common usage for a 'hand-held spindle'. As analyzed in Nirukt by Rishi Yask, the Vedic term *dheetibhih*[4] in the above RigVeda mantra (7/15/9) is for the use of the hand for writing. The term *didhiti* in the other RigVeda Samhita reference (5/1/23/1) implies the same meaning for creating fire with an *arani*. It is the use of the 'human fingers'-- use the fingers to hold a pen to write the script and use the fingers to hold an *arani* to create fire by attrition.

The above mantra tells us - 'use your hand' to generate writing of *akshara*[5] where the term *akshara*[5] means the letters of a script. The word *ŏopaakshara*[5] is in fact a euphonic combination of *ŏop* and *akshara*[5] which together mean the 'secondary letters' or the 'resulting letters'; which means that the letters are primarily created in the mind and then secondarily reproduced by hand on a medium. We all accept the fact that humans after inventing the script began writing the *akshara* with hand and therefore a human hand is the oldest instrument for word processing in the human civilization. A human hand in the Sanskrit terminology has five fingers; the thumb is also counted as a finger. Thus, the term *dheetibhihi* represents all five fingers in the human hand and is the naturally contrived biological instrument for writing.

The most important feature of the mantra is the instruction in it for going beyond the biological instrument, the hand. The mantra explicitly suggests speeding up the process of writing to *sahasrinee*[6], which appearing in plural means '*thousands*'. In common figure of speech, this expression is similar to '*a thousand times*' and therefore the mantra's emphasis is to achieve a much higher speed in the process of writing than the speed attainable by a hand. This also tells us that the mantra is not instructing one to write faster and faster by hand; rather it is explicitly telling us to devise the process to increase the speed of writing thousands of times faster than the speed of hand-writing. Additionally, the mantra instructs to achieve this in collaboration with other scholarly intellectuals.

First a theoretical aspect of the instruction in the mantra: The terms *ŏop*[1] and *yanti*[3] in the first line are separated unlike in the first compound term *ŏop-akshara*[5] of the second line in the mantra. Linguistically they are also in a euphonic combination. Therefore, when taken together the compound word *ŏop*[1]-*yanti*[3] would lead to the meaning - '*a well thought out resultant plan (or system)*'. When other root meanings of the word *yanti* are considered, one finds that, it also stands for '*in an organized method*' or '*in a controlled method*'. These closely related meanings obviously are for developing a plan or a process for speeding up the writing process.

Now let us look into the practical aspect of the meaning of *ŏop-yanti.* A derivative of the term *yanti* is the term *yantra*, which conveys the *yaugik* meaning of '*that which alleviates one's suffering*'. This is why, the term *yantra* in common usage is for 'instrument' that helps to reduce the pain and suffering of the physical body. The two words together forming a single term *ŏop-yantra* is also an alternative to the most commonly used term *ŏop-karan* in most regional languages of India that have their root in Sanskrit. The term *ŏopkaran* is commonly used for an instrument or for an instrumental aid. A telescope for example is a *ŏopkaran* (secondary instrument) to enhance the capability of the eye where the eye is a *karan* (the primary doer of action). A car is a *yantra* or *ŏopyantra* to alleviate the suffering in walking long distances. A machine in general is a *ŏopkaran* or a *ŏopyantra.* An instrument by definition enhances the ability of a sense and the Sanskrit term for sense is *karan.* The mantra is encouraging the practical aspect of speedwriting. Going beyond a hand is possible with a machine (*yantra*). Since not all scholarly necessarily have a technical bent of mind for developing and implementing technical *ŏopkarans* (instruments), the mantra additionally instructs the reader to achieve the goal of speed-scripting with the help of other talented accomplished people. To develop the process of writing further, the humans have used various methods, media and aids. Later the humans have gone way beyond and invented mechanized instruments for writing such as typewriters, printing-machines and computerized printers that have enabled the script writing a thousand times faster than a hand.

Now, if we were to revisit the mantra translation from a modern perspective, the mantra translation would not be too far from - *develop a fast mechanical system to carry out the process of writing a thousand times faster than the hand.* The Veda, which the Aryans consider the source of all knowledge, advises humans to invent the process of writing by hand and 'thousands of times' faster by mechanical means. One needs script for a machine to print!

An *akshara* (letter) can be taught using only the auditory method of listening and pronouncing. The *ŏopkarans* utilized for that are the voice and the 'sense of hearing'. Nevertheless, the real meaning and purpose of the *akshara* are enhanced when its distinct geometric shape is visually recognizable. This adds an additional *karan* to the process, the sense of seeing. As mentioned before the root meaning of the word *akshara* itself is '*a non-decaying entity*' or '*a non-diminishing entity*'. This in mathematical terminology is the 'invariant'. In most languages a letter is just a symbol to represent a sound where its components, 'le - tt - er' themselves have no deeper meaning in them. In Sanskrit on the other hand, the term *akshara* conveys the quantized power of an entity that for all purposes is abstract, with no subsistence in it. One could dismantle a word into many letters (*aksharas*) by separating all its *aksharas,* but an *akshara* itself continues to exist. This also means that just as matter at its fundamental state is quantized without a quantum being further divisible, the language is made up of words that are divisible into *akshara,* an entity not divisible further. In Sanskrit however the half-*akshara* are also used when an *akshara* is pronounced for shortened sound. This line of thinking on the part of the *rishis*, where giving meaningful nomenclature to an entity that reflected the properties of the entity, is science (*vigjyaan*). The structure of the Sanskrit language and the structure of matter are similar. Two hydrogen atoms and one oxygen atom form a water molecule with its chemical formula being H_2O. The water molecule is like a word dismantling into its 'non-decaying *aksharas*: H, H and O. Therefore, while naming a letter as *akshara* the *rishi* left a stamp on it that an *akshara* is a fundamental building block of words in a way similar to H, H and O are in a water molecule. What atoms are to a molecule the *aksharas* are to the word. This is the case of all *aksharas* in the Sanskrit script, which with given rules together form the words of the script. Sanskrit script, which is Devnagri script, has the built-in integrity of being phonetically consistent, unlike the English language in which for example the same *akshara* 'u' carries different sounds as in the words 'but' and 'put'.

Different languages of the world have differing numbers of *aksharas* (letters) in their scripts. For example, there are 63 in Sanskrit script including the vowels and the consonants, 24 in Greek, 23 in Latin, 26 in English and so on. Thus, as soon as an *akshara* appears, the writing process begins. It becomes clear from the above mantra references that as far as the Vedic reading and writing is concerned there has not been a separation of verbal recitation and the writing process. RigVeda the oldest book of the Aryans clearly advises to develop both, the Vedic tradition of oral learning as well as of writing by hand and writing with the aid of *ŏopkaran* (instrument).

It may be of interest for the reader to know that unlike the previous handwritten mantras, the last mantra above (RigVeda, 7/15/9) was typed in Devnagari (Sanskrit) font by this author using the keyboard and printed using an electronic printer. To accomplish this, a sheet of *bhoj-patra* was fed through the printer. This certainly expedited the process perhaps a thousand times, as the above mantra instructed one to do! A *bhoj-patra* is the naturally occurring writing material for the Vedic Aryans to write on and to preserve the written Vedic text. A few more words on *bhoj-patra* are in the following.

This section is concluded by presenting only the first line of a mantra from AtharvaVeda that should remove all doubts about the myth that the Aryans did not have the script to write in and had only the oral tradition of learning. The relevant part of this mantra states:

यस्मात् कोशादुदभराम वेदं तस्मिन्नंतरव दध्म । (AtharvVeda, 19/72/1)

Transliteration: "*Yasmaat koshaadudbharaam vedam tasminnantarav dadhma*"

Mantra translation:

" *Just as the book of Veda is taken out of the chest and returned to it after reading, the Vedic knowledge emanates from God and returns to Him* (on dissolution of the creation) ... " (AtharvaVeda, 19/72/1).

In a typical Vedic style, the mantra explains a phenomenon from the spiritual realm by using a simile from the practical daily activities of life in the physical realm. The Veda in the physical realm is a book of knowledge, a physical object that is brought out of a chest for reading and returned to it for safekeeping. In the spiritual realm, the Veda is the knowledge itself, a non-physical entity that emanated from the Omniscient and returns to Him at the time of the dissolution of the creation.

This mantra of AtharvaVeda completely negates the hypothesis that the Aryans had only the oral system of learning because one can take a book out from a chest and return to it, only when it was in the written form. This activity by an Aryan is possible when a script existed and the process of writing was in place, all the time right from antiquity. This mantra puts a 'double-strikethrough' in the conjecture of Max Muller et al. and all those who still insist that the Aryans only had the oral tradition of teaching and learning.

Whereas it was advisable for a student to learn to recite all Vedas in their entirety (in up to 13 different ways), he/she was also advised to complete a prescribed curriculum to finish his/her education. Depending on how much study one aimed to take up, the prescribed study period in the Vedic education system ranged up to the age of 25 to 35 years. Except for some geniuses, it was impossible for the average students to cram the vast Vedic Literature, both in prose and poetry, preserving the accuracy of the text and the sanctity of the knowledge. To learn by heart all the mantras from the four Vedas a student would have to cram 10,521+ 1,975 + 1,875 + 5,977 = 20,348 mantras (including some common mantras) in the same sequence as presented in the Vedas – a very difficult challenge indeed! Those who completed the task of learning by heart the 20,348 mantras from the four Vedas earned the high revered title of Chaturvedi. Likewise, the surnames of Trivedi and Dwivedi were conferred for mastering respectively three and two Vedas. Those who carry such surnames are the descendents of learned ancestors.

The strong emphasis on oral tradition of learning prevailed in the Vedic society for several good reasons and it continues in the present-day *guru-kulas*. A good reason is that the oral learning allowed a student or a teacher to carry a book in his/her mind and not in his hand. This practice also enabled a quick retrieval of a mantra or a reference to advance a discussion on a subject. The practice of having the Vedic text in a *rishi*'s memory was like having a powerful living and mobile computer from which the desired mantras or *sutras* accessed sequential or at random. Just as in the modern computer age, the recalling of references requires computer hardware or a hardcopy of the required information, the *rishis* in ancient times also knew the importance of preparing the hardcopy of their thoughts. They certainly wrote their thoughts down on writing media in a well-developed script and initiated a tradition to carry the script-writing practice forward for the future generations.

A civilization may erect material monuments to display and preserve its history. Some monuments go back millennia in the past. A monument being a material manifestation of the science and culture of the society, it is primarily the subject of the sense of seeing. Another civilization may go beyond a proud display of material objects and instead lay emphasis on the writing of the books of wisdom. The books of the Vedas for the ancient Aryan society are a manifestation of the pinnacle of intellectualism and spiritualism. When it comes to determining the age of a civilization, both, the material monuments and the books furnish information on the age of the civilization. In the case of a material monument, the legends and the written history about it and the carbon-dating would lead to its age. In the case of a book, the linguists would carefully study the parameters and applicable historical references to determine its age. The contents of a book itself may reveal the way of life and thought pattern of the people of that era. A necessary condition for the accurate determination of the age of a book from its text is to have the thorough knowledge of the language of the text and the history and culture of the society to which the book belongs. The Vedic civilization emphasized the pursuit of philosophy and spirituality. The Veda books and the books of the Vedic Literature in general are the spiritual monuments that the *rishis* left behind. The mantras are the invariant building blocks of this spiritual monument that do not decay with time.

Along with inventing a script and carrying out the writing work, one needs a medium for writing. One can start with the engraving on stone and etching on metal plates, but one cannot go too far with that. To write books and to keep written records, one needs paper and ink; barring the digital (virtual) copies. The ink is relatively an easy item to extract from vegetation while the process of producing paper was yet to be invented. It is said that the Chinese invented paper in 105 CE. However, the records of the foreign travelers to India during the period of Alexander-the-Great in the 4th century BCE express that the Indian businesspersons kept business transactions written down on paper locally available. If the Aryans had the writing tradition but had no paper until 105 CE, how did they carry out the writing of the vast Sanskrit Literature? This question is addressed in the following.

Writing Medium in the Aryan Civilization

The generic term for a writing medium in Sanskrit language is *patra.* The word *patra* means a thin flat surface like a leaf. For that matter, a sheet of paper for writing and printing is a *patra.* Several writing media have been used by the ancient Aryans, most popular being the *bhoj-patra, taad-patra, kapaas-patra* and *resham-patra.* According to an ancient tradition in Vedic Culture when a king or the society venerated a person, a plaque made of a thin copper sheet was presented with the distinction engraved (etched) on it. Such a medium of writing for important occasions is termed as *'taamra-patra'* (copper-paper). Therefore, while a *patra* means a leaf it also signifies a thin flat writing medium.

There is a common expression in contemporary Indian society that the ancient Aryans used to write on *bhoj-patra.* In this context, the *bhoj-patra* is not the leaf of the *bhoj*-tree. A *bhoj-patra* is a thin layer of the *bhoj*-tree's bark. In the case of a *taad-patra,* however it is not the bark of the *taad*-tree (palm-tree) but the broad leaf of the *taad*-tree. The writing of government land records on a thin pressed cotton-sheet (*kapaas-patra*) or on silk cloth (*resham-patra*) has been an age-old practice in the Indian countryside. Indelible ink was used on these mediums to prepare land ownership records with descriptive survey diagrams to preserve them for centuries. The village councils in south India, as late as the end of the twentieth century, still kept land records on *resham-patra* in parallel with the paper work. In case of the renewal of land records or the land-dispute, the village *patwari* (land officer) would bring out his land register containing land records written in black indelible ink on thin semi-transparent silk-cloth. Such old records with the village *patwaris* have served as the basis for the current land-data-base under the present state governments. This author has witnessed a *patwari*'s register and has acquired land-ownership records on silk-cloth for the land owned by the previous generation in the family.

In the context of the Indians keeping good records, Max Muller in the VII chapter of his book 'India-What Can It Teach Us?' has stated a historical fact related to Alexander-the-Great. Alexander's admiral named Nearchus sailed down the Indus River in 325 BCE. Records show Nearchus stating that when he frequented the maritime stations of India he met Indian merchants, who for commercial contracts and documents '*wrote letters on cotton that had been well beaten together*'. Three facts emerge from Max Muller's writing. First - the script existed in India in the fourth century BC; second - the writing was quite prevalent down to the level of business transactions; third - the Indians wrote on paper made out of cotton. Max Muller usually contradicts himself!

Before the invention of mass printing, the *bhoj-patra* was in ancient times an inexpensive medium for writing, which nature provided in abundance. Depending on the specific application, the *bhoj-patra* in varying bark thickness could be retrieved for writing; from a few millimeters down to thinner than regular printing paper, with a natural smooth surface quite like paper. Most gurus in ancient India lived in ashrams in forests. When a guru assigned a student to take up serious writing work for preserving a text, the student would go out and collect the naturally peeling off bark of a *bhoj*-tree or retrieve from the naturally fallen *bhoj*-tree. A *bhoj*-tree is Betula Utilis or Himalayan Silver Birch tree. One finds several separable laminations under the outer rough white surface of the unrolled birch bark. When the outer layer is peeled off, four to six successive layers of the bark underneath are quite smooth for writing. This bark is much stronger than a printer paper of the same thickness. A *bhoj*-tree (silver birch) sheds its bark naturally. This enabled the ancient *rishis* to do their writing work on *bhoj-patra* without affecting the health of a tree. In the present method of producing paper, the death of a tree is inevitable.

To demonstrate a sample of Sanskrit writing in a way quite similar to the ancient Vedic times, this author has hand-written the two earlier cited mantras (RigVeda, 6,53/7, RigVeda, 6/53/8) on bhoj-patra. To integrate this writing into the text of this book, the author has included the scanned portion of the hand-written mantras. The original handwriting is on the third layer from the outer white rough bark of *bhoj-patra*. The difference here of course is that these mantras are written on a *bhoj-patra* retrieved from a Canadian birch and not from a Himalayan birch. The thin laminae that one peels off from a piece of the naturally separated bark of the birch-tree are not only thinner than the regular printer paper but they are also waterproof to last longer than a regular sheet of paper. The shade of the successive lamina changes from brown to a cream color providing good contrast for the black ink as shown above in the two printouts. One could use thicker laminate of *bhoj-patra* to preserve a manuscript for centuries. As mentioned earlier, it was the physical form of RigVeda Samhita's transcript, hand-written on *bhoj-patra* by Acharya Sayana, sometime in 14th century that Max Muller received in England for his English translation of RigVeda.

Record-keeping and History Writing in Ancient India

The nineteenth century effort by the Western Indologists and neo-Sanskritists to rewrite the ancient history of India generally excluded the historical existence of kings like Vikramaditya after whom the Vikram Calendar began 57 years prior to the Common Era Calendar of the West. They also excluded the other kings who ruled millennia prior to Vikramaditya; to name a few here - Manu of Manusmriti, Rama of Ramayana, Krishna, Yudhishthir and Arjun of Mahabharata (texts listed in Table 4.1). For the Aryans (Hindus) of India, all these kings of the past have been part of their glorious ancient Indian history. Influenced by Puranik mythology, the Western Indologists deemed these significant persons of history as fictitious and mythical characters. The historical accounts in Ramayan and Mahabharata and the recorded genealogies of kings preserved in Sanskrit Literature were also deemed fictional. With such a starting premise there was hardly any substance left for anyone to write about the ancient history of India except to talk about the invasion of India by Greeks, Muslims and Huns. The fact is that compared to the long ancient history of the indigenous Aryans of India, the invasions by the outsiders and their rule over India, are relatively recent.

Even during the foreign rules, the ongoing ancient practice of record-keeping throughout India remained independent and unaffected. To get to know about the history and record-keeping practices in ancient India, one had just to read the available accounts of the foreign travelers and visitors to India. Max Muller in the first chapter of his book – 'India- What Can it Teach Us?' has narrated from the second century writings of Arrian, the pupil of Epictetus, on the practice of reporting administrative information by appointed public overseers and superintendents in ancient India:

> "*They oversee what goes on in the country or towns, and report everything to the king, where the people have a king, and to the magistrates, where the people are self- governed, and it is against use and wont for these to give in a false report; but indeed no Indian is accused of lying.*"

Although Arrian's above account is from the second century CE, the practice of reporting to the king and magistrates is a very ancient practice in India. Such a practice could be meaningful and sustainable only when a script existed in the society. Writing in a script and writing of the history had been a prevalent cultural practice in India from ancient times.

Old genealogy records of the dynasties of kings in different regions of India are still preserved. Writing and recording the history is not an imported practice in India. They are a result of the indigenous practice ingrained in the Indian civilization. A few examples of such ancient historical records are given here.

Based on a book written in 1725 CE the unbroken record of genealogy of the Aryan kings who ruled from Indraprasth (Delhi), going back right up to the Emperor Yudhishthir in 3140 BCE were presented in Table 3.3 in Chapter 3. Also discussed in Chapter 3 is the history book entitled Rajtarangini, written in 1148 CE by Kalhan that takes us three millennia BCE detailing the genealogy of the Kashmir dynasty.

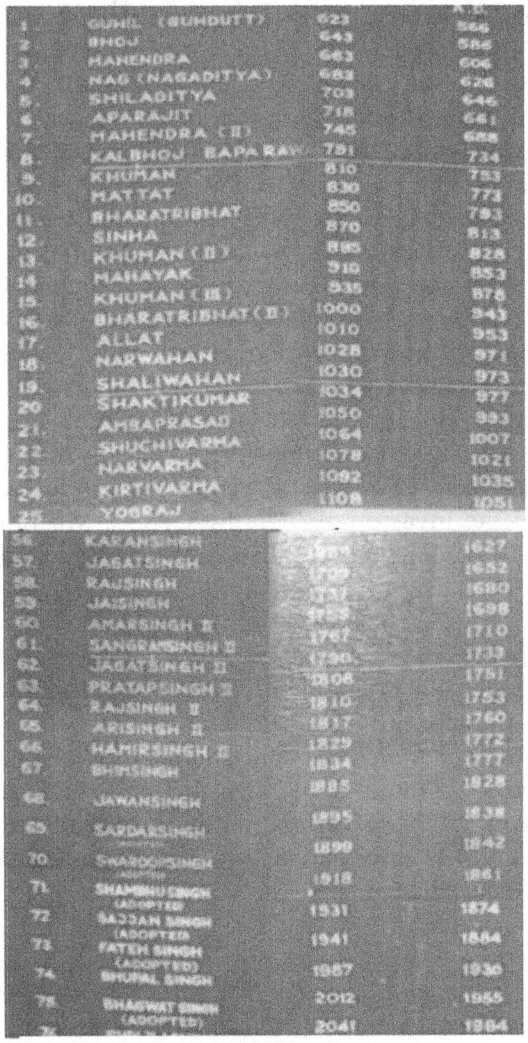

Udaipur Dynasty: Continuous record from 566 AD to 1984 AD

An example of a continuous *vanshavali* (genealogy) record, going back from the present time to 566 CE over 76 rulers of the Mewar Kingdom, is painted on the entrance wall of the Udaipur Palace in Rajasthan, India. This author has truncated the list for brevity. Before the British introduced the Common Era Calendar (CE) in India, the dating of written work including *vanshavalis* was carried out in one of the ongoing Indian *samvats* (calendars) discussed in Chapter 3. The Udaipur *vanshavali* begins with the first king Guhil (Guhdatt) who established the Kingdom of Mewar in the year 623 of Vikram Era calendar (566 AD). The painted genealogy of the Mewar kings presents the serial number, the name, the year of coronation and the number of years of that king's rule (not shown in this table). The Common Era year (A.D.) shown in this genealogy was added in recent times by the palace record keepers for the benefit of the tourists. Udaipur is one of the most frequented tourist spots in present day India. The fact remains that such a record will not go on a wall without it first being recorded in a ledger. Such a detail was not possible without having a tradition of written history. Not only the record-keepers of the kings preserved the records, the temples also independently maintained ledgers. These records have remained uninfluenced by the foreign rules or by the British methodology of record-keeping.

Maintaining historical records is a common practice all over India. The record-keeping continues at all levels of the Hindu society, from the kings to the common people. A simple example of common record-keeping is the preparing of the birth certificate. When a child is born in the present developed Western civilization, the birth certificate records the child's name, parents' names, place and time of birth. These details are also recorded in child's *'janm-patra'* (birth certificate), however, they are complemented with the astronomical data as well. A geometrical chart is drawn on the *'janm-patra'* that provides the planetary positions at the time of the child's birth. To top it all, the birth constellation of the child is also recorded. The birth constellation is one of the 27 constellations named indigenously in Sanskrit and it indicates the Sun's position in the ecliptic when the child is born. The present astronomical nomenclature for all constellations used in such preparations is the same as the one given in the AtharvaVeda, discussed in Chapter 11 Volume 2.

About two thousand years ago, the practice of fatalistic astrology crept into the art of preparing the *janm-patra* of the newly born. Those who prepared the *janm-patra* charged money to interpret the supposed effect of the planets on the newly born and performed *pooja*, a ritualistic ceremony, to reduce the adverse effect of the planets. Regardless of such astrological practices, the process of writing and maintaining of birth records has been a very beneficial tradition in the Indian civilization. Of course, the uneducated farmers did not worry about getting the *janm-patra* prepared.

The process of record-keeping was not limited to a locale; it extended even to far-away places of pilgrimage that connected India from the north to the south and from the east to the west. Remarkable about these specific pilgrimage places is the ancient practice of writing down a pilgrim's name, place and date of pilgrimage along with the amount of donation given to the brahman priest performing the ceremony to bless the pilgrim. These records continue to exist since centuries.

The *rishis* in ancient times introduced the practice of going on pilgrimage to holy places. The *rishis* in their wisdom established places of pilgrimage at the extremities of the country and advised people to travel to these places. The multiple benefits of undertaking a pilgrimage were obvious. It kept the history of the holy place alive and a pilgrim got to know the geography of the country and became familiar with different customs of the inhabitants in different parts of the subcontinent.

This author recalls vividly his visit as a child along with his parents to the pilgrim place called Haridwar at the foothills of the Himalayas on the banks of the River Ganga (Ganges). It was during the mid-1940s when India was about to get freedom from the British Raj. As we got off the train at Haridwar, a representative from the Haridwar priestly community greeted us. He asked for my father's name and place of birth. Right away this representative steered us to a small office where a priest sat in the lotus position at a slanted writing desk. The priest welcomed us, asked us to sit down and declared that he and his ancestors have been our family's priests for many generations during the pilgrimage journeys to Haridwar. He performed a short worship service by reciting mantras for the successful completion of our family's pilgrimage. Then he asked a young priest to hand him a specific register related to our district and our family. He opened the register and started reciting the genealogy of my family like a passage from the scriptures, going back many generations. This monotonous recitation included the names of my grandfather and great grandfathers, the dates of their pilgrimage in Vikram-calendar (Hindu calendar) and the amount of donations received. My father gave the priest some cash donation and we left for a dip in the River Ganga. I heard my father telling my mother that as far as he could recall the priest's recitation of our family genealogy was flawless.

The priest did not render special treatment to this author's family because our ancestors were privileged people or from a big town. The fact is that this author's parents and ancestors have been the residents of a small village that is several thousand years old, in the state of Rajasthan, some 500 km from this pilgrim place at Haridwar. Nevertheless, the practice of recording the dates of pilgrimage continues even now with the difference that the amount of donation to the priest is probably inflated and the record-keeping probably is computerized.

The practice of record-keeping is ingrained at all levels of Indian culture.

European Scholars on the Age of the Vedas

Historians have used several methods to estimate the age of the ancient civilizations, cultures, monuments and literature. There are several different methods utilized by the historians for differing subjects:

(a) Mythological Method compares the different mythologies to ascertain their period of development.
(b) Antiquarian or Historical Method examines and interprets the ancient records at hand to ascertain a historical period.
(c) Contemporary Method establishes a succession period of specific literary works by fixing their individual dates (period).
(d) Scientific Method subjects the ancient article to laboratory testing, conducts an analysis of the scientific data in the areas of astronomy, geology, chemistry, ecology, satellite observations, the genome studies and so on.

European scholars have utilized different applicable methods to determine the age of the Vedas that led to a wide range of inconclusive estimates.

Mythological Method and Puranas

An implicit assumption in applying the Mythological Method on the timeline of a society is that a society begins with myths and blind-beliefs that diminish with time when education and science increase. An attempt to trace back the timeline of a developed society may lead to the beginnings of that society. While there is no difficulty in accepting that education and science eradicate myths and blind-beliefs, the difficulty is in accepting that the timeline of a society is a bi-ended straight line, with the primitive state of the society at the first end and the developed state at the second. In this author's view, it is a curve with peaks and valleys rather than a straight line; peaks at the height of civilization and the valleys when the society is saturated with ignorance, myths and blind-beliefs. A society that once was at the cusp of civilization with a high level of education could go down into an abyss with time, the causes being external invasions, internal degradation or nature's calamities. The same civilization could rise again. To judge a society's up and down over a span of five-six millennia is not logical because the intelligent humans according to anthropology have been around at least a quarter of a million years.

An examination of the Vedic culture tells us that the Vedas have existed in the Hindu society first and have no mythology in them. The other major literary works based on the Vedas, written in the Aryan society of India prior to and up to the Mahabharata period 5,000 years ago, such as from the Upanishads right down to the six Darshans, have no trace of mythology in them. The mythology in the Hindu society begins after the society went through a dark period during a millennium or two after the Mahabharata War when the education in the society declined. The mythological stories took firm hold after the Puranas were written from about 500 BCE. This was also the time when the form-worship was introduced by Buddhism and Jainism to worship respectively the Buddha and Teerthankars when the priestly brahmans followed suit to counter the conversion of Aryans/Hindus to these two religious disciplines, which according to the Vedic philosophy, were atheistic. Aryans/Hindus also began building temples with their own deities in them. Archeological excavations reveal that even the most ancient Hindu temples are younger than about two and a half thousand years. The process of creating mythology continued in the Hindu society for a thousand years after the beginning of the Common Era when it became a part of Hindu life. In the present Hindu society, one would find both kinds of people, having a highly philosophical approach to religion without believing in the myths and those believing in myths, ghosts, animal sacrifice and so on. Therefore, the age of a society's literature cannot be determined only by the rate of decline of the mythology and the onset of scientific progress. From the study of human history, be it of Vedic India or of the Egypt of the Pyramids, one can observe that the cycle of progress and decline of societies is a long-term oscillatory phenomenon.

Antiquarian (scientific), Antiquarian (Historical) and Contemporary Methods

The Antiquarian Scientific Method when applied to a hardcopy of a manuscript of the Veda, hand written on *bhoj-patra*, would yield the age of the manuscript and not of the Veda. The Antiquarian Historical Method of crosschecking for the references of the Veda in other Vedic Literature indicates clearly that the Vedas are mentioned in other books and commentaries on the Veda mantras but in the four Veda books, no other books are mentioned. Historically speaking, the deduction drawn is that the Vedas are from an era long before the other books from the ancient past of the Indian Aryan Civilization. The Contemporary Method required a thorough knowledge of the Vedas and the Vedic Literature. The Western scholars of that time were thronged with pre-conceived notions and prejudices about the Veda and the post Vedic Literature.

Observing the mediocrity in applying the Antiquarian and Contemporary methods to determine the succession period in Vedic Literature, Pundit Guru Datta noted in "The Terminology of the Vedas and the European Scholar" (1880 CE): -

> "*One is indeed wonderstruck at the way in which European scholars mistrust Indian chronology, and force their hypothetical guess work and conjecture before the world as a sound historical statement of facts. Who, that has impartially studied the Darshan literature, does not know that the Darshans existed centuries before even the first word of Buddhism was uttered in India? Jaimini, Vyas and Patanjali had gone by, Gautama, Kanada and Kapila were buried in the folds of oblivion when Buddhism sprang up in the darkness of ignorance. Even the great Shankara, who waged a manly war against Buddhism or Jainism, preached nearly 2,200 years ago…Further, there is no event as certain in Indian History as Mahabharata, which took place about 4,900 years ago.*"

The six names in this quotation are those of the authors of the six Darshans (Table 4.2) that precede the great war of Mahabharata. The cross-referencing studies in Sanskrit Literature by contemporary Indian Sanskrit scholars (i.e. application of Antiquarian/Historical and Contemporary Methods, e.g. Pundit Udayaveer Shastri) establish that the other branches of Vedic Literature (items 1 to 7 in Table 4.1) preceded the six Darshans (item 8). Going deeper into Vedic philosophy one gets the revelation that the manifestation of the Darshan *sutras* and their structure had an underlying process that required the pre-existence of their source. Undisputedly, the four Vedas are the source of the six Vedic Darshans and all the other post-Vedic literary works listed in the Tables 4.1 and 4.2. The interpretive exposition of the Veda mantras in the books of Brahmanas and a foundational discussion in the Upanishads were a prerequisite for the development of the Darshans. Glimmering through the sutras of the Darshans are also the principles laid in the Smritis.

The 'great Shankara' in the above quotation is Aadi Shankaracharya who established the five Mathas (monasteries) between 499 BCE and 482 BCE in the five regions of India. A detailed discussion on Aadi Shankar was presented in Chapter 3. The period of Aadi Shankaracharya is highly relevant to the state and the fate of Budhism in India. The written historical material preserved in these monasteries is one of the best possible evidence for the Antiquarian and Contemporary methods. The European scholars hardly touched this material before they hypothesized about the history of the Vedic Literature or hypothesized about Aadi Shankaracharya's period. Those Indian scholars who tooted the horn of the European scholars also overlooked this valuable well-preserved and protected historical material in the Sanskrit language. Acharya Udayaveer Shastri (cited earlier) researched the Matha literature arduously while staying at the monasteries. The results of his studies in Shankaracharya monasteries were presented in Chapter 3. In the same chapter are also discussed the findings based on the Scientific Method that include the evidence related to astronomy, geology, ecology and the state of the art genetics, all of which place the Aryans, their culture, language and religion way back in the antiquity of modern human development.

Guru Datta in the above quotation over a century ago did not pull out of a hat the number 4,900 years for the historical time of the Mahabharata War. He based that time on the historical accounts from the ancient Indian literature and he questioned the *European scholars' mistrust in Indian chronology.* Swami Dayanand Sarasvati and Guru Datta Vidyarthi pointed out independently in differing ways, the basic flaw in the Europeans' approach to determining the age of the Veda but their advice went unheeded. On this subject, Max Muller had put forward several self-contradicting hypotheses.

The real milestones of the Vedic Civilization are the books listed in the Table 4.1. To apply the Contemporarian Method to these books, one has to be able to read and understand the written material in them and the written commentaries on them. Learning to read Sanskrit is not sufficient to understand the meaning of the writings on these milestones. For example when one is looking for a milestone such as the 'Six Vedic Darshans' (Table 4.2) one will find that they are six different milestones separated in time. When one arrives at one of these milestones and begins to read the *sutras* in it, one may not be able to apply the Contemporarian Method to determine the dates of their creation. The ill preparedness of the Indologists meddling in the Indian history and the history of the Sanskrit Literature is summarized by a well-known incident in Max Muller's life (noted from "Early Indologists & Indology of India, PART 2; by Swami B. V. Giri):

> *A brahman student from India went to England. When he heard that a great Sanskrit scholar named Max Mueller taught at Oxford University, he went to meet Max Mueller. When he came face to face with Mueller and spoke to him in chaste Sanskrit, Professor Max Mueller was dumbfounded, unable to respond. Mueller admitted that he could not understand 'what the gentleman was saying'*!

Based on Max Muller's mantra translations, his contemporary in India, a monk with mastery over the Sanskrit language and having immense Vedic knowledge, Rishi Dayanand Sarasvati observed that Max Muller had indulged in distorting the meaning and the knowledge in the mantras. In his book Satyarth Prakash, he compared the Sanskrit ability of Max Muller to - *'a toddler learning to walk'*.

About the same time the German philosopher Arthur Schopenhauer (1788-1860), who had studied the Upanishads well, also cast doubt on the ability of the Westerners claiming authority on Sanskrit. He noticed that neo-Sanskritist Europeans were meddling amateurishly in determining the age of the Vedas.

Swami Dayanand Sarasvati advised the Indologists and neo-Sanskritists where their deficiencies lay in translating the mantras. One simple example of a deficiency pointed out by Swami Dayanand was their ignorance of the elementary rule of Sanskrit grammar in the usage of singularity and plurality. These Indologists had not learned that in addition to a singular form of a word there were two plural forms of the word – dual plural and multi-plural. Max Muller while translating mantras had treated a dual-plural as plural causing distortion in the meaning of the mantra.

While translating the very first mantra of RigVeda and its first word *agni* in particular, Swami Dayanand Sarasvati gave references of several other mantras in the Vedas and the ancient statements from the Shatpath Brahmana, Nirukt and Manusmriti. He emphacized that the word *agni* stands for two meanings respectively in the spiritual and the physical realms and the context decides which meaning applies. A Veda mantra may at times have two meanings valid for each realm. It is under a specific poetic-adornment rule in Sanskrit called the *'shlesha-alankaara'*, a unique beautiful feature of the Vedic Sanskrit poetry when a single mantra renders more than one meaning, both valid. Swami Dayanand Sarasvati in his Hindi translation and commentary of the RigVeda pointed out, in the first few pages, that the Western Sanskritists and scholars of Indology namely Wilson and Max Muller did not understand this multidimensional poetic adornment. The consequence was their erroneous translation of the Veda mantras. The very first word of the very first mantra of RigVeda, the *agni*, remained beyond their grasp of understanding.

Swami Dayanand Sarasvati said that to understand a Veda mantra, having some vocabulary of the Sanskrit language, a handy Sanskrit dictionary and just being able to read the rules of Sanskrit grammar were not sufficient. One had to understand and learn the required methodology emphasized in the Vedic Literature by the ancient *rishis.* One had to understand the books of Brahmanas, the logic of the language, Vedic Sanskrit grammar (Rishi Paanini's Ashtadhyayi), understand Vedic terminology (Rishi Yask's Nirukt and Nighantu), other educational tools and finally understand the context of the mantra itself. However, the Indologists ignored his advice. Observing such a state of affairs Swami Dayanand Sarasvati expressed his opinion in the 11th chapter of Satyarth Prakash (1881 CE) as in the following (translation taken from Guru Datta Vidyarthi):

> "*The impression that the Germans are the best Sanskrit scholars, and that no one has read so much of Sanskrit, as Professor Max Muller, is altogether unfounded. Yes, in a land where lofty trees never grow, even a ricinus communis or the castor-oil plant may be called an oak. The study of Sanskrit being altogether out of question in Europe, the Germans and Professor Max Muller may there have come to be regarded as highest authorities....I came to learn from a letter of a Principal of some German University, that even men learned enough to interpret a Sanskrit letter are rare in Germany. I have also made it plain from the study of Max Muller's "History of Sanskrit Literature" and his comments on some mantras of the Vedas that Professor Max Muller has been able only to scribe out something by the help of the so-called tikaas, or paraphrases of the Vedas, current in India.*"

Some Western and Eastern scholars have drawn clues from the Veda mantras for quick and easy conclusions to establish the historicity of the Vedas. However, such clues have not yielded sound conclusions. Their ill-fitting linguistic clues were based on choosing certain words that appeared in some Veda mantras. One such word is *manu. Manu* is also the name of a king, the author of Manusmriti and the king who established the city of Ayodhya. Their argument was that because the term *manu* exists in the Veda, the Vedas have history in them and without any further historical analysis Max Muller baselessly hypothesized that Manusmriti was written probably after 400 BCE and therefore the Vedas must have been written in not too distant a past from that period.

Swami Dayanand has provided the meaning of the term *manu* and explained how it is derived from the "*munn gjyane*" *dhaatu* (root) for "*to think*". God is addressed as Manu because "*His knowledge is His glory*". In the context of a Veda mantra the term *manu* is for the "*all-accepting Creator*" and not a person named Manu who became a king. Not understanding the context of why the term *manu* appeared in the mantra and what its *yaugik* meaning is, the wrong conclusions were drawn by Max Muller and others about the historicity of the Vedas.

Vedic Principle for the Age of the Vedas

The Vedas neither have in them any specific calendar date for their creation nor do they carry any human history. They just have the formulations of thought, called the mantras. The Big Bang cosmologists believe that as soon as the process of the creation of the universe began and the matter started to form, the gravitational field started to coexist with matter as the property of matter. They label the gravitation in matter as an 'inherent property of matter'. In a similar way the Vedas emphasize that the knowledge is also an 'inherent property of the creation', available for living beings to tap into at will. The level of acquisition of this knowledge is however according to the ability and predisposition of the living being. This predisposition of a living being depends on its physical and spiritual evolution acquired through the karmic cycle. The Vedas for the Aryans of India are the revelation of knowledge, the term for which in Sanskrit is *gjyan*. The *gjyan* according to the Vedas and the Vedic philosophy is one of the properties of the universe (Nature) that has a relationship and interaction with conscious-beings and is not merely a chemical process in a living-being's brain. What mass is to matter the *gjyan* is to consciousness. Just as many entities and phenomenon of eternal and transient nature, such as Time, Space, energy, gravitation in matter and so on are revealed at the onset of creation, the *gjyan* energy field is revealed through the consciousness in life forms. It is an energy field unto itself that emanates from the all-pervading consciousness field called God; it persists through the universe to be availed by a conscious-being individual soul, according to the soul's preparation. This preparation is the soul's ability to acquire *gjyan* (knowledge). *Gjyan* begins emanating from the time zero of the creation and therefore assigning an age to it is the same as assigning an age to Time or assigning an age to the gravitational field of the universe. At the end of creation, the field of *gjyan* energy, along with all forms of matter and energy (Nature) retract back into their source, the Creator energy field. This is a short explanation without invoking details of the Vedic philosophy for the Vedic connection between the consciousness and the *gjyan* (knowledge)

It is natural to think that this concept of knowledge and its origin is un-provable and therefore a self-serving hypothesis. Fair enough. However, this human limitation is not limited to philosophy it is also a drawback of Big Bang Hypothesis as well. A hypothesis enables a beginning point for the search of knowledge, be it in the realm of matter or of consciousness. Therefore, a hypothesis attempting to prove a phenomenon in the realm of matter is not fit to reject a hypothesis pertaining to consciousness.

As soon as matter manifests in the creation the mass appears in it, which creates the gravitational field that embeds the universe in it. In the Big Bang Hypothesis, there is the all-permeating gravitational field because there is matter in the entire universe. In the theistic Vedic thought, there is the ubiquitous knowledge (*gjyan*)-field because there is the all-pervading Creator-energy field with the additional property of consciousness. In the Vedic concept of *gjyan* the consistency of logic would require that the knowledge field existed from the beginning of creation because for the Vedic concept the Creator entity exists from the beginning of creation. This is what the Vedas propound, the Upanishads explore and the Darshans conclude in their *sutras*.

No living creation can begin and survive without the spark of knowledge appearing in a living being. Therefore, the Aryans believe that because the knowledge exists from the very beginning of the universe, the Vedas are present right from the beginning of creation. The soul, a localized formless quantum of consciousness in a living physical body, endeavours through the physical body equipped with knowledge gathering functions. For the emanating Creator-energy knowledge-field, each soul serves as a reception antenna receiving the level of knowledge according to its own reception ability. This ability is a function of the soul's *karmic* balance. It is an antenna's attribute to intercept and receive the signal. The formless quantum of soul also has the attribute of interacting with matter to form its physical/biological body for implanting the knowledge senses. It is through the physical/biological body, equipped with the knowledge senses, that the soul transforms the formless knowledge into a visible script. This script in turn can manifest into a book.

When one goes beyond this abstract concept of knowledge and thinks in terms of the substantive form of the Veda, one begins to wonder at what stage of creation the Vedas came into the form of chapters (*mandalas*), sections (*suktas*) and verses (*mantras*). It is the physical form of the book or a manuscript that is generally assigned an age, 500 BCE or 1500 BCE as the Indologists did, erroneously. The presumption in assigning such dates is that the knowledge contained in the mantras of the Vedas and the physical book of the Veda came simultaneously. This assumption however breaks down when one comes across the reference of the Veda in a book much older than previously thought. All books in the vast and ancient Vedic Literature, anterior to Mahabharata (in Tables 4.1 and 4.2) are based on the Vedas and are inundated with references of the Vedas but do not make a single mention of the age of the Vedas. One is amazed with the fact that not a single reference of any texts, manuals, notes or dictionaries is found in the Vedas that might have preceded the Vedas. We do not have a single book or even a single phrase that a historian could find which might have preceded the Vedas. A follower of the Veda cannot venture upon assigning an age to the knowledge (*gjyan*) field that persists from the beginning of the creation. That is why the common expression - *the Vedas are ageless*. This implies that they are from the beginning of the creation. Since the Vedic Civilization in India is very ancient and the invariant Vedas have been the fountainhead of knowledge for the Aryans of India, the Vedas are put in a category of revealed knowledge from the beginning of creation. The *gjyan* energy field ought to exist before the first conscious-being living cell is to appear in the creation and survive its next moment.

Max Muller et al Credited and Discredited

There is no doubt that Max Muller's dedication to Sanskrit and the studies in Indology resulted in voluminous writing, over 3000 pages, covering the Vedic and Buddhist literature. He carried out editorial work on the RigVeda and the RigVeda *bhashya* (interpretive translation) from the original hand-written manuscripts that he received from India. He also wrote several books on the Vedic religion and culture. Many people in the literary field are in high admiration of him including some scholars of Indian origin who have even called him a *rishi.* Obviously, these scholars have no literary knowledge of the definition of a *rishi.* A *rishi* by definition is the '*revealer of the mantras and their recondite knowledge*' and not just a translator with a Sanskrit dictionary by his side.

Another writer called Max Muller the 'Vedvyas of Kaliyuga', Kaliyuga being the present era in the Vedic Calendar with onset on February 18, 3102 BCE. Rishi Vedvyas, the great philosopher and the author of the Vedanta Darshan (Brahm Sutra philosophy) and the book of Mahabharata cannot be compared with mediocre translators of Sanskrit text. Just to name a few themes expounded in Mahabharata alone are politics, science, duty to society, social ethics, character-building and philosophy. Rishi Vedvyas was also the writer of the interpretive commentary on Yog Darshan. Rishi Vedvyas' positive effect on the mind of the Aryans is immense. Max Muller on the other hand has been instrumental in negating what Rishi Vedvyas stood for. Those who have called Max Muller with the superlatives of '*rishi*' and 'Vedvyas' must spend more time reading and understanding Vedic Literature and shake off the adverse effect of the educational scheme of Lord Macaulay, Max Muller and the British Raj.

Max Muller's dedication to his Sanskrit studies and Indology, as we understand from his life history, was to earn a high status of living and to uproot the Vedic religion, the religion that can claim to be universal. Any damage to the non-sectarian philosophy of the Vedas, with the message of universality, is not the damage to the Vedas themselves, because the Vedas are the invariants in time and space. The real damage is to the whole world and a great loss to all intellectuals and those striving for spiritual enlightenment free of the crutches and dogmas of sectarian religions.

References of King Manu, the codes and principles in his book of Manusmriti, inundate the book of Ramayan (too many to cite here). Five millennia ago Rishi Vedvyas' Mahabharata composition is inundated with references of Ramayan (Mahabharata-Shanti Parv, 11/20-24) and Manusmriti (Mahabharata-Shanti Parv, 13/35; 18/11 Anushaasan Parv, 12/15;....). Yet, Professor Max Muller states in the III chapter in his book - "India –What Can It Teach Us?" about the ancient book of Manusmriti:

> *"As to the laws of Manu, which used to be assigned to a fabulous antiquity, and are so still sometimes by those who write at random or at second-hand, I doubt whether, in their present form they can be older than the fourth century of our era, nay I am quite prepared to see an ever later date assigned to them. I know this will seem heresy to many Sanskrit scholars, but we must try to be honest with ourselves. Is there*

any evidence to constrain us to assign the Manavdharmasastra (Manusmriti), such as we know possess it, written in continuous slokas, to any date anterior to AD 300?"

Max Muller in the above quotation imposes literary honesty on others and himself and says that Manusmriti could not have been written before 400 AD or 300 AD. Max Muller pulls both numbers out of his hat without furnishing any logical basis or a solid historical or literary reference, yet he had the audacity to challenge others to provide proof that Manusmriti was written before 300 AD. He obviously needed a lesson in logic. He ought to have provided some proof for his own dates before raising the rhetoric question in his quotation above.

Max Muller never visited India and India's literary institutions and he never developed the required foundation in Sanskrit grammar. He based his logic or lack of it, on his limited perspective developed in Oxford, England. He never subjected himself to even one *shastrarth* (a traditional literary and philosophical challenge/debate) to discover how far he was right or wrong. It would not be excessive to say that an old Sanskrit idiom applies to him - *koop-mandook*: a frog in a well that has little knowledge of what lies outside.

How could the simple logic escape him that in 326 BCE when Alexander the Great invaded India the orderly Indian society was already following Manu's laws from his book Manusmriti? The educated Indians commonly quoted Manusmriti and its *shlokas* and the Indian society was following the dharma based on the principles ascribed in Manusmriti. Max Muller himself writes in "India - What Can It Teach Us?" that "*Alexander stood speechless in front of the yogis and maneeshis*...". The *maneeshis* are the thinkers and followers of dharma, the dharma code of ethics that is based on *Manav-dharma-sastra* – the Manusmriti. Therefore, the book of Manusmriti had to be much older than 300 AD. Manu and Manusmriti are quoted in the books of Ramayan and Mahabharata. If Muller's logic were followed, then these two books would belong to a period after 300 AD. Absurd!

Max Muller's scholarship and professionalism are questionable. To get a clearer picture of his honesty and professionalism let us examine a few more of his statements. Why is the present author singling out Max Muller's incompetent work related to the historicity of the Vedas and his mantra translations? The reason is significant.

The primary aim of this book is to present the mantras on science thought and cosmology in the Vedas. The reader is going to find differences in the translations of the relevant mantras provided by Max Muller and this author. Most scholarly in the West and a large number of educated Indians unquestioningly accept Max Muller's conjectures about the Vedas and the Vedic Literature and his translations of the Veda mantras. The question is whose translation is correct? To translate a mantra this author has first provided the *yaugik* meanings of all the terms first and the composite translation in English is constructed based on the Vedic methodology mentioned earlier. Max Muller has used *rudhi* meanings and provided composite translations that at times have no relation to the mantras and at other times are meaningless and weird. The reader will notice this fact when two translations are compared. A sample of Max Muller's nonsensical translation of a mantra is presented under the sub-heading

"Max Muller's Scholastic Sanskrit Ability in Question" in the following where he wrongly translated 9 words out of about only a dozen in the mantra, RigVeda (1/162/2). In addition to the nonsensical translations of the mantras he is among the early Indologists who has done the most harm to the good name and the value of the Vedas and the Vedic knowledge. This author is making an effort to shake the negative impression that Max Muller has created all over the world with his well-publicized myths about the Vedic culture, Vedic history, Aryan invasion of India and Veda mantras, including those containing science thought and cosmology. This author's humble desire is to promote the *satya*, the 'truth' of the Vedas and question Max Muller's *a-satya*. The humble hope is for *Satyameva jayate,* meaning - *'only the truth prevails'.*

Max Muller did not consider withdrawing the erroneous translations even when pointed out. Many contemporary authors in the arena of Vedic writings (Swami B. V. Giri, Swami Vidyananda Sarasvati), who are referred to in this text, have carried out thorough research in the literature on Max Muller and presented quotations from Max Muller's books and letters that he wrote. This author has cited in the following just a few quotations from the published literature (see Ved Meemansa, publ. 1984 and Aryon Kaa Adidesh aur Unki Sabhyata, publ. 1989, by Swami Vidyanand Sarasvati).

After Max Muller learnt Sanskrit and carried out some translation work in Germany, he went to England at the age of 23 in search of a job. There he met Lord Thomas Babington Macaulay (1800-1859 CE), a member of the Supreme Council of India representing the British government in British India. Macaulay was in the process of implementing the British Education Plan for India. Soon after his arrival in 1834 CE, Macaulay commissioned Max Muller to translate the main religious book of the Hindus, the RigVeda, into English at a rate of 4 shillings per page. After that, he was appointed at Oxford where he translated and wrote many books on Vedic culture, Vedic Literature and Buddhist philosophy. Max Muller is therefore considered the champion for popularizing the Indian wisdom contained in the Vedas and the philosophies based on the Vedas. His letters to different people are available in printed form and tell us about his true state of mind. Following are some excerpts from 'Life and Letters of Fredrick Max Muller'.

(A) From his speech at the Christian Missions in Westminster Abbey in 1873:

> "*Hinduism was dying or dead because it belonged to a stratum of thought which was long buried beneath the foot of modern man... The worship of Shiva, Vishnu, and other popular deities was of the same and in many cases of a more degraded and savage character than the worship of Jupiter, Apollo or Minerva. A religion may linger on for a long time, it may be accepted by large masses of the people, because it is there, and there is nothing better. But when a religion has ceased to produce defenders of the faith, prophets, champions, martyrs, it has ceased to live, in the true sense of the word; and in that sense the old orthodox Brahmanism has ceased to live for more than a thousand years.*"

His first sentence in the above quotation that "*Hinduism was dying or dead …long buried beneath the foot of modern man…*", is in direct contradiction with his assertion in the third chapter of his book, 'India - What can it Teach Us?' where he said, "*I shall say even more, and I have said it before, …we can hardly understand how, at so early a date, the Indians had developed ideas which to us sound decidedly modern*".

(B) To his wife in 1866 about the translation of RigVeda and the Vedic religion:

> "*This edition of mine and the translation of the Veda will, here after, tell to a great extent on the fate of India and on the growth of millions of souls in that country. It is the root of their religion and to show them what the root is, I feel sure, is the only way of uprooting all that has sprung from it during the last three thousand years*". ('Life and Letters of Fredrick Max Muller', vol. I, chap. XV, p.34.)

(C) To the Duke of Argyle the Governor of India in 1868:

> "*India has been conquered once, but India must be conquered again and that second conquest should be a conquest by education…the ancient religion of India is doomed, and if Christianity does not step in, whose fault will it be*"? ('Life and Letters of Fredrick Max Muller', vol. I, chap. XVI, p.378).

(D) E. B. Puse a friend from Oxford wrote to Max Muller:

> "*Your work will mark a new era in the efforts for the conversion of India*".

(E) In 1876, while writing to a friend, Max Muller said that he would not like to go to India as a missionary because that would make him dependent upon the government. His preference was this -

> "*I would like to live for ten years quite quietly and learn the language, try to make friends, and then see if I was fit to take part in this work, by means of which the old mischief of Indian priestcraft could be overthrown and the way opened for the entrance of simple Christian teaching…India is much riper for Christianity than Rome or Greece were at the time of Saint Paul*".
>
> "*The rotten tree for some time had artificial supports …but if the English man comes to see that the tree must fall…he will mind no sacrifice either of blood or of land…I would like to lay down my life, or at least lend my hand to bring about this struggle"… "I do not claim for the ancient Indian Literature any more than I should willingly concede to the fables and traditions and songs of savage nations. I simply say that in the Veda we have a nearer approach to a beginning, and an intelligent beginning, than in the wild invocations of the Hottentotes and Bushmen,..*"

(F) Max Muller wrote and exhibits his arrogance towards an Indian intellectual, N.K. Majumdar:

"Tell me some of your chief difficulties that prevent you and your countrymen from openly following Christ, and when I write to you I shall do my best to explain how I and many who agree with me have met them and solved them...From my point of view, India, at least the best part of it, is already converted to Christianity. You want no persuasion to become a follower of Christ. Then make up your mind to work for yourself. Unite your flock - to hold them together and prevent them from straying. The bridge has been built for you by those who came before you. STEP BOLDLY FORWARD, it will break under you, and you will find many friends to welcome you on the other shore and among them none more delighted than your old friend and fellow labourer. F. Max-Muller."

(G) Max Muller repudiates his own hypothesis:
Max Muller wrote for Anthropological Review in 1870 to popularize the Aryan Racial Theory where he had classified the human race into seven categories on an ascending scale - with the Aborigines on the lowest rung and the "Aryan" type at the top. Those who have studied his work, find that repeatedly he pulled self-contradictory theories out of his hat without any logical basis and his theory on the human race with an ascending scale, putting the Aryans at the top, was one such example. To save his reputation as a Sanskrit scholar who flunked in understanding the meaning of the Sanskrit term *arya* (for Aryan) which actually meant '*a person with noble qualities and character*' or the "*ishvar putra (progeny of God)*" without any connotation of race, color or physical stature, he recanted his earlier hypotheses stating:

"I have declared again and again that if I say Aryan, I mean neither blood nor bones, nor hair, nor skull; I mean simply those who speak an Aryan language...to me an ethnologist who speaks of Aryan race, Aryan blood, Aryan eyes and hair, is as great a sinner as a linguist who speaks of a dolichocephalic dictionary or a brachycephalic grammar".

Based on the quotations presented above one should have no doubt about the condition of Max Muller's mind. As Pundit Guru Datta Vidyarthi had said - *It is necessary that the mind should, by an adequate previous discipline, be raised to an exalted mental condition,*...So is it with Vedic philosophy; Professor Max Muller's mind was nowhere near an exalted state.

Max Muller wrote several books and articles on India including, 'Sacred Books of the East', 'Laws of Manu' (a book on Manusmriti), 'Sanskrit Grammar for Beginners in Devnagari and Roman Letters', 'The History of Ancient Sanskrit Literature', 'India - What Can It Teach Us?' and many more in addition to his multi-volume commentary and translations of Veda mantras. The entire writing work was undertaken ill prepared and with a mind that was prejudiced with preconceived notions about the Vedic philosophy and infested with ulterior motives.

Max Muller an Epitome of Contradictions

Max Muller in the first chapter of his book – 'India - What Can It Teach Us?' refers to the historical accounts of several foreign writers and travelers to India. For the excellence of the justice system in ancient India, he refers to Ktesias, the famous Greek physician of Artaxerxes Mnemon, who was present at the battle of Cunaxa in 404 BCE. His period overlapped with that of Megasthanese who was Seleucus Nicator's ambassador in the court of the Indian king Chandragupt of Pataliputra (Sandrocottus of Palibothra, in Greek). Ktesias was the first Greek writer who wrote a special chapter titled - 'On the Justice of the Indians'. He quotes from the writings of Megasthanese that in India of that time - '*theft was extremely rare, and that they honoured truth and virtue*'Ktesias's account of the excellence of the justice system in ancient India precedes the year 288 BCE which marks the death of the great Emperor Chandragupta Maurya. Max Muller also quoted from the second century writings of Arrian (86-160 CE) a Greek philosopher, historian and a military commander who was the pupil of the Greek Stoic philosopher Epictetus (55-135 CE). As cited earlier, Arrian wrote about the systematic work of public overseers and superintendents in India of that time.

Max Muller continues to shower praise for the honesty and integrity of the Indians of ancient times citing the records of well-known Buddhist pilgrim travelers from China, foreign Muslim writers after the Muslim invasion of India and the records of Marco Polo, to name just a few. As mentioned above, he cited Nearchus, the admiral of Alexander the Great, to have stated in 325 BCE that for commercial contracts and documents '*the Indians wrote letters on cotton that had been well beaten together*'.

Hence the big question is – if Max Muller found credibility in the accounts of foreign travelers to India, that the Indians had high honesty and integrity, why does he not trust the ancient accounts of history in the ancient Sanskrit writings that Indians wrote? It is strange behaviour on the part of Max Muller that he would mistrust the historical accounts laid down by the honest *acharyas* and *rishis* of the ancient Sanskrit texts, which said time and time again that Mahabharata took place over 3000 BCE and that the books of Ramayan and Manusmriti were even older, the Vedas being the oldest. For these *acharyas* and *rishis*, the '*naanritam*' (truthfulness) was godliness and *asatya* (falsehood) was sin. Max Muller may not have understood Sanskrit and the Veda very well but at least he understood these values of the Aryans that he mentioned in his writings. He chose to mistrust the truthful scholarly people of India so that he could distort the ancient Indian history.

Max Muller's Scholastic Sanskrit Ability in Question

Max Muller was aware of Swami Dayanand's RigVeda *bhasya* (translation) being published in installments in an Indian journal. It is also obvious from his comments that he had read Swami Dayanand's *bhasya.* Within the first few pages of his RigVeda *bhasya,* Swami Dayanand invoked several ancient references including the emphasis on Nirukt etymology and Ashtaddhyayi grammar. In the same first few pages Muller must have also read of Swami Dayanand pointing out the mistakes in the mantra translations of H. H. Wilson (1786-1860), Max Muller, R. T. H. Griffith (1826–1906) and Sayanacharya. Muller compromised little and rather dismissively commented on Swami Dayanand Sarasvati -*"He has done some new translations of Vedic words and mantras, but nothing that is not available in the old tradition, I have not found some demonstration of new thinking."*

Although Max Muller has mentioned Nirukt and Ashtaddyayi in his subsequent writings, it is obvious from his Veda translation that he did not understand them. If he had read them with good understanding, he would not have passed a derogatory comment on a profound scholar of Sanskrit, Swami Dayanand Sarasvati. The fact was that Swami Dayanand Sarasvati neither invented a single word of Sanskrit nor did he come up with an arbitrary meaning of his own in the translations. In addition to the Vedic wisdom bestowed on him by his guru Swami Virjanand, Swami Dayanand followed Rishi Yask's Nirukt, Rishi Paanini's Ashtadhyayi with the science of word formation and Nighantu, an ancient word assemblage of Vedic terms. Instead of admiring Swami Dayanand Sarasvati's incredible scholarship, Max Muller wrote disparagingly of Swami Dayanand's translation: "*... I have not found some demonstration of new thinking.*"

Tired of Max Muller's arrogance emanating from his incompetence as a translator and interpreter of RigVeda, his contemporary Pundit Guru Datta Vidyarthi questioned him by analyzing his mantra translations. Based on Sanskrit grammar and etymology of Vedic terms, Guru Datta took up Muller's mantra translations and pointed out how *'incongruous, unintelligible and superficial'* his translations were. Some passages from "The Terminology of the Vedas and European Scholars", authored by Guru Datta are reproduced in what follows.

Guru Datta points out that in one mantra alone (RigVeda, 1/162/2) Max Muller wrongly translated nine words out of about only a dozen words that formed the mantra couplet. Following is Pundit Guru Datta's text (reproduced from "Works of Pt. Guru Datta Vidyarthi", edited by Dr. Ram Prakash, Professor of Chemistry (Punjab University) published in 1998.

Guru Datta Vidyarthi states:

"We come now to the second mantra which runs thus:

यन्निर्णिजा रेक्णसा प्रावृतस्य रातिं गृभीतां मुखतो नयंति ।
सुप्राङ्जो मेम्यद्विश्वरुप इन्द्रापूष्णो: प्रियमप्येति पाथ: ॥(RigVeda, 1/162/2)

Transliteration:

"Yannirnija reknsa praavritasya raatim gribheetaam mukhto nayanti |
Supraangjo memyadvishvaroop indraapooshno priyamapyeti paathh ||"

Max Muller's translation of this mantra is –

> "*When they lead before the horse, which is decked with pure gold ornaments, the offering, firmly grasped, the spotted goat bleats while walking onwards; it goes the path beloved by Indra and Pushan.*"

"Here again there is no sense in the passage. The bleating of the goat has no connection with the leading of the offering before the horse, nor any with its walking onwards. Nor is the path of Indra and Pushan in any way defined. In fact it is very clear that there is no definite specific relation between the first mantra and this, according to Muller's translation, unless a far-fetched connection be forced by the imagination bent to discover or invent some curious inconceivable mythology. And now to the application of the principle that all Vedic terms are *yaugik*. Max Muller translates *reknasas* into gold ornaments, whereas it only means wealth (see Nighantu, ii.10). *Rati* which signifies the mere act of giving is converted into an 'offering'; *vishvarupa* which only means one 'having an idea of all forms' is converted into 'spotted'; *aja* which means 'a man once born in wisdom, *being never born again*', is converted into a 'goat'; *memyai*, from root *mi* to injure is given to 'bleating'; *suprang*, which means from root *prachh* to question, 'one who is able enough to put questions elegantly', is translated as 'walking onwards', *pathah* which only means drink or food, is translated into 'path' and lastly, indra and pushan, instead of meaning the governing people and strong are again made to signify two deities with their proper names 'Indra' and 'Pushan'. Concerning the word *pathah*, writes Yask, vi.7. .. (a sutra cited from Yask).. *Mukhato nayanti*, which means, "they bring out of the organ of speech, or they explain or preach', is translated by Max Muller into 'they lead before".

"It is thus clear that, in this one mantra alone, there are nine words that have been wrongly translated by Max Muller, and all is due to this that the *yaugik* sense of the words has been ignored, the *rudhi* or the *laukika* sense being every where forced in the translation."

Based on the above word analysis Pundit Guru Datta presented the composite translation of the mantra as:

> "*They who preach that only wealth earned by righteous means should be appropriated and spent, and those born in wisdom, who are well versed in questioning others elegantly, in the science of forms, and in correcting the unwise, these and such alone drink the potion of strength and of power to govern.*"

Pundit Guru Datta Vidyarthi emphasized that when a Veda mantra is translated with the *yaugik* meaning of the Vedic terms with proper grammar rules, it yields a purposeful and sensible meaning. Additionally, it also reveals insight into the profound and extraordinary thinking of the *rishi* of the mantra. From Guru Datta's writings one gets considerable insight into the literary activity pertaining to the Vedas during the 18th and 19th centuries. The works of a few more prominent Western Sanskritists of that era are cited in the following.

H. H. Wilson (1850-1888) wrote a book entitled 'Religious and Philosophical Systems of India'. Based on Acharya Sayana's commentary on RigVeda Wilson also translated RigVeda into English. As pointed out earlier, Swami Dayanand Sarasvati in his RigVeda translation into the Hindi language pointed out the deficiencies in Wilson's RigVeda translation. Arthur Anthony Macdonell (1854-1930) was a student of Professor Max Muller. Even though there was a large pool of highly qualified competent senior Sanskrit scholars of Indian origin teaching in Kolkata University, Macdonell was appointed as the head of the Sanskrit Department under the British Education Plan. He wrote several books: 'History of Vedic Grammar', 'History of Vedic Mythology', 'A Sanskrit Grammar for Students', 'A History Of Sanskrit Literature', 'A Summary of the Deities and Myths of the Rgveda' etcetra. Obviously, like his guru Max Muller, he was a prolific writer. Based on the discussion in this chapter, one can imagine the literary pollution they together generated!

As an example, only one incident related to Macdonell's level of understanding of the Vedas is cited here. Vedic scholars of India pointed out that in his writings he grossly erred on finding polytheism (multiple gods) in the Vedas and that he and his teacher Max Muller were completely off base because the Vedas had only a unitary, formless universal spirit, the God-entity. Macdonell vehemently disagreed. Out of the many mantras stating that the Vedas only had one God, a particular mantra from the first chapter of RigVeda (1/164/46) was brought to his notice. The second part of the mantra states:-

एकं सद्विप्रा बहुधा वदन्त्यग्निं यमं मातरिश्वानमाहुः| RigVeda (1/164/46)

"*Ekam sadvipra bahudha vadantyagnim yamam maatrishvaanmaahu.*"

This means,

"*God is only one and owing to his virtues the intellectuals address Him with different names such as agni* (all-pervading energy), *yam* (all-controlling), *vaayu* (all-pervading dynamic force)".

Although the first line of this mantra is not presented here, it also mentions several other names attributed to the same unitary all-pervading spirit, referring to the virtues of the Creator.

To salvage his reputation as an authority on the Vedas and to continue to hog the undeserved lucrative position of Sanskrit Chair at the Kolkata University, Macdonell came up with an argument. He said that this mantra was not an original Veda mantra and was inserted later in the RigVeda. To save his credibility he did not hesitate to concoct this absurd hypothesis, not realizing that he was negating another of his own equally absurd previous hypothesis. He had earlier put forward a baseless hypothesis that 8 out of 10 *mandalas* (chapters) of RigVeda were written first and the last two were added later! In his weak defense, he forgot the basic logic that the structure of the Vedas is such that even if someone were to insert a mantra, it could be inserted only at the end of the last chapter or at the most in the last two *mandalas*, which according to his arbitrary theory were added later. However, the particular RigVeda mantra on monotheism, pointed out to Macdonell and Max Muller by the scholarly Sanskrit pundits is in RigVeda's very first *mandala* (chapter) and not in the last *mandalas*. As cited above by this author, it is the 46th mantra out of 51 mantras in the 164th *sukta* of the first *mandala.*

Even if Macdonell thought that the mantra in question was inserted later in the first *mandala* of RigVeda, the fact remains that through this mantra the RigVeda emphasized monotheism. Conveniently, he also forgot what his guru Max Muller had stated in the third chapter of 'India - What Can It Teach Us?' that the collection of the Vedic Hymns in the order in which they are found presently, existed from at least 1500 BC. Max Muller in the seventh chapter of the same book again emphasized that all the 10,580 RigVeda mantras, all in perfect meter, were composed about 1500 BC and handed down without any variation up until 1500 AD.

Macdonell, his guru Max Muller and all those under Max Muller et al. were in a quandary because accepting that the Vedas had monotheism would mean that all their translations and interpretations of the Vedas were a huge infringement and that would be disastrous for their academic reputations. Accepting their mistake would have equated their lifetime's literary work to zero. Between Macdonell and Max Muller it was a case of '*yatha guru tatha shishya*', meaning 'as is the guru so is the student'. As far as the Vedic culture and religion is concerned, the literature that this duo produced is as good as offal. The ego and superiority complex of these personalities would not let them accept their mistakes and grow in wisdom. They rejected the most understood principle in the Vedas and Vedic philosophy, that to achieve enlightenment and up-liftment of the consciousness (soul), one must let go of one's ego. No wonder that Max Muller et al. did not grasp the prime Vedic principle of monotheism, the principle of the unitary God-entity.

Yet another Western Sanskritist, Sir M. Monier-Williams who wrote 'Indian Wisdom', 'Sanskrit English Dictionary (1899)' and 'The Study of Sanskrit in Relation to Missionary Work in India (1961) said:

> '*When the walls of mighty fortress of Hinduism are encircled, undermined and finally stormed by the soldiers, of the cross, the victory of Christianity must be signal and complete'*.(taken from Vidyananda Sarasvati, "Ved Meemansa", 1984, p15).

Pundit Guru Datta Vidyarthi in a lecture entitled, "The Hymns of the Vedas" took up 13 points from Monier Williams' criticism of the Vedas and presenting mantras after mantras from Rik, Yaju and Atharva Vedas, he showed that the Vedas only had monotheism. He presented the summary of a set of five mantras from AtharvaVeda (13/4/16-18, 20, 21) showing that monotheism was the only theism in the Vedas:

> "*He is only one, there is no second, no third, no fourth God. There is no fifth, no sixth, no seventh God. Yes, there is no eighth, no ninth, no tenth God. In Him, the Unitary Being, all live, move and have their being*". AtharvaVeda (13/4/16-18)

In spite of the scholarly Vedic pundits pointing out the deficiencies in their translations of the Vedas, the Sanskritists such as Sir M. Monier-Williams continued writing and publishing. Unfortunately, what they wrote continues to serve as a resource for researchers in Sanskrit Literature. Included among Max Muller et al. is also Arthur Berriedale Keith (1879-1944) (author of 'History of Sanskrit Literature') who assigned the age of the Vedas to be 1400 BCE whereas H. H. Wilson and R.T.H. Griffith assigned it 2000 BCE. It is conceivable that a number of Ph.D. dissertations were written based on Max Muller et al.'s misinterpretation of the Vedas. The process unfortunately continues.

CHAPTER 5: The Primary Vedic Entities in the Creation of the Universe

Vedic Terms in Cosmology

Developments in contemporary cosmology have led to the development of new physics terminology. One does not expect this to happen in the most ancient Vedic philosophy, science and cosmology in particular. The self-sufficiency and value of the Vedic terminology can only be assessed from what already exists. The limited number of examples of the terms in Vedic cosmology presented in the previous chapters would indicate that the Vedic definitions of cosmology terms are unique and deep in meaning. The word-formation in Sanskrit being root-based, each term conveys the meaning that it is derived for. It also puts a restriction on an inappropriate usage of a term. Therefore, if a term is used in cosmology, it has to convey a meaning that relates to cosmology. Emphasizing, analyzing and interpreting a specific term in Vedic cosmology helps a reader to go deeper into the meaning of the term.

A precondition for Vedic creation is to recognize the pre-existence of certain real entities of eternal nature that take part in the creation process as 'general cause', 'material cause' and 'efficient cause'. Some of these entities remain unchanged and unconsumed in the process while the others change in their physical form and shape. For example the inert Nature, the matter conglomerate of fundamental particles, the 'material cause', pre-exists at the commencement of the creation and enters into a transient phase during the creation. The formation of the universe consumes it totally, exhibiting it in the form of observable and non-observable cosmos and energy, yet conserved in its totality during and after the creation. No substance in the universe is formed in isolation. It requires a minimum essential set of entities at causal level for it to come into existence. This is true for the universe itself.

The eternal cyclic order of creation and dissolution in the Vedic cosmology requires an eternal set of entities which serve as the stepping-stone for a creation and serve as the cause for the next steps to follow. The laws of nature must pre-exist to create the first physical systems in the universe and these first systems also under the laws of nature must serve as the cause in the next steps of creation. This chain in creation helps sustain the universe and prevents the universe from entering into chaos.

To express the process of creation of the universe one needs a smart set of terminology and the set of entities that partake in the creation. The necessary condition however is that all partaking entities are defined with clarity and there is no violation of the universal principle 'cause-and-effect'.

In the Vedic perspective, all things that we see around us and all that we are unable to see but observe the effects that they create, the substantive diversity of matter or energy forms, together make up the universe. The observed inherent transience in the nature of substances, with previous substances serving as the 'cause' for the new substances, leads us to think that all substances of the universe have come into existence at some point in time and collectively the entire universe also had come into existent at some point in time. As mentioned above, in the Vedic creation, inert Nature and certain real entities are eternal and do not come into existence at a given point in time.

The two components of the 'creation' in the Vedic framework, animate and inanimate respectively relate to the creation pertaining to the consciousness (spiritual) and the inert physical aspect of the universe. Both components come into action at a given moment in time. This may sound similar to what is in other religious traditions; however, the Vedic perspective on this differs significantly at the fine structure. The differences pertain respectively to two different realms of governing laws, *ritam* and *satyam*. According to the Vedas the *ritam* laws govern the consciousness realm of creation and the *satyam* governing the inert physical creation. The consciousness having strong relationship with the inert matter, the *satyam* laws are encompassed by the *ritam* laws.

Frequently used Vedic Cosmology terms are as follows:

सृष्टि (*Srishti*), सर्ग (*Sarg*), जगत (*Jagat*), प्रलय (*Pralaya*)

The English term 'creation' is for the phenomenon of forming the universe. The equivalent term in Sanskrit is सृष्टि (*Srishti*). When the root of the term सृष्टि (*Srishti*) is analyzed from the perspective of the Vedas and the Sanskrit language one learns that the term has strong relationship to a physics process than to a religious connotation. The etymology of the term सृष्टि (*srishti*) reveals that the universe has come into existence at a given point in time and it encompasses the advent of both realms of the universe - animate and inanimate. The other expressions such as 'creation of the universe' and 'Vedic Creation Principle' are all included in the Vedic term "*srishti*".

The root meaning of the word सृष्टि (*srishti*) tells us that it is सृष्टि (*srishti*) because it has undergone सृजन (*srijan*), the process of formation. Therefore the सृष्टि (*srishti*) is - *'the one which has been created'*. Implicit here is that it is not something that has its existence without a beginning, as hypothesized in some of the modern cosmology theories, discussed in Chapter 2. The Vedic term *srishti* represents the universe as an entity that began to come into existence at a given point in time. The connotation of the word *srishti* includes – "*one with a beginning and existence*" and "*not an ever-existing phenomenon*", latter being contrary to the 'Steady State Hypothesis' in contemporary cosmology.

The term सर्ग (*sarg*) encompasses the active phase of the universe during its existence and stands for the visible and invisible matter, in the form of cosmos, forces and energies that carry on their dynamic activity within the *ritam* and *satyam* laws. While the word *srishti* represents a cosmological event (the past) and its existence (the present), the word *sarg* represents –'*the present functioning universe*'.

The word जगत (*jagat*) for the universe is a fascinating term that tells us exactly how most contemporary cosmologist see it; that the universe has come into being, that it will disappear and that it is functioning presently between its beginning and end; yes, all three properties are concealed in the three syllable term जगत (*jagat*). At the same time the word *jagat* exhibits the marvel of the Sanskrit language where there is no dearth of proper nomenclature in all areas of this creation; natural and super-natural. This simple three syllable Vedic term *jagat,* though common in usage in all Sanskrit-based Indian languages, including the slang expressions of the Indian countryside, has amazing depth in it and therefore requires further analysis.

The Vedic Sanskrit grammar and philology reveals that the term जगत *(jagat)* originates from three separate roots. These three roots forming the conjoined word जगत are three single letters/syllables: ज + ग + त = *ja* + *ga* + *ta*. The three verbs for the three roots are:

ज = *ja* → जायते (*jayate iti jagat*) stands for '*one that is born* or *one that comes into existence*'; both apply.

ग = *ga* → गच्छति (*gachchhati*) stands for '*one that goes away* or *one that ends its presence*'

त = *ta* → तिष्ठति (*tishthati*) stands for '*one that stays* or *one that exists or one that persists*'

The Vedic creation therefore stands for a phenomenon that happens in reality wherein '*it is created*', '*it ceases to be*' and '*it exists*' between its two ends. The formation of the single word ज + ग + त = *ja* + *ga* + *ta* = जगत *(jagat)* from the first letters of the three different verbs is remarkable. The Veda mantras (RigVeda 10/129, 190) in the *suktas* on the creation of the universe, discussed in Chapter 6, further reveal that the *srishti* is not a one-time event; rather it is a cyclic event with a Veda assigned definite time period.

The common nouns for the three stages of creation are thus related to these roots as: in the term *jagat =ja-ga-ta,* the *ja* relates to *srishti, ta* relates to *sarg* and *ga* relates to प्रलय (*pralaya).* The word *pra-laya* in the above heading represents the 'dissolution' of the universe where the nature dissolves into its causal state, a non-recognizable state of matter where matter enters into an indefinable state, yet continues to exist in non-active form till the onset of the next creation. The word *laya* alone in the word *pralaya* is sufficient to define the 'dissolution' of the universe; however, the prefix *pra* to the term *laya* adds emphasis to the process of *laya*, indicating that it is the process that takes place with certainty. The terms *srishti* and *pralaya* are also used to represent the time durations of creation and dissolution phases respectively. When the term *srishti* is used in the context of the time period it represents the time for creation's activity and also termed as the *Brahmn-din* (God's day). Likewise the term *pralaya* being the time-period for creation's rest, it is termed as *Brahmn-ratri* (God's night). Two other terms expressing the opposite nature of *srishti* and *pralaya* are *kalp* and *vikalp*; defined below.

Kalp and *Vikalp*

Kalp is the creation phase of the universe lasting for a period of 1,000 *deva-yugas* of 4.32 BY (Billion Years, AtharvaVeda, 8/2/21) each, when the matter is in *Sata*-state, a state that is utilizable to form substances, objects, forces and energy forms of the universe. The *srishti, sarg* and *jagat* adjectives of the universe exist in the *kalp* phase. *Vikalp* is the period of the dissolution phase of the universe, also lasting 4.32 BY, the same length of period as in the *kalp* phase. The *pralaya* phase belongs to the *vikalp* period when the matter of the universe enters back into its indefinable *Asata*-state. This state of Nature is also termed as *avyakta* (*a-vyakta*) meaning- '*not explicable in words*'. Just as the period of a day is one day of humans, the Vedas have termed the period of a *kalp* as God's one day and just as the period of a night is one night of humans, the Vedas have termed the period of a *vikalp* as God's one night. Compared to this perennial cycle of *kalp* and *vikalp* in the Vedic principle, the modern Big Bang Hypothesis creates the universe only once; not knowing what happened before the Big Bang.

The Vedic creation process discussed in detail in Chapters 7 to 10 is complex and beyond the simplicity of the expression - "*Let there be light and the light was there*". The Nature in the Vedic creation process switches into the *Sata*-state at a given point in time, enters into the creation phase, prepares the primordial matter developing necessary heat energy over a period of one *deva-yuga* of 4.32 Million Years (1000th of the *kalp* measuring 4.32 BY) before creating a tremendous explosion in it hurling matter far and wide. Then a sequence of entities of matter (energies, forces, fields, physical and metaphysical systems, visible matter etc), referred to in numerous mantras to be cited later, are created which further branch off to create other visible and invisible, macro and minute forms of entities. These are created because they are deemed essential to sustain the existence of the universe and all things in it including life on Earth and throughout the universe.

The Nature has strewn splendor around us in its utmost complexity in the form of mountains, rivers, oceans and the part of the living universe itself. All this is the variety of matter with emblems that form features of a *kalp* that disappear in *vikalp* including a large number of invisible forces, systems and phenomena of the creation phase. This is equivalent of saying that in a *vikalp,* the matter of the universe ceases to carry forward any memory of a *kalp.* Are there entities if any in Vedic creation that continue their existence in both phases so that something more could be extracted before $t = 0$ sec of a Big Bang? It is not known from the contemporary Big Bang Hypothesis whether more information could be extracted from such entities if subjected to a mathematical treatment that could take us back in time, however, the Vedic terms for the entities that continue through both the phases of kalp and *vikalp* in Nature are said to be *anaadi* and *nitya.*

Anaadi and *Nitya* (अनादि, नित्य)

The word *anaadi* means '*one that does not have a beginning*'. The equivalent term for *anaadi* is eternal. In a perennial cycle of creation such as in the Vedas, the Time and Space entities preexisted and will continue to exist during the *pralaya* phase of Nature. Therefore, they are the *anaadi* entities of Vedic cosmology. When one leaves the physical realm defining *anaadi* entity to be '*without a beginning and end*' and enters into the realm of Vedic philosophy one invokes the framework of cause-effect. Then the *anaadi* is defined as:

'*An anaadi entity is not an effect of a cause but serves as a cause for other effects*'.

In modern physics, one can have a premise to create matter out of nothing and can even infer a proof for it. One can even have a fantasy of phantom particles that appear and disappear to create matter at a given point in space 'out of nothing'. Even then it is not said that the existence of a substantive matter particle consumed a piece of Time or a piece of Space. However, nothing in the universe is created without the pre-existence of time and space – in 'time' it is created and in 'space' it is placed.

A natural question here is - why the Vedas would assign *anaadi* (eternal) status to certain entities like Time and Space that have no cause for their own birth and existence. The logic in the Vedic perspective is that because the Vedic creation-dissolution process is cyclic and eternal the *anaadi* Time and Space entities have to be always available to serve as the *saadharan-kaaran* (ordinary-cause). The causes that might create Time and Space are not observed in the universe, but the existence of Time and Space is either observed or inferred from the observed effect that they are responsible for. Modern cosmologists accept the reality of Time and Space but cannot create in the laboratory a piece of either, no one can. The Vedas recognized that if Time and Space were consumed during the creation process, they would be exhausted and nothing would be left for the present and the future creations. The Veda mantra formulations have recognized Time and Space for what they are and assigned the category of अनादि (*anaadi* = eternal) to them.

Another important term in Vedic philosophy that relates the entities and substances to their existence is नित्य (*nitya*). The term *nitya* will become clear from a few *sutras* from the fourth chapter of Rishi Kanaad's Vaisheshik Darshan.

सदकारणवन्नित्यम् । (*sadkaaranavan-**nityam***, Vaisheshik, 4/1/1)
"*A substance in existence that has no cause for its existence is **nitya.***"
(Vaisheshik, 4/1/1), and,

तस्य कार्यं लिङ्गम । (*tasya kaaryam lingam,* Vaisheshik, 4/1/2).
"The action (cause) is its defining characteristic (lingam)"
(Vaisheshik, 4/1/2).
because,
कारणभावात् कार्यभाव: । (*kaaranbhaavaat kaaryabhaavah,* Vaisheshik, 4/1/3).

"The existence of cause leads to the existence of effect" (Vaisheshik, 4/1/3)

Rishi Kanaad is conveying here that those substances or entities that have no cause for their being but can be the cause for the other substances are *nitya* (eternal and ever-present) and that such entities must exist before they cause an effect. On the other hand a substance resulting from a cause (action) and is only an intermediate cause of a subsequent effect, is transient in nature and thus termed as *a-nitya* (non-*nitya*).

In a definition akin to physics, the *nitya* substances are invariants in Time and Space, inclusive of Time and Space themselves. The terms *anaadi* and *nitya* (eternal and non-transient) are synonymously used in Vedic philosophy. With respect to entities forming the universe, the difference and similarity between the Vedic and modern Big Bang concepts is quite obvious. Whereas the Time and Space are *anadi* and *nitya* in the Vedas, they are treated in Big Bang Hypothesis as *nitya* (ever-present) after the Big Bang Hypothesis's t = 0 sec. This is a contradiction in Big Bang Hypothesis because one cannot create an entity and make it an 'ever-present' entity for the rest of its life without a mechanism for its growth. Unless the Big Bang Hypothesis comes up with the cause of Time and Space, the hypothesis itself cannot be considered based on a sound illogic.

On one hand the effulgent variety of Nature is praised in the Vedic mantras and Upanishad formulations and on the other hand it is also declared to be an *anitya* phenomenon, transient on a short term basis and ending completely into an indefinable state at the time of *pralaya* dissolution, when even the laws of Nature under *satyam* disappear. That indefinable state of all matter of the universe will still remain an entity called *Prakriti* in the Vedas, which is the inert Nature in a suspended state for a period, waiting to spring into a new creation phase.

Four important entities of modern cosmology are: Space, Time, matter (including Dark Matter) and Dark Energy, discussed in Chapter 2. Since modern cosmology is unable to claim with certainty the mechanisms for the birth or death of these entities, one cannot assign them the label of eternal or non-eternal. The Dark Energy being energy, a degenerate form of matter, which by definition is an effect, would not pass as an eternal entity in the Vedic frame of reference. There are five eternal entities in the Vedic cosmology that participate in the cause-effect process. Surprisingly, three of them are the same as in modern cosmology. Since the Vedic creation follows an eternal cycle, these five entities are considered to be eternal. They are defined in the sections to follow after the terms *ritam* and *satyam* are defined.

ऋतम् (Ritam) and सत्यम् (Satyam)

There are several constants and invariant laws of Nature that enabled Nature to form the universe and continue to provide continuity to the universe. An example from this set is the gravitational constant G and the law of gravitation discussed in Chapter 2. Any change in these constants and laws would create chaos. The Vedic theistic approach for the advent of the universe stipulates that the creation takes place under the specific laws that the conscious-being Creator imposes on the preexisting Nature. All laws governing the animate and inanimate creation come under the *ritam* and *satyam* laws.

If the Creator is taken out of the picture the Nature has to create its own laws. There is no indication in Nature for a trial and error methodology for evolving its laws. The matter of the universe is created under Nature's laws but under what laws the Nature (matter) created the laws that create matter? The universe in a theistic reference frame is created by the will of a conscious-being Creator. For example, when a human-being generates a thought in his mind to write a sentence on a piece of paper the non-substantive thought alone is not capable of writing the sentence down. To implement the writing of the non-substantive thought that the mind created, the mind also creates a thought that instructs the substantive hand of the living-being to carry out the instruction. The logic in the Vedic premise is that the formless all-pervading Creator has the will-power and thought to create the universe, but it needs hands to carry out the job. The *ritam* and *satyam* laws are formless Creator's formless, invisible, non-substantive perfect hands. The 'mind over matter' is the underlying premise.

The modern cosmology assumes that there is only the physical realm of the universe that comes into existence under the physical laws. The Vedic Cosmology on the other hand emphasizes that there are two realms of the creation where two sets of laws apply, the spiritual (consciousness) realm and the physical realm. All laws pertaining to the consciousness creation go under ऋतम् (*ritam*) and all the laws pertaining to the physical creation go under सत्यम् (*satyam*). To express it differently, the ऋतम् (*ritam*) laws and the सत्यम् (*satyam*) laws respectively govern the animate and the inanimate creation. The ऋतम् (*ritam*) and सत्यम् (*satyam*), pronounced with soft 't' are some of the most fundamental terms of the Vedic cosmology and philosophy.

To elucidate the point further let us take the example of Newton's third law which states - "*To every action there is an equal and opposite reaction*". This law applies to an 'action' in the physics realm. The Sanskrit term for 'action' is 'karma'. In the context of the gravitation of the Earth, Rishi Kanaad stated in his Vaisheshik Darshan (5/2/1, 5/2/2) that the property of gravity in the Earth is its अदृष्टकारितम् (*adrisht-**kaaritam***), where the term *adrisht* means 'invisible' and ***kaaritam*** means 'karma' (action). The karma in this case is generated by an inert entity, the Earth. The 'action' under the cause-effect rule is the cause and the 'reaction' is the effect (result). Dealing with everything physical (material) this rule comes under Creator's सत्यम् (*satyam*) system.

When one moves from the physical realm to the consciousness realm the 'karma' is generated by a living-being. A living-being can do karma that is pure physical in nature and inconsequential. Also a living-being owing to the free-will can do karma that affects favourably or adversely the other conscious-beings. Such a karma is respectively labeled as 'good karma' or 'bad karma'. But because it is also 'karma' (action) it should generate a right 'reaction' (result). There is no law in physics to cover such an "action-result" equation because physics does not deal with the animate creation (consciousness). According to the Vedas all laws governing the consciousness realm (animate-creation) including the balance in "karma-result" of an individual conscious-being, come under the Creator's ऋतम् (*ritam*) system. The physics laws that have been discovered by physicists and those remaining to be discovered, do account for the physical universe under *satyam* but the metaphysical part of the universe remains hidden and unaccounted for. If this is not taken care of the system is not fair. As this author has grasped it from the Vedic philosophy, the ऋतम् (*ritam*) system encompasses all implicit rules of the creation, physical and metaphysical, that never change and the सत्यम् (*satyam*) system of rules encompasses all explicit rules governing the created physical universe that also do not change during the creation phase of the universe. Just as the 'truth' is truth which never changes with time, never transits into something else, the *ritam* and *satyam*, encompassing unchanging rules of creation and dissolution are together addressed as Creator's unchanging *ritam* and *satyam* systems.

With regard to the invariance of the ऋतम् (*ritam*) system the unchanging karma principle binds the conscious-being soul such that the doer of the karma bares the effect (consequence), good for good and bad for bad. The karma principle operates at spiritual level to which physics laws do not apply and the soul has no choice but to function within this foreordained unchanging system. As in Newton's third law on action-reaction applicable to the physical realm, there is equivalence between the karma and its consequence in the metaphysical realm.

The *ritam* and *satyam* relate to each other yet have their application at two distinct levels. The comprehension of the term *ritam* is at a qualitative and philosophical level to relate to queries such as - why is the creation of the universe initiated, why it is in a given form, who is behind this creation, why is there consciousness in the creation and the other questions of this nature. The comprehension of the term *satyam* on the other hand is at an obvious level because it applies to the matter that has come into existence in the form of observed Nature. Since the matter (Nature) of the universe comes into existence under *satyam* its sustenance and the process of dissolution also come under *satyam*. For a modern physicist all transience in Nature in the form of physical phenomena from the beginning of the creation to the present, pertain to Nature alone and are seen directed by Nature itself. The physicist however does not answer the basic question – if the Nature created all physical laws with perfection at the first attempt during the early moments of the Big Bang, without any sign of evolving and finalizing any of the physics laws, why does the process of creating laws not continue for the rest of the universe's life? Many such questions remain outstanding.

Some modern physicists have suggested that the universal constants of Nature, such as the gravitational constant G, may not be constant and instead changing with time. Under the Vedic principle of *satyam*, the physical laws of

Nature remain invariant through the creation. Also, because the Vedic creation is a continuum cycle of *srishti* and *pralaya* (creation and dissolution) there have to be rules that govern the Nature even during the dissolution phase. It is the *ritam* of the Creator that continues even through the dissolution phase of Nature when the physical laws under *satyam* governing it, disappear along with the universe.

The first mantra of the 190th *sukta* of the 10th *mandala* (chapter) of RigVeda simultaneously states the terms ऋतम् (*ritam*) and सत्यम् (*satyam*) in relation to the Vedic principle of creation. A detailed discussion on the mantra is in Chapter 6; here the meaning of the first part of the mantra is presented in the context of *ritam* and *satyam*. It states:

ऋतं च सत्यं चाभीद्धात्तपसोऽध्यजायत । (RigVeda, 10/190/1)

Transliteration: "*Ritam cha satyam chabhiddhattapaso addhyajaayata* ।"
Mantra Translation:

> "*Under All-illume Creator's* **ritam** *and* **satyam** *rules and His* ***tapasaha,*** *the heat form of energy, the Nature comes into its day-epoch of creation....*" (RigVeda, 10/190/1).

The *Eekshana* Force

Eekshana is a force created in the mind of a conscious-being by the consciousness. When the conscious mind creates a thought it also creates a quantum of instruction in the brain to implement that thought. For implementation of the instruction the brain uses the *eekshana* force to orders a limb in the physical body of the conscious-being to generate the desired mechanical action. The term *eekshana* of Vedic philosophy can be translated into common English as 'desire' or 'free-will'. However, the terms 'desire' and 'free-will' alone do not represent the essence of the *eekshana* force, a force that is of metaphysical nature, enabling the creation of tangible objects. All subtle forces operational through the living physical body of a conscious-being according to the Vedic creation are physical or metaphysical but the entity exerting *eekshana* that generates the force to create action is neither physical nor metaphysical, it is spiritual. It is the *atma*-entity the soul. It is also written and expressed as *atman*. A section is devoted in the following to this entity that serves as the general-cause in the creation of the universe. It is the inner force of a conscious-being that through *eekshana* translates a creative idea into observable tangible physical object. Four forces of physical nature were discussed in Chapter 2, namely, 'Strong Force', 'Weak Force', 'Electro-Magnetic Force' and 'Gravitational Force'. These forces exist in Nature but the *eekshana* force, different from these physical forces, is created by the consciousness in the living physical body of a conscious-being. The conscious-being utilizes it to create mechanical action. A short discussion on *eekshana* is presented below.

While answering questions on the creation of the universe, the Vedic philosopher Swami Dayanand Sarasvati commented on *eekshana* in the eighth chapter of Satyarth Prakash (1881 CE) written in Hindi language. His specific definition of *eekshana*, translated by this author into English is:

"Eekshana is the power of the mind with the combination of three factors; visualization, thought and desire, by which a conscious-being is able to create tangible things".

This trait in conscious-beings is absent in inert objects. Since the Creator is considered to be an all-pervading force, with consciousness as His additional attribute, Creator's *eekshana* power is also all-pervading. Because the *eekshana* creates impetus to create tangible things, Swami Dayanand Sarasvati ascribed the *eekshana* virtue in the unitary universal spirit, the Creator, to supply necessary impetus to the inert Nature, the building material of the universe, to launch into action. This action renders the effulgent tangible universe. It is the recondite *eekshana* force that forms the link between the conscious-being Creator and the created inert matter of the universe.

In contrast with the all-pervading *eekshana* of the conscious-being Creator the quantized conscious-beings *atmas* (souls), localized in their individual physical bodies, having been endowed with limited *eekshana*, are limited in their ability and power to exert their influence in a limited sphere of action. Conversely the presence of *eekshana* in a living matter-conglomerate is the proof of the presence of consciousness in that mass – the living body.

To ascertain the presence of consciousness in a physical body, one should be able to observe the triplet - *visualization, thought* and *desire*; primarily in the physical body that created a tangible object and secondarily the reflection of the triplet in the created tangible object. It is the triplet-virtue of *eekshana* that leads to an intelligent design. No intelligent design on Earth, ranging from the creation of a piece of script by an author to a wonder of the world is possible without the three components of *eekshana.* Only one or two components are not sufficient to create a tangible object with intelligent design. The living-beings create small tangible objects and the all-pervading *eekshana of the Creator* creates the enormous universe. The universe, observed by the Vedic *rishis* as a tangible object was deemed as the result of all-pervading *eekshana* with *visualization, thought* and *desire*. This would differentiate the Vedic philosophy from the other philosophies which also advocate intelligent design. Included in Vedic creation is the intelligent design of the whole universe with unwavering physical laws of Nature playing their role in the physical realm. This also includes the physical evolution of humans and the rest of the living-beings under both, the physical laws of Nature and their spiritual evolution under the laws in the consciousness realm. From the Vedic perspective the existence of all invariant physical laws governing Nature, such as the perfect law of gravity, are part of an intelligent design.

In Vedic thinking it is the *eekshana* in a living-being that causes the evolution of the physical body in which it resides. The dexterity in human hand evolved because the *eekshana* rising in living-being's mind directed the hand to evolve in that way and the living-being as a whole evolved to adapt to its own environment because it analyzed and visualized through five knowledge senses linked to *eekshana* to get tangible results. The *eekshana* in Vedic perspective is not a slave of time for the evolution to take millions or billions of years, as advocated in the Darwinian hypothesis. It is the slave of the conscious-being who does have limitations to get quick results.

When a man builds a house for his family members, he is the efficient-cause for the house and the building material its material-cause. He uses the *eekshana* power with the *visualization, thought* and *desire* to build the house, a tangible object. The relationship between the Creator and all the souls of the universe is similar to the man building a house for the activity and shelter of his family members. To create things is in the nature of conscious-beings, be it God to create the universe or a human-being to create a house. This nature is a general rule of the creation where not just humans but living beings at all levels implement their *eekshana* to create their homes in a best possible way. Ahead of the arrival of chicks a weaver-bird builds an elaborately woven nest which is a wonder of engineering, planning and execution. The insects with tiny bodies create their dwellings with application of chemistry and physics, some of which are difficult for humans to replicate. The *eekshana* with *visualization, thought* and *desire* is seen in action throughout the living creation. The consciousness and recognition of *eekshana* in it is part of the Vedic cosmology. To label it as a mechanical behaviour or instinct in living-beings is low-grading a universal phenomenon deep seated in the universe and exhibited only in living beings.

In Vedic thinking, the creation of the complex bodies of human-beings and the rest of the living-beings is part of the complexity of the universe itself, the complexity that we observe in all physical phenomena. This complexity in the universe is not recent. In the beginning of the universe the scales of physical phenomena were different but the universe was as complex within seconds of the onset of creation as it is now in the form of galaxies, solar systems, biomass on Earth, living bodies and each cell forming a living-being. It did not evolve over billions of years to get more complex. If that early complexity of the physical universe did not take the evolutionary long period of billions of years, then the hypothesis that life began with the first uncomplicated cell evolved over millions of years on its biological path into complex humans, should be questionable. The preclusion of complexity of the human body from the beginning of the creation is equivalent of saying that the universe in Big Bang began with a single nuclear particle and evolved later into complex galaxies. With the same argument, the complexity of a human physical body with its preexistent spiritual stature could be a part of the design of the physical universe with *eekshana* in living beings doing its own part. In Vedic thinking, the obvious complexity of the universe and of the different phenomena within it is only a minute part of the whole creation because much more unobservable, undetectable and ungraspable complexity still exists beyond cognition.

In Vedic philosophy, it is the inherent *eekshana* and its level (power) that dictates a change in the body of a living-being to adapt to the surroundings. The power of *eekshana* in a living being depends on its individual pre-evolution at the metaphysical level (not the Darwinian evolution). In the Vedic philosophy the inheritance comes not only from the parents (DNA) but also beyond the parents, brought in by the immortal quantum of consciousness in the form of metaphysical traits that make each living-being different. Consideration of the phenomenon of non-substantive consciousness directing the creation of tangible things is currently absent from the discussion on cosmology even though it is also a result of cosmology. This discussion forms the subject of another book entitled "**Consciousness and Matter – A Vedic Perspective"** to be published soon by this anuthor.

The Vedic cosmology and philosophy together offer a list of inert and spiritual entities that enable a perspective for unifying the physical and metaphysical creation of the universe. Same as in contemporary cosmology the Vedic philosophy considers the Nature-entity (matter) as the "material-cause" and the Time and the Space entities as the "general-cause" (ordinary-cause). Additionally, the Vedic cosmology gives consideration to the Creator-entity as the "efficient-cause" and the discretized field of consciousness as the "general-cause". It is this general-cause, a collection of discretized entities called '*atmas*', that are the 'atoms' of a discretized field of consciousness, that individually and in association with others (parents) take part in the evolution of their own bodies that they occupy while functioning. The five eternal entities mentioned above are defined below.

The Five Eternal Entities of Vedic Creation

(1). The Creator-Entity (*Purush-tatva*)

Owing to many virtues, the Creator is addressed with many names. One of the most common names is *Purush*. The term '*purush*' *means – the one pervading all*. The author calls this all-pervading conterminous consciousness as the consciousness of "First-kind", the "Second-kind" being the quantized consciousness – the *Atma* (Soul). To differentiate *Purush* from *Atma* (soul) the Creator is addressed as *Param-atman* (the highest *Atma*). In Vedic philosophy the *Paramatma* and *Atmas* having the common virtue of consciousness and ubiquity are also addressed together as *Purush*.

Other common expressions for the *Purush*-entity are: Omnipotent, Omniscient, Omnipresent, formless, conterminous unitary entity, the supreme consciousness and so on. In the context of cosmology this author has addressed the Creator as the Creation Energy-field.

Definition of the term '*purush'*

The primary name of the Creator in the Vedas is OM formed with three syllables o u m. When a detailed interpretation of the term o u m is looked up, one finds that the Creator is addressed as OM because it encompasses all creation. When the Creator and Nature are addressed together as a duo the expression is *Purush-Prakriti*. In the compound word *Purush-tatva* the second term *tatva* means 'entity'. RigVeda's 10th *mandal*'s 90th *sukta* and YajurVeda's 31st chapter are devoted to *Purush-devata* (*purush* subject). Swami Dayanand Sarasvati in his book entitled 'Rigvedaadi Bhaasyabhumika' (Publ. Arsh Sahitya Prachar Trust) has presented the interpretive commentary in Sanskrit and Hindi languages on all 22 mantras of the Purush Chapter of YajurVeda. He has substantiated the etymology of the term *purush* citing Nirukt by Rishi Yask.

While interpreting one hundred different Vedic terms for the Creator-entity Swami Dayanand Sarasvati in his book entitled 'Satyarth Prakash' explains the root-meaning for the term *purush* to be पृ पालनपूरणयोः (*pri paalanpoornayoh*) *to protect and to fill* (complete) and defines the term as यः स्वव्याप्त्या चराऽचर जगत् पृणाति पूरयति वा स पुरुषः (*yah svavyaaptya charaachar jagat prinaati pooryati vaa sa purushah*). This translates into: '*one who is the indweller of all substances of the universe, animate and inanimate*'. The meaning of the term also includes '*one who pervades the entire universe and its substances big and subtle*'.

The Creator-entity's virtues are defined in large number of Veda mantras, *shlokas* of Smritis, statements in Upanishads and the aphorisms of the Darshans. Any commentary on all that is beyond the scope of this book. A representative mantra from the YajurVeda summarizes the virtues of the *Purush* (*Brahmn*),

स पर्यगाच्छुक्रमकायमव्रणमस्नाविरं शुध्दमपापविध्दम् ।

कविर्मनीषी परिभूः स्वयम्भूर्याथातथ्यतोऽ अर्थान्व्यदधाच्छाश्वतीभ्यः समाभ्यः ।।

(YajurVeda 40/8)

Transliteration:

"*Sa paryagachchhukram-akaayam-vrinmasraviram shuddham-apaapviddham* ।

Kavirmanneeshee paribhuh swayambhoo-r-yaathaatathyato-a-arthaan-vyadadhat-chchhaashvateebhya samaabhyah ।।"

Mantra Translation:

"*The Brahmn is swift and all-powerful (Omnipotent), without a body (formless), without holes and non-permeable, conterminous, all-pervading, omnipresent, omni-directional, unbound by the vice of the system of nerves and arteries, pure without defects, Omniscient, knowing thoughts of all minds, disapprover of sinners, self-existent, eternal entity, revealer of knowledge, creator of all substances for all in His creation. He and He alone is to be worshipped.*" (YajurVeda 40/8).

The compact disposition of a large number of terms in this single YajurVeda mantra defining a single entity certainly delineates a linguistic marvel. Out of some twenty different virtues of Brahmn cited in this single mantra, there are about half a dozen in the context of the physical world and the rest are in the spiritual context. The Vedas give credit to Brahmn who uses his *eekshana* power to create the universe from the preexisting material-cause.

Saankhya Darshan of Kapil Rishi provides a composite set of the physical entities of the Vedic creation where certain energy fields and forces are created as the byproducts of Nature. As stated in his Darshan the *Purush* is the efficient-cause of creation:

स हि सर्ववित् सर्वकर्त्ता । "S*a hi sarva-vith sarva-karta*" (Saankhya, 3/56)

Sutra Translation:

"*He is Omniscient and the Creator of the universe in its entirety*"

(Saankhya, 3/56).

Rishi Kapil deals with the age-old physics question–

"*Can one prove the existence of the Creator*"?

Recall from the discussion under 'Immeasurability of Entity' in the beginning of the Chapter 2 where it was inferred that when a field is all-pervading and infinite and is well-beyond the universe, it remains immeasurable and non-provable. Such is the situation with the Creator-entity. Recall the so-called physics *sutra* by this author, formulated in Chapter 2 that, *"The true knowledge of an entity is possible if the instrument of knowledge is able to interact with the entity"*. Under this deduction the detectability of a physical phenomenon hinges on the fact that the instrument used for its detection is made of matter and it interacts with that phenomenon. Since the instruments made out of matter are unsuitable to interact with the Vedic defined Creator a physical detection of the Creator can never be possible. The Creator being the all-pervading consciousness, an instrument made of consciousness by logic could sense the presence of the all-pervading consciousness. However, as was deducted in the Chapter 2, if an entity is not detectable in the physical realm one cannot conclude that the entity did not exist and therefore the supplementary physics *sutra* was laid down by this author - "*Futility of detection of an entity is not a proof of the absence of the entity*". The detectability of one consciousness with another consciousness is a subject beyond the scope of this book.

Let us consider an example from the physical realm. A photo-detector is the instrument to detect light photons. It is made up of a flat surface of photo-sensitive material. When a light photon falls on its surface there is interaction between the light photon and the surface of the photo-detector. In this interaction each photon (corpuscle of light) releases one electron from the molecule of the photo-detector material. The measurement of this electron in the circuitry of the photo-detector yields the indirect detection of the photon. If one exposes the photo-detector surface to the rain the detector fails to detect the number of rain-drops falling on it because the photo-detector surface interacts with light photons and not with the rain-drops. Obviously the photo-detector is able to prove the presence of light and is unable to prove the presence of rain even when the rain is falling on it. This is the gist of 'Immeasurability of Entity'.

All photons falling on the detector come from outside the detector one by one and detected by the detector proving their existence. However, if the detector is immersed in the light field, with photons already inside it and outside it, the detector would not be able to do its job even for detecting the photons. Therefore, when certain entities are all-pervading and infinitely large in their expanse, the immeasurability of such entities is a reality. The provability of such entities runs into difficulty because the provability depends on measurability. The Vedic Creator-entity being infinite and all-pervading, in and beyond the universe, remains non-provable for the same reasons.

A reasonable question is generally asked - if the Nature is observed as a reality and found deterministic by intellectuals (cosmologists) why the Nature is not assumed as the Creator of the universe? Implication of this question is that the Nature is the material-cause as well as the efficient-cause of the creation. A related question was raised eons ago by Kapil Rishi in his Saankhya Darshan (*Shastra*) *sutra* (1/56) as - '*why Ishvar (God) cannot be assumed to be the material cause of the creation*'? Rishi Kapil's answer to this question is in his next-very *sutra* (1/57) wherein he stated that ईश्वरासिद्धेः (*Ishvarasiddheh*), meaning - '*Such an Ishvar cannot be proven'*. What the *sutra* means is that if the Nature is considered to be the *Ishvar* (Creator), such a Creator remains unproven because a material-cause has no initiating properties to create its own universe. Rishi Kapil has emphasized this to draw a distinction between *Ishvar* (God), a conscious-being and the Nature an inert entity treated in the Veda as the material-cause of the universe.

Rishi Kapil emphasized this again in the third chapter of his Saankhya Darshan stating in the *sutra* (3/56) that the *Ishvar*, "*is Omniscient and the Creator of the universe*" and then he concluded his argument by formulating the next *sutra* in the same chapter stating,

ईदृशेश्वरसिद्धिः सिद्धा । "*Iddrisheshvarsiddhihi siddha*" (Saankhya, 3/57)

Sutra Translation

"Ishvar thus defined has been proven with certainty in this Shastra (Darshan)". (Saankhya, 3/57).

The references from the Vedic philosophy are presented here to show how the Vedic philosophy has addressed the subject of logic and cause-effect. The underlying argument of the Vedic philosophy is that the matter of the universe by itself is no different from a dead-body of a living-being without activity in it and without *eekshana* to initiate creation of tangible objects. The same material-body with the presence of a conscious force of a soul in it was a creator of abstract thought and tangible things in its own limited way. The argument of the Vedic philosophy is that if it were left to the inert matter of the Nature, the material-cause, it would not have created anything. To distinguish between the *Prakriti* and *Purush* (*Ishvar*), Bhagvadgita (3/46, 47) states that "*Purush is beyond Nature and nothing is beyond Purush*".

Just as the physicists are credited for discovering Nature's laws that preexisted under what is *satyam* system in Vedic philosophy, the ancient *rishis* in their own way recognized in the mantras the hidden laws under *ritam* and *satyam* and reformulated them in the form of *sutras* and *shlokas* in the Darshans, Smritis and Upanishads. An example of this in *satyam* realm is the Vedic definition of Nature that at its most fundamental level it is quantized in three types of fundamental particles named *sata, raja* and *tama.* Creator-entity is credited by the Vedas for an initiation in the fundamental *Asata*-state of Nature in its undefined material state to send it to the *Sata*-state in its defined matter state. This is a rule under *ritam* under which each particle of the tri-property Nature becomes active to manifest as matter from the *Asata*-state to the *Sata*-state. This section is summarized with a *sutra* from Saankhya Darshan where the extent and nature of Creator's interaction with Nature is stated,

तत्सन्निधानादधिष्ठातृत्वं मणिवत् ।

("*Tatsannidhaanaat adhishthaatritvam manivat"*), Saankhya 1/61)

Sutra Translation:

> *"Activity in Nature is inspired and governed by His (Creator's) interaction with it, as in a mani"* (Saankhya, 1/61).

Physicists would vouch that all activities in Nature are governed by its unchanging laws. What Saankhya states here is that all these laws of Nature are not created by the inert Nature itself, rather the prime-cause for all actions in Nature is the Creator. The Creator's all-pervading *Eekshana* yields the creation of the tangible entity, the universe. Instead of saying that 'that is the way Nature is', Rishi Kapil in the above aphorism states that the conscious-being Creator interacts with Nature (matter) in a manner similar to light interacting with a *mani* (crystal) without being consumed in the crystal. Whatever laws of nature that we discover, that are necessary for the universe to come into existence and to remain sustained without chaos, may not be too many to count but they are not straightforward and simple. They are abstract and unobvious for all to grasp. They are too complex for inert Nature without brain to create. To create a tangible object one needs a pre-thought with *visualization, thought* and *desire*. This is absent in inert entities. The explicit simile of the light and the crystal in the *sutra* for God and Nature respectively, is indicative of the fact that two separate entities are involved: one is the energy (light) and the other is the substantive matter (crystal). The usage of a crystal for a simile not only tells us about an observation in physical optics by Rishi Kapil, it also exhibits a perfect example for others to understand the nature of the relationship between two ungraspable entities, the Creator and the Nature.

Since the Nature is an inert entity, the interaction between the Creator and the Nature is one sided; the Nature remaining passive in this interaction. This is also the reason why Rishi Kapil chose the example of a *mani*. The *rudhi* (common) meaning of *mani* is jewel (a crystal). From the perspective of its physical properties a *mani* by definition is an "*object capable of concentrating and revealing"* more than what is obvious. An example for a *mani* in common usage, since ancient times, is the *sphatik-mani*. The Sanskrit term *sphatik-mani* stands for a quartz-crystal. A *sphatik-mani* accepts the light and reemits it without being affected by the light. The other example of a *mani*, in a non-optical application, is *ayaskaantam-mani* which is a magnet. Owing to the magnetic properties in the *ayaskaantam-mani* it creates the concentration of the magnetic field which is demonstrated by the alignment of ferrous material (iron filings). The iron-filings align themselves along the magnetic lines of force owing to their association with the magnetic field of the *ayaskaantam-mani* where there is no consumption of material. In the *sutra*, the example of a *mani* implies that just as the light inspires a crystal to glow and a magnetic field inspires dynamic behaviour in a piece of iron, the Creator inspires the Nature for all activities without being affected Himself. The Creator can be viewed as a Creation Energy Field just as the several other un-provable hypothetical entities of modern cosmology, discussed in Chapter 2. The Creator (*Brahmn*) has perfected the laws of Nature because unlike the assumed singular event of Big Bang for the creation of the universe, He has created and dissolved the universe, infinite number of times before!

In this book when God is addressed as the first person, this author with the limitation of English language has addressed God as He. God viewed as a formless entity, conscious-being, all-pervading force, cannot be addressed as He, She or It. The Sanskrit language or the Vedic philosophy having no such limitation, addresses God as God. God having many virtues is addressed with many adjectives including 'father' as well as a 'mother'.

(2). The Nature-entity (*Prakriti-tatva*)

The general properties of the Nature-entity are: Material-cause of the universe, an inert entity, a conflux of matter and its forms, the Vedic defined Nature at its fundamental level is quantized in three types of fundamental particles:
satva*, *raja and ***tama.***

The Nature in the Vedas and in post-Veda ancient Vedic Literature, though an inert conglomerate of matter, is addressed as an entity and has many synonyms and adjectives related to its function and context. Acharya Udayaveer Shastri in 'Saankhya Siddhant' (1991) has provided a lengthy discussion on the subject of *Purush* and *Prakriti* with valuable references from ancient writings. This author has benefited from his in-depth philosophical exposure of the complex Vedic concepts and recommends the Hindi readers to explore his writings on all six Darshans.

The *Prakriti* has been expressed variously in the Vedas and Darshans, all leading to the same meaning, that the *Prakriti* is the material-cause of the universe. In Saankhya philosophy (see Saankhya Siddhant by Acharya Udayaveer Shastri, 1991) the term '*prakriti*' for Nature is defined based on its *dhatu* and *pratyaya* (root and affix) as:

प्रकर्षेण क्रियते सर्वं जगदनया इति प्रकृति:

(*prakarshen kriyate sarvam jagadnaya iti prakritih*),
which translates into:-

> "*With which the entire universe is pre-eminently created is termed as Prakriti*".

The analysis of the word *prakriti* thus makes it obvious that the universe is a phenomenon created out of *Prakriti* and has a beginning. Because *Prakriti* displays large number of properties and functions it has large number of Sanskrit synonyms. Nomenclatures for *Purush* (Creator) and *Prakriti* (Nature) are differentiated by the *vibhakti* rules of grammar. In a grammatical construct for example the term *prakriti* appears in *tritiya vibhakti*, indicating with certainty that *Nature is tertiary to the Creator* and the Nature and the Creator are *two separate entities.*

The Vedic cosmology terms for Nature by themselves tend to illustrate the Vedic creation story. Whereas the term *Prakriti* indicates the necessity of raw material for the creation and that the process has a beginning, the other important Vedic term for Nature entity being *Pradhan*, indicates that the story has a definite ending. The term *pradhan* is for प्रकर्षेण धीयते अन्तर्लीयते सर्वं जगत् यस्मिंस्तत्

(*prakashen dheeyate antarleeyate sarva jagat yasminstat*) which translates into:

> '*Because it pre-eminently exists in creation phase and dissolves the entire universe within itself, that is why (it is Pradhan).*

These two terms for nature in fact establish the primary Vedic principle that the universe has a beginning and an end. Further along in the Veda mantras is the emphasis that the creation and dissolution is a definite perennial cycle that is imposed on Nature under the Creator's *ritam.*

There are two different Vedic terms used for Nature entity in two successive mantras of RigVeda's *Bhavavrittam sukta* on the history of creation of the universe, discussed in detail in Chapter 6. The term *Svadha* (RigVeda, 5/34/1; 10/129/1, 5; 1/164/38; 1/165/5, 6) stands for '*the entity with eternal self-existence*'; implicit in it is its causal status that it is not an effect of a cause. The second term *Salilam* there means '*the entity with fluidity with all its properties dissolved into uniformity*'.

An argument based on the meaning of the term *Svadha* could be raised by a modern cosmologist that if the Nature is '*the entity with eternal self-existence*' then it could, without Creator's help, self-generate into a universe under its governing physical laws and dissolve itself under the same laws into an indefinable state. The disagreement from the Vedic philosophy relates to physical laws of Nature and the nature of Nature.

To create a universe the matter must follow some physical laws. If the laws are created first where do they exist if the matter is not created yet? If the matter and the laws of Nature are created simultaneously what mechanism created the laws and what makes them perfect and critically balanced? Therefore, the Nature alone is not able to create its own laws to create a universe.

Contemporary cosmology is in agreement that the matter of Nature is quantized at its fundamental level. If the matter has not developed properties yet (before creation begins) each component of matter would behave independently to do its own thing to lead to a permanent chaos. No chance of forming a universe.

For a matter conglomerate without preexisting physical laws in it each matter particle is in an inactive equilibrium. This equilibrium extends to the entire matter of the pre-universe. Therefore to have the state of matter changed an extrinsic trigger of energy is needed. In the Vedic Cosmology this trigger is provided by the *Eekshana* of the Creator-Energy-Field. The Creator's *Eekshana* imparts the perfect physical laws (*ritam* and *satyam*) to the system and all properties such as the mass and charge to the matter to begin the creation process.

All thoughts, Vedic or otherwise, consider all substances of the universe to have some property. The difference is in the emphasis on primary versus secondary in that the substance is primary and the property is secondary. The Vedic philosophy puts it as - *the property is sheltered in the substance, is inseparable from the substance and cannot be harnessed out of a substance*. An example of this is the sweetness in sugar. The term *Salilam* for Nature when treated as a substance also means '*a physical entity in dissolved state*'. Implicit in this is that at the starting point of creation the whole matter is non-diverse. In the

Veda mantras, the Nature is also addressed as *Aditi* (RigVeda, 10/72/1-9, entire *sukta* is on *Aditi*). The Sanskrit root of the word *aditi* yields - '*the entity for utilization' or* '*the entity for consumption'*. This means that the Nature is utilized in creating the universe and the creation is for utilization. It is utilized in creating cosmos and it is utilized by living beings for forming their bodies and for their survival in general. A RigVeda mantra beautifully expresses that it is the diverse *Aditi* that is seen all around us in different forms.

अदितिर्द्यौरदितिरन्तरिक्षमदितिर्माता स पिता स पुत्रः ।
विश्वे देवा अदितिः पञ्च जना अदितिर्जातमदितिर्जनित्वम् ॥

(RigVeda, 1/89/10)

Transliteration:

"*Aditirdyauraditirantariksham aditirmaataa sa pita sa putrah* ।
Vishve deva aditih panch jana aditirjatamditirjanitvam" ।।)

Mantra Translation:

"*Dyu, the brilliant space around the Sun, is Aditi* (Nature); *Antariksh, the spread-out vast space is all Aditi. Aditi is the mother of all substances, Aditi is the father and Aditi is also its own progeny. All devas are Aditi and five knowledge senses are Aditi. All that is in the universe and all that is to manifest is from Aditi*" (RigVeda 1/89/10)

The Veda in this mantra has encompassed a big range of the creation from the vast space and cosmos to different forces within living beings in the form of knowledge senses and the process of generation from mother to child and so on. All that is in the universe and all new things that will come to be are also *Aditi*. The *Aditi* produces diverse substances but ultimately they all are *Aditi* itself. The thought in the mantra is remarkably simple yet defines uniquely the *Aditi*-entity.

The term *Avidya* for Nature means '*not possessing knowledge*' and the term *Tamah* means an '*inert entity*' and it also means a '*non-knowledge entity*'. Both terms, *Avidya* and *Tamah* for *Prakriti* indicate that *Prakriti* the material-cause of the universe is an inert entity without having any knowledge and its own initiative; unlike-living beings. Such a rich nomenclature in the Vedas for a single entity is an outstanding example in the subject of cosmology. In his Vedic studies, this author has come across numerous synonyms for the Nature-entity, however only 39 of these are compiled in the Table 5.1.

Table 5.1: Based on its attributes and functions the 39 Sanskrit synonyms of *Prakriti* (Nature)

Item No.	Synonyms of *Prakriti* (Nature)	Word Meaning	Reference
1	Aabhu	Nature transited out of its causal form	RigVeda (10/129/3)
2	Aayu	An aged-entity	
3	Achetan	without consciousness	Shvetashvar Upanishad

4	Aditi	*aadyate iti aditi*, utilizable entity	RigVeda(10/72, 1/89/10, 10/64/5)
5	Agyaan	Nonknowledge	Shvetashvar Upanishad
6	Aja Ajamekam	Unborn, unitary-eternal	Taittireeya Aranyak (10/10/5) Shveta Upanishad (4/5)
7	Akshara	Non-decaying with time	
8	Alinga	Non-descript, neutral gender	Yoga sutra Vyas-bhasya (2/19); Mahabharat, Shanti (303/47)
9	Anaksha	Blind, devoid of knowledge	RigVeda (10/27/11)
10	Andham	Blind	RigVeda (10/27/11)
11	Avyakta	Non-definable (during dissolution phase of creation)	Saankhya; Shvetashvar Upanishad; Kathopnishad (1/3/11); Bhagvadgita
12	Avidya	Inert, not possessing knowledge	Shvetashvar Upanishad
13	Bhogya	Utilizable (matter for cosmos, utilized by souls)	
14	Chhaya	Image, Shadow, Darkness	Shvetashvar Upanishad
15	Goh	Cow: provider of nourishment to creation	Vaayu Puraan (23/55)
16	Jyeshtha	Olden, older than all other objects	RigVeda (10/121/1); Nirukt (13/37)
17	Kshetra	Field: area with defined space	Shvetashvar Upanishad; Bhagvadgeeta (13/1-5)
18	Leen-Vastu	Substance that dissolves everything in it.	Saankhya Darshan (1/56)
19	Mahan	The Great	Mahabharata, Shanti (238/26)
20	Maaya	Complex to understand, an intended design	RigVeda (6/47/18); Shvetashvar Upanishad (4/10)
21	Mool	Root-cause of creation	Saankhya (1/32)
22	Parinamini	Subsistent	Saankhya literature
23	Pradhaan	Principal cause	Mahabharata, Shanti (238/26); Shvetashvar Upanishad (1/10); Saankhya (2/1)
24	Prajapati	Progenitor of population (cosmos), Nature in Creation phase	AtharvaVeda (10/7/7, 8,4 0, 41; 10/8/13); several more
25	Prakriti	The specific entity that serves as the material cause for	Vedas, Saankhya + many references

		creation	
26	Prasav dharmini	Embryo-genesis	Saankhya literature
27	Salakshma	Of uniform properties (pralaya state of nature)	RigVeda (10/12/6)
28	Salilam	Fluid in nature, pervasive in universe, uniform in properties (pre-creation state of nature)	RigVeda (19/129/3); AtharvaVeda (10/7/38)
29	Satva	Reality (not an illusion), state in creation phase	RigVeda, Pangi Upanishad
30	Shakti	Energy	Shvetashvar Upanishad
31	Svadha	self-sustaining entity, not being created, cause and not an effect	RigVeda (10/129/2, 5)
32	Tamas/ Tamah	Darkness: non-knowledge entity, inert entity	YajurVeda., Manusmriti (1/5) ; Shvetashvar Upanishad
33	Tridhatu	Tri-matter or tri-property (sata, raja, tama)	All through Vedic literature
34	Tri-gunatmika	Tri-property	Many references
35	Vun	Presents variety in creation. Utilizable in creation. Diverse as in biodiversity	YajurVeda (1/20)
36	Vidhanam	Orderly, with rules and physical principles	Deval Dharma Sutra
37	Vishu Roopa	Of numerous forms	RigVeda (10/12/6)
38	Vriksha	Tree: It grows, appears in numerous branches, leaves and fruit (as in creation phase). It is cut (as in dissolution phase)	RigVeda (10/31/7, 10/81/4, 1/164/20); YajurVeda (17/20); Taiti San. (4/6/2/5), Geeta (15/1)
39	Yoni	Material-cause, Womb of all creation,	RigVeda (10/8/3); Shvetashvar Upanishad (1/2)

Each term in this table generates a different thought about this single entity, Nature. From this pool of words, one can appropriately choose the one that best expresses the underlying thought to create a picture in the reader's mind. In the Table 5.1, a brief meaning of the Sanskrit term is given and references wherever possible are also given in the last column.

Rishi Kapil in his Saankhya philosophy has extensively dealt with the subject of *Prakriti* and the creation of the universe. Rishi Kapil's reference appears several times in the book of Mahabharata (Figure 3.1, Timeline-A, 3140 BCE). The Shanti-parv chapter (63/152) of this book presents a reference of an ancient Rishi Yagyavalkya who praised Rishi Kapil and his Saankhya Darshan as - *"Naasti Saankhyasamam Gyaanam"* which translates into – *'There is no knowledge like the Saankhya knowledge'*. It is also stated in Mahabharata that - *"Gyanam Saankhyam param matam"*, which translates into – *'The Saankhya knowledge is the profound principle'*. Upanishads, prior to Mahabharata, have also praised Rishi Kapil and his Saankhya philosophy. This author has cited earlier a number of *sutras* from Rishi Kapil's Saankhya Darshan. Contextually the *sutra*, मूले मूलाभावादमूलं मूलम् (1/32) (*mole moolaabhaavaad-moolam moolam*) was discussed in Chapter 1 explaining the different meanings of the term *mool*. The *sutra* translation is –

"The mool is without mool due to nonexistence of mool in the mool"

and its meaning is –

"The Nature is without a cause due to nonexistence of its cause."

Implicit in Rishi Kapil's expression is that the Nature is an eternal entity and not an effect of a cause. This is the nature of Nature in the Vedic thought. It is the root-cause of all matter, the *Mool*. The term *mool* is commonly used for the term *jada* which commonly stands for inert substances. The *Prakriti* (Nature) in Vedic science is an inert entity.

As mentioned before, the two states of *Prakriti* are indefinable *Asata* during the pralaya phase and definable *Sata* during its creation (*srishti*) phase.

Prakriti (Nature) a Conglomerate of Tri-property *Aapah* Particles

Aapah the fundamental particles of matter in three categories *sata, raja* and *tama* were introduced in Chapter 1 and 2. *Aapah* is one of the most important terms in Vedic cosmology. *Aapah* are the main ingredient of Nature in both its phases, creation and dissolution. *Aapah* are to Vedic physics what is fundamental particles to modern physics. A detailed discussion on *aapah* is presented in the following with a pictorial depiction to differentiate the two states of matter, *Asata* and *Sata*. Mysteriously, the Vedic *rishis* have proclaimed that matter at its fundamental level is a tri-property corpusclized entity, the three properties of the three categories being *sato-guna, rajo-guna* and *tamo-guna* where *guna* suffix to the three kinds of *aapah* stands for 'property'. A brief discussion on the three properties is also included in the following.

Aapah (आपः): the fundamental Tri-property unit of Matter

The Vedic creation of the universe is from the fundamental form of matter called the *aapah*. The *aapah* in the Vedas appear as a pluralized term in female gender (RigVeda, 7/47/1-4). Therefore, *aapah* term in this text is not pluralized by adding 's' to it. *Aapah* are pluralized in the Vedas perhaps because even the formation of a single atom of matter needs more than one *aapah* and the female adjective to *aapah* perhaps is because they together give birth to the whole creation when impregnated by the *Eekshana* energy of the Creator.

An *aapah* particle of the Veda is with independent existence in the likeness of a quark of modern physics that are formed right in the beginning of the Big Bang and are consumed in forming the nuclear particles leading to the formation of atoms and molecules. Formation of the smallest atom, Hydrogen, with two 'up-type' and one 'down-type' quarks was depicted in Figure 2.3 in Chapter 2. The Vedic philosophy stipulates that no particle is formed without the *aapah* particles. This implies that a negatively charged electron orbiting around a nucleus is formed with some *aapah* particles and even a photon of light presently thought as a mass-less particle is also formed with some *aapah* particles. An electron being only 659th of the mass of a quark mass it is likely that an *aapah* particle is even smaller than a quark or even smaller than an electron.

Notice that the *Sata*-state of matter in the creation phase and *sato-guna* (*sata*-property) of matter use the same term '*sata*'. To keep a distinction between the state of matter and the property of matter, this author has used first syllable in upper-case for the *Sata*-state of matter and the lower-case for the *sata*-property of *aapah* particle. In the Vedas, the context determines whether the state of matter is being addressed in a mantra or the properties of matter are the subject. For example in the RigVeda *sukta* (10/129/1) on Creation of the Universe, both terms *Asata* and *Sata* have appeared together in the same Veda mantra, where the context is the state of matter transiting from *Asata,* the indefinable state of matter, to *Sata*, the defined state of matter. In context with the property of matter the word *sata* or *satah* is for the property of matter, the other two being *rajah* and *tamah*. As mentioned above, these properties when hyphenated with the term *guna* (property) their Sanskrit transition is: *sato-guna, rajo-guna* and *tamo-guna.*

A coarser example of *guna* (property) from the observable world is the sweetness in sugar. When a tiny sugar crystal is broken into smaller pieces, each piece still tastes sweet. Even a single sugar molecule will taste sweet. A sugar molecule is made up of three types of atoms, Hydrogen, Carbon and Oxygen. None of these individually taste sweet. Therefore, if a sugar molecule is broken into Hydrogen, Carbon and Oxygen atoms, the sugar molecule loses its identity and its sweetness too. Other chemicals are also formed with differing Hydrogen, Carbon and Oxygen atoms but they have other *guna*, yielding other tastes or smells (properties). It is a special chemical arrangement of these atoms that creates a specific *guna* in a substance. What this coarse example of sugar tells us is that the *guna* of a substance accompanies the substance right to its final minuteness. The three *gunas* stated above take us well beyond the *gunas* of taste and smell at atomic and molecular level. The *sato-guna, rajo-guna* and *tamo-guna* are the *gunas* (properties) of the most fundamental form of matter beyond the atoms and molecules (*paramanu* and *anu*) of elements and therefore beyond the subject of knowledge senses. An *aapah* particle with three *gunas* in it indicates that it is a trio with three types of particles in it.

Vedic principles stipulate that during the creation phase of Nature, differing proportions of *sata, raja* and *tama* create substances (atoms and molecules) with differing properties. This is something similar to differing proportions of nuclear particles in the nucleus form different atoms.

The observable active matter/energy of cosmos is said to be dominant in *rajo-guna*. It is the *aapah* of *rajo-guna* property that serves as the seed for the formation of observable matter in the creation. This is why in common language, the fine sand grains are addressed as *raj* (sound as in nudge). The basic property of matter being gravitational attraction, it is the *rajah*-property of matter that creates gravitation in matter. The orbital motion in cosmos is also credited to *rajo-guna*. This motion in physics terminology is owing to the kinetic energy. The radiational energy (light) is also owing to the *rajo-guna*. The resplendence cosmos in motion are addressed as "*yo antarikshe rajaso-vimanah*" (YajurVeda, 32/6) which translates into, "*those matter-objects in Space, in flight like energetic plane*".

In Vedic science, the intellect in living-beings is a separate entity acquired by living-systems from the pervading field of *aapah* with *sato-guna*. Since the intellect entity is a byproduct of *aapah* it is a physical entity.

Matter made of *aapah* with non-active comatose characteristic is said to be dominant in *tamo-guna*. Keeping this in mind a reference from Kapil Rishi's Saankhya Darshan on the state of *Prakriti* (Nature) in its dissolved state is briefly discussed.

The *Saamyavastha Prakriti*: State of Neutrality in Nature

Rishi Kapil lays down a canon defining the state of *Prakriti*. The first part of the *sutra* that deals with the *Prakriti* and its products is,

सत्त्वरजस्तमसां साम्यावस्था प्रकृतिः,...
"S*atva-rajs-tam-saam saamyavastha prakritihi*..." (Saankhya, 1/26)

Sutra Translation:
"The saamyavastha (balanced, with neutrality) state of satva, rajah, tamah is Prakriti...". (Saankhya, 1/26)

Rishi Kapil in this formulation states that the *Prakriti* is in its dissolved state when three properties of matter, are in a uniform balance, indistinguishably mixed together. This state is when the three properties of matter exhibit no diversity, no separation or no activity in any fashion. This is Nature's *Asata*-state which is the dissolution phase in a non-diverse state of the three *gunas* (properties). This is matter's indefinable state. This uniform and non-descript is the state from which the Nature enters into the creation phase. To understand Rishi Kapil's *saamyavastha* (balanced state) of Nature a parallel from modern physics is drawn in the following.

A system in physics is said to be in static equilibrium when no movement is observed in it or in its parts. This would occur when there are no forces acting on the system or all forces acting on the system in different directions cancel out to zero. Assume this system to be a container with some gas in it. By itself sitting on a table the container is in static equilibrium. Inside the container however the story is quite different. The gas molecules inside the container are not stationary, they are moving around at high speeds. The cause of this motion is the temperature of the system. In Chapter 2 while discussing temperature of a substance the 'absolute zero temperature' (0° K) was explained. This is the temperature when motions of all atoms and molecules of the substance cease to exist. Such a state will render static equilibrium of even higher degree than the one mentioned above. The question is whether this really is the highest state of static equilibrium of a substance? Certainly not and this is still not the *saamyavastha* of *Prakriti* in Rishi Kapil's definition either. Further discussion on this makes it clear.

Consider a stationary atom in the system at static equilibrium. This stationary atom in fact has tremendous dynamics because the tiny electrons are ceaselessly orbiting at very high speeds around its nucleus; reaching over 2000 km per second. This means that apparently a system in static equilibrium at 0° K still had a high degree of dynamics hidden in it. The difference in this static equilibrium and Rishi Kapil's *saamyavastha* equilibrium is that there is no motion of electrons or the existence of any activity in Nature (matter) even at its fundamental level. The *'saamyavastha'* state of Nature in Rishi Kapil's *sutra* is not an ordinary equilibrium state; rather it is the state when even the fundamental particles of matter have no motion of any kind whatsoever. By the Vedic definition, any motion in matter represents the presence of *rajo-guna,* which if existed, would disturb the *saamyavastha* state of Nature. In the realm of particle physics and the Big Bang Theory, it could be the state before Plank Time. It is that point of equilibrium beyond which, as in the rest of the *sutra* (not presented above), the *Prakriti* enters into creation phase producing a set of 23 entities of matter at the most subtle matter and energy forms. These first-born distinct subtle forms of energies serve as the foundation for creation of the Vedic universe which includes both, the animate and inanimate creation.

The emphasis in the sutra is the 'balanced state' of *satva, rajah, tamah* in the inert conglomerate. This state is also expressed as the inactive state of *Prakriti* that should continue to persist unless an extrinsic spark of energy is put into it.

The infusion of energy into the balanced state of matter (*saamyavastha* of *Prakriti*) by the *Purush*-entity at a given point in time causes a reversal in the non-active nature of matter into becoming active.

This Vedic concept of Nature wherein it is an ensemble of three types of particles at its fundamental level is unique to the Vedas. Just as in modern cosmology the universe starts as an uncharged neutral singular blob of matter in its early state and later it exhibits a charge separation as in an atom where its nucleus is positive and electrons orbiting around it are negative. If all charges of the universe, positive and negative, are summed up, it is a neutral entity. This separation of charge is the nature of the universe in creation phase. Likewise the separation of three *gunas* to form substances dominant in specific *gunas* is the Vedic concept of the universe. A simplistic example for the *Asata*-state of matter in Nature's balanced state in dissolution phase during *pralaya* is a sesame seed. Just as the three ingredients, sugar-oil-chaff, are uniformly hidden in a sesame seed, the three properties *sata-rajah-tamah* of matter are uniformly hidden in Nature. The diversification of sugar-oil-chaff is observed when the sesame seed is crushed by extrinsic force. This is in the likeness of the creation phase of the universe where the balanced state is disturbed revealing its diverse properties.

Etymology of the term *Aapah*

The great grammarian of Sanskrit, the scholar and reviver of the Vedas, Swami Dayanand Sarasvati (1825 - 1883 CE) has provided the root and the word formation of the term *aapah* in his book entitled 'Panch Maha Yagjya Vidhi'. He states:

" ..The word aapah is derived from अप (*ap*), which *originates from the root* आप्लृ व्याप्तौ (*aaplri vyaptau*) *and in the realm of physical creation it is always feminine and plural..."*

Swami Dayanand Sarasvati further elaborated upon the term *ap* (pronunciation as in 'up') when he took up the translation and interpretation of an AtharvaVeda mantra. The root meaning of the term *ap* is contained in the second part of the root *vyaptau* of *aaplri vyaptau*. This means that *aapah* are 'all-pervading'. The term to represent something which is all-pervading/ubiquitous, is व्यापक (*vyapak*), also derived from the root व्याप्तौ (*vyaptau*). The term व्यापक (*vyapak*) itself is formed with वि आप क (*vi aap ka*) with अप (*ap*) being at its core. Owing to the property of all-pervasion, the *Brahmn* (Creator-entity) is addressed with the adjective of सर्व-व्यापक (*sarva-vyapaka* = all-pervading) with *ap* at the core of the second term. Even though the Vedic Creator-entity is an all-pervading singular unitary spirit, the *Brahmn* is also addressed as *aapah* in singularity. When the context is the physical aspect of the creation, the *aapah* are addressed in plurality with feminine gender. The usage of the word in a sentence determines the context.

The *rudhi* meaning (ordinary or colloquial) of the word *aapah* also stands for water because fluidity is water's basic nature. The word *aapah* appears very frequently in the Veda mantras on cosmology where early stage of the universe is depicted. There was no water formation in that early stage of creation and water is not all-pervading in the universe. Therefore, the *yaugik* meaning (root-meaning) of the term must apply.

Aapah the Viscosity of Space

The term *aapah* having generated from the *dhatu* (root) आप्लृ व्याप्तौ (*aaplri vyaptau*) necessitates that it is ubiquitously present at every point in space. Based on this definition and from other references in the Vedas this author has understood that out of the seven progeny of *Aditi* (Nature), discussed in Chapter 5 and 8, the *aapah* is one of the first born *devatas* with the property of ubiquity. Therefore *aapah* by itself is an entity and no corner of space is devoid of it. The space in between the stars and galaxies that one thinks is empty is in fact populated by the background *aapah* field. This field acts as the viscosity of empty space and moderates the motion of all moving objects from massive celestial bodies to high speed fundamental particles like photons. The reason a photon of light travels with a fixed speed of $3x10^8$ meters per second in vacuum and not more is due to the ubiquitous field of *aapah*. A photon is made of *aapah* and therefore must have some mass. Therefore, it also must exhibit gravitational attraction to mass. Its speed through the *aapah*-field limited to $3x10^8$ meters per second is indicative of the fact that the mass distribution in space within the *aapah*-field affects the motion of the photon.

The nature of *Aapah*

Here is a reference from Taittireeya Aranyak, written after Saankhya Darshan, where the same three inherent eternal properties of Nature, *satah-rajah-tamah* as being the cause of the creation, are expressed beautifully with the adjective of colours.

अजामेकां लोहितशुक्लकृष्णां बव्हीं प्रजां जनयन्तीं सरुपां |
अजो ह्येको जुषमाणोऽनुशेते जहात्येनां भुक्तभोगामजोऽन्य ||

(Taittireeya Aranyak, 10/10/5)

Transliteration:

"*Ajaamekaam lohit-shukl-krishnaam bavhim prajaam janyanteem sarupaam*|
Ajo hyeko jushmaano anushete jahaatyenaam bhuktbhogaamjo-anya"||)

Translation:

"*With the* (fundamental) *properties of lohit-shukl-krishn is one and the only one Aja* (Nature) *that itself never takes birth but from it are created numerous diverse populations that inherit the same properties. Among the eternal-beings* (souls) *one* (kind) *utilizes Aja to enjoy materialism and continues to be trapped in it and the other* (kind) *utilizes it to gain access for the enlightenment of the Brahmn, thus liberating himself from it and attaining salvation."* (Taittireeya Aranyak, 10/10/5).

In the above *shloka* the *lohit* (red), *shukl* (white) and *krishn* (black) respectively correspond to the three fundamental properties of Nature, the *satva, rajas,*and *tamas*. The three fundamental properties of matter are the same as stated before. This shloka also unfolds that the tri-property Nature is one entity and is utilized to create the cosmos to populate the universe with all objects of Nature inheriting the tri-property of the parent. The second line of the shloka refers to the eternally existing conscious beings (third invariant to be defined below) that are born in physical bodies in Creator's creation to utilize Nature and its creation. As mentioned before, Nature acquired the synonym of *Aditi* because it is for utilization by conscious-beings. The Taittireeya statement cited above states figuratively that one kind of conscious-being, while utilizing the Nature, is caught up in it and the other kind becomes cognizant of the fact that the Creator and the creation (Nature) are two different entities. This realization by the enlightened conscious-being frees him/her from the attachment to matter (Nature) and leads him/her to salvation, the so-called 'eternal bliss'. Simply put-the materialism is a hindrance to knowledge and enlightenment and therefore to the salvation.

The first word *Aja* of Taittireeya *shloka* is a noun and stands for an entity that never takes birth, meaning that it is an eternal entity. *Aja* of Taittireeya is the Nature-entity and the three adjectives to it (red, white, black) are its three causal properties of *satah, rajs,*and *tamah* respectively that in Kapil's sutra are listed *sata, raja* and *tama*. The difference here is that the writer of the Taittireeya Upanishad expressed them with adjectives of colours. Since the colours are present only in inert matter and not in the *atma* (soul) of a conscious-being, the entity with three colours (three properties) is inert and produces many populations out of these three fundamental properties of matter that are inert like itself. Acharya Udayaveer Shastri in his book in Hindi language, 'Saankhya Darshan Ka Itihaas', has elaborated upon this topic with detailed etymology of different terms.

Pandit Bhagvaddatta in his book 'Ved-Vidya Nidarshan' (1959) summarized from the fifth chapter of Vayu Purana shlokas 9 through 13 that draw a sketch of Vedic creation of the universe. The English translation of the same is given below:

> "*The pralaya* (dissolution) *is when the gunas* (three properties of matter: *sato-guna, rajo-guna, tamo-guna*) *are in a non-differentiating balanced state of dissolution and srishti* (creation) *is when the balance is perturbed. The rajo-guna in pralaya exists in indefinable dissolved state, sheltered in tamo-guna and sato-guna like oil in sesame seed and butter in milk. Parameshvar* (the Supreme Being) *with immense yoga* (action with conscious effort) *entered the egg* (primordial matter embryo), *created perturbation in the Pradhaan* (Nature) *and the awakening in All-pervading Purush self. The perturbation in the Pradhaan manifested the rajo-guna. It is the rajah that creates exalted action in all, like water sprouting a seed. Under the governance of Purush, incongruous state of the gunas* (properties) *creates the srishti* (universe)".

Although the Puranic references are not as old as the references from the Vedic literature of pre-Mahabharata period 5,000 years ago, the ideas contained in them have descended from ancient Sanskrit Literature and remain indigenous to India where Nature continues to be defined as *tri-guna-tma-ka* (tri-property entity). As far as the above statements of Vayu Purana on creation are concerned the underlying thought is not different from that in the Vedas and the Darshans. According to the Vayu Purana the equilibrium state of the three properties of Nature is when there is a complete lull in Nature. It becomes active upon inspiration from the Creator.

Of the three properties the rajo-guna that was masked earlier by the other two, becomes free to create dynamic state in matter. This new dynamics in the primordial matter creates various subtle forms of matter some of which with the nature of ubiquity. Rishi Kapil has enumerated the forms and the sequence of these forms. One matter form dominant in rajo-guna is Ahankaar (ego), a force generating activity in all, animate and inanimate.

It is the *rajo-guna* that imparts the property of mass in matter leading to the property of gravitational attraction with motion at large scale in heavenly bodies and at small scales in atoms to rotate the electrons around nuclei.

The Nature in the above statement being compared to an egg indicates that in its early state the Nature was a three-dimensional entity in space in the shape of an ellipsoid. Large number of Veda mantras and Brahmana statements, presented in Chapter 7 and 8, emphasize that the primordial matter of the universe (*Prakriti* = Nature) was initially in the shape of an ovate in *paryaplav* (rotation in floatation) before it exploded in the Vedic Big Bang. This is contrary to the modern Big Bang hypothesis where the whole nature is in a dimensionless point; a non-physical state only possible in mathematics. The *aapah*-trio with the three properties of *satah, rajah* and *tamah* are pictorially in Figure 5.1). The colours white, red and black for *satah, rajah* and *tamah* particles are arbitrary; however, they are adopted from the Taittireeya Aranyak citation presented above.

This author's concept of *Asata*-state is shown on the left in Figure 5.1. The active tri-property *aapah* particle in Nature's creation phase is shown on the right of Figure 5.1. Based on the Vedic knowledge grasped by this author, all forces of Nature are absent in its *Asata*-state of dissolution phase and they come into existence when *Asata*-state switches into the *Sata*-state of creation. The figure on the left is devoid of any kinetics in *aapah,* external or internal, leading to all *satah, rajah* and *tamah* to come together and confine in *a* diminutive space, reducing the space around them (shown by outer circle in the diagram). In the creation phase on the other hand, the differing forces (attractive and repulsive) emerge as byproducts of Nature (represented by arrows) that disturbs the balance in Nature leading to separation between the *satah, rajah* and *tamah*, stretching the Space-entity occupying increased volume (shown by outer circle).

***Asata*-state)** (inactive balanced state) (*sata, raja, tama*)	***Sata*-state** (*active aapah*) (*sata, raja, tama*)

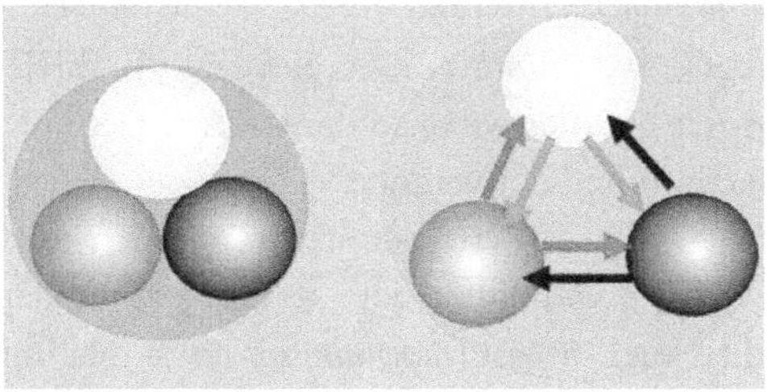

Dissolution Inactive Phase	**Creation** Active Phase

Figure 5.1: *Prakriti* - *Asata* to *Sata* state, Inactive to Active*: satah* (white), *rajah* (red), *tamah* (black). The outer circle represents the minimum space occupied by three kinds of *aapah.*

The concept presented in the above diagram for two states of *aapah* particles when extended to the whole Nature, leads one to conceptualize matter of the universe confined in a smaller space during its dissolved phase and it swelled up in its creation phase. This is what is expressed in the RigVeda mantra (1/129/3) in the *Bhavavrittam sukta* (History of Creation of the Universe) that the *Prakriti* (Nature) in its *Asata*-state is confined *in 'tuchchhayen'* (diminutive) space. This is further discussed in detail in Chapter 6. The matter particles in the creation phase transit from 'introvert' to 'extrovert' due to the *tapah* (heat energy) of the Creator Energy Field (RigVeda, 1/129/3; 1/190/1). It is therefore intuitive to suggest that the energized *aapah* would stretch the Space of the embryonic universe.

A nuclear scientist coined the word 'flavor' for describing the property of a quark (Table 2.0) even though a quark has no taste in it. Still the Figure 5.1 draws a parallel with physics in that the up-type and down-type for quarks is something similar to the up and sown arrows differentiating *satah* (white), *rajah* (red) and *tamah* (black) particles. In reality these three particles are colourless.

The *aapah* are said to be the un-exhausting fuel of the creation, the onset of which is said to begin with the onset of vibrations in *aapah*. These thoughts are presented in the RigVeda mantra:

आपो यं व: प्रथमं देवयंत इन्द्रपानमुर्मिम[1]कृण्वतेळ: ।
तं वो वयं शुचिमरिप्रमद्य घृतप्रुषं मधुमंतं वनेम ॥ (RigVeda, 7/47/1)

Transliteration and Word Translation:

आपो (*aapo) = aapah (pluralized)* , यं (*yam*)= to which , व: (*vah*) = your, प्रथमं (*prathamam*) = the first, देवयंत (*devayanta*) = desiring *devas,* इन्द्रपानम् (*indrapanam*) = remunerative to *Indra deva* (electricity) , ऊर्मिम्[1] (*urmim*) = waves or vibrations, अकृण्वत (*akrinvat*) = to prove or to draw inference with contemplation, इळ: (*ilah*) = voice, action with controlled deliverance ।
तं (*tam*) = that , वो (*vo*) = for you , वयं (*vayam*) = we , शुचिम (*shuchim*) = pure, अरिप्रम् (*aripram*) = without defect, अद्य (*adya*) = today , घृतप्रुषं (*ghritprusham*) = saturated with fluidity, fuel, मधुमंतं (*madhumantam*) = elating substance, वनेम (*vanem*) = with specificity॥

Mantra Translation and Interpretation

> *"Today let us be contemplative to draw inference on aapah the elating substance that is fundamentally pure without defect and is the un-exhausting fuel with fluidity (ghrit-prusham). The controlled first vibrations[1] (ilah-urmim[1]) in aapah were for the desiring devas and were remunerative to the Indra deva (electricity)".* (RigVeda, 7/47/1).

The reference to *devas* in the mantra with the adjective of 'first' is for the first specific energy and force fields created in the Vedic creation process that lead to the next steps of creation in the universe. One of the seven *devas* created first in the beginning of creation, discussed in detail in Chapter 8 is the *Indra-devata*, the all-pervading electricity of the universe.

First of the 7 *devas* is *Aapah-devata* itself the matter at its most fundamental stage which is the subject of this mantra. The adjective of 'pure' to *aapah* particles forming the *Aapah-devata* means that it is an unadulterated primordial matter. The 'purity' of *aapah* here means that it is made of itself without components and before any degenerate states of matter are formed out of them. Since *aapah* are the source of all other forms of matter and energy, observable and unobservable, they are addressed with the adjective of घृतप्रुषं (*ghrit-prusham*), the unlimited fluid fuel of the universe; the *Indra* being one of the first beneficiaries of it to manifest as electricity.

An important and mysterious term in the mantra is ऊर्मिम्[1] (*urmim*) in the primordial *aapah*. The term *urmim* stands for both, waves and vibrations. As it will become clear later from the other mantras that the cause of the first vibrations in the fundamental matter of the universe is *tapah* the heat energy from a source extrinsic to *aapah*, this mantra hints that the initiating mechanism for the creation of the universe is through the vibrations. These first vibrations give rise to the first seven *devas* which become the fundamental properties/nature of matter. These seven *devas* are the subject of the *Devata sukta* of RigVeda (10/72/1-9) discussed in the Chapter 8.

When the Vedic term '*indra*' is subjected to etymological analysis, one learns that its root represents properties that represent several entities. In the context of *devas*, the *indra* among seven inert systems of creation represents the all-pervading electricity of the universe. From a practical perspective, one knows that the electrons, the representative of electricity, pervade through all matter of the universe and even the energy (photons) is produced by electrons when they transit from higher exited states of atoms to lower states. Therefore, an electron and differing forms of electricity are in the category of *indra-devata.*

The controlled ऊर्मिम् (*urmim*), the waves and vibrations in *aapah* leading to the creation of *indra-devata*, the electricity, is the subject of this mantra. The waves and vibrations in all-pervading electricity are also the subject of the sixth mantra of the *Devata sukta* (RigVeda, 10/72/6). Stated therein is - 'यद्देवा[1] ... नृत्य[2]तामिव[3] तीव्रो[4] रेणु[5]रपायत[6] (*yad-deva*[1] ...*nritya*[2]*tamiva*[3] *teevro*[4] *renu*[5]*rpaayata*[6]), which means – "*The devas*[1]...*with their each primordial fundamental quantum*[5], *spread*[6] *out forcefully*[4] *far and wide through this universe moving like*[3] *a dancer*[2]". The dance of particles in the early minute form of matter is ऊर्मिम् (*urmim*), the waves and the vibrations mentioned in this mantra. Implicit here is that the first vibrations in *aapah* the inexhaustible fuel of the universe create the first *deva*-systems. The other mantras of the *Aapah-sukta* of RigVeda (7/47) have also re-emphasized the ऊर्मिम् (*urmim*) in the *aapah*. In RigVeda (7/47/2) the mantra states that the ऊर्मिम् (*urmim*) the vibrations in the *aapah* cause आशुहेमा इन्द्रः (*aashuhema indrah*) meaning *'a tremendous increase in indra* (*the electricity*)*'*. When it comes to the creation of the universe, the thoughts presented above and other related mantras are mysterious yet quite explicit in picturing the process and relationship of entities playing a role in the early stages of the universe. This author has made an effort to draw attention to the processes and relationships of the entities that are involved in the Vedic Cosmology scenario.

With regard to the Nature as a tri-property entity several Upanishads have followed Rishi Kapil and his Saankhya Darshan. In Chapter 1 a reference of the students of Rishi Shvetashvatar, the author of the Shvetashvatar Upanishad, was cited wherein they had posed a question to their teacher asking which entity is most important in creating the universe. In Shvetashvatar Upanishad (5/2, 4/5), the Vedic *maatatli* (cosmologist) Rishi Shvetashvatar expressed high respect for Rishi Kapil and his Saankhya Darshan and proceeded defining the Nature as the tri-property entity, in a way quite similar to the one presented by Rishi Kapil. The same is in Chhaandyoga Upanishad (6/3/3, 4 and 6/4/7). So there is no dearth of discussion and consistency on Nature's nature in the Vedic literature. Not only the Upanishad references put Rishi Kapil in the distant past of the Upanishad period in the Vedic history, the Upanishads, Darshans, Mahabharata and the other ancient books propound Rishi Kapil's philosophy on the fine structure of the Nature-entity. Rishi Kapil of course has based his philosophy on the Vedas.

Yogiraj Krishn in Bhagvadgita stated that Rishi Kapil's Saankhya principle based on the Veda, that the Nature is a tri-property entity, is the '*all-accepting principle*'. Bhagvadgita's '*tri-gunya-vishaya veda*' (tri-property-subject of Veda) in its shloka (2/45) is also a reference to Veda's tri-property Nature where Krishn preaches to Arjun to separate himself from *Prakriti* to strengthen his spiritual vitality. This is exactly the same as the message in Taittireeya Aranyak shloka for salvation. Krishn in the Bhagvadgita shloka (13/23) further emphasizes the subject of '*Purush-Prakriti*' (Creator-Nature) very close to what was in the shloka of Taittireeya Aryanak. He preaches about the hidden secret for attaining salvation. This secret is to attain freedom from being trapped in the material body with limitations, through the cycle of reincarnations which is known as the bondage of Nature (matter). In the present context of a discussion on the tri-property Nature, it is interesting to learn that Lord Krishn stated to Arjun in Bhagvadgita shloka (14/5) explicitly the effect of matter on the soul. The shloka translation reads thus:

> *"O Mahabahu!* (Arjun, the warrior with great arms) *the satva, raja, tama, properties of the Prakriti* (Nature) *tie the imperishable conscious soul by bestowing a physical body to it".*
>
> (Bhagvadgita, 14/5).

Krishn's message to Arjun may sound as though the Nature entity has its own mind and registers control to tie down an eternal-being soul to the material pleasures of the physical body, thus diverting the soul away from eternal bliss of salvation. It has been emphasized severally in this text that Vedas treat Nature, its byproducts and its systems (*devas*), as inert entities. Krishn makes a distinction here and throughout the Bhagvadgita that the soul is conscious and is an imperishable entity while the physical body derived from *Prakriti* is transient. The inert entity like Nature and its other inert systems like air, energy, Sun, Earth and so on are also inert.

Vedic Triune Atom

According to modern physics the inherent nature of matter is corpuscular. An atom may have several neutrons and protons forming its nucleus. Each proton or neutron is formed with three quarks each; a proton formed by two up-type and one down-type quark and a neutron is formed by two down-type and one up-type quark. When an electron is 659th in mass of a quark it cannot be said that a quark is matter's last stage. An interesting observation in modern physics is that a triplet of matter (3 quarks) gives rise to stability in forming matter.

A remarkable feature of the ancient Vedic science is that the corpuscularized (made up of fundamental particles) Nature at its fundamental level is made up of three types of particles *sata, raja* and *tama*. However, based on the discussion on matter in the Vedic Literature, descending down from the ages, this author understands that each component of this Vedic triplet is subtler than the stable particles observed and measured by modern physics. This author posits even the mass-less photon, a particle much smaller than a quark, must have at least one *rajah* particle in it. We have not seen yet what is inside a quark leave alone what is inside a photon.

For a moment let us assume that a quark is not the fundamental particle but it is divisible into smaller particles of matter. A quark weighs about $6x10^{-28}$ kg (6 divided by a number with 28 zeros after 1) and an electron weighs about $9.11x10^{-31}$ kg (about 659th of a quark). We can at least divide a quark, hypothetically, into 659 little pieces with each piece weighing the same as an electron. But why stop at that. Another real particle in modern physics called neutrino weighs only about $1.78x10^{-37}$ kg which is about 3.37 billionth of a quark. The Big Bang Hypothesis uses the time division up to 10^{-43} seconds (Plank-time). Therefore, if time-division that small can make sense in physics, no mathematical limit could be imposed on dividing matter well beyond the size of a neutrino. Thus if one continues to divide a quark to reach a trillionth of its mass, one ends up with a tiny piece of matter with some matter still left in it. So, we are left with two questions:

(a) are the *sata-raja-tama* of the Veda the same as the neutron-proton-electron of the contemporary physics? Or

(b) are the *sata-raja-tama* of the Veda the same as the three quarks of contemporary physics that form a neutron or a proton? Or

(c) are they the last stage of matter yielding the formation of quarks?

These are difficult questions to answer. One other pertinent question is – how did the Vedic *rishis* reach the conclusion that the fundamental particle of the Vedas is a triplet of three particles of different properties?

The Vedic term for an atom is *paramanu (param+anu)*. The prefixed adjective *param* to the suffix *anu* means the ultimate reality of *anu* or 'beyond an *anu*'. In the Vedic concept the unit of *aapah* is said to be *tri-gunatmaka* (tri-property), indicating that the separation in the three properties is a possibility. A *paramanu* on the other hand is said to be the last state of matter. Therefore a Vedic *paramanu* is either a *sata* particle or a *raja* particle or a *tama* particle. In Vedic concept a *paramanu* by definition is that entity of matter which is not divisible further. More than one *paramanu* form an *anu*. This distinction is similar to the one used for an atom and a molecule in modern physics.

The Sanskrit term *anu* stands for the antonym of 'fat'. This means that an *anu* is not fat yet with extra matter in it. It is implicit in an *anu* that it is a three dimensional substance and a unit by itself without any fatness in it. By fat here is not meant butter. It means that it is the least substantive. This also means that it is not a conglomerate of many such. For example a small quantity of water is not an *anu* because it can be reduced down to smaller portions till only one molecule of water is left – no more extra fat (weight) left in it. Therefore an *anu* is a single molecule of a substance made up of several individual fundamental *paramanus*. This is the understanding throughout the Vedic literature. The euphonic combination of *param* + *anu* forming *paramanu* therefore refers to the last stage of *anu*. Nature as a whole is eternal and exhibits continual change during its creation phase forming atoms and molecules and their aggregates in the form of cosmos. However, *paramanu* being the last minute state of matter, it is a *nitya* entity, which does not change with time. These unchanging smallest units of matter make the *Prakriti* as a whole, an eternal entity.

With regard to the physical shape of a *paramanu*, Rishi Kanaad in Vaisheshik Darshan has stated the nature and feature of a *paramanu* as, नित्यं परिमण्डलम् (*nityam parimandalam*, Vaisheshik, 7/1/20). This translates into - "*eternal and spherical*". The eternity (*nityam*) is the nature of a *paramanu* and sphericity is its physical shape. In Figure 5.1 an *aapah* trio of *sata, raja* and *tama* is depicted. One cannot form a परिमण्डलम् (*parimandalam* = *spherical*) particle with two or more spherical particles. Therefore, Rishi Kanaad's definition of a *paramanu* necessitated that it is a single spherical entity. Therefore, it becomes obvious that *sata, raja* and *tama* each are the state of a *paramanu*, the last stage of matter. Implicit in Rishi Kanaad's definition of *paramanu* being "*nityam parimandalam*" is that through the eternal cycle of creation and dissolution of the universe when Nature goes through change of state, *Sata*-state to *Asata*-state, a *paramanu* does not change. For a *paramanu* to be *nitya*, a *sata* or a *raja* or a *tama* particle has to be indivisible. Thus these three are the ultimate building blocks of Nature and the Vedic defined Nature is equally divided into three types of particles with three individual properties.

In the Vedic philosophy larger *anu* (molecules) of different substances are formed by combining differing proportions of *sata, raja* and *tama*. In the context of describing elements, the Vedic philosophy uses the terms *parmanu* and *anu* in the same way as the terms 'atom' and 'molecule' in modern physics. The difference however is that the Vedic definition pertains to a finer stage of matter than in modern physics. The term 'atom' in Modern Physics does not represent the *nitya* virtue (eternity) of the matter particle because in reality all atoms of elements are newly formed during this creation process of the universe and will ultimately dissolve into fundamental particles.

The term *guna*, the property in a substance, indicates that it is associated with the substance that already exists. When the term *guna* is used in the realm of the creation of the universe the word *guna* in the context of metaphysics and philosophy is synonymous with the particle (matter) itself. In the Vedic creation of the universe the material-cause for the onset of creation is Nature as a whole, but at a finer level it is the triplet of *gunas* (properties): *sato-guna, rajo-guna* and *tamo-guna* that reside in the three types of particles *sata, raja* and *tama* forming the Nature-entity and therefore these *gunas* are addressed as the cause for creation. In a euphonic conjugation these three *gunas* (properties) are written as *satoguna, rajoguna* and *tamoguna*. This is equivalent of saying that the matter at its fundamental state is triune.

Physical Aspect of the Triplet

The Vedas recognize energy forms in both realms, physical and metaphysical. The examples of the enrgy forms in the physical realm are *agni or taapah* (heat-energy), *indra* (electricity) and *vaayu* (kinetic-energy) and so on. Contextually the *eekshana* force/energy associated with the consciousness was briefly discussed in this Chapter and will be elaborated upon in Chapter 6 Volume 2. The Vedic philosophy considers a whole set of energy forms in metaphysical realm (to be discussed in a separate book). These most fundamental subtler energy forms are byproducts of matter, dominant in *satoguna* (*satah-aapah*) are addressed with adjectives such as *devata, rishis, hotas, indriyas* and so on; even though they are all inert energy functions. These energies exert their own forces in specific fields wherein they function. Vedas and the Vedic philosophy propound that the five *tanmatras* of *Ahankaar-tatva* (Ego-element) lead to the creation of matter and physical energy forms. It is the dominance of *rajoguna* that creates most activity in the matter leading to the observable form of matter. In Vedic philosophy the dynamic nature in observable universe is owing to the dominance of *rajoguna.* It is the *rajoguna* that creates observable matter that has *roop* (form and shape) and *guna* (property) like the mass in matter. For example the Vedic way of expressing the Sun-Earth system in terms of *rajoguna* would be – Sun's *rajah* keeps the Earth attracted to it. Therefore, although the gravitational force is minute enough to be termed as a *devata*, it is still said to be due to *rajoguna* in matter.

Even though the life on Earth and on other *lokas* (the other worlds in the universe) according to the Vedic creation is said to have appeared later in the creation, the physical creation of celestial objects is considered to be secondary. The physical creation is only a preparation for life to emerge in it. Also since the conscious-beings need physical body (matter) to function in this creation the tri-property in matter must relate to the living creation also. In the Vedic philosophy therefore, the overall nature of a conscious-being is determined by given proportions of *satoguna, rajoguna* and *tamoguna* in his/her persona. This is briefly discussed further.

Spiritual aspect of the Triplet

The basic nature of matter does not alter whether enormous cosmos and solar systems are formed out of it or small living bodies grow out of it. Hydrogen atoms forming the stars of the Milky Way Galaxy are similar to those forming water and other organic molecules in a living cell of a living-being. The amazing thing in all this is that the complexity in chemical compounds formed in an enormous star is not enormous compared to the inordinate complexity in forming and functioning of a living cell of a small living-being. The automatic functioning of the thousands of different sub-functions in a living body and the maintenance of complex chemical structures, including DNA differentiating one from the other, are completely absent in the enormity of a star. The complexity of the living body disappears at its death. The presence of the complexity in the living body is due to life in it and this life's conscious effort, while it is present in the body. In Vedic expression the complexity of the living body is due to the presence of the quantum of consciousness in it otherwise known as the spirit (*atma*) and the development of enormous complexity is the minimum requirement for the ownership of the material body by this quantum. The individuality in a conscious-being is due to the specific cultivation of three properties of matter, the *sata, raja* and *tama.*

In Vedic philosophy the effect of the three properties is seen through the ability of the physical body and through the metaphysical functions that generate thought and exhibit *eekshana.* The *satoguna* represents purity and its cultivation by the conscious-being enhances mind-function and cognitive ability. The *rajoguna* represents medium property and its cultivation enhances physical activity and the *tamoguna* represents inertness.

The minutest and innermost physical entity nearest to the soul is the *Buddhi-tatva,* the Intellect-entity. The composition of *Buddhi-tatva* is dominant in *satoguna*; differing amounts of *satoguna* creating different intellects with different elevations of cognitive power. The coarseness increases in all other functions down from *Buddhi-tatva* to *Ahankaar* (Ego-entity), knowledge senses and the karma functions in a living body. The coarseness is indicative of less *satoguna* and more of *rajoguna* or *tamoguna*. The next after *Buddhi-tatva* being the *Ahankaar-tatva,* (Ego-entity), it is said to be dominant in *rajoguna*. A detailed discussion on these entities, based on Saankhya Darshan, will form the subject of another text under the title. 'Vedic *Devas*: A Link between Inanimate and Animate'. Here it would suffice to say that it is the *Ahankaar-tatva* that forms the physical body and keeps it going by creating the desire not to die. Higher level of *rajoguna* creates higher level of ego in the person. It is equivalent to the instinct of survival in modern science.

Higher level of *tamoguna* in a conscious-being is indicative of inertness, non-achieving behaviour and inferior development of intellect. Manusmriti in its 12th chapter has related the three properties to a person.

"Satvam gyanam tamoagyanam raag-dveshau rajaha smritaha"
(Manusmriti, 12/10).

This translates into -

"The knowledge in a person indicates satva, the foolishness tamoguna and attachments to material things and jealousy indicates rajoguna".

Manusmriti further relates the three properties to the state of mind of the person. The *satoguna* relates to happiness, the *rajoguna* relates to infelicity and instability of mind and *tamoguna* relates to non-intellectual activity, business in sensuality and materialism. Contrary to this perspective the modern science deals with matter in a separate compartment unattached to the basic nature of the person. The basic nature of the person for modern science relates only to the chemical composition of the physical body and the chemical balance in it.

Vedic versus Modern Science on Nature

The approach in Vedic science/philosophy for defining Nature may not be the same as in modern science, nevertheless it is thought-provoking and well above the religious dogma. It has many features common with modern science. The most important and unexpected commonality between the two is the corpuscular nature of matter at the basic level of matter's physical composition. The other equally important commonality is the triplet forming a fundamental particle that leads to the formation of matter. The difference between the two systems is that the *satah, rajah* and *tamah* distinction in the Vedic cosmology is an eternal property of matter of the tri-property Nature while in the modern cosmology the quarks appear in the creation phase of the universe after the so-called Inflation Era when Quark-Lepton Era begins at 10^{-32} seconds and ends at 10 seconds after the onset of Big Bang. Not clear from the Big Bang Hypothesis is what causes different flavours in quarks and whether quarks existed in matter before they appear in Quark-Lepton Era. When *satah, rajah* and *tamah* are considered as a unit of matter the common term used for them is *aapah*. The similarity here is that a stable nuclear particle made up of three quarks is in the likeness of an *aapah*. The geometric definition of *satah, rajah* or *tamah* is a *paramanu*. Modern science discovered the basic structure of matter by theory and experiment; however, it should be puzzling for a modern physicist as to why and how the Vedic *rishis*, eons ago, arrived at a triplet as matter's most fundamental composition.

In modern physics, Einstein's matter-energy equivalence principle ($E = mc^2$) relates the two forms of matter - mass and energy. The matter in mass m is in coarse form which after transformation into energy E is in subtle form. In this transformation, it is the matter that loses its identity including its identity to exhibit mass. When a hydrogen atom with three quarks in its nucleus made of one proton and one electron orbiting around it is transformed into energy, it transforms into a certain number of energy packets called photons. In modern physics a photon is not known to have any quarks in it or they would add weight to a photon. In the Vedic principle, though untested yet, any transformation of matter that might change the matter's form to equivalent energy would continue to possess either all three types of fundamental particles *sata*, *raja* and *tama* in a given proportion, or at least possess one of the three particles. Therefore it seems that the *sata*, *raja* and *tama* are at a much subtler level of matter than even quarks and their combinations that create mass in matter. Summarily, based of the Vedic definition of matter, this author's thinking is that a quark is too big to be a fundamental particle – it should be divisible further into smaller particles *named sata, raja* and *tama.*

Vedic thought defining Nature's fine-structure in terms of three fundamental properties was not necessary to promote a religious dogma. The conclusion according to the Vedas that there are three separate fundamental particles/properties at Nature's fundamental level puts Vedic thought straight into the physics arena. Once in physics arena, it would not be scientific on part of a modern physicist to say that the Vedic thought of triplet forming matter is remarkable but it could just be a lucky guess by a Vedic *rishi*. It is not just the idea of a matter triplet discussed here; put forward millennia ago by the Vedic *rishis*, it is the multitudes of thoughts in Vedic science that would put a Vedic reader in amazement.

Ancient *rishis* knew how to grind substances into fine powders. This however was not sufficient to see three components of matter *sata, raja* and *tama*. They had clear understanding of three phases of water –solid, liquid and vapour. But, that could not have led them to deduce the triplet nature of matter. Therefore, the Vedic *rishis*' suggestion that if one were to go to the final division of a substance it would turn out to be in three distinct quanta, not two not four but three, should remain an intrigue indeed! There is no evidence that the Vedic *rishis* had 'particle smashers' of modern day physics. Did they have a different methodology to determine the fundamental nature of matter? This would remain a puzzle for intellectuals of modern times.

An ordinary answer to such a question is – there must be more ways to understand fundamental nature of Nature. To understand the fundamental structure of matter the subject of particle physics has evolved from its simplicity over a few centuries to experimenting with sophisticated particle smashers. The Vedic *rishis* who drew conclusions to put forward the Vedic concepts for Nature's variants and non-variants had a way to understand nature that we are yet to understand. Until then some of the features of the Vedic science will remain a mystery. A prejudgment that the Vedic ideas could just be imaginations and speculations by the Vedic *rishis* is up for several surprises because this author will present a number of Vedic science thoughts in this and the forthcoming chapters that are quite intriguing. Some of the examples to substantiate this author's contention related directly to physics where observation, measurements and calculations by ancient Vedic *rishis* led to mind-boggling discoveries with high accuracy, remarkably close to that achieved by modern experimental science.

Purush and *Prakriti* : Togetherness of Creator and Nature

Just as the consciousness and material body pair up to create the reality of a living-being's existence, the Vedic philosophy sees *Purush* and *Prakriti* as a pair for the reality of the creation. It will become clear in the following that in the Vedic cosmology, even though ample recognition is given to the *Purush*, the emphasis is not always to glorify the *Purush* to fulfill the need of a worshiping human-being. When it comes to describing the creation of the universe or the function of a cosmological body itself, a Vedic mantra can be at a high level in poetic expression, yet it delivers contextual knowledge with clarity. If it were only to glorify the Creator to increase faith in a Supreme-being, the mantras that deal with creation of the universe need not have included physics details of creation that would interest only those pursuing knowledge in cosmology. As an example the Veda's assigning a total life-span to the universe of 4.32 BY compared to short life-span of 100 years for humans does not necessarily create greater faith in the Creator. These Veda prescribed numbers may be smaller than those suggested by the contemporary cosmology (13.8 BY); nevertheless, a reference to incredibly large mathematical figures in mantras is primarily for dissemination of cosmological knowledge rather than for creating religious fervor. The mantras furnish the meaning of the term Veda (revealed knowledge) and therefore the Veda mantras should be viewed in that purport. The spirituality in a mantra is for spirituality and the knowledge about the universe and the other physical phenomena revealed in the mantra is for enhancing knowledge which leads to enhancing the intellectual potential. It becomes obvious from the mantras cited in this book that the Vedas have included various necessary steps that are required to build the cosmology scenario for Vedic creation which are not needed for mere worship of the Creator. The Vedic knowledge provides benefits for all, including an atheist.

In the Vedic cosmology scenario, there are no other parallel universes to the one existing. There is no need for many universes when this one is complex enough and large enough to fulfill all purposes.

(3). The Soul-entity (*Atm-tatva*)

The third eternal Vedic entity named the *Atm-tatva*, expressed as *Atma* or *Atman* is a quantum of consciousness. There are differences and similarities in the two conscious entities, the *Purush* and the *Atmas*. Both have *Eekshana* power to create tangible things with their respective abilities. The differences stem from the singularity of *Purush* versus plurality of the *Atmas*, area of influence, potency, science (knowledge) and many more factors to list just a few here.

To differentiate the *Purush*-entity from an *Atma*-entity, the *Purush* is also addressed as *Param-atma*, meaning the 'primary *atma*' (consciousness). *Atmas* are individualized energy packets independent from each other, all having the virtue of consciousness in them. The term *purush* stands for the all-pervasive virtue of the *Purush* and the term *purush* is also used for a soul because the souls, though discretized, pervade through the universe and available in all places to take birth throughout the universe. An *Atma* is a mass-less non-perishable eternal entity that persists through the creations, indulging with matter by assuming physical body or remaining free from the bondage of matter, depending on its state of spiritual evolution. It is an individualized energy packet like a light-photon with the difference that an *Atma* is with the virtue of consciousness and a light-photon is inert. *Atma* is addressed as the soul or as the spirit in other theistic frameworks of philosophy.

The souls in the creation process serve as part of the 'general-cause'. Striving for knowledge is inherent in a soul. The knowledge in a soul is inversely proportional to the ignorance in it. The enlightenment of a soul gives it freedom and the ignorance becomes a vice, the former taking the soul towards bliss and the latter towards its indulgence with matter (Nature). To gain knowledge of the universe (Nature) and of the Creator, a soul has to indulge with matter by taking a physical body equipped with the knowledge senses and the karma-functions. This is possible when the Nature is in the creation phase with its matter in usable state. The availability of the creation fulfills the soul's desire to gain knowledge of the pleasurable yet complex inert Nature and of the all-bliss Creator, leading the soul toward higher level of its spiritual evolution. This evolution, not of the physical body but of the soul, leads the soul in its birth sequence to become an enlightened scholarly person (pundit), a scientist or a cosmologist and it enables to gain knowledge of the 'self' and the Creator, leading towards self-realization and bliss. The trend in modern science is to go for cultivation of knowledge and not for self-realization. The Veda mantras advise one to strive for both goals and not to leave the job only half done. Not working for both is said to be the missed-opportunity. In other words emphasis in the mantras is – do satisfy the need of the brain as well as of the spirit (knowledge of matter and consciousness).

Considering that the souls in the universe being infinite in number it is expected in such a Vedic system that souls at different levels of development with free will are going to be at all levels; some at a primitive level and others at human level. Among humans a wide spectrum of levels exists with emphasis ranging from materialism to spirituality. Discovering new knowledge by a pre-enlightened soul in a physical body of a physicist is also expected under this system.

Owing to its minuteness and corpuscular nature the *Atma* is mobile in space when not bound by the material body, latter bound by time and space. In the Vedic system the inert matter acquires the status of animate owing to the presence of *Atma* (consciousness) in it. In Vedic cosmology the presence of this eternal *Atm-tatva,* ubiquitously distributed through the universe is the cause for discretized life-systems on Earth. The Veda mantras reveal that the life-systems are not only on the Earth but they are throughout the universe on the planets of other solar systems. It is this soul-entity that directs the intellect and other faculties of the living body to discover the laws of physics or to create a tangible thing.

In modern cosmology/physics there is no separation between Nature (matter) and consciousness, the *Atm-tatva*. The Vedic system with the existence of the utilizer of this nature, the *Atm-tatva*, discretized in infinite numbers, provides a holistic approach to understand the meaning of the existence of cosmos. The Vedas have provided the connection between the physical world that can be sensed, observed, measured and examined by tools of science and the consciousness; the Creator to create the universe and the *Atm-tatva* as its limited utilizer. In the Vedic Creation Principle, just as the matter of the universe pre-existed in indefinable matter state of *Asata* to serve as the material-cause, the *Atm-tatva* is a pre-existing eternal entity distributed throughout the universe to serve as the 'general-cause'.

In terms of creativity, the Creator is credited for creating the entire universe and the *Atma*, using its functions, senses and the *eekshana* (desire), creates small things within its limited area of influence. There are several other terms for *atma*. The common ones are *chetan-tatva* (conscious-entity), *jeev-atma*, *jeevatma-tatva* or just *jeev*. The underlying meaning in the term *jeev-atma* is that it is the cause for life in the physical body and its presence in it exhibits action and activity.

A concise definition of the *jeev-atma* (soul) is in a RigVeda mantra (1/164/30):

अनच्छये तुरगातु[1] जीवमेजद्ध्रुवं मध्य आ पस्त्यानाम् ।
जीवो मृतस्य चरति स्वधाभिरमर्त्यो मर्त्येना सयोनि: ॥
(RigVeda, 1/164/30)

Transliteration:

"*Anachchhaye turgaatu*[1] *jeevamejad-dhruvam madhya aa pastyanaam*।
Jeevo mritasya charati svadhabhiramartyo martyena sayonih" ।।

Mantra Translation:

"Jeev-named entity is swift[1]*, immortal and endeavouring. It dwells stably, localized in the house of mortal body and is a repeated visitor of the mortal world".* (RigVeda, 1/164/30)

These are some of the important characteristics of *jeevatma*. The soul in Vedic philosophy is considered to be the minutest entity such that when in free-state without a physical body, matter does not pose a barrier to its movement. When it is in bound-state in a physical body its movement is limited by the body. The soul with its *eekshana* (will-power) directs the physical body to do actions in its limited sphere. When it is free from the bondage of Nature, its *eekshana* can instantaneously transport it to any point in the universe. This is the meaning of the second word of the mantra, *turgaatu* (तुरगातु). It is swifter than a photon of light, a photon having the maximum known speed among the physical byproducts of Nature.

The soul is not a byproduct of Nature (matter) but unlike the *Purush*-entity it interacts with matter by taking a physical body to do karma, the desired actions. It is the freedom to do karma that makes the *jeevatma 'a repeated visitor of the mortal world'.* If the soul has not attained complete freedom from karma and therefore the freedom from the bondage of matter, the law of karma forces the *jeevatma* to begin creating a new physical body in a new mother's womb right after the death of its previous body. The Creator's purpose of creating a universe is for the souls to do karma for improving their status towards freedom from the bondage of matter and therefore by doing further good karma to attain the ultimate eternal bliss. The birth re-birth cycle continues until it becomes completely knowledgeable of itself (self-realization) and of the universe (in a limited sense, not in the omniscient sense). This state of freedom is soul's bliss because it permits a soul to contemplate at will the entire creation and its beauty up close. If such a soul wishes to take birth again it brings the firsthand knowledge of the creation and the Creator and expresses it in the form of mantra formulations.

While the soul is empowered to own the physical body, it provides impetus to the functions of the physical body to gather knowledge with the combination of 13 metaphysical and physical distinct implements (functions), including the 5 well-known knowledge senses. The physical body, especially at the human level, equipped with the highly developed 13 implements called *karans* in Vedic philosophy, provides a special privilege to the soul to do karma (action) with the soul's free-will. A discussion on the 13 *karans* is beyond the scope of this book. However, it is the effect of its actions (karma) that sets the soul's direction towards the next stop after it leaves a body. It is bound by the *ritam* laws of the Creator. Yogiraj Krishn in Bhagvadgita preached to Prince Arjun about soul's (*Atma's*) minuteness and immortality. English translations of only two Bhagvadgita *shlokas* on this subject are presented here:

> *"Neither it (**Atma**) is born nor it ever dies. It is not that if it came* (to take birth in a body) *once it will never come back again. It is eternal, unchanging perpetual and ancient. It is not killed by killing the body, it lives on"* (Bhagvadgita, 2/20).

> *"No weapon can cut the Atm-tatva, no fire can burn it, water does not wet it and the air does not dry it"* (Bhagvadgita, 2/23).

There is considerable discussion on the *Atm-tatva* throughout the Vedic Literature. The emphasis has been that it is a non-decaying and non-defective consciousness quantum with divine characteristics. This discussion (not included here) includes *Atma*'s location in the living body, its method to control all actions by its living body, what comes with the *Atma* at the time of Creation of the universe, what changes are inherited by it when it leaves one body to take another, what ultimately remains with it when it attains the freedom (salvation) from matter and the cycle of birth and death and so on.

A Philosophical Perspective: Rishi Vaarshganya of Ancient times, Aadi Shankaracharya of 500 BCE and Prof. Hawking of modern times

In the context of philosophical discussions pertaining to the eternal entities and the creation of the universe, several references from the ancient Vedic and non-Vedic philosophies are found in the post-Vedic literature. The non-Vedic philosophies have been closer to atheism and pantheism. The whole universe for an atheist is inert Nature which maneuvers itself under its own laws to create this vast and diverse universe. For a pantheist the whole universe is out of a single entity, the Nature, or out of a single conscious-being Creator. In ancient India there have been a few thinkers and philosophers, who subscribed to such non-Vedic philosophies. The freedom of thought and speech has always been a progressive feature of the Vedic religion and philosophy. To name one here, a philosopher non-concurrent with the Vedic philosophy, during the ancient Vedic period, was Rishi Vaarshganya. Owing to his scholarship he is addressed respectfully as a *rishi*. He hypothesized that the inert Nature alone was sufficient and responsible to create the universe and the existence of God was not needed (discussion in "The Vedant Darshan Ka Itihas (1991)" by Acharya Udayaveer Shastri)

In the Vedic cosmology arena it would not come as a surprise when the renowned cosmologist Stephen Hawking in his recent book entitled "The Grand Design" declared that God was not needed to create a universe and that the universe would create itself out of nothing and sustain itself owing to gravitation (this text was written before Professor Hawking passed away). Hawking being a mathematician/physicist, his declaration is based on the elaborate mathematical hypothesis. Whichever way one puts it, such an argument ultimately precipitates into saying that the inert Nature (matter) by itself creates the universe. This was also said by the ancient *rishi* Vaarshganya eons ago. The logic behind the argument is that the gravitation in matter is due to mass in it and mass in matter is due to the existence of matter, and all of this is out of 'nothing' - the other name of which being the "space-time singularity" of modern Big Bang Hypothesis. Rishi Vaarshganya in ancient times and Hawking and others in modern times do need Nature to create the universe.

The Advaita Philosophy (non-duality) of Aadi Shankaracharya (509 BCE - 477 BCE) was alluded to in Chapter 3 and 4 which stipulates that everything in the universe - all souls, all cosmos and the entire matter of the universe, is nothing but God. In this philosophy the difference between *Purush* and *Prakriti* or *Purush* and *Atma* is non-existence. God creates the universe out of Himself and the observable universe is not a real entity but a mere illusion (*maaya*). That everything in the universe is the Creator is similar to the philosophy of pantheism where everything is Nature. A soul-entity in Rishi Kapil's Saankhya Sutra and Rishi Kanaad's Vaisheshik Sutra is an individual and independent entity with consciousness as one of its virtues. It is different from the other souls and different from the Creator of the universe. Even the Vedanta Sutra (1/2/11) of Rishi Vedvyas has stated that the God and the soul are two different entities.

In the Saankhya Darshan *sutra* (1/61) presented earlier under "The *Purush-tatva*", Rishi Kapil states that God is the controller of Nature. In the successive *sutra* he also states that in specific actions the *jeevs* (souls) are also the controllers:

विशेषकार्येष्वपि जीवानाम् । (*Visheshkaaryeshvapi jeevaanaam*)
Saankhya, 1/62)

"Souls are also the controllers in specific actions" (Saankhya, 1/62).

The word *jeevaanaam* in this sutra represents the plural of the word *jeev*, indicating plurality of *jeevs* (souls). The singularity of God and plurality of souls in Saankhya also represents the clear distinction between the two separate conscious entities. The soul in the Vedas, Saankhya Darshan and other Vedic Darshans is an individual entity responsible for its own actions and is a controller of its own area of influence to a limited degree. This is why *Atm-tatva* (*Jeevatma-tatva* or soul-entity) was expressed as being a part of a discretized field of consciousness embedded within a single conterminous field of consciousness, the *Purush*. In the Vedic philosophy there is a clear separation between the components of the Vedic-trinity (*trishtup*): God-Nature-Souls.

As mentioned before, owing to the ubiquity of *Atmas* (souls) through the universe they are also addressed as *purush*. Based on observation and reason, Rishi Kapil in his Saankhya Darshan emphasized the plurality of the *purush* (souls),

जन्मादिव्यवस्थात: पुरुषबहुत्वम् ।
("*Janmadi-vyavasthaateh purush-bahutvama*") (Saankhya, 1/114)

Translation: *"The system of births and rebirths reveals that the souls are many"* (Saankhya, 1/114).

This principle of Saankhya Philosophy is supported by Rishi Kanaad in his Vaisheshik *sutras*, 3/2/20, 21).

Aadi Shankaracharya's Advait Philosophy that everything is God is based on his interpretation of the last Darshan, Brahm Sutra authored by Rishi Vedvyas close to historic milestone at 3140 BCE. The fact is that the Vedanta Sutra does not support the Advait Philosophy. Just one of several *sutras* is presented from the Vedanta Sutra,

गुहां प्रविष्टावात्मानौ हि तद्दर्शनात् ।
(*Guhaam pravishtaavattmaanau hi taddarshnaat*) (Brahm Sutra, 1/2/11)

Word Translation:

गुहां (*guhaam*) = in the cavity (of mind), प्रविष्टौ (*pravishtau*)= (two entities) entered, आत्मानौ (*atmanau*) = two *atmas* (*two souls*), हि (*hi*) = with certainty, तद्-दर्शनात् (*tad-darshanaat*) = due to their revelation there.

Sutra Translation:

"In the cavity (of mind) *reside two souls* (*the Atma* and *the Parmatma*), *where both are revealed with certainty"*. (Brahm Sutra, 1/2/11)

In Vedic thought the place of the soul (*Atma*) in the living body is in the cavity of the mind. In context with the locale of the soul in the living body the Veda, Upanishads, Darshans and the rest of the Vedic literature have all stipulated that the soul resides in the interior of person's head in a specific small locale addressed as गुहा (*guhaa*) which means a cave (cavity). This cavity is said to be the size of the tip of the thumb. Depending on the context it has also been addressed with other terms such as *gyan-granthi* (knowledge-gland), *antehakaran* (interior mind function) and so on. In the yoga practice and meditation the place for invocation of God and self realization is also the mind.

The singular term for *Atma* according to Sanskrit grammar is आत्मा (*atma*), its dual-plural is आत्मानौ (*atma-nau*) and its plural is आत्मानाम् (*atma-naam*). Addressing the presence of the souls in the cavity, Rishi Vedvyas's usage of the word is unambiguously the आत्मानौ (*atmanau*) for dual-plural. This tells us that there are two separate *Atmas* (souls) in the single cavity of the mind. Since the God is thought to be all-pervading it would be found in all cavities, Rishi Vedvyas expresses in his *sutra* with "certainty" the presence of the *Atma* and *Paramatma* (the Creator) in a single locale. The *sutra* by itself suggests that the God and the soul reside in the cavity of mind *where both are revealed with certainty*. The two entities in one place are possible when they are two different entities. In the Brahm Sutra the *Purush* and *Atma* are two distinct entities and are also eternal. Therefore, the *Advaita* Philosophy does not change the Vedic premise on the list of eternal entities.

In the context of the creation of the universe, whether it is the ancient atheistic philosophy of Rishi Vaarshganya wherein everything is matter, or Shankaracharya's theistic philosophy wherein everything is God, or it is the modern atheistic philosophy wherein matter and its gravitation serve all three 'causes' for the birth of the universe, as advocated by Professor Stephen Hawking and others, they all in fact are saying the same thing with differing terminology. This author has put Hawking's theory into the category of philosophy because Hawking has hypothesized the creation of the universe based on his mathematical equations and has talked about the uselessness of God in the process of creation without having at least a hypothetical God in his equations. When one talks about Nature and its gravitation as the creator of the universe without proving his deduction in a physics laboratory, one is a philosopher first and a mathematician and a physicist next. In these non-Vedic philosophies, the material-cause is its own efficient-cause and vice-versa, call it Nature, God or gravitation. Several other non-Vedic atheistic philosophies have also emerged from the Indian soil such as the Buddhism, Jainism and Chaarwak (materialism).

The Vedic Trinity (Vedic *Trishtup*)

The Vedic Religion has a trinity which is addressed as the *trishtup*, meaning - the three together. This *trishtup* is formed of the three entities defined above - *Purush, Prakriti* and *Atmas*. Similar to the Vedic trinity is the trinity in Christianity formed of Father, Son and the Holy Spirit - all three spiritual entities. The origin of the Sanskrit term *trishtup* is in the far antiquity of the Vedic Literature. The Vedic trinity is formed by two spiritual entities and one inert entity – the *Prakriti* Nature). There is a causal relationship between components of the Vedic *trishtup*: *Purush* the efficient-cause of the universe, *Prakriti* the 'material-cause' of the universe and *Atmas* the 'general-cause' of the universe.

The causal relationship of the three causes extends to the cause-effect rule in the creation and it becomes a tool for enhancing knowledge in both realms of the creation, animate and inanimate. Kanaad Rishi in Vaisheshik *sutra* (2/1/24) stated: "*It is observed that the properties in the effect are accordant of the cause*". The implication of this statement is that the properties in matter are those inherited from the original material-cause, the Nature. In the Vedic creation and therefore in the Vedic *trishtup*, the three entities (*Purush*, *Prakriti* and *Atmas*) play their own roles but do not convert into each other.

The Vedic *trishtup* is very beautifully depicted by the RigVeda mantra (1/164/20) in a poetic adornment called रूपकालङ्कार (*rupakaalankaar)* wherein the subject matter is expressed in an allegory of two birds with beautiful wings perched on a tree, one being the *atma* (the soul) and the other being the *paramatma* (Supreme Being) and the tree being the creation (Nature). The mantra translation is:

> *"There are two conscious entities* (soul and the Supreme Being) *possessing divine qualities, coexist in the same creation* (the Nature) *as intimate friends in companionship. One of them enjoys the fruit of his action (indulges) and the other simply observes from all sides without indulgence."* (RigVeda, 1/164/20).

(4) Time-Entity (*Kaalah-Tatva*)

The fourth eternal entity of the Vedas is Time. Vedic term for Time-entity is *Kaalah-tatva* or *Kaal*. First part of this section deals with the definition of Time and the second part deals with measures of Time. The Vedic definition of an abstract entity such as Time reveals the depth of thought and grasp of understanding by the Vedic *rishis*. The concept of Time used in modern cosmology is millennia old with its origin in the Vedas. The measurement of Time with well thought-out time-scales reflects the ability of the Aryans and the Aryan society in experimentation and application. The elaborate time-scales of the Aryans relate to practical astronomy encompassing time scales of billion-years in *yuga* units down to mind-boggling and experimentally derived *nimesh* time unit in millisecond. An AtharvaVeda mantra delineates the complex concept of time:

कालो ह भूतं भव्यं च पुत्रो अजनयत् पुरा ।
कालादृचः समभवन् यजुः कालादजायत ॥ (AtharvaVeda, 19/54/3)

Transliteration: *"Kaalo ha bhootam bhavyam cha putro ajanyat pura.
Kaaladrichaha sambhavan yajuh kaalaadjaayat".*

Word Translation:

कालः (kaalah) = the Time-entity , ह (*ha*)= is itself, भूतं (bhootam) = past, भव्यं (bhavyam) = future, च (cha) = and, पुत्रः (putrah) = son, अजनयत् (*ajanyat*) = given birth to, पुरा (pura) = the one born before (father)।

कालात् (*kaalaat*) = from Time, ऋचः (richah) = knowledge bestowing mantras, सम अभवन् (*sam abhavan*) = are created ,यजुः (yajuh) = YajurVeda (or knowledge of right actions), कालात् (*kaalat*) = from Time, अजायत (ajaayat) = is born of ॥

Mantra Translation:

"Kaal the son is its own father (the past) *and will be born as its own son* (the future). *Richas, the knowledge bestowing mantras, are born out of Kaal and Yaju the knowledge of righteous actions is born out of Kaal".* (AtharvaVeda. 19/54/3).

The *devata* (subject) of this *sukta* is *Kaalah-tatva*, the Time-entity. The first part of the mantra was also presented in Chapter 1 of this book. The continuity in Time-entity's flow could not be expressed in a better way than this riddle! In real life a father gives birth to a son and the father and the son represent respectively the past and the present. Even though their life-spans generally overlap on the timeline, the irreversible Time puts them on their respective time segments. When it comes to Time-entity itself, the past is created only because the present is born. There cannot be an overlap between the past and the present. Therefore, the *putro* (the son) of Time gives birth to its own *pura* (father) because a past is not born without the present being born.

The mantra completes the riddle in its first line by stating that it is the Time-entity which itself is born as its own son the *bhavyam* (future). The next mantra (19/53/4) in the same sukta, presented below, also repeats the same theme that Time is its own father and its own son. There the mantra explicitly uses the terms, पिता (*pita*) and पुत्र (*putra*) respectively for father the past and future the son. Therefore, is there a better way of defining Time than it being its own father and its own son? The mantra also reveals the thought process of the *rishi* to come up with a unique way of relating the past, present and the future of the Time-entity. This is his way of defining a real continuum.

The second line of the above mantra states that the *richas* are also the result of time. *Richas* are the mantra compositions revealing hidden knowledge of the creation and all subjects within. They could not have come about without the existence of Time. After knowledge comes action. Therefore the mantra gives special reference to the word *Yajuh* that encompasses 'knowledge of all right karma'. Implicit here is that YajurVeda (second Veda), which is the book of knowledge for right and righteous actions, has also come into being because the existence of Time. Therefore, the emphasis in the AtharvaVeda mantra is that one needs the existence of the Time-entity to acquire the knowledge of the past, present and the future.

Some well-known cosmologists like Roger Penrose have suggested that the mind-entity that perceives time should also be studied and understood before one can know what Time is. Time in Big Bang cosmology is used but not understood. It is thought to be a real entitity which takes birth at t = 0 sec, about 13.8 BY ago and will continue to grow bit by bit in future. An entity that has a begining can grow only bit by bit and not knowing its nature and the mechanism for its growth one could not say that it could not come to an end. If it does not come to an end it will grow to infinity and would affect the functioning of the universe in some unknown way. The bit by bit growth means the entity is discontinuous at its fine structure. Different bits of such an entity cannot have togetherness in them forever; one part may affect the universe one way and the other part may affect in another way leading to chaos.

Kaal (Time-entity) in the Vedas is an all-pervasive eternal entity covering the entire creation and flowing irreversibly forward nonstop as a continuum in three dimensions. *Kaal* in the Vedas is a real entity. The Vedic argument for the pre-existence of Time from even before the present creation is straightforward. The Vedic creation being a recurring phenomenon, with matter changing its state cyclically from indefinable *Asata*-state to definable *Sata*-state and back again needs the Time to be in existence in both states. Of course a state of matter, *Asata* or *Sata*, needs a container wherein to exist. This container of matter is the Space-entity. Because the two differing states of matter (*Asata* or *Sata*) occupy differing Space, the Space-entity has to be flexible and non-snapping. Let us assume that the matter is in the *Asata*-state of dissolution confined in the diminutive Space (*tuchchhayen* of RigVeda, 10/129/3) and there was no Time-entity. Because all events take place in Time and if Time is not there the matter will remain in diminutive dissolved state of *Asata* or in whatever state it is. In other words the matter of the universe without Time will remain in a status quo. Therefore both, Time and Space, are needed to pre-exist for the creation of the universe. The eternity of Time-entity in the Vedic creation is also a prerequisite in the 'Cyclic Universe Theory" proposed by Steinhardt and Turok (Science, May 5, 2006). The basic premise of this theory is the same as in the Vedic creation that the universe is in a perennial cycle of creation and dissolution.

The concept of the Time and the Space as entities is not an invention of contemporary cosmologists. Considering that the Time and Space are defined in the mantra formulations wherein their properties are revealed, the concept of Time and Space is as old as the Vedas. The Vedic *rishis* had recognised Time and Space as an entity in the process of creation of the universe and captured it into mantra formulations well before the birth of "space-time continuum" in modern cosmology.

In addition to numerous Veda mantra references on *Kaal*, the AtharvaVeda has devoted two successive *suktas* 53 and 54 of its 19th Chapter with a total of 15 mantras stating the characteristics of Time and its association with other entities. For the sake of brevity, direct English translations of some mantras, without their Sanskrit text and word-for-word translation are presented in the following. Interpretations of the mantras are presented where necessary. In modern literature one comes across expressions and similes for Time, such as the 'arrow of time' and 'like the flow of a river'. In this respect the Vedas have not been abaft. Some of the ancient adjectives for Time that have appeared in the mantras will be pointed out. The emboldened Vedic terms in mantra translations are interpreted further because these terms establish the relationship of Time with the hidden thoughts in the mantra.

Mantras from *Kaal Suktas* (53, 54) of AtharvaVeda's 19th Chapter

(AtharvaVeda mantra 19/53/1)

> *"Kaal is like the potent un-ageing horse in perpetual motion (flow) on which ride the* ***sapt-rashmi*** (seven-colour-rays) *and the* ***sahashra-aaksha*** (thousand-eyed) *Sun. On it ride the knowledgeable and intellectuals. All* ***bhuvanani*** (created celestial bodies) *ride on its* ***chakra*** (wheel or cyclic order)*".*

This mantra lists time-dependent physical phenomena with several adjectives associated with Time. One unique adjective for the Time-entity is the '*potent un-ageing horse*'. A horse symbolizes power and motion. The expression that Time is the *potent un-ageing horse* means that this entity is unstoppable and eternal which continues on its eternal journey. A familiar fact about this entity is that when the 'time comes', the event that is to take place will take place. Additionally, this mantra is a concentrate of thoughts expressed explicitly and implicitly on time-dependent physical phenomena. These thoughts are remarkably adorned by compact Vedic Sanskrit terminology. The ***chakra*** of Time in the mantra means the cycles and the periods observed all through Nature. The specific reference to Sun in the mantra is important. The Sun relates to all ***chakra*** in the solar system including the ***chakra*** of day-night and year on the Earth. The Sun's reference also includes its rays which are a time-dependent dynamic phenomenon, solar rays reaching the Earth in 8 minutes. The mantra puts an explicit designation of ***sapt-rashmi*** on the rays of the Sun which translates into 'seven-rays of the Sun', the major seven colours in the solar spectrum. Not only the physical optics is a subject here that the *rishi* of the mantra recognizes but he is pointing to the fact that solar rays are a time dependent dynamic phenomenon.

Humans have been seeing the multi-coloured rainbow since they have seen the Sun. A rainbow remained a mere abstract painting far away on the canvas of the sky, out of reach. However, the assignation of number 7 to the count of colours in the sunlight, as in the above mantra and the rainbow as '*Indra Dhanush*' (Sun's bow) in other mantras, is an example of deductive optics by the Vedic *rishis*. The Vedas had recognised that the colours seen on the screen of the sky were in fact those in the rays of the Sun. The reference of ***sapt-rashmi*** (seven-rays) for the sunlight has appeared in the Veda mantras many times (e.g. RigVeda, 1/50/8, 1/164/2, AtharvaVeda, 7/107/1, 9/9/2,....). More discussion on the solar rays is in the section on Sun in Chapter 8. Even the term *mani,* which means a crystal dispersing solar rays into colours, has appeared with reference to the solar rays. The modern optics began with the ***sapt-rashmi*** designation of the

Sun by the Vedas. Sir **Isaac Newton** (1642-1727) is credited to have assigned 7 colours to sunlight. Obviously he was not the first to do so.

The other adjective for the Sun in the mantra is the ***sahashra-aaksha*** which translates into – *the Sun with thousands of eyes.* Eyes are to see objects. Most living beings with just the two eyes see in a considerably wide field of view without turning their heads. If humans were blessed with a thousand eyes they would see all around them at the same time. What the eyes receive from the objects that they see is the solar spectrum reflected from the object. Therefore, the metaphor of Sun having thousands of eyes means that the Sun illuminates all objects all around it. The Sun in other Vedic mantras is the illuminator of all ***bhuvanani*** around it. The term ***bhuvanani,*** plural of ***bhuvan***, means '*astronomical objects of terrestrial nature*'. Dynamics being a feature of all time-dependent phenomena in nature, implicit in the mantra is the time-dependent relationship between the *Kaal* (Time) and the Sun's rays and the Sun itself. A basic deduction from the mantra is that the rays of light take time to go from one point to the other.

A powerful horse is generally associated with a chariot with ***chakras*** (wheels) which continue to turn in their own cycles while the horse ahead of them keeps moving in one direction. Therefore the horse-simile for Time is drawn to show its continuity within which many time-cycles associated with diverse cosmological bodies and phenomena persist. Humans move faster when they ride a horse. The ***sapt-rashmi*** of the Sun riding on the horse of Time implies that the Time moves even faster than the solar rays.

The second line of the mantra stipulates that, in addition to physical systems where time is of the essence, even the intellectuals and the wise cannot function without the sense of time. Time is of essence for an intellectual idea to happen. The knowledge of time itself is put on a high pedestal by the Vedas. The term for the scholar of time-science is '*maatali*'. A cosmologist of today is a *maatali* of the Vedas.

(AtharvaVeda, 19/53/2)

> *"The Kaal rotates the* **sapt-chakran** (the system of seven wheels) *around their respective seven* **naabhis** (foci/centres) *with just one* **amritam-aksha** (immortal axle). *The Kaal with certainty is the* ***first*** **deva** *that reveals all objects of creation".*

It was the time-dependent *sapta-rashmi* (seven rays) of the Sun in the first mantra. In this mantra the Time is correlated with a mysterious *sapta-chakran*, a system with seven cycles, each cycle being around its own focal point (***naabhi***) yet the system has only one immortal-axis (***amritam-aksha***). Since the subject relates to wheels and cycles, the axis in the mantra is the axis of rotation. The condition of a single axis serving all seven wheels is fulfilled only when each wheel is confined into the next without their circumferences crossing each other. This is similar to the orbits of the planets around a common axis, the Sun. Usually an axle of a wheel wears out with time. But the axle that the mantra is talking about is such that it never wears out - thus maintaining its immortality. All substantive objects change with time and so should the common axis of the seven wheels mentioned in the mantra. But the mantra states that this particular axis is not a transient axis and it does not wear out with time and therefore the adjective of ***amritam-aksha*** for the axle. This means that this un-wearing-down axis is the axis of rotation and not a substantive physical axle.

The major terminology of the mantra being in the realm of mechanics with reference to astronomical bodies, it is not far from the reality to say that the mantra is alluding to the solar system with seven observed planets forming the ***sapta-chakran*** with one common ***aksha*** (axis) of rotation through the Sun, an axis that does not show any sign of wearing out. Several more mantras in Chapter 11 in Volume 2 also support the astronomy nature of this deduction.

The Time in this mantra is addressed as the ***first deva*** because it exists before the beginning of the creation when the other *devas* are created. The *deva* or *devata* systems in the Vedic cosmology are said to be the first progeny, the physical systems of Nature that enable the creation and its sustenance till the end of creation. Time in this mantra being adorned by the adjective of '*the first deva*' could also be translated as '*the deva before all devas were created*'. Among a total of 33 inert *devas* in the Vedic creation are the first seven *devas* known as the 7 *Aditi-putraha*, the Nature's progeny (discussed in Chapter 8). The Time therefore is one of the important entities in the Vedic creation.

(AtharvaVeda, 19/53/3)

> *"On Kaal is sustained the brimful-kumbha* (pitcher), *the kumbha that we of the present observe in numerous ways. Kaal continues to move ahead of all substances of creation. It is this Kaal that is understood and deduced by intellectuals from the understanding of param-vyom* (vast space)".

A *kumbha* is the spheroidal earthen-pitcher used for bring water home from a well which generally is some distance outside the village. The career carries the brimful-pitcher balanced on the head. The relationship of a brimful-pitcher of water and the Time is quite mysterious. *Kumbha* has been the main object in Vedic philosophy for exemplification of the cause-and-effect where the clay of the pitcher is the *upadaan-kaaran* (material-cause) which is Nature and the pot-maker is the *nimitta-kaaran* (efficient-cause) the Creator. In this mantra the *kumbha* is the Nature in its definable form.

The mantra expresses that *we of the present observe Nature in numerous ways*. A *kumbha* faraway looks small which continues to grow as the pitcher's carrier approaches nearer and nearer. This approach is time depended. The mantra through the simile of Nature as the brimful-pitcher delineates an observational scenario where the pitcher is perceived differently from different distances and at different times. Like a *kumbha,* the objects in *param-vyom* (vast Space) are seen in a relative perspective. It also suggests that to understand Time's nature one has to understand the objects of Nature far and near and within the perspective of Time. The mantra further propounds that to learn about Time the intellectuals extend their observation of Nature from close quarters to *param-vyom*, the farthest of the deep-space. The pairing of Space and Time is the obvious thought in this mantra.

Although the mantra does not give us hard numbers on observations of the vast-space (*param-vyom*) that could be used to deduce the specific nature of Time, it does tickle one's imagination to think of the principle of relative position in space and time of the moving objects. We may not be aware of it but we follow this principle in daily life. The time of creation in billions of years was arrived at by Vedic *rishis* inspired by similar mantras when they observed the *param-vyom* which embedded the system of *sapta-chakran* (seven-wheeled system) of planetary motion with the Sun being their single axis (*aksha*). It is the observation of distant objects in motion that tell us more about time and space. Let us take up a contemporary example related to studying substances in Space (*param-vyom*).

Space stretches from the Earth to the farthest galaxy. Beginning with the Earth, a study of the oldest rocks on Earth by scientists of the University of McGill in Montreal (in 2008 CE) concluded that the Earth is at least 4.28 BY old. Canadian geologists also found (March 1, 2017) rocks with microbes petrified in them about 4 BY ago. The age of the Earth is estimated to be 4.6 BY. Therefore, the life evidently began on Earth within 600 MY of the formation of the solar system including the Earth. Another example of scholarly studying substances and placing time stamp on them is extra-terrestrial object, a meteor. The age estimate of a meteor yields 4.6 BY, which by reduction is the age assigned to the entire solar system. Red-shift in the spectrum of light rays (*sapt-rashmi*) coming from the edge of the *param-vyom* tell us that the universe is 13.8 BY old and expanding. Thus the observation and knowledge of objects of *param-vyom* teaches us about the nature of Time. By definition the term *param-vyom* is the highest or deepest of space that according to Veda "*preserves all that exists there*".

(AtharvaVeda, 19/53/4)

> "*It is the one (Kaal) that has nourished the cosmos, it is the one that has completely and uniformly encompassed the cosmos. While being the* पिता (*pita* = father) *of the cosmos it has provided continuity by being the* पुत्र (*putra*) the *son of itself. In this creation none is older than Kaal*".

Notice the riddle of Time being the father and the son is repeated in this mantra where the continuity of Time is of essence. *In this creation none is older than Kaal* makes this entity pre-existing the creation.

(AtharvaVeda, 19/53/5)

> "*Kaal has created the dyu (illuminated region around the Sun) and the Earth. In Kaal is its past, with certitude in Kaal is cast its future*".

(AtharvaVeda, 19/53/6)

> "*Within Kaal the matter comes into being, within Kaal the Sun exists to give off heat energy. All cosmos are in Kaal and in Kaal the eye sees variously*".

These mantras are self-explanatory. Full translations are not provided for the remaining mantras in the two *Kaal suktas* of AtharvaVeda, only the essence is presented. These mantras state the importance of measuring *kaal* (time) from the

Kaal-entity. They state the relation of *Kaal* to other phenomena and systems of creation that relate to the physical as well as metaphysical creation.

(AtharvaVeda, 19/53/7-10)

"Kaal encompasses all entities of the creation that relate to forces of Nature that sustain the living system in the creation, provides opportunity for karma, enables the knowledge of Veda and the nomenclature of all substances of creation therein. Kaal is the revealer of the creation and the Brahmn the Creator. Kaal has fathered all celestial praja, the heavenly population of this creation, and it is the nurturer and binder of all creation. In it are stably sustained all cosmos".

(AtharvaVeda, 19/54/1, 2, 4, 5)

"Kaal is the measure of the Kaal-entity (Time-entity*), the time, which gives us knowledge of all. In Kaal are born the disha (directions). The Sun rises in it and sets in it punah* (again and again). *All motions exist within kaal including movement of the wind. The Prithvi in it has become mahi* (the Earth has grown colossal) *and the dyu* (the illumine space, the abode of Sun) *has been placed there. All lokaha* (terrestrial planets) *flying in space and all that non-decaying stability of the whole cosmos is due to Kaal-entity".*

Lastly, it states that,

...ब्रह्मणा कालः स ईयते परमो नु देवः ॥ (AtharvaVeda, 19/54/5)
("*Brahmna kaalah sa eeyate parmo nu devah*").

This translates into:

"...Biggest of all devas, the Kaal, coexists with Brahmn the Creator and continues to move fast" (AtharvaVeda, 19/54/5)

This mantra is explicit in defining the *Kaal-tatva* (Time-entity) to be creation's eternal entity. Coexistence of *Kaal* with the eternal entity *Brahmn* the Creator and the efficient-cause of the universe, makes it automatically an eternal entity. The additional knowledge in the last part of the mantra is that the *Brahmn* and the *Kaal* are two distinct entities that coexist in this creation. The *Brahmn*-entity is not created and therefore Time-entity in Vedic thinking is not created at any particular event like t = 0 sec of modern Big Bang hypothesis. This limited discussion shows that the Veda has gone deeper in defining and describing the nature and properties of the Time-entity than modern science has.

Time and space in Theory of Relativity are knotted together as space-time continuity. As this author has emphasized before, none of these entities could be measured absolutely; only their differences are measured. Of the five eternal entities being discussed here, two being consciousness do not permit their quantitative measurement. The other three permit limited access for their measurement. Nature allows us to measure mass in matter but the Time does not allow us to measure its 'present' because the present is infinitely minute. But it allows us to measure its linear span in its past. The Space allows us to measure it in three dimensions. All these measurements together help to further our knowledge within our reference frame which to some degree is extendable to the reference frame of the universe. The recognition of the importance of measurement of these entities has been an ancient practice. Before presenting the Time and Space measurement scales of the Aryans let us look at the fundamentals of Aryans' approach to complex concepts.

Measurement (*Maatra*) leads to Enhancement of Knowledge

Aryans have looked at the Time-entity in its totality. A unique, remarkable and insightful evidence of this is in the 10th *shloka* of Surya Siddhant (The Sun Principle). A treatise on indigenous practical astronomy, "Surya Siddhant", was referred to several times in Chapters 1 and 3. Recalling this shloka from Chapter 1, one finds that the Aryans have viewed the Time-entity for what it really is – unfathomably eternal entity and inconceivably minute as well as measurable to some extent in between the two extremes. Surprisingly, the terminology in the *shloka* is as if it has been drawn from a modern physics text book. The *shloka* translation is repeated in the following:

> "*Time is in two classes. First, the cause of creation and dissolution* (being eternal) *is unknowable and the second is knowable. The second-class is in two categories – one is moort* (comprehensible and measurable) *and the other is sookshma* (immeasurably minute)." (Surya Siddhant, 1/10).

The depth of thought in this shloka on the nature of Time is obvious. It is the comprehensible and measurable *moort* time that is the subject of this section. When a measurable entity or a phenomenon is adroitly defined, as Time-entity in the AtharvaVeda mantras and in the *shloka* of Surya Siddhant, its measurement is inevitable by those who have grasped its reality. Considering the mind-boggling level of science development that has taken place in the past two-three centuries a measurement methodology need not take millennia to develop. The time measurement and the time division of the Aryans discussed below would make it obvious that the Aryans had already achieved a high level of success in time measurement and in devising time-scales for advancing the knowledge for their future generations.

One method for advancing the knowledge is to devise a measurement scale for measuring an entity. Once a stable measurement scale is worked out, it continues through future generations and sometimes through future civilizations. For example the scientists of the present civilization have benefited from the measurement scales invented in the past civilizations; popular ones being hour-minute-second for time, inches-foot-yard-mile (British) or centimeter-meter-kilometer (metric system) for distance and ounce-pound-ton (British) or gram-kilogram-quintal (metric system) for mass. With the progress in science it became necessary to extend both sides of the preexisting scales and add to the roster some new units and scales of distance, force, energy, electricity, magnetism and so on.

No measurement of any entity is in isolation. When an entity or a part of it is measured the other entities serve as a 'general cause'. Therefore, a simple dictum in measurement could safely be postulated by this author as:

"No measurement of Space, Time or Matter is independent of the other two"

When a cube of matter is measured in its length, width and thickness the space gets measured and it is not independent of 'time'. When a person makes a time measurement it has a beginning of time and the end. During this time the person has aged changing the matter in his body which also changes the space occupied by the person, be it a few cells. Therefore, what is being measured and what else gets measured needs to be recognized. In this recognition if one is measuring an entity the person must know what are the other associated entities taking part in this measurement and recognize where the entity belongs - in time, in space or in matter? To resolve this philosophical imperative the Vedic philosophers have attempted to put a handle on it. This handle is termed as the '*linga*' or '*lingam*' of the entity. It was mentioned in Chapter 1 and in the Vaisheshik *sutra* (4/1/2) presented in this chapter that the word *linga* stands for the 'defining symbol' or the 'defining property' of the entity/phenomenon. Any measurement of an entity required a pre-recognition of the *linga* of that entity.

Kanaad Rishi in his Vaisheshik Darshan has examined phenomena, substances and entities from this perspective and laid down the principles (*sutras*) for future knowledge enhancement. For the Time-entity he states (Vaisheshik, 2/2/6):

अपरस्मिन्नपरं युजपच्चिर क्षिप्रमिति काललिङ्गानि ।
(Vaisheshik Darshan, 2/2/6)

(*Aparsminnparam yujpachchir kshipramiti Kaal-lingaani*)

Translation:

"The knowledge of younger, older, simultaneity, later, sooner are the lingas of Kaal (Time)" (Vaisheshik Darshana, 2/2/6).

For Rishi Kanaad to have put together in one *sutra* all *lingas* of Time is uniquely remarkable. To the best of this author's understanding, not one factor can be added to or subtracted from his list. These *lingas* of the Time-entity are self-explanatory. A time related phenomenon would have at least one or more of these *lingas* present in it, differentiating the measure of time from the measure of space or matter.

Expressing a substance in terms of a unit of measurement is inherent in humans. In Vedic practice the generic term for a measurement unit is 'maatra', amazingly close in sound to the word 'meter'. The maatra or the unit of measurement in the Vedic civilization is as old as the Vedic civilization itself. Considering the antiquity of the Vedas, the range of scales for measuring time, space and matter is surprisingly quite extensive on both, the lower as well as on the higher ends of the scale. The Vedic counting from zero upwards is summarized in Table 5.2.

These units, with their original Sanskrit nomenclature, as it appears in the Vedas and the Vedic Literature, are listed in the second column. This author has added in the first column of the tale the numerals in the standard form with zeros after 1 or in the scientific notation with the number of zeros expressed as the exponent of 10. In the third column are added the terms in the common usage in English.

A YajurVeda mantra (17/2) was cited in Chapter 1 which teaches the counting of large numbers in metric system from 'one' to '*paraardh*' (10^{18}) (with 18 zeros after 1) in 19 steps. After listing these steps of the metric system of counting, the long *chhanda* (mantra) of YajurVeda terminates with an interesting conclusion in its second part. It states –

"...*When you think of counting bricks, think of the vast Space and those worlds out there that are observable* (धेनवः लोके = *dhenavah loke*)[1] *and non-observable* (धेनवः अमुष्मिन् = *dhenavah amushmin*)[2] *to which also these numbers apply, where you may be in rebirth* (अमुत्र = *amutra*)[3]".

An AtharvaVeda mantra (10/8/24) in a different context also gives a count of 100 (*shatam*), 1,000 (*sahasram*), 10,000 (*ayutam*) and 100,000,000 (*nyarbudam*).

The remnants of Vedic education system still existed in the county-side of India when this author started schooling. In the elementary classes, right after learning the numerals the guru taught the multiplication-tables up to 40 and additionally taught the times-tables for quarter, half, three-quarters, one-and-a-quarter and so on till four-and-a-half. Then the guru taught the large range of numbers and the nomenclature of units to go with it right up to 10^{19} (19 zeros after 1 which is ten-billion-billion, one step more than in the YajurVeda mantra) which is still in this author's memory for recital in the correct order.

Table 5.2: Numerical counting system of the Aryans

No.	Vedic Nomenclature	Western Nom.	No.	Vedic Nom.	Western Nom.
0	***shoonya***	**zero**	**10^9**	***arbudam***	**billion**
1	***Ekam (ikaai)***	**unity**	**10^{10}**	***dash-arbudam***	**ten billion**
10	***Dasham***	**ten**	**10^{11}**	***kharab***	**one hundred billion**
100	***Shatam***	**one-hundred**	**10^{12}**	***dash-kharab***	**trillion**
1000	***Sahasram***	**thousand**	**10^{13}**	***neel***	**ten trillion**
10000	***dash-sahashram (ayutam)***	**ten-thousand**	**10^{14}**	***dash-neel***	**one hundred trillion**
100000 (10^5)	***Laksh***	**one hundred-thousand**	**10^{15}**	***padam***	**million-billion**
10^6	***dash-laksham***	**million**	**10^{16}**	***dasha-padam***	**ten-million-billion**

10^7	***kror***	**ten million**	**10^{17}**	***shankh***	**hundred-million-billion**
10^8	***dasham kror (nyarbudam)***	**Hundred million**	**10^{18}**	***dash-shankh***	**billion-billion**
			10^{19}	***maha-shankh***	**ten-billion-billion**

A question arises as to why such large numbers existed in the Vedas and in Vedic civilization and why were they taught? There are several parts to the answer. One answer is in YajurVeda mantra cited above. For that answer to be acceptable one has to believe in rebirth. It is however interesting to note that the discovery of thousands of exo-planets in astronomy proves the second part of the mantra that there is life on some of the exo-planets, even if there is no rebirth system. The second part of the answer is that the Vedic tradition could not do without mathematics that they derived from the Vedas, including the invention of zero without which the world would not be the same today! The Sanskrit terminology of the numbers, with 7 zeros (kror) and up to 9 zeros (arab or arbud, a billion), still continues in India today in commercial and government budgetary expressions. The large numbers were also taught because at one point in the Vedic civilization the professional sciences were a part of the school curriculum in the *gurukulas* (centres of learning at gurus' *ashrams*). The term assigned to each unit in the above table shows the development of science and mathematics in the Vedic society for an extended period of time.

Some of the units of Vedic mathematics in the first column of the Table 5.2 with nomenclature in the second column relate to the well-known units in the third column used in the western civilization like thousand, million, billion and trillion. The same question when posed to the present civilization why such big numbers needed, yields the answer - *'necessity is the mother of invention'*. It is obvious from the Table 5.2 above that the nomenclature right up to 10^{19} was developed in the Vedic civilization in the far antiquity of the Indian civilization that even had finer steps with proper nomenclature than the million-billion-trillion scale, jumping forward with three zeros at a time. The necessity must have been the mother of invention of the Sanskrit nomenclature too.

It is a generally accepted fact that the concept of zero originated from the Aryan Civilization in India. Zero is an entity that has no value of its own but it increases the value of a number by association. Aryans invented the concept of zero because it must have become a necessity for counting and dealing with large numbers. Even if they pulled it out of the Vedas, the credit for introducing zero in mathematics goes to them. The obvious fact here is that the Aryans generated acronyms of the units and utilized large numbers. It is further clear from the AtharvaVeda mantra (8/2/21) that states the period of creation. The first part of the mantra is presented below,

शतं तेऽयुतं हायनान् द्वे युगे त्रीणि चत्वारि कृण्मः ।...॥(AtharvaVeda, 8/2/21)

Word Transliteration and Translation:

शतं (*shatam*) = one hundred, ते (*te*) = for you, अयुतं (*ayutam*) = ten thousand, हायनान् (*haayanaan*) = years, द्वे (*dve*) = two, युगे (*yuge*) = to add, त्रीणि (*treeni*) = three, चत्वारि (*chatvaari*) = four, कृण्मः (*krinmah*) = given to you/done for you, put sequentially ।

In the context of the mantra the translation of this part of the mantra is *"One gets the age of creation by placing two, three and four before one hundred 'ayutam' years"*. (AtharvaVeda, 8/2/21).

In a typical Vedic mathematical way of expressing large numerals, the mantra requires one to build the number by proceeding from the lower place value to the higher (from right to left). An *ayutam* occupies 5 places of numerals and one hundred *ayutams* occupy 7 places of numerals. Since no specific numerals are assigned for the seven places, the instruction is to place 7 zeros first to occupy one hundred *ayutam* spaces and then place 2, 3 and 4 in that order going towards higher orders to get the required number. The result is 4,320,000,000 years. This, 4.32 BY (billion years), is the total age of the creation phase that the universe will last for. Although the present age of the universe, based on astronomical observations is determined to be 13.73 ± 0.12 BY, it is much bigger than the Vedic assigned age of the universe. Nevertheless, a figure such as 4.32 BY coming out of the ancient Vedic discipline of thought in science and mathematics is in the same order of magnitude. This remarkable formulation on the creation period of the universe in billions of years is presented in AtharvaVeda in a beautiful poetic mantra structured in a unique meter

The figure of 4.32 BY as the Vedic age for the universe could not have been arrived at by mere chance or by fanciful imagination. The Vedic sage could have given a round figure such as 10 thousand or a million or even a billion years but the mantra formulation enunciates a specific number, the 4.32 BY. There must be a logical explanation behind this precise number.

The next sections under the Time-entity are devoted to its measurement.

Measure of Time-entity: Large Time Scale of the Aryans

Detailed Time Division of 4.32 Billion Years

The interesting feature of the Vedic age of the universe is that it is not just a lump-sum mathematical number of 4.32 BY; it is further divided into specific epochs of time called *yugas*. The *yuga* periods are not just another set of lump-sum numbers, the Vedic scholars suggest that the *yuga divisions are* like the seasons on Earth having their distinct natural character. Because the number 4.32 BY itself relates to the long-term astronomy this author's thinking is that they must have specific relationship to the long-term astronomical periods associated with the Sun's traverse and position in the constellations. An important unit of time is a *deva-yuga* period lasting for 4.32 Million Years, 1000th of the reation period of 4.32 Billion Years.

In the Vedic literature the detailed division of this period have been presented in quite a few ancient references, including the Manusmriti authored by the great King Manu of the far antiquity of Vedic civilization anterior to Mahabharata (3140 BCE) and the books written after Manusmriti such as the Bhagvadgita (3140 BCE) and Surya Siddhant (The Sun Principle, author and time of writing not known). There are differing procedures used in each of these references to arrive at the same length of a *yuga* (a defined period of time) and the same final total of 4.32 BY. The details of time-division presented in this book are based on the book of Manusmriti. Manusmriti's time division begins from *nimesh* as the smallest unit of time which equals to 0.1778 second and goes right up to 4.32 BY. For the sake of brevity Manu's *shlokas* on time division are directly translated from Sanskrit to English without detailed word for word translations. The time division from one day and upwards is presented first. The *shloka* (1/65) of Manusmriti states that:

> *"The Sun enables the division between human's day-night and the devic* (of the *devas*) *day-night where the day is for activity and endeavour by living beings and the night is for sleeping"*. (Manu, 1/65).

This is indicative of the time division by the natural system of Sun-rise and Sun-set. The important feature of this statement is that there are two kinds of day-night periods: humans' and *devas*'. This difference is stated in Manusmriti shloka (1/67):

> *"One year of humans is one devic day-night. The day-period of devas is when the Sun is udgayanam* (north of equator) *and the night-period of devas is when the Sun is dakshinaayanam* (south of equator)*"* (Manu, 1/67).

This remarkable shloka laid down by Manu right in the beginning of the Vedic civilization speaks volumes. It deals with the science of physics and geography to establish the time division. The time division on Earth relates to Earth's geography and the Sun's latitudinal position in the sky with respect to Earth's equator. The terms *udgayanam* and *dakshinaayanam* in this shloka specifically relate to the Sun's geographic latitudinal region or Sun's abode in the sky with respect to the equator of the Earth: *udgayanam* is for the Sun being in the northern hemisphere and *dakshinaayanam* is for the Sun being in the southern hemisphere. The term *udgayanam* in subsequent ancient astronomical Sanskrit also appears as *uttaraayan* with *uttar* for geographical north with the common suffix *ayanam* representing the place or abode. This also tells us that the creator of the nomenclature, *udgayanam* and *dakshinaayanam*, Manu in this case, knew the dividing line between the two. This dividing line is the equator. The Sanskrit name for it is *vishuvat-rekha*. When one looks into the word meaning of the term *vishuvat-rekha* alone, one is pleasantly revealed that the ancient Aryans had tremendous grasp of physics, astronomy and geography. The terms *vishu* stands for "in two parts", *vat* stands for "similar to but not actual" and "*rekha*" stands for "line". When put together "*it is the line that divides in two similar parts but not real parts*". Obviously, it is the line dividing the Earth's globe into equal halves. The other Sanskrit dictionary meaning of *vishu,* in context of astronomy is "similar to equinox".

One wonders why Manu introduced two different time units in a single *shloka* (1/67), one human's and the other *devic*. It is obvious from the Vedas and Manusmriti that the ancient *rishis* including Manu had the knowledge of the Earth, that it was round, spinning around its axis and that the solar angle changed with latitude through the year, sending the Sun into *udgayanam* and *dakshinayanam* (northern and southern hemispheres). Manu realized that defining a full day by two equal halves (daylight and night-time) was not absolute because such a definition changes with time of the year and latitude of the place when the day and night are not equal. Whereas a day in the equatorial region of the Earth is defined to be the daylight period between sunrise and sunset, a day at the geographic pole is six months long and a night there is also of six months long. To have his definition applicable to the entire globe, pole to pole, Manu intentionally introduced two time units and related them within one *shloka*. This is the meaning of "*One year of humans is one devic day-night"* in the above *shloka.*

Appropriate terminologies have been employed to minimize any confusion in the definitions. To differentiate a day from a night the Vedic expression is *aho-ratra* (day-night), however, in the count of days in a Vedic year the 360 *aho-ratra* form one year which in common usage is 360 day *saur-varsh* (solar-year). Thus in the count of 360 solar-days at the equator, a day is inclusive of a day-night pair. All these time-units, time-scales, their nomenclatures and definitions prescribed by Manu in his Smriti have their root in the Vedas; which are considered by the Vedic people to be the ultimate reference. The Vedic *rishis* based their knowledge and derivations on the Veda mantras. The *shloka* itself makes it glaringly obvious that Manu had full grasp of Earth's geography and astronomy to have introduced astronomy terminology *udgayanam* (Sun in northern hemisphere) for six months and *dakshinaayanam* (Sun in southern hemisphere) for six months. This *shloka* of Manu provides a time conversion factor: **1 *devic* day equals 1 human year**. It is also obvious that the Earth for Manu was a round globe as it has been in the Vedas. He had full knowledge of the daily variation in Sun's position causing change in the length of the day and night through the year.

To determine the onset of *udgayanam* and *dakshinaayanam* one needs to ascertain the dividing line between the two, the *vishuvat-rekha* (equator). Therefore Manu must have known where the equator on the Earth was. A brief discussion with references was presented in Chapters 1 and 4 on Manu being a globe-trotter with grasp of knowledge on global geography, global ecology and languages of societies exterior to Aryavart (India). The consideration of the Sun's position with respect to the equator indicates advanced astronomical knowledge of Manu and therefore of Aryans of the far antiquity. The deductions contained in the *shlokas* require knowledge of north and south sides of the globe and the dividing line. Manusmriti and the Vedic literature define a day between the sunrise and the sunset and a night between the sunset and the sunrise to make a calendar day, as all do.

After relating the human year and *deva* day, Manu in the next *shloka* introduced the unit of Brahmn-day. While doing so he urged the scholarly *rishis* in attendance before him 'to pay attention'.

> *"Be attentive that now I will present to you in brief the measure of Brahmn's day and the measure of yugas one-by-one, kramashaha* (sequentially)!" (Manu, 1/68).

A *Brahmn*'s day is a Creator's day, a period of time during which the creation is in its active epoch of definable matter. Similarly, a *Brahmn*'s night is the period of time when the creation is in dissolved slumberous state. In the terminology introduced earlier, a *Brahmn*'s day is *srishti* (creation) and His night is *pralaya* (dissolution). The length of *Brahmn*'s day based on the AtharvaVeda mantra is 4.32 BY. A *Brahmn*'s day can be expressed in terms of human-years, *devic*-days or *devic*-years. A *Brahmn*-day is expressed in yet another unit called *yuga*. The meaning of the word *yuga* is the 'period' of time, however, a specific period of time has specific name for the *yuga*.

The word *kramashaha* (sequentially) in Manu's above *shloka* is for one to know four different *yugas* sequentially in the order of their duration. These four *yugas* from the onset of creation successively are:

(1) *Kritam-yuga*, its more popular name is *Sata-yuga,*
(2) *Treta-yuga*
(3) *Dvapara-yuga*
(4) *Kali-yuga*

This unit of four *yugas* together sequentially cycles through one thousand times during one *Brahmn*-day which is the full age of creation. This means that the creation begins at the onset of the first *sata-yuga* (*kritam-yuga*) followed by the next three and at the end of the first *kali-yuga* begins the second *sata-yuga* and so on, till they all sequentially cycle through one thousand times in one creation.

Manu has considered a year to be of 360 solar days and not of 365 solar days as in the present. It is possible that the Earth did have a 360 day-year at the time of Manu and therefore the scale of a year was considered to be of the 360 days. This also stems from the fact that RigVeda mantras consider a year to be of 360 days. The significance of 360 day-year and its division into 12 months in relation to the division of Earth's orbit around the Sun being 360° will be discussed in the following. Here it would suffice to say that all numbers presented in terms of human, *devic* and *Brahmn* days and years are based on a 360 day-year. This means that while a human-year is of 360 solar days a *devic*-year being of 360 *devic*-days (at the North Pole) is 360 human-years long.

After defining one *devic*-day in the *shloka* (Manu, 1/67) as 360 human-days (one year) Manu in the *shloka* (1/69) presents the mathematical definition of a *sata-yuga* in terms of the main period plus the two union periods either side of it.

***Sata-Yuga*:**
"In terms of those, devic-years, a sata-yuga is said to be of four thousand years, added to it is the sandhi (union) period before and after its fraction of four hundred years,". (Manu, 1/69)

Therefore, according to Manu, along with two union periods, a *sata-yuga* is:

Sata-yuga = *sandhi* + 4,000 *devic*-years + *sandhi* = 400+4,000+400 =4,800 *devic*-years.

From Manu (1/67), presented above, 1 *devic*-day (a day-night at the North-Pole) is 1 human-year at the equator with a Vedic definition for a year being 360 days, the above number can be represented in terms of human years.

The *sata-yuga* period in terms of *devic*-days = 4,800x360
=1,728,000 *devic*-days
=**1,728,000** human-years

Therefore, a ***sata-yuga*** = **1,728,000 human-years.**

Based on Manu's shloka (1/70) the other three *yuga* periods are calculated as following,

Treta-yuga, Dvaapar-yuga and Kali-yuga:

Manu's definition of these three yugas is:

"For the remaining three yugas one gets their measures by subtracting sequentially a thousand from their main period and a hundred from their union periods". (Manu, 1/70).

Its Mathematical summary is:
From *Sata-yuga* one gets ***Treta-yuga*** = (400-100)+(4,000 -1,000) + (400-100)
= 300+3000+300
= 3,600 *devic*-years
= 3,600x360 *devic*-days
= **1,296,000 human-years**

In the same manner from *treta-yuga* one gets,
Dvaapar-yuga = 200+2,000+200
= 2,400 *devic*-years
= 2,400x360 *devic*-days
= **864,000 human-years**

From ***dvaapar-yuga*** one gets ***Kali-yuga*** = 100+1,000+100
= 1,200 *devic*-years
= 1,200x360 *devic*-days
= **432,000 human-years**

Manu must have felt the need for clarification on what has been said in the above *shlokas* and therefore he elaborates further:

> "*This, what has been enumerated here for the chatur-yugam (four-yugas), is twelve-thousand divya (devic)-years. It is devas' one yuga*". (Manu, 1/71).

Manu is stating that one *chatur-yugam* (four-*yuga*-period) of humans is equivalent to 12,000 *deva-years* which is termed as one ***deva-yuga***. This is 12,000x360 = 4,320,000 human years. One gets the same figure by adding the measures of all four *yugas* presented above,

1,728,000 + 1,296,000 +864,000 + 432,000 = 4, 320,000 years = 1 *devic-yuga*

It is obvious from this addition of four *yugas* that each is a multiple of the ***Kali-yuga*** as,

4x432,000 + 3x432,000 +2x432,000 + 1x432,000
= 4,320,000 human years = 4.32 MY

The next shloka of Manu (1/72) defines the last unit of Vedic time-scale for cosmological estimates and *Srishti*-calendar,

> "*The number determined from one thousand deva-yugas is termed as one day of Brahmn and the same length of time is one night of Brahmn*". (Manu, 1/72).

The mathematical representation of this statement is,

1 *Brahmn*-day = 1000 *deva*-yuga = **4, 320,000 x 1000 human years** = 4,320,000,000 years = **4.32 Billion Years (**BY)

Different nomenclatures and their relationships are summarized below:

(1) A set of four successive *yugas* is termed as a *chatur-yuga* or a *chatur-yugi*.

(2) In terms of the names of the *yugas* a *chatur-yugi* is:
Sata-yuga + *Treta-yuga* + *Dvaapar-yuga* + *Kali-yuga* = *chatur-yugi* = 4.32 MY

(3) *Chatur-yugi* = *Maha-yuga* (*maha* means big).

(4) *Chatur-yugi* = *devic-yuga, deva-yuga, divya-yuga* (all five synonyms for a *chatur-yugi* , each with the span of 4.32 MY.

This time period of 4.32 MY, the *devic-yuga*, is important in Vedic Cosmology because, as brought out in Chapter 8 on Vedic Creation Scenario, this is the time period needed for:

(a) the primordial matter (*Maartaand* = Mortal-ovate) to develop sufficient heat energy to explode in Vedic Big Bang and break into *koti-koti* (uncountable) smaller *Maartaands*,

(b) This is the time period needed after the Vedic Big Bang to establish order in the creation and for the living matter with root-system to first appears on the Earth.

This number, 4.32 BY, is the same as in the AtharvaVeda mantra (8/2/21) cited earlier. In that particular mantra a total for the number of years is given without time division in terms of *yuga* periods of different names and measures. Elsewhere in the Vedas some of the units of time mentioned above appear in differing contexts. Manu is found to be the first in the Vedic Literature to have presented a break-down of the total period of one *Brahman*-day, the full span of the creation phase of the universe. The above *shloka* (Manu, 1/72) also states that after each creation (*srishti*) follows Brahmn-night (*pralaya*) of the same length. In one of the shlokas Manu said that human-day is for activity and the night for sleeping. A *Brahmn*-day is for the *Prakriti* (Nature) to be in an active phase (awakened state) and after that recede into stupor during *Brahmn*-night. In a different context, Yogiraj Krishn also states in Bhagvadgita (8/17) that,

> *"Those who talk in terms of the definition of day-night and are scholars of the science of Time know that one day of Brahmn is of one thousand yugas and one night is of the same duration"*. Bhagvadgita (8/17)

One thousand *yugas* in this shloka are one thousand *devic-yugas* which is the same as one thousand *maha-yugas* as defined above.

The ancient astronomy book "Surya Siddhant" (The Sun Principle) has given in its first chapter the basic definitions of the *yugas* and their relationships to each other in terms of numerical values and many other details. The basic definition and the numerical values of the *yugas* are the same as in Manusmriti. For example in the Surya Siddhant shloka (1/15) in particular, it has stated that, "*One chatur-yugi is equivalent to 12,000 devic-years which is equivalent to 4,320,000 solar years*". It has gone further talking about the orbital periods of the planets and how to arrive at the age of the universe and how many *yugas* and part thereof have elapsed since the creation of the universe began. Drawing more details here from the Surya Siddhant will distract the context of this chapter.

Table 5.3: Vedic Time-scale for cosmology and important epochs in the history of creation (Manusmriti 1/69-73).

	No. of years	Unit Nomenclature	Comment
1	1,728,000 years	1 *Sata-yuga*	4x *Kali-yuga*
2	1,296,000 years	1 *Treta-yuga*	3x *Kali-yuga*
3	864,000 years	1 Dvapara-yuga	2x *Kali-yuga*
4	432,000 years	1 *Kali-yuga*	1x *Kali-yuga*
5	4,320,000 years (sum of four above)	1 *Chatur-yugi* (1 *devic-yuga*)	Also called a *deva-yuga* or a *deva-samvatsar*
6	Cycle of 1 to 4	→	1,000 times during one

	repeats		creation
7	1,000 *Chatur-yugis* (4,320,000,000 years)	1 *Kalp* (*Brahmn*-day) (1000 *samvatsar*)	Creation period (beginning to the end of creation)
8	1,000 *Chatur-yugis* (4,320,000,000 years)	1 *Vi-Kalp* (*Brahmn*-night) (1000 *samvatsar*)	Opposite to *Kalp* = dissolution phase of the universe (end of creation to next beginning of creation)
9	311.04 Trillion Human-years	1 *paraant-kaal*	100 *Brahmn*-years

There is one more unit of time which relates to the period of salvation of the soul. According to the Vedic philosophy, the soul's privilege to be in 'eternal bliss', free from the vice of *Prakriti* (matter) and free from the birth-death cycle and the pain and suffering that come with it, is not forever. It is a predetermined period after which the souls in salvation come back to take birth in the creation. Swami Dayanand Saraswati in the 9th chapter of Satyarth Prakash discusses the subject of salvation (*mukti*) with a reference from Mundak Upanishad (3/2/6) and states that the salvation period is one *paraant-kaal,* also known as *maha-kalp.* This period is 100 *Brahmn*'s years.

1 *paraant-kaal* = 100 *Brahmn*-years → 100x360 *Brahmn*-days (day+night) → 100x360x4.32 BYx2 human-years → 311,040,000,000,000 human-years (311.04 Trillion human-years. This is shown in the last row of the Table 5.3. This number in Sanskrit mathematical terminology is read as: 31 *neel* 10 *kharab* 40 *arbud.* All calculations pertaining to the time units used in Vedic science and cosmology are summarized in the conversion Table 5.3.

It is important to note from the table that the *chatur-yugi* or 1 *deva-yuga* was equated to yet another term, *samvatsar.* There is reason for naming yet another term for the same time period because each term conveys a slightly different meaning. The generic term *samvatsar* is for a cyclic order. As an example the term *samvatsar* in context of short term cycle is for one orbit of the Earth around the Sun, which is the period of one calendar year. It was mentioned in Chapter 3 that the Vikram-samvat started from King Vikramaditya, 57 BCE and the present Kali-samvat started from February 18, 3102 BCE.

The term *samvat* is short for *samvatsar.* In the context of the Earth, its *samvatsar* represents a perpetual cycle with the periodicity of one year made up of 12 months. In the context of the universe, its *samvatsar* represents a perpetual cycle of 4.32 MY made up of four different successive *yugas* that cycle through 1,000 times. All these perpetual cycles appear during the creation phase and disappear during the dissolution phase and the perpetual cycle of creation-dissolution or *srishti-pralaya* also continues with its periodicity of 8.64 BY.

Let us take a few more steps to look into the Vedic cosmological time-line, where the 4.32 BY period has been further divided into different epochs of time that relate to the history of the Earth and its future.

Manvantar Period and the Big Flood

The Earth due to its position on its orbit around the Sun cycles through well-defined seasons. Does the universe go through any cycles? Modern cosmology assigns an age to the universe but does not suggest that universe goes through or does not go through any long-term cycles. The only variation in universe's character according to Big Bang Theory is onetime change as five epochs discussed in Chapter 2 of which the first three end within 10 seconds of universe's life. No cyclic behaviour in the universe's life is observed or predicted in the modern Big Bang Hypothesis.

In Vedic cosmology, the universe began with its well-defined first era of *sata-yuga* to last for 1,728,000 years, followed by three other eras of specific periods. According to the Vedic cosmology, considerable resetting in the universe takes place at the end of each *maha-yuga* of four successive *yugas* totaling 4.32 MY. This resetting affects the Earth and thereby the history of the universe relates to the history of life on Earth. For the astronomical reasons presently unknown, the Vedic *maatalis* the scholars of time-science, related the history of the universe to life on Earth and grouped the age of the universe into bigger epochs of time giving them specific individualized names. The names are not only given to the epochs that have passed but given to those that are yet to come.

Vedic *maatalis* have grouped the 1000 *maha-yuga* (*chatur-yugis*) of the creation into 14 successive sub-divisions called *manvantars* of equal duration. In doing so the first three and the last three *chatur-yugis*, a total of 6, are separated as the *manvantar-sandhi* (union) periods and the middle 994 *chatur-yugis* are divided into 14 equal periods of 71 *chatur-yugis* in each *manvantar*. Therefore each *manvantar* period is 306.72 MY long. The specific names given to these 14 *manvantar*s are listed in the Table 5.4. A union period of 3 *chatur-yugis* (4,320,000 x 3) of 12,960,000 years each is shown at the beginning and at the end of the *manvantar* list in the table. The Vedic creation begins with the *Svayambhuva-manvantr* and ends with the *Bhautakya-manvantr*.

The two large union-periods in the Table 5.4 for the whole creation are different from the smaller sub-union periods that were used by Manu for formulating the *yuga* periods to arrive at the four different *yuga* periods. Life with root-system (vegetation) according to Vedic thinking (RigVeda, 10/72/3) first appears on the Earth in the first *deva-yug*a of 4.32 MY after the onset of creation and it takes two more *deva-yug*as of 8.64 MY for humans to appear. During the last union-period the situation on Earth and in the rest of the creation is said to be chaotic and disorderly, leading towards the dissolution of the universe. Although the Vedic calendar keeps track of the total 1000 *maha-yugas*, the thinking in the Vedic literature is that the useful period of the creation is during 14 *manvantars,* which is 994 *maha-yugas* (*chatur-yugis*).

The RigVeda mantra (10/97/1) from the "Vegetation" sukta is the basis for the three devic-yugas being used for the first union period. This RigVeda mantra states:

"I know with certainty the names and places of 107 medicinal vegetation of deep red colour that were born first in three deva-yugas prior to the birth of humans".(RigVeda, 10/97/1).

Important point to note is that the Veda did not say that the humans were born first who sowed the seeds giving birth to vegetation. Therefore, the sequence assigned by the Veda to the birth of vegetation and humans is logical, scientific and therefore is remarkable.

The Vedic rishis have upheld that a Big Flood takes place on the Earth after each *manvantar* which destroys most of the establishment and civilizations on the Earth, forcing a new beginning in the order on Earth. This Big Flood is termed as the "**Jal Plaavan**". This author has not been able to find from the Vedic literature a scientific basis for the cause of the **Jal Plaavan** occurring with the frequency of a *manvantar*, 306,720,000 years each. The Biblical story of a Big Flood is well-known when Noah saved the humans and animals in his ark. The Biblical Big Flood according to the Bible occurred about 3,000 years BCE and there is no mention that such Big Floods occur periodically with specific frequency, as given in the Vedic time-line of about once every 307 million years.

It is intuitive to suggest that the source of this *manvantar*-periodicity bringing a Big Flood on the Earth could be within the solar system, originating from either the heat-cycles within the Sun raising the waters of the oceans or the astronomical positioning of the planets and asteroids. The flood occurrence having a long period over 307 million years also inspires one to think of a cause beyond the solar system.

Table 5.4: Fourteen *manvantar* names and their durations

No.	**Manvantar Name**	**Duration (years)**
	First sandhi-period	**12,960,000**
1	**Svayambhuva**	**306,720,000**
2	**Svarochish**	**306,720,000**
3	**Ottami**	**306,720,000**
4	**Tamas**	**306,720,000**
5	**Raivat**	**306,720,000**
6	**Chakshush**	**306,720,000**
7	**Vaivasvat**	**306,720,000**
8	**Saavarni**	**306,720,000**
9	**Daksh Savarni**	**306,720,000**
10	**Brahm Savarni**	**306,720,000**
11	**Dharma Savarni**	**306,720,000**
12	**Rudra Putra**	**306,720,000**
13	**Raucha**	**306,720,000**
14	**Bhautyaka**	**306,720,000**
	Last sandhi-period	**12,960,000**
	Total Creation Period	**4,320,000,000**

Just as the duration of time required for the Earth to orbit once around the Sun is termed as the 'terrestrial year' or the 'solar year', the duration of time required for the solar system to orbit once around the centre of the Milky Way Galaxy is termed as the 'galactic year', also known as the 'cosmic year'. The duration of a galactic year being about 250 MY is of the same order as the manvantar frequency of 307 MY. At the present time it is not known as to what all hurdles of cataclysmic proportions are in the orbit of the Sun carrying its solar system, while completing an orbit around the galactic centre?

The smooth rotation of the Milky Way Galaxy with all stars running in their respective tracks may suggest that there cannot be any possible hurdle in the track of the solar system to cause catastrophic event resulting into a Big Flood on the Earth. However, there is a possibility that on each orbit around the Milky Way our solar system passes through some severe gravitational hurdle owing to another star or another galaxy causing the Big Flood on Earth. Let us analyze the long-term periods.

The Big Bang is said to have taken place 13.8 BY ago and the Sun was formed only 4.14 BY ago. If one assumes that the Earth was also born along with the Sun, it took a while before the oceans collected water on Earth. It is estimated that the oceans appeared about 3.937 BY ago. Now if we use 250 MY for each solar orbit around the galactic centre, the solar system orbited around the galactic centre about 3.937 BY/ 250 MY = 15.75 times. If the solar system faces at least one catastrophic hurdle on each orbit causing Big Flood on the Earth, there would be about 16 Big Floods in about 4 BY. This is too many compared to the Vedic count of 14 in the entire life of the universe.

Even though there is no relationship between the modern cosmology/astronomy and the Vedic cosmology/astronomy, surprisingly the order of magnitude of the age of the universe is in billion years in both. The Vedic numbers stipulate 14 *Jal Plaavans* (Big Flood) in 4.32 BY period of the life of the Vedic universe. The number 14 in the Vedic stipulation and 16 over a period of about 4 BY with one hurdle on the orbit of the Sun is interesting to note.

Whatever the cause for a Big Flood, if it has any relation to the orbit of the solar system around the galaxy and if the Vedic contention is right that the Big Flood frequency is about 307 MY, it should be of interest to geologists to look for a 307 MY periodicity in the Earth's crust and if possible in the crusts of all planets of the solar system. According to Vedic calculations 7 *manvantars* have elapsed yielding 7 Big Floods. Geologists should look for at least 7 scars in the crust of the Earth separated by 307 MY.

One could think of several reasons for one-time or occasional Big Flood in the life of the Earth. A big meteorite impact in the ocean could create a gigantic global tsunami. However, the *Jal Plaavan* phenomenon mentioned in the Vedic literature being a long-term periodic phenomenon with the frequency of occurrence of about 307 million years, the probability is that the cause for Big Flood would be found outside the solar system.

Prediction of the Subsequent Big Floods

According to the Vedic *srishti* calendar, we are in the 7th *manvantra* of Vaivasvat. The time elapsed since the beginning of the Vedic creation is **1,972,949,118** (≈ 2 BY) years in 2018 CE. The 6th Big Flood since the beginning of creation was at the end of Chakshush-manvantar (Table 5.4). By accounting for one union period of 12,960,000 years, the last Big Flood occurred 145,589,118 years ago in 2018 CE. In the coming period we may get some devastating floods which occur unpredictably every now and then, however, according to the Vedic prediction, the next Big Flood (*Jal Plaavan*) will occur 161,130,882 years in the future at the end of the present manvantar. It is worth examining on a scientific basis whether such a catastrophic event as the Vedic ascribed Big Flood, with a periodicity of 306.72 MY, is a real event due to the Earth and the solar system being at a specific position within our galaxy.

Contemporary astronomers have suggested that almost all observed galaxies are older than 2 billion years with their arms formed of stars gone around their axes over 8 times. The Vedic time division of creation into 14 *manvantars* of which 6 *manvantars* have already been lived means that there have been 6 Big Floods of cataclysmic proportions leading to seven rejuvenations of living creation on the Earth. This closeness of 8 times rotation of most galaxies and 6 times renewal of the living creation on Earth after a Big Flood is quite intriguing to note.

Long-term Seasons in the Creation

In addition to the Vedic prediction of a periodic occurrence of Big Floods several questions on the time division of the Aryans arise. Notice that the Vedic creation period of 4.32 BY is divided into numbers that are the even-fractions of this big number, a *maha-yuga* (*chatur-yugi*) being 1000th of it. Then a *maha-yuga* of 4.32 MY is further divided into four *yugas* not of equal duration but in multiples of 2, 3 and 4 of the *kali-yuga* period of 432,000 years. The obvious question is regarding the specific and differing duration for the four *yugas* put in sequence and why the same period of 4.32 BY was divided into the 14 *manvantars* of equal and specific duration? There is considerable indication in Vedic literature about the answer to this question, even though presently that answer only remains academic.

One observes in Nature that it is inundated with periodicity at all its levels beginning from a tiny electron orbiting a nucleus of an atom to the Sun orbiting around the centre of our galaxy carrying with it the entire solar system. Therefore, it is possible that the Big Flood is part of the periodicity that occurs for certain astronomical reasons. The Veda has divided Earth's year into 6 seasons which have been named according to the specific changes they bring on Earth. The 14 *manvantars* separated by the periodicity of 306,720,000 years may have something to do with our solar system's position on its orbit around the Milky Way Galaxy. This is like the existence of sub-harmonic waveforms within a resultant waveform. What the Vedic system has certainly done here is to arouse the curiosity in the enquiring minds to examine these Vedic ideas employing modern science tools and methodology. One has to wonder why the *rishis* chose to divide the period of billions of years into specific periods. They certainly did not indulge in this incredibly complicated mathematical maneuvering for fun!

Deductions

- The creation period is divided into *yugas*, cycling through 1000 times. Whole creation period is further divided into 14 *manvantars* of 306,720,000 years each (306.72 MY).

- Earth suffers from a Big Flood at the end of each *manvantar* and periodically renews the living creation on it.

- Since the onset of the Vedic creation, the last 6th Big Flood occurred about 145.59 MY ago and the next one will occur in 161.13 MY.

Time-keeping by the Aryans

If the Aryans have carried forward the four Vedas intact for thousands of years, with no change even in one syllable of a mantra, the Aryans were also capable of carrying forward their time-counting process. Based on their astronomical observations, discussed in Chapter 11, they have been preparing calendars for yearly, monthly and daily use with the accuracy of *kaashtha* (3.2 seconds) in time (Table 5.5) and within seconds of arc (1/3600th of a degree) in angular positioning of the celestial objects on the celestial sphere, observed all around us as the dome with fixed constellations on it. They have been aware of the fact that the *Kaal* (Time) brings changes in the society and cataclysmic events create drastic discontinuity in the smooth functioning of the society, affecting its learning process and the knowledge base. Along with extracting knowledge from the Vedas and bringing it to a level that is common to all in the society, the Aryans of Aryavart devised means and ways, to preserve the Vedas and the Vedic practices. Much can be written on the safeguards devised by the Aryans to protect the Vedas and the Vedic knowledge; just a few points related to time-keeping by the Aryans are noted here.

In the ancient Aryan tradition that still continues in India, an auspicious ceremony generally **begins by invoking a specific *shloka*,** called the '***sankalp-shloka***'. This *shloka* is to establish the geographic and time coordinates of the event. An example of a *sankalp-shloka* is in the following when this author on August 4, 2013 recited it to begin his lecture on physics in the Vedas as:

ओ३म् तत् सत् श्रीर् ब्रह्मणो द्वितीये प्रहराध्र्दे, वैवस्वते-मन्वन्तरे अष्टाविशंति-तमे कलियुगे, कलि-प्रथम-चरणे, एक वृन्द: - षण्णवति कोटिः, अष्ट लक्षः, त्रिपञ्चाशत् सहस्रे, चतुर्दश - उत्तर - शततमे (1, 96, 08, 53,114) सृष्टि सम्वतसरे , कलि सम्वतसरे पञ्च सहस्र चतुर्दश उत्तर शततमे (5,114), भूमण्डलेस्थिते उत्तरपाताल-उपखण्डे, कैनाडा देशे तस्मिन्-स्थिते टोरोन्टो नगरे तत्रस्थ स्कारबोरो-उपनगरे, उत्तरायणे पुनर्वसु नक्षत्रे, वर्षात्रृतौ आषाढ़ मासे कृष्ण (शुक्ल) पक्षे तृतीयायां तिथौ रविवासरे प्रातः(सायं) काले, शुभमुहूर्ते, अहम् शिवराजपालः, भौतिकविज्ञान-विषये सर्वविज्ञानार्थं स्वकीयान् विचारान् भवत् भद्र-स्त्री पुरुष समक्षे प्रकटयामि।

Transliteration:

"*Om tat sat shreer brahmno dviteeye prahraardhe, vaivasvate-manvantare ashtaavishanti-tame kaliyuge kali-pratham-charne, ek vrindah-shanvati kotih, asht lakshah,tripanchashat sahashre chaturdash-uttar - shatatame* (1, 96, 08, 53,114) *srishti samvatsare, kali samvatsare panch sahasra chaturdash uttar shatatame* (5,114), *bhoomandale-sthite uttar-pataal-*

upkhande, Canada deshe tasmin-sthite Toronto nagare tatrasth scarborough-upnagare, uttarayane punarvasu nakshatra, varsha ritau aashaadh maase krishna (shukla) pakshe tritiyayaam tithau ravivasare pratah (sayam) kale, shubhmuhurte, aham shivrajpalah, bhautic- vigjyan-vishaye sarvavigjyanaartham svakeeyaan vicharan bhavat bhadra-stree-purush samakshe prakatyaami."

Translation:

Om - O Eternal Brahmn with you in my heart, in the second quarter of your creation day, in the Vaivasvat Manvantar period of the creation history, in the 1st step of the twenty eighth Kali Yuge, in the Creation Calendar year of 1,96,08,53,114 (1.96 BY), *in the Kali Yuga calendar year of* 5,114, *in the Scarborough suburb of the city of Toronto, situated in the country of Canada in the North American subcontinent on the globe, when the Sun is in the northern hemisphere in the constellation of Punarvasu during the rainy season on the Sunday on the third day of the dark lunar phase of the month of Aashaadh, I, Shiv Raj Pal, recite this 'sankalp-shloka' to commence this auspicious occasion to express my thoughts on physics for your knowledge, respected ladies and gentlemen.*

Notice that in this *Sankalp Shloka* the date of the *sankalp* is expressed in the year of the Srishti Calendar (Natural creation calendar of the Aryans) and the day of the *sankalp* is expressed in terms of the position of the constellation within a *muhurta*, which is approximately on an hourly basis. The solar-*ayan* (hemispherical) position is assigned to the day. The day is also expressed in terms of the position of the place on the globe and is also related to the lunar phase (bright or dark) with the ongoing social calendar of the place, in the Vikramaditya Era calendar. Not only that, the name Vaivasvat of the *manvantar* period (7th in Table 5.4) is recited with the number of the currently ongoing 28th *Maha-yuga.* The details recited in the shloka are quite elaborate and scientific in nature. Therefore this system is remarkable. It is amazing that this practice of reciting the *sankalp-shloka* prevails with the priestly brahmans throughout India, even deep in the countryside. Swami Dayanand Sarasvati in his book, the 'Rigvedadi Bhasyabhumika' presented a sample of such a *Sankalp Shloka* for those readers desirous of learning it and preparing it for special occasions. There he also stated that- "*... the Aryans of India have been incrementing and recording the Srishti-Calendar in the same manner as the ledger books are kept up to date in the society*".

Measure of the Time-entity: Small Time Scale of the Aryans

Large numbers with the scientific notation and the nomenclatures in the Vedic system and the Western system were presented in Table 5.2. The smaller units below one are also expressed in the science notation, millisecond and micro-second being the examples. The scientific notation is used with negative exponent to the base 10 when the numbers are extremely small. Did Aryans have smaller time-scales to represent small time units? Yes they did. Based on the discussion presented below one would find it mind-boggling as to how did they arrived at some units with the accuracy that could not have been possible without an accurate and difficult experiment in physics; the unit of *nimesh* being an example!

In addition to the large units of numerals and time presented in Tables 5.2 and 5.3 the Aryans have also dealt with the minuteness of the *Kaal* (Time-entity) and developed time units below a day, hour, minute and second. Modern physicists have been able to measure time in the order of femto-second (10^{-15} sec) going down to atto-second (10^{-18} sec, 1 divided by a number with 18 zeros after 1). The Vedic units do not go as low as these science units, however, they exhibit that the Aryan approach for dividing the time in specific proportions was very scientific. Manusmriti's *shloka* from its Srishti (Creation Principle) chapter presents the nomenclature of five time-units and gives the conversion factor between adjacent units. The shloka begins with the smallest practical Vedic unit of time named *nimesh* and ends with *aho-raatram* (day+night).

निमेषा दश चाष्टौ च काष्ठा त्रिंशत्तु ताः कला ।
त्रिंशत्कला मुहूर्तः स्यादहोरात्रं तु तावतः ॥ (Manu, 1/64)

Transliteration: "*Nimesha dash chaashtau cha kaashtha trinshattu taah kala. Trinshatkala muhurtah syaadahoraatram tu taavatah*".

The shloka is translated directly into the first five rows of the Table 5.5.

Table 5.5: Vedic Unit of Time from nimesh to one day and to one year

	Unit of Time	Conversion	Common unit	Comment
1	1 *nimesh*		1 nimesh = 8/45 sec = 177.777 millisec	Derived from time taken by 1 blink of the human eye
2	1 *kaashtha*	18 *nimesh*	3 sec +200 millisec	
3	1 *kala*	30 *kaashtha*	1 min and 36 sec	
4	1 *muhurt*	30 *kala*	48 min	
5	1 *aho-ratra* (*divas*)	30 *muhurt*	24 hr (day+night)	Shatpath Br. (10/4/313-24) 10800 muhurt in 360 days
6	1 *paksh*	15 *aho-ratra*	½ month (15 days)	
7	1 *maas*	2 *paksh*	1 month	
8	1 *ayan*	6 *maas*	½ year (6 months)	Uttarayan –Sun in N-hemisphere. Dakshinayan – Sun in S hemisphere
9	1 *varsh* (*manav*)	2 *ayan*	1 year (human's)	1 deva-day (at the pole)

The nomenclature of time units with conversion factors between each successive unit is the same as in Manu's *shloka*. Since the units are sequentially recited by Manu with conversion factors, a conversion from one to any other would also be straightforward.

The first five items in this table being those in Manu's *shloka*, this author calls it as Manu's table. The details of the conversion factors in Manu's table show that the ancient *rishis* were quite advanced in physics and astronomy where such fine time divisions were needed; or why else did they develop such time conversion factors?

The columns 2 and 3 present the Vedic names of the time-units and their relationship to the next higher unit. The column 4 gives corresponding modern equivalent to the Vedic units. The common unit of 1 day in the 5th row of the table provides direct relationship between the Vedic unit of 1 *aho-ratra* (*divas*) = 30 *muhurt* = 24 hr = 1 day (day+night). Using this correlation the value of 1 *nimesh* is exactly 8/45 second (about 178 milliseconds).

It may be mentioned here that in addition to the time classification laid down by Manu, several other small units of time have also been used in Vedic time measurement and applied to practical astronomy. They are: *truti* (0.5 *nimesh;* about 0.09 sec), *praan*, *pul, vipul* etc. Based on the definitions of the time units in other *shlokas* in Manusmriti and those in the ancient text 'Surya Siddhant' (Sun Principle), this author has added the last four rows in the Table 5.5.

It is obvious from the table that the Aryans developed a very comprehensive list of time units covering time-scales smaller than a day. These units have been in use since ancient times for astronomical calculations to prepare astronomical tables. These units were also used to prepare calendars and to record special historical events in the society. An important feature of the Table 5.5 is the smallest time unit of *nimesh* that takes us to about $1/6^{th}$ of a second and more accurately to 177.777 milliseconds.

The Mind-Boggling Case of *Nimesh* (निमेष): the unit of time

Just as the present civilization divides a 'day' into four units– day, hour, minute, second with the factors of 1, 24, 60, 60 the ancient king Manu in his book of Manusmriti furnished 5 time-divisions from a day: day (*aho-raatra* = day+night), *muhurt, kala, kaashtha*, *nimesh* with the factors of 1, 30, 30, 30, 18. Observing the Vedic time-scale of Manu one finds that an unmatched number 18 is part of the scale for the last unit of *nimesh* and not 30 to make it uniform. Why not a uniform time division of 30 *nimesh* in a *kaashtha*?

The discussion in this section will show that the time unit of *nimesh* (निमेष) is a derivative of the applied physics and not a theoretical factor of 30. The *nimesh* is being labeled here as the 'mind-boggling case' because unlike hoards of other science thoughts and principles that one comes across in the Veda mantras, the word *nimesh* would have been formulated following Sanskrit grammar but the time unit of *nimesh* could not have become possible without going through arduous yet accurate experimentation in science involving time measurement and human physiology. Since the time unit of *nimesh* is an incredible ancient outcome of applied science it is analyzed in the following to show what all incredible possibilities existed in the distant past of the Vedic Civilization.

The word *nimesh* appears in the Vedas often for the blink-function of an eye (e.g. YajurVeda, 23/3) and considered to be a vital physiological function in a living being. As we know it now that the blinking of an eye is involuntary action generated by the physiological necessity to moisten the eyeball. For this important physiological function of the human body even a prayer for its health is offered in the mantras.

The Vedic term *nimesh* (निमेष) represents a verb for creating specific motion of the upper eyelid. The prefix *ni* (नि) in it represents 'down'. The conjoined word *ni-mesh* means – '*down-ward motion of eyelid*'. The term *mesh* (मेष) also includes the action of striking. So the alternate meaning of *nimesh* is – '*down-ward strike of eyelid*'. The word *nimesh* conveys with clarity the downward strike by the eyelid.

The term *nimesh* does not represent both actions of the eyelid – closing and opening. Because the closing and the opening of the eye are opposite actions, the Sanskrit being a perfect language has two separate terms for them. While the eyelid striking down is *nimesh* the reversal of the eyelid to open the eye is *unmesh* (उन्मेष) with the prefix *un* (उन) replacing the prefix *ni* (नि). The sound of the word *unmesh* (उन्मेष) is in fact closer to *oonmesh.* It is also worth noting that the prefix *un* before a verb generally reverses the meaning in languages other than Sanskrit. This usage may have gone from Sanskrit grammar to other languages for undoing an action; examples being – kind and unkind, known and unknown etc. In Vaisheshik Darshan (2/1/4), its author Rishi Kanaad lists 12 signs of the presence of the soul in a living body. In that reference Kanaad Rishi states that *nimesh* and *unmesh* are two separate actions in a living being. The *nimesh* is the time taken by the eyelid to strike the lower eyelid and not the blink rate of the eye.

The main action in this physiological necessity is the strike of the eyelid; its recovery being secondary. The action of the eyelid is quite similar to the strike of a hammer wherein the action starts from hammer's topmost position and ends when it makes contact with the object being struck. In this scenario, the time taken for the strike is of essence rather than the time taken to raise the hammer. The time of the strike by the eyelid is the time it takes for the one-way motion – the downward motion. With regard to *nimesh* becoming a part of the timescale of the Aryans, the ancient *rishis* must have first discovered that a *nimesh* of the human eye was smaller than a *kaashtha* (item- 2 in Table 5.5), which is 3.2 seconds long. Then they must have found out that there are 18 *nimesh* in a *kaashtha*. This makes a *nimesh* to be exactly 8/45th of a second. Considering that *nimesh* is an ancient unit of time, for all practical purposes it is a remarkably small time resolution that the ancient *rishis* practically achieved.

For a measurement of blink-time one needs a starting reference time and a time-device to click time exactly when the upper lid strikes the lower. How did Manu do this experiment? All these questions are in the realm of physics yet the answers to these questions are not easy. This is why the unit of *nimesh* is 'mind-boggling'.

For measuring a *nimesh*, even today one needs high-tech sophisticated instruments and timing devices. It is easy to count how many times a normal human eye blinks during one day or in one hour but measuring and assigning absolute time to the blink itself is another story. The *nimesh* is the time taken by the upper eyelid to strike the lower in one blink.

Compared to a *nimesh* being about 178 milliseconds, let us check out derived values of the blink-time of an eye based on modern methodology using advanced technology. There are differing methods that the modern scientists have employed to determine the blink-rate and the blink-time has been a byproduct of the experiment.

Blink-Rate and Blink-Time Measuring Technique and the Nimesh

Without reviewing different methodologies for determining the blink-time, cited here is one of the most recent techniques that employs photo-detection with digital cameras with known frame rate to detect movement of the eyelid. The computer interface of camera images yields a statistical measure of both, (a) the blink-rate of the human eye derived from the blink count over a period of time and the blink-time of human eyelid (see e.g. Grauman et al. 2001, 2003; Chau and Betke, 2005) by resolving the short time duration of the blink. In these and all other modern experiments, the blink-time is defined to be the time taken by the upper eyelid from the start of the upper eyelid, striking the lower eye-lid and receding back to its original position. This definition of blink-time, including both the actions of the eyelid, is different from what the *nimesh* (eyelid coming down to close the eye) is. Once the difference between the modern and Vedic definitions of the term 'blink' is established, let us now look at the relationship between the Vedic *nimesh* and the time measurement of the blink by modern technologies.

The laboratory measurements yielded a statistical range for the blink-time (*nimesh* + *unmesh*) for a normal human eye to be about 300 to 400 milliseconds; about twice as much as the Vedic *nimesh* of 178 milliseconds. For practical purposes the time elapsed either side of complete closing time of the eye is symmetric and is about the same. Therefore a *nimesh* with eye-closing time is about half of the total blink-time measured in these modern and accurate experiments. Therefore the measured statistical range for *nimesh* is 150 to 200 milliseconds, giving an average of 175 milliseconds. Within statistical errors of modern measurements Manu's *nimesh* of 178 milliseconds is exactly as it is supposed to be. **This is not an ordinary result.** Whichever way one looks at it, the numbers that the modern science has generated and the number that Manu has provided are remarkably on the dot. Obviously, Manu was a good physicist. One can draw an easy conclusion here that the case of *nimesh* is intriguing and that the ancient *rishis* knew more than we think!

Commentary on Vedic Time Scales

A close look at the Vedic time units in the practical timescale (Table 5.5) makes it obvious that the units of day (day-night), *muhurt, kala* and *kaashtha* are not ad hoc; they are products of experiments and invented for practical use. The Table 5.2 based on the Vedas indicates that the Aryans carried out their counting in the decimal system going from zero to extremely high levels of billion-billion. The division of the unit of 'day' into *muhurt, kala* and *kaashtha* with a factor of 30 is meaningful and indubitably well-advanced. King Manu and his Manusmriti are considered by the Hindus to be among the most ancient sources of knowledge (details in Chapter 3). Therefore the time-division is extremely ancient anterior to the historic milestone on February 18, 3102 BCE. Even the age assignment for the creation being 4.32 BY is an unexpected figure.

Why did the ancient Aryans finalize a time scale from a day down to *muhurt, kala, kashtha* in multiples of 30? There may be other explanations for that particular division; this author suggests the following possibility.

It was expected that the Aryans would extract knowledge regarding time and space from the Vedas and further develop it. To have it accepted and finalized throughout Aryavart must have necessitated a discussion among the scholarly people as to how to ultimately put such a knowledge base in practice for the greater good of the society. To solve the problem of how to divide time and make it acceptable to all, they must have looked into the Vedas which were before all writings, including the Manusmriti. There are numerous mantras in the Vedas that would have guided the ancient Aryans of India in developing the foundation of their timescale on. There are several relevant mantras in the 164th *sukta* of the Chapter 1 of RigVeda; however, just two mantras are presented here, RigVeda (1/164/11, 48). The author suggests the reader to also explore AtharvaVeda's section (10/8/4). The RigVeda mantras have been extensively interpreted by the ancient Rishi Yask in his Nirukt Shastra (4/27). The translations of the two RigVeda mantras reflecting Rishi Yask's interpretive commentary are as following:

> *"Know from this creation the un-weakening wheel of samvatsar* (year) *circling (chakram-pari) the Sun and having twelve sectors where ritam* (the creation principle) *in communion with the samvatsar, has produced seven hundred and twenty offsprings"* (RigVeda, 1/164/11).

> *"Know it by discussion and logic that it is the single-most wheel of samvatsar* (year) *with three foci that has twelve pradhi* (circum-sectors in togetherness) *and three hundred and sixty saakam* (spokes) *firmly implanted. They 'move' with time as one system and yet 'move not'"* (RigVeda, 1/164/48).

Once we know that the Aryans' practical *samvatsar* (year) was of 360 days, the meaning of the two mantras becomes obvious. According to Rishi Yask's interpretation, the 720 offsprings in the first mantra are 360 pairs of day and night and the 12 sectors of the wheel are 12 months of the year. This is made clearer by the second mantra in that the *samvatsar* is made up of 360 days (day+night) with 12 months out of a *pradhi*. The word *pradhi* stands for circumference of a circle. When used in plural the term *pradhi* represents sectors of the circumference. In the case of a wheel when *pradhi* is used in the plural sense, as in the mantra, the 12 *pradhi(s)* are 12 sectors of the rim forming the whole wheel. While providing etymological interpretation of these mantras Rishi Yask in a typical style of posing a question to himself or to his disciples, asks - '*why pradhi*?', then he provides the answer, '*pradhi, because they move together as one system*'. A wheel is not strong if all *pradhis* are not joined by the spokes to the central hub of the wheel. Therefore, the mantra states that it is the 360 *saakam* (spokes) that hold the wheel together. The emphasis in the mantra is that these 360 spokes are 'firmly planted' in the wheel. Rishi Yask refers to Brahmana books, in his Nirukt and states that these 360 spokes are symbolic of 360 solar days in the *samvatsar*, which add up to one year. These 12 *pradhis* and 360 *saakams* in the *un-weakening wheel*, functioning under the *ritam* (Creator's principle inherent in Nature) thus form invariants of the *samvatsar* (year).

The three foci in the mantra are interpreted by Rishi Yask and the Brahmana books as the three main seasons: summer, rainy and fall. In the second mantra the meaning of '*move*' and '*move not*' of the system of the *saakams* and *pradhis* (spokes and sectors) is now obvious. They '*move*' because they are in perennial motion (cycle) and they '*move not*' because they do not become loose. Implicit here is that the process of the yearly cycle is like a rotating wheel without the need for service or repair unlike an ordinary physical wheel fraught with decay.

The Veda mantras and Brahmana books of the Aryans led them to establish a working time division for the year. The 360-day-year on a circle (the *chakram* in the mantra) corresponds to one degree of arc for one day. The division of a circle into 360 degrees can be found in Vedic literature far back in ancient literature including Vedvyas's book on Mahabharata 3140 BCE, in which he stated himself that from the bank of the River Sarasvati, he was making observations 'day in and day out' of the planetary positions. This was possible only when fine angular resolution of astronomical objects was derivable. Some details on this subject were presented in Chapter 3 under the sub-heading of "Sarasvati and Astronomy in Mahabharata". Observations of such astronomical events were not possible without the knowledge, practice and instrumentation in geometry and trigonometry, including the division of angles in an arc of a circle. Since no books are found to be older than the Vedas, a high possibility is that the concept of 360 degrees in a circle originated from the Vedas and travelled to the other Western regions of the globe.

The *schloka* couplet (1/28) in the ancient book on astronomy, the Surya Siddhant (The Sun Principle) states explicitly the angular division of a circular orbit of a planet as:

> "*A complete circle* (orbit of a planet) *is* 1 *bhagan;* 12 *raashi* in 1 *bhagan*; 30 *ansh* in 1 *raashi*; 60 *kala* in 1 *ansh*; 60 *vikala in 1 kala"*. (Surya Siddhant, 1/28).

Translation of the same in terms of modern terminology is:

> *A complete circle is 1 orbit, 1 orbit is divided in 12 sectors of* 30° *each, 60 minutes* (of an arc) *in* 1° *and 60 seconds* (of an arc) *in 1 minute* (of an arc).

The 12 *pradhis* and 360 *saakams* of the RigVeda mantra are thus the same as Surya Siddhant's 12 *raashi* in 1 *bhagan* with each *raashi* being 30° which is same as 360° in 1 orbit (*bhagan*). The term *bhagan* in common language is "the run" which is in context of a planet. Therefore, 1 *bhagan* means 'one run' of a planet which is 1 orbit of the planet. The Veda and the Surya Siddhant make it abundantly clear that one circular path of a planet is divided into 360 small sectors of 1° each and 12 sectors of 30° each with their Vedic nomenclatures.

Therefore there is clarity in the choice of 30 *muhurta* in a day (not 24 hours). It begins with the above cited Veda mantras wherein 12 *pradhis* (sectors) of a circle correspond to a sector of 30° each, which correspond to 30 days in a month of 360 day-year. The factor of 30 day in a year is extended to 30 *muhurta* in a day and so on and other divisions are also carried out with a factor of 30. Using a judicious mixture of time-units and angular units, the Aryans have been predicting angular positions of astronomical objects and even now printing in yearly *panchaangs* (five-limbed almanac).

The astonishing case of *nimesh* discussed above lends further support to this author's view that the ancient Aryans had a strong experimental base and devised ways and means to derive precise time and angular units. Their experiments certainly covered a wide range of subjects from the cosmos to the phenomena very near. The 18 *nimesh* in one *kaashtha* based on blink-time of the human eyelid is an evincing example of experimental superiority that existed in the far antiquity of Vedic civilization. The details of the methodology for measuring *nimesh* remain a mystery.

(5) The Space-entity (*Aakaash-tatva*)

The Space-entity in Vedic cosmology is as important as it is in modern cosmology. The difference however is that one gets to learn more about the properties of the Space-entity in Vedic cosmology. A good amount of discussion on the Space-entity is preempted in Chapter 1 and 2. The idea of Space being an entity by itself, separate from the Nature-entity (matter-energy) and the Time-entity is relatively young in modern science. Well before the fog cleared from the idea that the Space was an entity when the space-time continuum and the Theory of Relativity by Einstein was introduction in 1916 CE, the great Vedic philosopher and thinker of the 19th century, Swami Dayanand Sarasvati quoted in his book of Satyarth Prakash (1881 CE) a statement on Space-entity that has been preserved in the far antiquity of the Vedic Literature. The statement is from the ancient Taittireeya-Upanishad and his commentary on it expressed with clarity that the Space (*Aakaash)* in the Vedas is a real entity and is one of the most important participants in the creation of the universe. The English translation of the relevant first part of the Taittireeya Upanishad statement and Swami Dayanand's commentary on it are presented in the following:

"…..तस्माद्वा एतस्मादात्मन आकाशः सम्भूतः "।

("*tasmadva etasmaadaatman aakaashah sambhutah*").

The commentary on this Taittireeya Upanishad's statement by Swami Dayanand is in Hindi language. This author has translated his commentary from Hindi to English as in the following:

> "*The aakaash-avkaash is a real causal substance (entity) spread out ubiquitously and its collection as a whole makes it Aakaash (Space-entity) the existence of which is revealed from the Creator and the creation (Nature).*"

He elaborates further on the Taittireeya statement asking a rhetoric question,

> *"...the reality is that the Aakaash* (Space-entity) *is not a created entity, because without a pre-existence of it where can Nature and its atoms stay..."?*

This statement shows that the concept of Space being an entity is extremely ancient. Taittireeya Upanishad was written in the far antiquity from the time Swami Dayanand wrote the book Satyarth Prakash and the concept of *Aakaash* (Space-entity) came into Taittireeya Upanishad from the Vedas before that. All Upanishads have derived their principles from the Vedas. The fact that the Space is an entity, pre-existing this creation, has been consistently stated and discussed in Brahmanas, Smritis, Shastras and the rest of the Vedic Literature and this is well before the development of thought on this subject in the Western philosophy and science. The earlier Western philosophers and the scientists continued to think that Space was an empty place called the 'sky' where all things and cosmos stayed and functioned and that it had no property of its own and therefore had no influence on the things that existed in it. The same thinking prevailed for the Time-entity. It is since a few centuries ago that the thinking about Time and Space started to change and the concepts are ultimately brought out in the open by the space-time continuum of General Relativity put forward in mathematical formulations of Einstein. The fact is that the concepts of Time and Space are not the recent adjunct to the developing physics of modern times; they are an endowment of the past. A large number of synonyms of *Aaakaash*, the Space-entity that this author has collected out of many more in the Vedas are also an endowment of the Vedas.

The ancient book of commentary on YajurVeda is the Shatpath Brahmana. A statement in this Brahmana on *Antariksh* (Space-entity), a synonym for *Aakaash*, states clearly what exactly the *Antariksh* is. The transliteration of the Shatpath Brahmana statement reads thus –

> "*...yatha heveyam pratishthavirasminloke prathivyvamu hevesha pratishthavirmushmin lokaha idamantariksham. Sa yadiha sannmum lokam na pashati tasmadahuha paroakshmsau loka iti.*" (Shatpath Brahmana, 11/1/6/22),

Its translation is:

> "*...just as the Earth is the support of all substances in this region of the universe, likewise the Antariksh is in that region of the universe. People sitting here are not able to see that region of space and this is the reason why they call that region as paroksh.* (Shatpath Brahmana, 11/1/6/22).

The reference here was for the deep-space beyond observation. Its simile with the Earth is to explain that just as the Earth is the support of all terrestrial objects that one can visualize and experience, the *Antariksh* is the support of all objects even though they are faraway in deep space 'beyond observation' (*paroksh*). Therefore, the concept of Space being an entity and its function being the support for cosmos is right out of the Vedas, the oldest books in human history. The Space has been understood by the ancient Aryans of India for what it is.

The concept of the Space being an entity is difficult to grasp because one cannot see it, smell, taste, touch it or cut a piece of it to study later. One is not able to put a measuring tape directly on it or experiment on it like we are able to on a piece of matter from the Nature-entity. The best one can do with the Space-entity is to measure the distance between two points and ascertain that the two points are separated in space by a given measure. But the Space is not confined in a plane, it spreads out in all three dimensions. The space can also be measured in three dimensions that is trapped by the volume of the object. But the object is not able to core-out its own trapped space from the rest of the Space outside it. Space by itself is not available to help prove its own existence. Sensing of Space-entity, as well as of Time-entity, by our senses is not possible because they do not interact with our senses. Now even the modern science says that these entities have real existence. This is what the Vedas have said since the far antiquity.

As mentioned before, the Space-entity like the other important entities of the creation has large number of synonyms in the Vedas, each reflecting Space-entity's specific function, its coverage of a specific region of the universe and its properties. Again, this exhibits the richness of the Sanskrit language. Just as a person in a small corner of a room of a big house is also said to be in that house, the term *antariksh* in some Vedic references is used for the specific space in which the solar system is housed. Just as the term 'earth' is used for the matter (soil) that forms the planet Earth, the space in a smaller region is addressed as *antariksh*. This author has come across in the Vedic Literature a large number of synonyms of *Aakaash* (Space-entity), including those listed in the Nighantu part of Yask Rishi's book of Nirukt Shastra (2/3/1). The 22 of these are listed in the Table 5.6. The meanings of a few most commonly used words for *Aakaash* are included in this table.

Nirukt, the ancient Vedic treasure of philology and etymology together, takes up words that occur in Veda mantras and provides interpretation of their meaning, giving the root of the word formation that leads to the right translation of a mantra. Of the 22 names of *Aakaash*-entity that this author has presented in Table 5.6, Nirukt has provided the interpretation of the 5th, 13th and the 15th terms, the *Antariksham* (or *Antariksh*), Naakah and *Pushkaram* - all synonyms for the same Space-entity. The Nirukt states:

> "*Antariksham is formed of antara+kshaantam* (inter-space + spreadout) *and because it exists between the Earth and the Dyu, or because it is antar+akshaya* (interspersed+non-decaying) *and exists within living body without suffering from change like the change and decay in Earth and matter"*. (Nirukt, 2/3/1; translation from Sanskrit to English).

The term Dyu in the statement stands for the brilliant region of space where the Sun's abode is.

Thus without a doubt, the Nirukt defines the Space to be an all-pervading physical real entity. The commentators on Nirukt have described *Aakaash* or *Antariksh* to be *mahaanavayava* (*mahaan+unavayava*). This conjoined word of two parts stands for an entity that is *mahaan* (great) and *unavayava* (without limbs or branches, or parts or shape or form). All substances in the creation are avyavi – having parts or limbs in them. A piece of iron in the shape of a sphere is observed to have no limbs in it but in Vedic philosophy it is an *avayavi* with limbs because it is made up of individual iron atoms, which are its limbs.

The *Aakaash* (Space-entity) addressed by the adjective of *mahaanavayava*, "great without limbs" puts it altogether in a different category not recognized yet in the contemporary Big Bang Hypothesis, that it is a one single entity, conterminous without a structure formed by quanta. It is a continuum in 3-D and is a great entity.

The important deduction here is that the Space-entity is not a quantized entity like the Nature-entity. This excludes a modern thought by some cosmologists who think that the Space has its own corpuscles.

The word *antariksham* for *Aakaash* (Space-entity) has come in Veda mantras a large number of times (e.g. RigVeda 1/110/6, 10/90/14, ...). The RigVeda mantra (10/90/14) states that when the creation commences the *Antariksham* arises from the *naabhi* (naval) of the Creator. The naval here represents the centre of activity. This gives Space-entity yet another synonym of *Nabha* in item (12) in the Table 5.6.

Table 5.6: The 22 Synonyms of the Vedic Space-entity

	Space-entity	Comment
1	Aakaasham	creates place for objects
2	Aapah	pervading in universe like *aapah*
3	Adhvaram	Nighantu, 19, in Nirukt (2/1/3/1). One that accommodates without getting injured (without getting altered)
4	Ambaram	place for aapah the material cause for creation
5	Antariksha m	space between and within cosmos. Ref: RigVeda 1/110/6, 10/90/14, 10/121/5, AtharvaVeda, 10/7/(several mantras).
6	Arnava	The sea of space, many mantras
7	Barhih	Whose form is minute like energy
8	Bhuhuh	that with characteristics of supporting the substances
9	Bireet	Place of fear for it is an entity unsupported by anything-else or the place for *vayansi* the objects that fly - animate and inanimate (Nirukt, 5/4/9)
10	Dhanv	vast and spread out like a desert
11	Kuru	A place of action, vast farm for cultivation and growth, YajurVeda (23/56)
12	Nabha	rises from the naabhi (naval, centre) of Purush (Rig. 10/90/14)
13	Naakah	A container or a holder, AtharvaVeda (10/7/43)
14	Prithvi	property to hold substances like Earth
15	Pushkaram	Pushkaram antariksham- Care provider for all in existence in it, Rig.6/16/13; Nirukt, 5/3/2/2,

		Nighantu 48.
16	**Sagarah**	**deep and extended like a sea**
17	**Samudrah**	**the sea of cosmos**
18	**Shenah**	**all-pervading, Rig.(10/144/4-5)**
19	**Svayambhu**	**being in existence by itself, no cause creates it**
20	**Vimanah**	**the place for objects to move and fly. Rig 10/123/1+many more**
21	**Viyat**	**which is seen spread in all directions containing different forms of objects (Nirukt, 2/1/3/1)**
22	**Vyom**	**Ubiquitous encompassing all matter of the universe by providing place for it, RigVeda (10/129/1)**

The other meaning of *Nabha* is '*one that expands*'. The most common Sanskrit term for the Space-entity is *Aakaash* and it is *antariksh* (YajurVeda, 31/13) in applied science thought where the word *antariksh* means "*the space between the cosmological bodies*".

Space the Flower of Creation

The job of science is to present dry facts about a phenomenon without worrying about its beauty which is in the eye of the beholder'! Physics/cosmology has ventured upon hypothesizing the nature of the Space-entity whereas the facts about its true nature still remain to be ascertained. As mentioned earlier, a cosmologist uses Space with Time to explain the nature of the physical phenomena of the universe but without a clue of Space's basic nature – is it a continuum or made up of tiny corpuscles? A modern cosmologist would say that Space took birth in the space-time singularity of Big Bang and expanded with the universe later in time, not knowing with certainty if it pre-existed holding the universe within its shrunken status. The physical nature of Space remains a mystery even for the advanced modern cosmology.

Because Space-entity is invisible, the beauty that is seen all around us in nature is not perceived as the beauty of the Space-entity (*Antariksh-tatva*); rather, the Space-entity is seen as the enabler of the beauty of the Nature-entity (*Prakriti-tatva*). The Vedic cosmology presents a root-derived term for *Antariksh* (Space) that delineates its properties and also reveals the unseen beauty of the Space-entity itself. This term for Space is '*pushkaram*' listed as item (15) in the Table 5.6. The term *pushkaram* for Space by itself is a meaningful science term that reveals the physical properties of the Space-entity but '*pushkaram*' is also a unique metaphor to manifest the beauty of the Space-entity. The term '*pushkaram*' is analyzed in the following and the reader would appreciate the fact that the Vedic cosmology presents the characteristics of the Space-entity with dexterity that have not been imagined as yet in modern times.

The Nirukt (5/3/2/2) provides the etymology for the term *pushkaram* where it states that *pushkaram* is *antariksham* the Space-entity. The term *pushkaram* is derived from a single *dhatu* (root) and is a multi-meaning term. The root for the word formation that leads to the term *pushkaram* yields several related meanings and functions associated with it. These meanings of *pushkaram* or *pushkar* are presented below in the characteristic style of Rishi Yask's Nirukt:

- *Space is pushkar because it does poshan of all substances of the universe, animate and inanimate, by benignly providing space for their existence.*

The term *poshan* stands for '*provides care*'. The virtue of benignancy in the Space-entity is emphasized by the book of Nirukt because a care-provider in the true sense is a beneficent giver without expecting anything in return. The Space-entity in Nirukt's definition being a cradle for all substances includes matter and energy. Therefore, Nirukt's definition of Space does not permit Space to be formed out of matter and energy. The Vedic Space-entity therefore is not made of matter or of matter in energy form. In the logic of the word-formation in Vedic system if the Space consumed matter of the universe for its own existence, the Space could not be termed as a benignant provider. This is why the *Antariksh* (*Pushkaram*), the Space-entity of the Vedas, is said to be an entity which is not an 'effect' of a 'cause'. The Space-entity of the Vedas is an eternal entity in the fellowship of the Time-entity and the other three eternal entities that were discussed earlier. The Nirukt continues listing other virtues of *Pushkar* (Space-entity).

- *Space is pushkar because it is pujakaram; it adorns in it all devas (fundamental properties and functions of Nature) for invocation* (not living-beings).

It was mentioned in Chapter 2 and discussed in Chapter 8 in Volume 2 that the Vedic *devas* are 'first-born' fundamental properties/systems/energy-forms that enable the creation and sustain it through the life of the universe. *Devas* are the Nature and the Nature is the *devas* that are held safely within *pushkar* the Space.

- *Space is pushkar because it rises above matter like a lotus flower rising out of mud* (not made of matter), *yet remaining beautiful and unblemished.*

A lotus flower is known for its beauty and its detachment from mud in spite of its association with it. In a Vedic spiritual discourse a lotus flower is usually cited as an example for one that exists in its efflorescent beauty without having attachment to its material surroundings: the mud and water. The Space is *pushkar* because it holds matter without getting attached to it. Therefore, the Space-entity's adjective is *pushkar*!

- *Space is pushkar because it is vapushkar; its character and form are not affected even when utilized.*

This property of Space primarily differentiates it from matter. To understand this, let us consider the utility of matter and the utility of Space. Take for instance two solid spheres of the same diameter, one made of iron with the density of 7.9 grams per centimeter cube and the other made of aluminum with the density of 2.7 grams per centimeter cube. The physical nature (composition) of the Space-entity according to the Vedas being minute, it pervades uniformly through both spheres of differing densities. The Space remains unaltered when inside the matter or outside of it or inside a dense material or inside a light material. This property of Space is *vapushkar*.

The *vapushkar* property of Space stipulates that reshaping of substances does not reshape the space occupied by the material objects embedded in it. The Space occupied by an object at a given place is not available to another object next to it but the Space is not consumed by the object and it becomes available as soon as the object moves to occupy another Space.

Let us now move from mechanics to pure physics where a modern cosmologist deals with Space. A cosmologist is often found saying or writing that 'in space-time continuum the time affects space and vice-versa'. In view of the Vedic concept of Space that Rishi Yask elaborated upon in his ancient Nirukt Sutra, the cosmologist should be given credit for being half correct. This is because there is no deterministic physics mechanism in which time changes into space or space into time, or the space ties to time or time raps around space. May be the mathematics plays the role in creating the space-time continuum but it cannot be proven that physics or chemistry does. What the modern cosmologist in fact is talking about is what actually is happening in his mathematical equation that he has set up, wherein the time and space are conjoined as two mathematical operators. The fact is that the Time and Space being two different eternal entities in the universe are not even capable of being conjoined with each other or with something else. A cosmologist in his mathematical construct (a gift from Einstein), imagining that one affects the other is unwittingly violating the *vapushkar* nature of the Space-entity.

The theoretical and observed curving of light rays passing close to massive bodies that Einstein explained on the basis of the Space developing excessive curvature, may not even be due to the curvature in Space. It could just be due to the gravitational effect on the invisible background *aappah* field in the vicinity of the massive body as delineated in Figure 9.11 in Chapter 9 under the sub-heading "Curved Space near Massive Bodies has a Different Meaning". The space becoming heavily curved in one region means becoming concentrated in that region and remaining rarified in rest of the void of the universe may cause a problem – it would create uneven islands of space that would move with massive bodies leading towards chaos. The *vapushkar* property dictates that the matter is embedded in Space and the Space does not alter it.

To understand the in-depth meaning of the *vapushkar* attribute of the Space-entity, let us conduct an imaginary experiment.

Consider a hollow steel sphere with wall thickness of say 2 millimeters with the same air inside it and outside. Now create a perfect vacuum in the sphere by sucking the last air molecule out and ask- what is left inside the sphere? The expected answer is – 'nothing'. This answer according to the Vedic thinking is invalid because it violates the Vedic definition of *Aakaash*. In the Vedic thinking, the sphere is devoid of air but there is still the presence of a real entity in the void of the sphere, the *Aakaash*, the Space-entity of modern cosmology. This entity was still conterminous in the entire volume of the sphere. The Space at any point inside the sphere was exactly the same as inside the 2 mm thick steel wall and outside the sphere. The wall with any thickness made even with the heaviest metal does not create discontinuity in Space. This is what was said by the Vedic *rishi* in a single Sanskrit term – *vapushkar*.

The nature of Space is unaffected inside and outside of the sphere when it is stationary or in motion. For example when a pitcher throws a baseball the Space that the ball occupies near the pitcher does not travel with the ball or the ball does not create and tunnel in Space as it travels between the pitcher and the batter, it leaves the space behind that it occupied before and occupies new space at each point on its path. In other words the space was always inside the ball but was not being captured by the ball or consumed by it. This is the meaning of the Vedic term *vapushkar*, which basically tells us that '*the Aakaash keeps its form even when in use by a substance of any form*'. Modern cosmologists use the Space in their field equations and deal with it mechanically but they have possibly not contemplated on its properties that have been pointed out by the Vedas. The *vapushkar* is inherent to *Aakaash* and *Aakaash* is an eternal entity not requiring evolution. The ancient term, *vapushkar*, representing an important attribute of the Space-entity, is unique to the Vedic physics.

In the Vedic concept of creation, the preexistence of the two entities, Space and Time, is a necessity. They are to be available at the onset of the creation when the first fundamental particle of matter takes birth in the creation or as Swami Dayanand Sarasvati had put it rhetorically – "*...without its* (Space-entity) *pre-existence where could the Prakriti* (Nature) *with its paramanus* (atoms) *stay....*". The Space has to be subtler than the smallest fundamental particle of matter because only then it can permeate inside each matter particle to hold it. Additionally, the Space has to be available beyond the edge of the universe to hold the entire matter and energy of the universe to serve as an overall container of the universe. It also has to have the property of stretching without snapping to accommodate the expansion of the universe. If it did not have these attributes, the matter would interact with it, capture it, enslave it, consume it, distort it or displace it ensuing chaos. This is how the Space is viewed in the Vedic philosophy. Since the Space-entity and the Time-entity are the eternal entities, they are non-provable; however, the modern cosmology is obliged to prove them because the Big Bang Hypothesis created them only 13.8 BY ago.

Some cosmologists have asserted that the time in Einstein's space-time continuum is affected by the space and vice versa. This assertion is not of great value if it happens in a field equation but not in the reality of Nature. If the space and time are variables in an equation a mathematical change in one variable would change the other variables.

An important outcome of general relativity is that the space is curved in the vicinity of heavy celestial bodies; heavier the body greater the curvature in Space. This outcome enabled improvement over Newtonian classical mechanics and explained better the motion of celestial objects. The observed bending of light rays towards the massive gravitating bodies in space is interpreted as the proof of this phenomenon. As mentioned above and discussed in detail in Chapter 9, this can be explained on the basis of the ubiquitous *aapah*-field in the so-called vacuum of space. In Vedic thought the background *aapah*-field also serves as the viscosity of Space serving as the velocity moderator for all objects including the photons in light beams. It is this viscosity that sets the speed of light at $3.0x10^8$ meters per second. Under this reference frame the fundamental matter particles smaller than the photons of light would have speeds higher than the speed of light.

The attribute of *vapushkar* in the Space-entity could be considered as a limited case of the relativistic mechanics; however, in the Vedic concept the Space is not consumed or altered by Time or matter or by gravitation. This is not to say that the entire space of the universe is not curved. Even if the *vapushkar* attribute of Space is seen valid in a limited regime, the recognition of the property of *vapushkar* in the Space-entity by the Vedic *rishis* indicates tremendous insight into an entity that continues to be as abstract in the minds of modern cosmologists as it was in the Vedic science and philosophy millennia ago.

- *Space is pushkar because it has the property of pushpate, one that swells.*

The word *pushpate* (*pushp*) is derived from the root that yields the '*action of swelling*'. The corresponding commonly used noun for *pushp* is flower because it is the result of swelling. This brings us to the relationship between the Space-entity, the *pushkar* and *pushp,* the flower.

The Vedic mantras credit the Creator for spreading the Space and the universe and the embedded celestial objects with it (e.g. RigVeda 10/56/5). This mantra is fully elaborated upon in Chapter 9 under the sub-heading of "**Expanding Universe - a Vedic Discovery Eons Ago**" in Volume 2. Even if the Creator is taken out of the equation, the residuum of this mantra still emphasizes that the universe has expanded with celestial objects going apart from each other, as in the part mantra - "*...Thus the diverse population of cosmos of the universe spread out in togetherness bound by Nature's rule"*. The Vedic principle that the Space-entity has the property of *pushpate* (one that swells) thus, in any shape or form, is no different from what now is the part of the observational astronomy and therefore a part of the Big Bang Hypothesis; that the universe is expanding.

The simile of the flower and Space was briefly discussed in Chapter 1 exhibiting the *pushpate* property of the flower and the Space.

Pushkar the Space-entity of the Vedas starts as a *tuchchhayen* (minute/diminutively small) as in the RigVeda mantra (RigVeda, 10/129/3), to be discussed in Vedic Creation Principle (Chapter 6), when the Nature in its dissolved phase is in non-descript matter form. Here a bud of course is only a metaphor because the initial size of the universe in the Vedic thought is diminutively small relative to the size of the universe but not small like a bud. The *pushkar* property is assigned to the Space entity and not to the matter of the universe. The *Prakriti* (Nature-entity) and *Aakaash* (Space-entity) for the Vedas are two distinct eternal entities. Based on Vedic references it will also be discussed that in the Vedic creation of the universe, the Nature starts as an ovate of matter and flattens out like a flower creating the cosmos. The Space assuming the shape of the Nature continues to swell along with the matter of the universe. The Space, otherwise a dry subject of physics, is expressed in the Veda in terms of the indescribable beauty and action of a flower; a unique definition of an invisible entity!

The Vedic delineation of expansion of Space through the metaphor of a flower (*pushp*) is remarkably simple, unique in modern perspective and close to the reality of the expanding universe. There are several fundamental dissimilarities in the two principles. The universe in a Big Bang of course starts from a dimensionless space-time singularity and in the case of a flower it starts from an invisible to the eye first specific plant cell that multiplies creating differing molecules to create later observed diversity within the flower. A flower certainly is not 'out of nothing'. Its cells multiply or replicate owing to the *eekshana* desire in the living plant. The Vedic wisdom here is admirable that brings a complex subject for common person to grasp. A flower represents beauty. The flower metaphor for Space creates a thought - if one had the ability to see the Space-entity and its expansion, it could even be a beautiful thing to observe. In the following are presented two successive AtharvaVeda mantras that illustrate yet another important character of *Antariksh,* the Space-entity.

Antariksh - the Vedic Space is Curved

Antariksh (Space) of a substance in Vedic expression is that substance's container, such that it holds the substance by pervading through the substance itself. In this perspective the *Antariksh* of that substance is in the same shape as the substance. The Earth's personalized *Antariksh* is spherical because the Earth is spherical and the solar system's *Antariksh* is in the shape of the solar system shaped like a discus, with Earth's *Antariksh* within it and part of it. Likewise, the *Antariksh* of the universe by reduction is in the shape of the universe itself which in the Vedic concept is a rotating wheel-shaped system of galaxies with a central axis of rotation (discussed in Chapter 9). A nook of the universe may be in so-called vacuum devoid of matter, but it is not devoid of two things, the *Antariksh* (Space-entity) and the ubiquitous undetectable *aapah* field, latter serving as the viscosity of space to regulate speed of substances including photons. In what shape the universe is and therefore in what shape the Space-entity holding it is, will be discussed further in the Volume 2 of this book.

Asking a question and answering it subsequently or later in another *sukta* (hymn) is a style in the Vedas and in the post-Veda literature. Following are two mantras, one with a question and the next with the answer. The question is primarily in the realm of cosmology and the answer is within the Vedic cosmological frame work. In the first mantra the question relates to three objects: *bhumi* the Earth, *dyau* (synonymous with *dyu*) defined as the brilliant region of space where the Sun resides and *Antariksh* the Space.

The question:

केनेयं भूमिर्विहिता केन द्यौरुत्तरा हिता ।
केनेदमूर्ध्वं तिर्यक् चान्तरिक्षं व्यचो हितम् ॥ (AtharvaVeda, 10/2/24)

Transliteration: "*Keneyam bhumir-vihita ken dyaur-uttara hita.
Kenedam-urdhvam,* ***teeryak*** *cha-* ***antariksham vyacho*** *hitam*".

Mantra Translation:

"*Who has created and prepared well the Bhumi* (Earth), *who has placed the Dyau* (illuminated region, the abode of the Sun) *high up and who has established the Antariksh* ***(Space-entity)*** *aloft which is vast and* ***spread-out curved like the path of an arrow****?"* (AtharvaVeda, 10/2/24)

The answer:

ब्रह्मणा भूमिर्विहिता ब्रह्म द्यौरुत्तरा हिता ।
ब्रह्मेदमूर्ध्वं तिर्यक् चान्तरिक्षं व्यचो हितम् ॥ (AtharvaVeda, 10/2/25)

Transliteration:

"*Brahmna bhumir-vihita, Brahmn dyaur-uttara hita.*
Brahmedmurdhvam, ***teeryak*** *cha-****antariksham vyacho*** *hitam*".

Mantra Translation:

"Brahmn the Creator has created and prepared well the Earth, Dyau the illumine region has been placed high up and Brahmn has established the ***spread-out curved-space****."* (AtharvaVeda, 10/2/25)

The Veda could have stated the above thought purely for spiritual gain as it does throughout the Vedas by stating that Brahmn the Almighty is to be worshipped because He has created the Earth, the abode of the Sun and the illuminated space. But by talking about a demarcation between the Earth and *Dyu* and by defining the character of Space to be curved like the path of an arrow, the Veda here is clearly imparting knowledge than advocating spirituality. The basic intent in these mantras is to impart knowledge by drawing distinction between the three major entities, the well-crafted Earth, the brilliant cosmos and the *Antariksh* (Space-entity) with the specific feature of being vast, *spread with curvature* in it.

It is important to note the two properties associated with the Space-entity cited in the mantras: it is *curved like the path of an arrow* and it is *spread-out* well beyond the Earth and the solar system. A third property of the Space-entity stated by another RigVeda mantra (10/129/3), to be discussed in Chapter 6 under the Vedic Creation Principle, is the initial confinement of the Space-entity in a small locale during the Nature-entity's *pralaya*-phase (dissolution). The adjective used by the RigVeda for that initial state is *tuchchhayen–aabhu*, means the diminutive Nature. The action of Nature spreading out in all directions at the onset of creation is also stated there in the successive mantras. This would make the Vedic-defined Space a flexible entity.

The physics contained in the तिर्यक् *(teeryak)* property of Space, emboldened in the transliterations of both mantras need some elaboration. In the compound word *teeryak* = *teerya*+*k*, the first word means 'an arrow' which is a small physical object. The *k* suffix to *teerya* transforms the word into a pure adjective yielding – *like the path of an arrow*. Therefore the mantras are not talking about an arrow itself; rather they are talking about the property of an arrow, that its path is curved. The mantras' contention therefore is that the *Antariksham* (Space) is curved. Also stated in the mantras is another property of *Antariksh*, the व्यच:

(*vyachah*). The term *vyachah* stands for '*vast and spread-out*'. Thus the Space-entity is not represented by an arrow itself or a linear curve formed by the path of an arrow, instead the *Antariksh* according to the Vedas is *curved, vast and spread-out.*

There are several other Veda mantras that are consistent with the fact that the Space-entity is an independent entity. The curved Space of the Vedas may be different from what Einstein conceptualized in his space-time continuum. However, the concept of space being curved started in the ancient Vedas. Some Vedic scholars have even remarked that Albert Einstein, who used to read Vedic Literature, may have gotten the basic idea of curved space from the Vedas! Many scholars translate *Aakaash* to be the 'nothingness'. The Vedas and Shastras (all 6 books of philosophy) and the rest of the ancient Vedic Literature define the *Aakaash* as a real entity, the Space-entity, which also is the basis for the modern cosmology.

In addition to discussing gravitation of the Earth, the Vaisheshik Darshan of Rishi Kanaad has dealt with *Aakaash-tatva* (Space-entity) assigning yet another property to the Space-entity. Just as Rishi Kanaad enumerated the *lingas* (defining signs or characteristics) of the Time-entity that were listed earlier in this chapter under Time-entity, he has stated in the *sutra* (2/1/20) the *lingas* of *Aakaash* as well:

निष्क्रमणं प्रवेशनमित्याकास्य लिङ्गम । (Vaisheshik, 2/1/20)

Transliteration: "*nishkramnam praveshanmityaakaasya lingam*".

Sutra Translation:

> *"The property permitting the 'exit from and entry into' is the defining sign of the Aakaash"* (Vaisheshik, 2/1/20).

The statement, '*exit from and entry into*' sounds quite simple because one can enter into and exit from a medium other than *Aakaash* (Space) as well, examples being air and water. While on Earth a person is already entered into air and can exit it riding a rocket out of the Earth's environment. To enter water one can jump into a swimming pool and exit from it too. If the *sutra* was that simple the *lingam* (defining sign) of air and water would be the same as that of the Space. This would render Rishi Kanaad's *sutra* fallacious because a *lingam* by definition is uniquely specific to a specific substance. Rishi Kanaad is a giant of Vedic philosophy. Considering how masterfully he framed the *sutras* on gravitation of the Earth that were cited in Chapter 1 and discussed further in Chapter 11, Rishi Kanaad's *sutra* (2/1/20) on the *lingam* of Space-entity has a deep meaning regarding the '*exit from and entry into*'.

The status of the pool-water changes when someone enters a swimming pool. The water equivalent to the volume of the body of the person is displaced. The same is true for air which is displaced when a body enters in it. The Space on the other hand does not get displaced when a substance '*exit from and entry into*' it. The implication of Rishi Kanaad's *sutra* is that when things enter into the Space-entity and exit from it, the physical state of the Space-entity neither changes nor is it consumed in the process. It permits the action of exit and entry for all substances without the substance or the Space being affected.

The Vedas have recognized the Space-entity with another adjective, the *vibhu*. The term *vibhu* is for an entity that is all-pervading through the universe without a break in it and without having branches and limbs which is the same as Nirukt's definition of *Antariksh* (*Aakaash*). In Rishi Kanaad's Vaisheshik Darshan, 8 sutras later (Vaisheshik, 2/1/29-31) the Space is considered to be a single and separate entity by itself.

Rishi Kanaad has also associated *linga* to some important substantive objects as well. In the Vaisheshik *sutra* (1/2/6) he states that 'action' is the *linga* of substantive objects having shape and discontinuity in them (made of separate atoms). He emphasizes the contrast between substances generating action (karma) and the three non-substantive entities: Direction, Time and Space as,

> *"Direction, Time and Space, having their dharma incongruous with the substances having dharma of karma (action), are action-devoid entities"* (Vaisheshik, 5/2/21).

This means that all actions (karma) seen in Nature are imperatives of Nature and not of Direction, Time and Space. This also means that the inherent property of 'action' in matter (Nature) is matter's *dharma*. The *dharma* in context of an inert substance is substance's expected inherent property that it displays on interacting with other substances. For example it is seed's action (dharma/karma) to sprout. In the context of living being, a person's *dharma* is to follow the rules of the society. In general, matter's *dharma* is to follow Nature's physical laws that the physicists have rediscovered in recent times. As we know that the gravity creates most observable karma (actions) in the form of motions of objects of the universe.

Newton in his third law of motion stated that '*to every action (karma) there is equal and opposite reaction (karma)*'. What Rishi Kanaad established eons ago in his *sutra* on action is what actually creates the action itself. Knowing that there are other pervading entities in the universe taking part in a complex manner Rishi Kanaad emphasized that all actions (karma) emanate from matter/energy (Nature = matter +energy) and not due to other entities. Then a natural question is – what is the big deal in Rishi Kanaad's *sutra* because all things happen in Nature anyway including action? The big deal is in the first part of Rishi Kanaad's concise sutra where he affirmatively states that the property of action is NOT PRESENT in or created by the three real entities - Direction, Time and Space, which are also present in the universe along with the matter. In other words, the *dharma* of each of these three entities is different from the dharma of Nature. Coming back to the context of this section, the Space-entity according to the Vedas is an action-less entity.

One other adjective that Vaisheshik Darshan (7/1/22) has associated with the Space-entity is विभव (*vibhav*). The adjective *vibhav* means '*to be interactive with all substances at the same time*'. This means that it is pervading through all substances at all times. The Space-entity with its all-pervading nature is an integral part of the Vedic science and cosmology because in the Vedic principles it is said to be the binding agent of matter during matter's indefinable state (*Asata*) in the non-creation phase and also in the definable state (*Sata*) of matter in creation phase.

Rishi Kanaad presents more *sutra* formulations to differentiate *Aakaash* from the other four elements of philosophy *Vaayu, Agni, Jal* and *Prithvi* (Air, Energy, Water and Earth). He isolates a basic property of *Aakaash* which also serves as its defining sign (*linga*). It is *shabd.* The properties of smell, warmth, taste and shape are absent in Aakaash. The ordinary meaning of the Sanskrit term *shabd* is 'sound' or an 'uttered word'. The deeper meaning of the term *shabd* is 'vibration'. The vibrations being action (karma), they can be produced only by matter but can be carried forward owing to the continuity of Space between the discontinuities of atoms of intervening matter. This means that matter could create vibrations but could not transmit or sustain them in the absence of *Aakaash*.

Space-entity is also termed as *Svayambhu* (17th term in the Table 5.6). The *Svayambhu* is the term for an entity that has its existence by itself and is not an effect of a cause. For an entity to be all-pervading like Space, it has to be more minute than the fine-structure of all substances immersed in it including the smallest fundamental particle and energy form of matter. In the Vedic definition, the Space is minuter than the smallest particle of matter, be it a quark or anything smaller particle than that, because it pervades inside such particles as well. For the Space-entity to be *Svayambhu* it could be interactive with all substances at all times (*vibhav*) but non-reactive with substances, making Space a no-decaying and non-consumable entity. This property of Space is made clear in Bhagvadgita as well, wherein Yogiraj Krishn preached to Prince Arjun about the nature of the Soul-entity. The Bhagvadgita's message to Arjun just before the coronation of Arjun's older brother Prince Yudhishthir in 3140 BCE in Hastinapur (ancient capital near Delhi), is not as old as the Vaisheshik Darshan and the Vedas. Nevertheless the Bhagvadgita establishes the consistency in Vedic thought and perspective on the concept of Space as thought by Krishn and all other *rishis* before him. In Krishn's message to Arjun the emphasis was on the nature of the Soul-entity where the Space was only an example. The property of Space comes through in his message as a byproduct of knowledge but the concept of Space in Krishn's message is no different from the one in Vaisheshik's *sutra* (7/1/22). Krishn's message is:

यथा सर्वगतं[1] सौक्ष्म्यादाकाशं नोपलिप्यते[2] ।

सर्वत्रावस्थितो देहे तथात्मा नोपलिप्यते ॥ (Bhagvadgita, 13/32)

Transliteration: "*Yatha sarvagatam*[1] *saukshmyaadaakaasham nopalipyate*[2] ।
Sarvatraavasthito dehe yathaatma nopalipyate"॥

Shloka Translation:

"Just as the all-pervading Aakaash (Space), owing to its minute nature, does not attach to any object in it, the soul-entity in the living-being also does not attach to the matter of the physical body." Bhagvadgita (13/32).

Brief commentary on this was included in Chapter 1.

It is clear from this statement that the *Aakash* (Space) for Krishn 5,000 years ago was a real entity. Krishn in this *shloka* of Bhagvadgita invokes the Space-entity to illustrate a common property between the Soul-entity and the Space-entity. This common property is expressed in the term नोपलिप्यते[2] (*no-api-lipyate* = non-interaction or non-attachment) with matter. The deeper sense of the *shloka* is that the soul-entity is non-consumable in chemical and physical processes of the physical body, where it lives unattached like the Space-entity that holds all substances of the universe yet remains unattached with the universe and all substances in it and remains unutilized by substances embedded in it. The bonus of the *shloka* is the mention made by Krishn of yet another property of Space, this property being सर्वगतं[1] (*sarvagatam*), meaning '*one that has gone into all*'. This means that the Space-entity is an all-pervading conterminous entity. An excellent definition of the Space-entity!

The Space in Vedic Metrology

A primary measurement of space is to measure shortest distance between two points separated in space. This is the same as a measurement on a straight line. Linear measurement is of great importance in all societies. The standard for measurement in Vedic civilization was a *hasta* (about 44 cm). Longer rope or a chain was calibrated by the smaller standard of length measurement. Markers and milestones were installed by societies to avoid repeated effort in measuring distances between different towns. A yard or a meter became most common measures of the present society.

The Vedic metrology standards for measuring space are briefly reviewed indicating that the scholarly of the society had developed the fundamentals for space metrology from the knowledge provided by the Vedas.

Case of the word mitra →metre (meter)

This section is to delineate how some of the fundamentals having origin in the Vedas became practices in the Vedic society. The unit of length measurement in metric system is one such example. It begins from the word *mitra* the common meaning of which is 'a friend'. The word appears in several Veda mantras and has deeper meaning as well. While interpreting a RigVeda mantra (1/2/7), the 19th century Sanskrit scholar and the Vedic writer in English, Pandit Gurudatta Vidhyarthi, quoted several times in Chapter 4, wrote thus:

> *"The first literal meaning of mitra is the measurer. The name is given to a substance that stands, as it were, as a measure or as a standard substance. It is the measurer of density, or of value, otherwise known as quantivalence. The word mitra is formed by adding unaadi suffix ka, to the root mi, according to the sutra 'amichimishsibhyah kr', un. 4/164. The meaning is* ***'minoti manyam karoti mitrah or one that measures or stands as a standard of reference****"* is a *mitra*.

From the same root is reflected the property of *mitra* – '*one that is most sought out and one that sets a standard for measuring up to*'. For its applicability in practical usage the term *mitra* is provided by the RigVeda for developing a standard for measurement. Yet another word from the same root with similar sound is *maatra*, which means 'quantity'. This also conveys the meaning of measurement. The word *maatra* is generally used qualitatively by common people even today in the Vedic society when the quantity of a substance is referred to. Surprisingly a term sounding similar to the Vedic terms *mitra* and *maatra* is 'metre' in the modern metric system for length measurement. The term 'metre' also fulfils the definition of the Vedic term *mitra* elaborated upon by Pandit Gurudatta Vidhyarthi in the above quotation. One observes from the open literature on Indian Culture that the Vedic society used extended range of length measurement. However, there is little surprise that the Sanskrit words *mitra* or *maatra* have migrated to other languages and civilizations and changed in pronunciation to metre or meter which ended up being used as a unit of length. The subdivision of the unit metre in the Western system also provides ample indication of word migration. The deci-meter and centi-meter are not too far in pronunciation from the original Sanskrit words *dashi* (tenth) and *shati* (hundredth) respectively, having the same meanings as in Sanskrit.

Beyond the measure of a hand was *danda* (stick), *rajju* (rope), multiples of rope-length, *krosha* and finally the measure of *yojan*. *Yojan* has been a common length measurement unit in Vedic Civilization for measuring large distances. The *yojan* for distance is straight from the Veda. In praise of *Rudra*, the hidden form of *Agni* (energy) in living beings including vegitation, the YajurVeda mantra (16/54) thus states poetically -

> *"Situated on Earth are uncountable discrete living systems with Rudra energy working through their respiratory function. The bow of their Rudra energy is stretched out to thousands (innumerable) of yojan".*

Not only the counting in this mantra on living systems implies the discrete nature of living systems, the *Rudra* energy form in them is said to have extended to large distances. Obviously the YajurVeda is talking about the biomass on the Earth and beyond. The *yojan* in this mantra is explicitly a measure of distance. The Shatpath Brahman (9/1/1/28-30) has provided an interpretation of Rudra's bow, stating that wherever one chose to examine, whether in the oceans on Earth or in the oceans of Space, the *Rudra* is present with its bow stretched. The meaning of the Brahmana statement's stretched bow is that the *Rudra* system is energized throughout the universe. When a bow is stretched the stretcher stores energy in the bow which is imparted to an arrow when released. In a typical Vedic Sanskrit expression 'thousand-yojan' generally means thousands of yojan or uncountable units of distance. A *yojan* in Swami Dayanand Sarasvati's interpretation of the above cited YajurVeda mantra (16/54) is equal to 4 *krosha* which is indisputably accepted by all.

The *krosha* measurement for distance and *krosha* engravings on milestones existed in some regions of India when British started to rule India. Since there was no confusion between the relationship of *krosha* and *yojan* the *yojan* unit of distance remained in limited practice. Disregarding the indigenous culture and language, the foreign rulers also made concentrated effort in imposing their own culture and language. The British were no exception. They introduced English and their system of length measurement. The Aryan units of length measurement including *krosha* and *yojan* were thus replaced by inch, foot, yard, furlong and mile. The British system mainly remained limited to urban India and along the railway system that they established. In the countryside however, the unit of *krosha* continued for all practical purposes till after the government of independent India adopted and implemented the metric system with kilometer replacing the mile. Although there is no confusion between the relationship of *krosha* and *yojan*, the confusion persists in *yojan*'s relationship to the modern units of measurement and the confusion also persists on its value cited in the Sanskrit literature developed in the past a few millennia.

The disagreement on converting a *yojan* into a kilometer stipulates that 1 *yojan* is somewhere between 5 km and 15 km. The majority of traditional Indian scholars have used a figure between 13 km and 16 km for a *yojan*. Religious scholar A. C. Bhaktivedanta Swami Prabhupada of Hare Krishna Mission gives the equivalent length of a *yojan* as 13 km throughout his translations of the Puranic scriptures. This author has accepted the value of 13 km for a *yojan* for a specific reason given under the next section. The reason is discussed in relation to the **speed of light in Vedic literature**.

As mentioned above the conversion factor of 13 km to a *yojan* is going to be the final one for this book. The *yojan* unit of distance is also associated with the spread rate of sunlight scattered from the Earth's atmosphere before sunrise. This is in the first chapter of RigVeda's Usha Hymn.

The Case of Vedic *Usha*: the Morning Twilight

RigVeda has devoted two successive *suktas* (hymns), 123rd and 124th, of its first chapter to *Usha Devata* with 13 mantras in each. The Vedic term *usha* is for the phenomenon of the solar illumination on the Earth's surface ahead of the sunrise. In common English language it is morning-twilight. Twilight is the indirect sunlight scattered from the upper layers of the Earth's atmosphere and received on the Earth's surface once in the morning before sunrise and once in the evening after sunset. The RigVeda has assigned the status of *devata* (subject) to *Usha* because of its phenomenal permanence in the creation and for bringing special benefits to the living creation on Earth. *Usha* began with the creation of the solar system and will continue through the creation phase of the universe. Thus, as far as Earth is concerned the phenomenon of *usha* is an invariant of the creation. This is why the RigVeda gave the importance of a *devata* to *Usha*, an inert invariant atmospheric phenomenon. The RigVeda basically deals here with a physics phenomenon.

The emphasis in these *suktas* is on the morning *Usha* because with it begins the 'new dawn' the new onset of activity in the creation on Earth. From a literary perspective the description of *Usha* in RigVeda is nothing short of a poetic marvel. Brevity restricts this author to present full translation and interpretation of all the relevant mantras. Only some adjectives, physical characteristics and actions credited to this phenomenon in 26 mantras of the two *suktas* are highlighted. From the physics perspective this alone will show the Vedic *rishi*'s interest in observing and studying the phenomenon of atmospheric optics. The two *suktas* on *Usha* phenomenon are credited respectively to two rishis named Deerghtamah Putrah Kakshivan, the son and Deerghtamah Kakshivan, the father.

Among the physics facts associated with *Usha* (morning-twilight) the fact relevant to the space measurement is contained in the RigVeda mantra (1/123/8). It states that *the onset of Usha appears 30 yojan from the place of Sunrise*; which indeed is an observation in atmospheric optics. This is discussed in the following in relation to Earth's geometry.

An atmospheric physicist (this author being one) would tell us that twilight (*Usha*) is a phenomenon in light scattering from the air in atmospheric layers immediately above the surface of the Earth that are laden with aerosol, heavy particulate matter and cloud droplets; all favorable for light scattering in the atmosphere. Depending on the light intensity levels, which depend on the solar angle below horizon, atmospheric physicists divide twilight into three periods, the most bright period closer to sunrise being called the 'civil twilight'. This definition is based on the time just before Sunrise and just after Sunset and is not based on light intensity measurements which would vary considerably depending on the atmospheric condition, latitude of the place and the time of the year. In this period of morning-twilight the objects become visible and the horizon is clearly discernible with unaided eyes. Lacking standardized unit for light level measurement, the observability of the objects and the horizon in the early twilight period remains quite subjective. However, this kind of subjective observability is accepted as scientifically defined 'civic twilight' at the present time.

In the following are listed a limited number of statements from the two *Usha suktas* (1/123, 1/124) of RigVeda which provide some insight into the Vedic poetic description of *Usha,* the twilight phenomenon in atmospheric optics.

- *'Usha lifts the darkness of night and is the banisher of diseases....'* (1/123/1);
- *'Like in all her antecedents the Usha comes as the first beautiful young woman, revealing all objects on the Earth even those that are high...'* (1/123/2);
- *'O Usha you come as Sun's sister and appear as Varun's (Atmospheric water-Air-Ocean system) young daughter...'* (1/123/5),
- *'With your fire-like illumination you illuminate from zenith* (उदाविष्कृण्वन्ति = *udaavishkrinvanti* = revealing from above) *the substances that are otherwise hidden under the cover of darkness...'* (1/123/6),
- *'When the chariot of the Sun arrives, the Sun with its brilliance removes the darkness of the night while you Usha illuminate substances of the creation with your* अन्यत् अद्यौत् *(anyat adyaut) the secondary illumination, while residing between night and the day that flow in the perennial cycle of Nature...'* (1/123/7),
- *'The onset of Usha appears 30 yojan from the place of sunrise...'*, (1/123/8),
- *'With the adjective of 'the first', O Usha you come without fail bringing with you the pure whiteness of your illumination, a new name for the day...'* (1/123/9),

- *'Usha moves in reverse yet comes back every day following the rule of Nature…'* (1/123/12);
- *'Staying in the lap of your creators (Earth's atmosphere) O Usha you are adorned with sunlight, you spread illumination in the half side of the Earth…'* (1/124/5);
- *'You (Usha) reach every place illuminating even a man fallen into a deep dark pit (dried up deep well)…'* (1/124/7).

These statements, otherwise in the beautiful Sanskrit poetry, no doubt create a vivid description of the twilight phenomenon. Additionally, the mention of associated optical effects and physical measure of its extent lends a touch of physics as well. The last bullet picturizes a man walking in the darkness of night and accidently falls into a deep *andh-koop* (a deep dark-dried-well). *Usha* illuminating even the dark deep pit raises the hope for the man fallen in the well. The mantra's poetic expression if presented in an unembellished physics statement, one would read it as – if the Sun is not overhead, the direct solar ray does not reach the bottom of a deep dark pit and the only light reaching there is twilight scattered due to the air in Earth's atmosphere. The sixth bullet (RigVeda, 1/123/6) exhibits twilight's one of the most important properties that it illuminates substances from above. This is delineated by the term उदाविष्कृण्वन्ति (*udaavishkrinvanti*) in the mantra which means 'revealing from above'. This is why this author used the term 'zenith' for *udaavishkrinvanti* in the sixth bullet. These mantras are therefore a source of considerable physics observation on the light phenomenon in the Earth's atmosphere.

Now the full translation and discussion on the RigVeda (1/123/8) mantra, which ties down *Usha*'s geographic extent on the surface of the Earth, is presented. This mantra also reveals that the hard number in *yojan* distance ascribed to the *Usha* phenomenon is a difficult experiment in physics that must have been conducted by the *rishis*. The mantras in these *suktas* suggest extensive indulgence of ancient people in scientific observation on the twilight phenomenon. The RigVeda mantra (1/123/8) and its full translation is presented below:

सदृशीरद्य सदृशीरिदु श्वो दीर्घं सचन्ते वरुणस्य धाम ।
अनवद्यास्त्रिंशतं योजनान्येकैका क्रतुं परि यन्ति सद्यः ॥ RigVeda (1/123/8)

With word-conjugation taken care of, the transliteration and word-meaning is as following:

Word transliteration and Meaning

सदृशीः (*sadrashih*) = unaltered, अद्यः (*adyah*) = today, सदृशीः (*sadrishih*) = unaltered, इत् (*it*) = it, or the emphasis on 'it be known', उ (*u*) = or, otherwise, श्वः (*shvah*) = next day (tomorrow), दीर्घं (*deergham*) = vast, big, great, सचन्ते (*sachante*) = caused by or results from or born out of, वरुणस्य (*varunasya*) = that of Varun (the atmospheric water - air - ocean system), धाम (*dhaam*) = home (for Varun's home the Earth's atmosphere) ।

अनवद्याः (*anvadyaah*) = demonstration of qualities, exhibiting its effect, त्रिंशतम् (*trinshatam*) = thirty, योजनानि (*yojanani*) = *yojans*, एकैका (*akaika*)= in concatenation or in succession, क्रतुं (*kritum*) = creating action, परि यन्ति (*pari yanti*) = attains repeatedly, सद्य: (*sadyah*) = swiftly ॥

Mantra Translation

> *"The Usha with its immutable occurrence is born daily in the vast home of Varun (atmospheric water-air-ocean system) and moves swiftly demonstrating its qualities in concatenation up to 30 yojans"*
>
> (RigVeda, 1/123/8).

The repeated theme in this and other mantras of these two *suktas*, e.g. "*....you come without fail...*" in (1/123/9) is *Usha*'s daily occurrence. In addition to this the mantra states that this phenomenon is born daily in the vast home of *Varun Devata*, the *devata* of waters. Since waters are in oceans and in the Earth's atmosphere in the form of ubiquitous clouds, the *Varun Devata* in the Vedic definition is the *atmospheric water-air-ocean system*. Because it is born daily in the vast atmosphere, the mantra (1/123/5) cited under a bullet above called it the *young daughter* of *Varun*. The daily birth of the twilight phenomenon in the vast atmosphere as depicted in the mantras is exactly how the twilight phenomenon is defined in modern atmospheric physics; with the exception that the Vedic definition is more beautiful and poetic with several unique observations never narrated before in physics.

The other point in the above mantra and in the other mantras (bulleted items) is *Usha*'s virtues where the illumination of the objects of the Earth being important. A remarkable observation in the above mantra is its 30 *yojan* long swath of illumination that keeps moving fast on the surface of the Earth westward of sunrise point in an एकैका (*akaika*) fashion. The term *akaika* means 'one-after the other'. This was appropriately translated by this author as the concatenating illumination where a 30 *yojan* long illuminated swath keeps moving successively fast towards the west as the Earth turns. This process exhibits one of *Usha*'s dynamic properties.

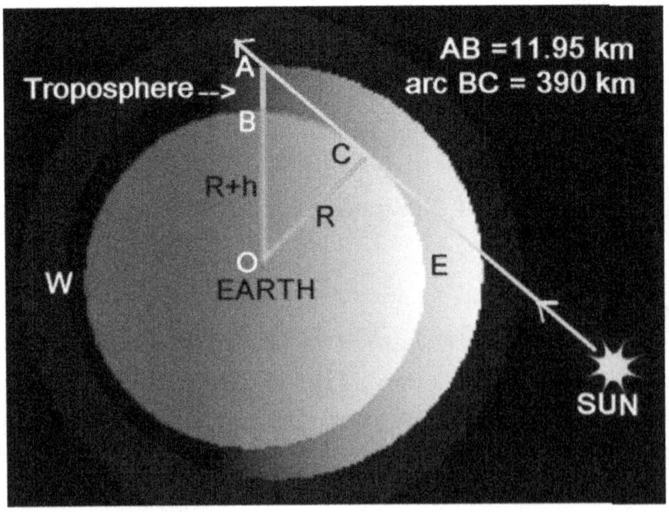

The sketch of the Earth and the solar rays (not drawn to scale) depicted the geometry of the twilight (*Usha*) phenomenon. The circle represents the spherical Earth which spins at a high speed from West to East. When the direct solar ray reaches tangentially at a point C, it is sunrise at C. The same ray keeps penetrating the atmosphere above the Earth's surface west of C. This ray on interaction with the air molecules and aerosol in the atmospheric layer scatters diffused sunlight towards the Earth's surface shown by illuminated atmospheric layer in the form of an arc above the Earth surface and around it. Now consider a point B due west of C directly below the point A on the ray at a height h above ground. All points on the ray between C and A send scattered light to the point B and to all other points between C and B on the curved surface of the Earth. This indirect light that illuminates the objects on Earth, west of C is twilight, the *Usha* of the Veda. This RigVeda mantra states that the distance between the sunrise point C and the point receiving first twilight at B, west of C is 30 *yojan*. Using 1 *yojan* equal to 13 km as discussed earlier, the distance BC is 390 km due west of C the sunrise point.

The arc distance BC can be measured by measuring the time duration between the onset of twilight at B and the onset of sunrise at B, provided one has the knowledge of spherical geometry/trigonometry, the spin rate of the Earth, the circumference or the radius of the Earth. One spin of the Earth takes 24 hours which is equivalent to the circumference of the Earth 40066.4 km. The 390 km sector of Earth's circumference corresponds to 14 minutes. Therefore the onset of twilight at B in the diagram is 14 minutes before the sunrise at B. This is extremely close to the definition of the morning/evening twilight. During this time the indirect sunlight (scattered sunlight) received from the height of the atmosphere is BA = 11.95 km.

The height of the scattering region of the atmosphere BA 11.95 km is interesting because on an average, in mid-latitudes, this is the tropopause-height above sea level. Tropopause is the upper boundary of troposphere. Troposphere is the atmospheric layer which is laden with particulate matter and aerosol and where all weather phenomena occur. As an atmospheric physicist this author is well aware of the fact that most of the solar radiation observed in twilight experiments is due to scattering in troposphere. Therefore, the physics experiment by the *rishi* assigning 30 *yojan* due west for the first appearance of *Usha Devata* before sunrise indeed is of considerable historical and physics significance. It is quite remarkable. This also leads to the fact that the conversion factor of 1 yojan equal to 13 km is quite reasonable.

A rishi having the knowledge of spherical geometry/trigonometry, the circumference/radius of the Earth is remarkable by itself and shows a high level of advancement in the Vedic Society. To carryout an experiment today to determine the twilight swath to be 390 km in the west of the sunrise point one needs more than one observer with their clocks synchronized and equipped with possible long-distance communication, separated east-west by a few hundred km to begin with. The fact remains that the mantra has given a hard number for the distance BC. It did not give 5 *yojan* or 50 *yojan* but it is 30 yojan, implying that a measurement was involved. It is mind-boggling!

This RigVeda mantra and many other mantras in the four Vedas, that present mysterious deductions on physical phenomena with real concrete numbers, make it glaringly obvious that the knowledge in the Vedas is an outcome of highly developed intellect among those who formulated the Veda mantras.

Under the eighth bullet listed above (RigVeda, 1/123/12), a unique expression for *Usha*'s movement is cited – '*Usha moves in reverse yet comes back every day following the rule of Nature...*'. The fact is that the Earth's daily spin would cause *Usha* to successively move in 'reverse' at a high rate. From the daily spin and known geometry of the Earth one finds that the *Usha*'s illuminated swath moves at a speed of about 464 meters in one second which is faster than the speed of sound in air; latter being 343 meters per second. Presumably no known vehicles or horses existed during the Vedic times that are known to travel at this speed, which could have been employed for in-situ measurements on twilight. The only conclusion on *Usha* starting at 30 *yojan* west of sunrise point is that the *rishi* had the knowledge of atmospheric physics and the speed of illumination on the surface of the Earth. This is indeed a mind-boggling realization of the knowledge imparted in an ancient Vedic mantra.

Another mantra's amazing observation under the fifth bullet, that – "*...your अन्यत् अद्यौत् (anyat adyaut) the secondary illumination, while residing between night and the day that flow in the perennial cycle of Nature...'* (1/123/7) further highlights the physics of the phenomenon which is none other than the phenomenon of twilight with the diffused illumination being secondary illumination. A bigger mind-boggling science observation from the Vedic Literature is given below.

The Mind-boggling Case of Space -Time Relationship: the Speed of Light in the Vedic Literature

Of the five Vedic entities presented here that form a set of causes in the creation of the universe, the Time and Space are of direct significance in developing physics of natural phenomena and advancing science knowledge in general. Under the sub-heading of "Measure of Time-entity: Small Time Scale of the Aryans" the sub-section entitled 'The Mind-Boggling Case of Nimesh' was presented. There it was said that the practical unit of *nimesh* could not have been derived by ancient *rishis* without a rigorous experiment in physics, here is yet another mind-boggling case where Space and Time, both are involved.

The simplest application where space and time are related is through the motion of an object. For example the speed of the Earth around the Sun being about 30 km/sec (= space/time) is the same as 0.41 *yojan/nimesh* assuming 13 km in a *yojan* and 8/45 sec in a *nimesh*. This conversion was straightforward because the conversion between a second and a *nimesh* was established in Table 5.5 and between a kilometer and a *yojan* was established under the sub-heading "Case of the word mitra → metre". The mind-boggling mystery is in the *shloka* written by Acharya Sayana (14th century CE) that was composed as the interpretation of a statement in Taittireeya Brahmana. Taittireeya Brahmana is one of the most ancient books on interpretations of mantras in the Krishn-YajurVeda. Taittireeya Brahmana belongs to the far antiquity of the Vedic civilization anterior to the Milestone established in Chapter 3 as February 18, 3102 BCE. Acharya Sayana's *shloka* is on the reading of Taittireeya Brahmana statement (2/8/7/51) on "*Doore Arthah Tarani-r-bhrajmanah*" which means the "*Meaning of Distance in Sun's Illumination*". A related mantra in Taittireeya Samhita ((1/4/31) is under "*Sooryo Atigreha*" (Sun the Special (extreme) Object of Astronomy); Soorya and Tarani being synonyms for the Sun. Hence, Acharya Sayana's *shloka* is on the speed of light. The original Taittireeya Brahamana statement is in prose to which Acharya Sayana gave a poetic form in the 14th century CE and the original thought of Taittireeya Brahamana is preserved in Acharya Sayana's *shloka* (see references in Ved Vidya Nidarshan, a Hindi publ. 1959 by Pandit Bhagvaddatta). The shloka by Acharya Sayana is:

योजनानां सहस्रे द्वे द्वेशते द्वे च योजने ।

एकेन निमिषार्धेन क्रममाणो नमोऽस्तु ते ॥

(Sayana / Taittireeya Brahmana, 2/8/7/51)

Transliteration: "*Yojanaanaam sahesrey dve dveshate dve cha yojane* ।
Eken nimishaardhen kramamaano namoastute" ॥

Translation:

"*The sunlight spreads through a distance of 2202 yojan in one half of a nimesh.*" (Taittireeya Brahmana, 2/8/7/51).

There is no other or different translation or meaning of Acharya Sayana's *shloka* (couplet) than what is presented in this sentence. Whichever angle one takes to translate the *shloka*, one arrives at the fact that the Sun's *bhrajmanah* (illumination) spreads (moves) at the rate of 2202 *yojan* in 0.5 *nimesh.* Sun's apparent disc itself is never seen spreading and certainly not through 2202 *yojan* and that the two numerals respectively represent the distance (*yojan*) and the time (*nimesh*) the Taittireeya Brahmana's statement relates to the rate of spread of sunlight. It is obvious that Saayan did not perform calculations for the speed of light as taken up by this author. Implicit in the 'spread' of sunlight is the 'speed of sunlight' because the two numbers given by the Taittireeya statement are distance and time. It is just a difference in expression. For example a meteorologist announces in the metaphor of 'Sun rise at 06:00 hours' knowing that the Sun never rises with respect to the Earth and it is the Earth that rotates to reveal the sunrise. Also in Rishi Taittireeya's statement, transformed by Acharya Sayana into a couplet, the spread of the Sun's *bhrajmanah* (illumination) means the spread of the solar rays and not the spread of the unchanging solar disc. Additionally, the spread of the Sun light is not the motion of the Sun in the sky. It is obvious that Sayan did not perform calculations for the speed of light as provided by this author.

Let us now examine the numbers cited in the shloka. A first glance at the number 2202 yojan in half the *nimesh* tells us that this unrounded number is not the result of a guess-work. The number 2202 *yojan* being 28,626 km is a large number about two and a quarter times the diameter of the Earth. Glaringly obvious here is that it is a number measured either in an experiment or it is the result of a calculation. This number being quoted as the speed of Sun's illumination in one half of a *nimesh*, further strengthens this author's contention that it is a measured speed of light by the *rishi* who was capable of measuring the fraction of a *nimesh*, the one half of a nimesh being only 4/45 second (about 89 milliseconds). No calculation related to the Sun can be performed in isolation without the input of some other measurement. Therefore this number decidedly is an outcome of observation followed by calculation. In a number such as this, an accuracy of at least 1 yojan per half *nimesh* is preserved. The second important feature of this statement is the half *nimesh*. A *nimesh* itself is about $1/6^{th}$ of a second and to assign half a *nimesh* to the distance of 2202 *yojan* confirms that it is a measurement of the speed of solar rays that was recorded in the book of Taittireeya Brahmana.

One *yojan* being equal to 13 km and one *nimesh* being equal to 8/45 seconds (Table 5.5), the speed of light based on the Taittireeya data can be calculated as,

2202 *yojan*/0.5 *nimesh* = 2202x13 km / (0.5x0.177778) sec = **3.22 x 10^5 km/sec**.

This is an extremely high speed. This speed is well beyond the speed of any known physical object. This is the speed of light. The obvious question to a physicist who works with space and time and the velocity of light in particular is:

> "*How is it that the speed of light in the ancient Taittireeya Brahmana turns out to be 3.22 x10^5 km/sec, which is so close to the speed of light 3.00 x 10^5 km/sec measured by modern physicists*"?

There is only one answer to this question - the number 3.22 x10^5 km/sec is so close to the speed of light because it is **nothing else but the speed of light,** the difference being that it was measured by an ancient *rishi*, well before the modern 'history of speed measurement of light' began. When the Brahmana statement associated the Sun's illumination with this speed, it is implicit that the Rishi Taittireeya is talking about the speed of the solar rays. Hence this author feels that the Vedic Space-Time relationship just cited here is **nothing short of mind-boggling.** The time assignment of one half of a *nimesh* (about 89 milliseconds) to this measurement also suggests that accurate space and time measurement techniques were at hand of the *rishi* who logged the numbers 2202 *yojan* and 0.5 *nimesh* in the Taittireeya Brahmana eons ago before the modern science was even conceived. Also, the *shloka* of Acharya Sayana in the 14th century is well before any activity in the West on the measurement of the speed of light. For this author, the Vedas continue to present wonder-upon-wonder.

Analysis

Light being the fastest physical entity in the universe is difficult to measure accurately. Its speed represented by the letter c is now accepted by scientists as 2.99792458x10^5 km/sec in vacuum. It is a physical constant for all wavelengths of radiation and known at the present time as the greatest speed that is known to exist in the universe. Regardless of the speed of light the conversion of 1 *yojan* to 13 km is generally accepted by scholars. When this author used the same conversion factor it yielded the speed of light extremely close to the real value arrived at by modern physics. Conversely therefore, because this conversion factor between *yojan* and kilometer gives a very realistic speed of light, 13 km in 1 *yojan* should be deemed an accurate relationship between the two distance units. Even if one took 8 km equal to 1 yojan, the calculations for the speed of light using Taittireeya data will yield 2 x10^5 km/sec, a speed not reached by any known object other than a photon of light, however, quite inaccurate for the speed of light. Since the Taittireeya Brahmana in this regard talks about nothing else but the Sun, the speed determination of 2202 *yojan*/0.5 *nimesh* yielding 3.22

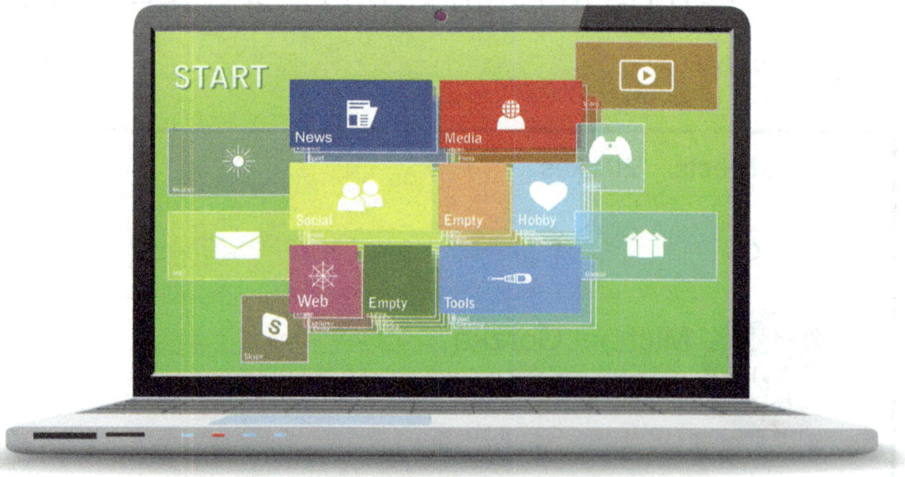

U

V

W

x 10^5 km/sec is the speed of light determined by a *rishi* far in the antiquity of Vedic civilization.

A Historical Note on Light Speed Measurement in Modern Era

Accurate determination of the speed of light in the Western civilization is not very old. Not only the original reference on speed of light, in Taittireeya Brahmana, is indisputably extremely old, even Acharya Sayana' *shloka* based on it, composed in the 14th century is at least three centuries older than the meaningful effort in the West made for measuring the speed of light. The reader may be curious to know when and how the Western civilization developed the methodology for accurate determination of the speed of light. From this history one could assess and compare the level of wisdom that existed in Vedic times, the wisdom that was adversely affected after the Mahabharata War five millennia ago.

The modern history of understanding and measuring of light in the Western civilization is said to have started with Aristotle (384 - 322 BCE) who thought that the light is not something that moves but is present. This cannot be considered as a start for measuring the speed of light. The other Greek mathematicians like Euclid (300 BCE) known as the father of geometry, Heron of Alexandria (10-70 CE) known as the 'greatest experimenter of antiquity' and Ptolemy (90-168 ACE) the geographer, astronomer and astrologer, all theorized that light is something emitted by the eyes and its speed must be infinite because all objects near and at infinity were revealed by the eye instantly. This also cannot be considered as the beginning of the process for measuring the speed of light. From all standards of time scale, this period in history is quite recent compared to the Vedic period of Taittireeya Brahmana where light is a measurable phenomenon with motion at 2202 *yojan* per 0.5 *nimesh*.

It was an Iraqi physicist Alhazen (965-1040 CE) who had the first scientific basis to suggest in 1021 CE that light starts from the objects and reaches the eye which detects it, making light an entity with finite speed. Still there is no sign of measurement of the speed of light! Then came Acharya Sayana's *shloka* in the 14th CE on the statement in Taittireeya Brahmana (2/8/7/5) formulated in the far antiquity of Indian civilization. This ancient statement gave the speed of light very close to its real value.

Acharya Sayana's period in the Indian history is well-known. A brief discussion on Acharya Sayana was included in Chapter 4 under the sub-heading of "Puranas". As the Prime Minister of the Vijayanagaram Empire with the empire constantly facing the attacks from the Moslem invaders coming from the North India he remained a prolific writer authoring over 25 volumes of interpretive commentaries on ancient books in Sanskrit Literature including five original books of his own. It was Acharya Sayana's hand-written RigVeda manuscript that Max Muller received in Cambridge, England, for English translation.

Numerous Veda mantras could be cited here to show that light for the Vedas is a real entity. The *sapt-rashmi* is the Vedic adjective to the sunlight with seven colours and it is emitted by objects and fire and it travelled with a high speed from the source to the sense of seeing, the eye. The brilliant Acharya Sayana put the Taittireeya Brahamana statement into a metered *shloka* couplet, presented above. Two important points are to be noted here: (a) that in the 14th century CE, there was nothing worthwhile in the area of light speed-measurement for Acharya Sayana to copy or extract from and put it down in the *shloka* couplet and (b) that no one outside India of 14th century CE had knowledge of relating *nimesh* to second and *yojan* to kilometer for Acharya Sayana to compose the couplet on. Since the original Taittireeya Brahamana principle on the speed of light was written eons before Sayan and Aristotle (384-322 BCE), the measurement of the speed of light 2202 *yojan* per 0.5 *nimesh*, which is 3.22×10^5 km/sec, is indigenous to ancient India.

The debate on finite versus infinite speed of light continued in the West for three centuries after Acharya Sayana. Well-known European physicists of the 17th century, Johannes Kepler (1571-1630), Francis Bacon (1561-1626) and René Descartes (1596-1650) concluded that the speed of light was infinite. After some experiments in optics by Isaac Beeckman (1588-1637), Galileo (1564-1642) and Robert Hooke (1635-1703) it was determined that the speed of light was finite. From Ole Romer's (1644-1710) astronomical measurements, Huygens (1629-1695) came up with the speed of light to be 2.2×10^5 km/sec, a value that was 27% lower than the actual speed of light and Sir Isaac Newton (1668-1704) reported the speed of light to be 2.1×10^5 km/sec that was 30% less than the actual value. Finally Bradley's (1693-1762 CE) value of 2.98×10^5 km/sec, determined by using star aberration and speed of the Earth around the Sun came very close to the real value. Other serious optical experiments followed yielding differing values in the same ball-park. In 1862 CE the speed of light determined by Fizeau (1819-1896 CE) was 3.13×10^5 km/sec (4% higher). Faucault's (1926-1984 CE) value was 2.98×10^5 km/sec. Revisions of the value of the speed of light continued in the 20th century reaching the most accurate value of 2.99792458×10^5 km/sec.

This was the Western historical perspective on light. First the light was a subject of philosophy, then a real phenomenon traveling with a finite speed and finally it was deemed as a measurable phenomenon. In the ancient Vedic civilization it may have gone through the same process of iteration to reach the value of 3.22×10^5 km/sec which is an over estimation by only 7% compared to the scientific measurements in modern times. This is remarkable and mind-boggling!

Index

References/Citations

Apte, V. S., "The Student's Sanskrit-English Diction..", publ. New Bharatiya Book Corp. (India).

Arp, Halton Christian, "1966 Atlas of Peculiar Galaxies".

Bailey, J. Martyn, Nature, 223, 1969.

Barbour's public lecture (Perimeter Institute for Theoretical Physics in Waterloo, Ontario, Canada, June 6, 2012.

Bhagvaddatta, "Ved Vidya Nidarshan", publ. Itihas Prakashan Mandal, Delhi, 1959.

Bilaniuk, Deshpande, and George Sudarshan, 1962.

Cai, Rong-Gen and Tuo, Zhong-Liang, (Chinese Academy of Sciences in Beijing),"...re-examined the data from 557 supernovas throughout the Universe and re-crunched the numbers".

Chao, E. C. T. et al., Journal of Geophysical Research, 75, 7445, 1970

Chau and Betke, 2005

Cliff et al., Journal of Geophysical Research, 77, 2007, 1972.

Feinberg, Gerald, 1967.

Flandern, Tom Van, (Meta Research), "The Top 30 Problems with the Big Bang", 1900.

Fred Hoyle, Hannes Alfven, Geoffrey Burbidge, Halton Arp and some thirty other scientists in British science magazine, the New Scientist, New Scientist, May 22, 2004.

Gerstenkorn, H. Z. , Astrophysics Journal, 36, 245, 1955.

Gilbert, 1893., Grauman et al., 2001, 2003., Hartman , 1966.

Grauman et al. 2001, 2003.

Hawking, Stephen, "A Brief History of Time", publ. Bantom Books, 1988.

Hooper, Dan, "New Scientist", Vol 244, No. 3256, 16 November 2019.

Kopal, Zdenek, Nature 183,169,737, 1959. Kuiper, G. P. J. of Geophys Res, 64, 1713, 1959.

Kopal, Zdenek, Nature, 210, 188, 1966.

Krasinski and Brumberg in 2004

Kuiper, G. P. Journal of Geophysical Research, 64, 1713, 1959.

Larry King, CNN TV host, script - September 10, 2010.

Max Muller, F., "India - What Can It Teach Us?", publ. Munshiram Manoharlal Publishers,1991.

Mitchell, William C. "The Big Bang Theory under Fire", Phys. Essays Vol.10, 2, June 1997.

Monier-Williams, "Sanskrit-English Dictionary", publ. Clarendon Press, 1899.

Narhari Achar, B. N., (Department of Physics, University of Memphis (USA)

Narlikar, J. (In his theory he rewrote Einstein's field equations), Annals of Physics, 107, 325, 1977.

Oldham, C. F., "Recent Research on the Sarasvati River", www.stephen-knapp.com.

O'nion and Pankhursts , 1972.

Papannastassiou and Wasserburg ,1971a, b.

Radiometric data for South Asian Archaeology (Univ. Pennsylvania, Philadelphia, USA) and British Archaeological Reports, Oxford, UK, 1982.

Roedder, E and Weiblen, P. W. , Science, 167, 641, 1970.

Rubin, Vera and Ford, Kent, (Carnegie Inst. of Washington), Amer Astronomical Society, 1975.

Sagan, Carl, "Cosmos", publ. The Ballantine Publishing, NY, 1980.

Sahoo[1], S. et al., "A prehistory of Indian Y chromosomes: Evaluating demic diffusion scenario", Proceedings of the National Academy of Sciences of the United States of America, Vol. 103, no. 4, 843-848, January 24, 2006.

Sengupta[5], S. et al., "Polarity and Temporality of High-Resolution Y-Chromosome Distributions in India Identify both Indigenous and Exogenous Expansions and Reveal Minor Genetic Influence of Central Asian Pastoralists", The American Journal of Human Genetics, Vol. 78, 202-221, February, 2006.

Singh, Udaya Narayan, "Soorya Siddhant", publ. Ramlal Kapoor Trust, Sonipat, 1986.

Sitaram Shastri, "Niruktam Interpretation", publ. Parimal Publications, Delhi, 2005.

Smith, Quentin, ""Did the Big Bang Have a Cause?", British Journal for the Philosophy of Science, pages 649-668, 45, 1994.

Smoot, George and Davidson, Keay, "…our galaxy is moving with a peculiar velocity of 600 km/sec…" , p137 "Wrinkles in Time", publ. Avon Books, NY, 1994.

Sreevastava, Mahaveer Prasad, "Soorya Siddhant" Translation and Interpretation in Hindi", publ. Dr. Ratnakumari swaddhyaya Sansthan, Vdya Mudran Greha, Ilahabad, 1987.Srivastava, Mahaveer Prasad, "Soorya Siddhant", publ. Svaddyaya Sansthan, 1987.

Steinhardt, Paul J. "Cyclic Universe Theory",2002.

Steinhardt, Paul J. (Princeton University (USA), Scientific American, April, 2011.

Surendra Kumar, "Vishuddha Manusmriti", publ. Arsh Sahitya Prachar Trust, January, 1996.

Taneja, Subhash, (Head of the Sanskrit Department, Rajasthan University, India.

Turok, Neil, (Cambridge University), "Cyclic Universe Theory", Science, May, 2006.

Udayaveer Shastri, "Vedanta Darshan Ka Itihas" (History of Vedanta Darshan), publ. Govindram Hasanand, Nai Sadak, Delhi, 1991.

Urey, 1951.

Urey, Harold C. " Physics and Astronomy of the Moon", Academic press, New York, 1962.

Urey, Harold C. and MacDonald, 526, 1971

Urey and MacDonald (1971) 526.

Urey, Harold C. , A short letter dated January 16, 1973.

Varma, Pundit Jiyalal, "Ved Shastra Talika", 1958.

Vaidiya, K. S., Resonance, pages: 19-28, May, 1996.

Veneziano, Gabriele, "The Myth of the Beginning of Time", Scientific American, May 2004.

Vidyanand Sarasvati, Swami, "Aryon Ka Aadi Desh Aur Unki Sabhyata", publ. International Aryan Foundation, Mumbai, 1989, cited Count Bjorustjourna, "The Theogony of the Hindus", publ. John Mu, London, p. 82, 1844.

Vidyanand Sarasvati, Swami, "Ved Meemansa", publ. Govindram Hasanand, 1984.

Vidyarthi, Pundit Guru Datta, "The Terminology of the Vedas and European Scholars", 1888CE.

Vogt, Steven, (Prof. of Astronomy and Astrophysics, University of California at Santa Cruz), Exo-planet Gliese 581, announcement, September 29, 2010.

Yashpal, B. Sahai, R. K. Sood and D. P. Agrawal (Proc. Ind. Acad. Sci., Earth Planet. Sci, Vol. 89, pages 317-331, 1980.

Limited List of Vedic Terms often cited in the Book

Aapah	An all-pervading tri-property fundamental state of matter
aatma, atman	soul
acharya	Teacher and scholar of high calibre
Antariksh, Aakaash	Space-entity, container of matter/energy of the universe, varies with eternal existence
anaadi	Entity without beginning, eternal
Asata-state	state of matter devoid of matter properties, indefinable matter state confined in diminutive space
avi-vritsan	(अविवृत्सन्): repetitive circular motion as in the orbital motion of a planet
avyakta	indescribable
bhuta	matter/energy, *bhutaani* (plural of bhuta, includes all matter in both realms: animate and inanimate. *panch-maha-bhootas* (five-major-elements of philosophy)
chakraprasaran	represents two actions together: rotating around an axis and spreading radially keeping its circular shape
Brahmn	The Creator, purush - one who pervades all (details in Chapter 5)
chakra	wheel, an orbiting system
chakrampari	Circular motion on a perimeter as in an orbiting celestial body
darshan	philosophy
deva, devata	One with brilliant properties – a conscious-being or inert system of Nature essential to sustain creation
dhoorshu	(धूर्षु): In the axis of rotation
hiranmayam-aandam	(हिरण्मयमाण्डं): *hiranya-garbh, hiranyand*, golden-embryo, premordial ovate/spheroid of Nature (matter), spheroid with golden brilliance
Inddraprasth,	Hastinapur, Ancient names of the capital now Delhi
Kaalah, kaal	Time-entity (irreversible)
kali-yuga	Time period of 432,000 yrs
kaaran	cause: *nimitta-kaaran* (efficient-cause), *upaadaan-kaaran* (material-cause), *saadhaaran-kaaran* (ordinary-cause)
kaarya	effect, action, karma
Krishn or Krishna,	or *Shree Krishn, Yogiraj Krishn*: All names of Lord Krishn who delivered the message of Bhagvadgita to Arjun
maatali	cosmologist, a sage with specific knowledge of creation, space, time and cosmos
Mahabharata	War between Kauravas and Pandavas near 3,140 BCE, an epic book
mantra	A metered poetic couplet

Manu, Manusmriti	Manu the first ancient king of Aryans who wrote the book of Manusmriti
Munn	Mind-entity
Nirukt	Book by ancient Rishi Yask
nitya	ever-existing
paramanu	fundamental particles of matter indivisible further, round in shape
pari-kraman	defines the 'step-wise motion on a circle'
Prajapati	Nature the master of cosmos (material realm), God as the master of all cosmos (spiritual context)
Prakriti	Nature
pralaya	dissolution of creation
paryaplavan	large scale circular rotation around an axis of a system in floatation as in hurricane
rishi	a highly enlightened person, revealer of Veda mantras and the hidden meaning in them
Sarasvati	a multi-meaning word: intellect, ancient extinct river, title for a Vedic scholar of high calibre
Sata-state	real definable matter, the state of matter to form universe
sato-guna, rajo-guna and *tamo-guna*	fundamental *aapah* particle made up of three different particles with three properties (*guna*) (details in Chapter 5)
Sanskrit	A highly developed, grammar-rich ancient language, the language of the Vedas
skambha	Pillar, support of the universe (Creator, physical structure of matter and energy at universe's centre
sneha	property of bonding matter or living-beings
srishti	Creation, one that does not happen by accident
sukta	group of mantras with common theme, a hymn, section of a chapter
Sutra	treatise, a book of philosophy
sutra	an aphorism, a principle
swami	master, highly revered person, a monk
tatva	substance with real existence, entity
tuchchhaya	space holding universe in dissolution phase in diminutive space when matter is in devoid of dynamics and properties
urmim	vibrations or waves
Vedas	The books of knowledge, four Vedas are RigVeda, YajurVeda, SaamVeda and AtharvaVeda
varootham	Cluster of atoms (a cosmic objects)
vitatah	(वितत:): instantaneous expansion as in an explosion
yojan	Ancient unit of distance (1 *yojan* = 13 km)
Yudhishthir, Arjun	Two of five Pandava brothers in Mahabharata

About the Book

Could there be another rational cosmology theory elsewhere in another civilization before the modern Big Bang was conceived by Georges-Henri Lemaitre in 1927 CE? Yes there is. It literally exists since antiquity, in 'chapter-and-verse', in the book of RigVeda, encased in eons old mantra formulations under the heading of *Bhavavrittam* which means "*The History of Creation of the Universe*". The author, Dr. Shiv Raj Pal a physicist, named the process the "Vedic Big Bang" because the universe in Vedic Cosmology begins with an impetuous explosion in the enormous primordial matter ovate termed *Hiranya-aand* (brilliantly hot Golden-spheroid). It is amazing that these two cosmologies, eons apart, are similar in some aspects and different in others. The uniqueness of the Vedic Big Bang is that a colossal rotating central-bar (pillar) of matter/energy initiates the Big Bang and sustains the universe in perpetual rotation and expansion in which the matter, energy and all physical processes are real and none hypothetical such as the hypothetical Cosmic Inflation and Dark Energy of the modern Big Bang.

The Vedic definitions of Time, Space and Nature (matter, energy) are superior and detailed. The Vedic Cosmology Scenario is followed by the Vedic theory of the formation of the solar system revealing intriguing features of the planets. The book takes the reader to India's historical period anterior to 5,000 years when the rich Vedic Literature was developed. Mind-boggling deductions from the mantras indicate an incredibly elevated thought-system prevailing in the Vedas. Prepare for an illuminating voyage to the beginning of the cosmos.

About the Author

Dr. Shiv Raj Pal is a distinguished retired Adjunct Professor of Physics & a Senior Research Scientist from the Physics Department, York University, Toronto, ON, Canada & Centre for Research in Experimental Space Technology (CRESTech), a Centre of Excellence under the initiative of the Province of Ontario.

Dr. Shiv Raj Pal was born and brought up in a small farming village in the State of Rajasthan, India. He completed BSc and MSc (Physics) from Rajasthan University and the PhD from the Physical Research Laboratory (PRL)/Gujarat University. The National Research Council of Canada awarded him a Post-doctoral Fellowship to carry out research at the Environment Canada in the subject of Twilight Scattering. The author joined the Physics Department, York University in 1972. Research work at York University included the Remote Sensing studies of aerosols, clouds, ozone layers at Toronto and in the Canadian High Arctic at Eureka in Elsmere Islands employing powerful LIDARS (Laser Radars). Serving as the in-charge of the LIDAR Laboratory the author participated in several international LIDAR missions.

Dr. Pal has published a good number of peer-reviewed research papers in physics journals, presented papers at conferences, published in conference proceedings and produced a large number of research reports. For more information please read author's website www.VedicBigBang.com. E-mail Address: shiv@vedicbigbang.com

Reviews

"A staggering erudite tome of explosive brilliance !"

Dr. Richard N. Dubinsky, Sky Council Link

Top review from the United States

Nitin C

★★★★★ **Comprehensive & Detailed account of Vedas as Science**

Reviewed in the United States on March 8, 2021

This is a very well written, well thought, well articulated book with tremendous gravity in terms of its sheer breadth of knowledge both in Physics & Vedas.
This is a comprehensive book that puts many theories both Western & Eastern to the test. From Prologue to the Epilogue you will be rewarded with gems of wisdom and rediscover that which was said many millennia ago.
I give my best to Dr Pal and hope to read more from him in the future.

Dear reader:

Please do log-on to the book on Amazon and evaluate it by assigning the number of stars.

Made in the USA
Monee, IL
06 November 2026

68030316R00240